Biological and Molecular Approaches in Pest Management

Biological and Molecular Approaches in Pest Management

Editors

Balwinder Singh

Director of Research

Ramesh Arora

Senior Entomologist

Department of Entomology

S S Gosal

Director of Research (Retd)

Punjab Agricultural University

Ludhiana – 141 004, India

Published by:

Scientific Publishers (India)
5 A, New Pali Road, P.O. Box 91
Jodhpur 342 001 (India)

Branch Office
Scientific Publishers (India)
4806/24, Ansari Road, Daryaganj
New Delhi - 110 002 (India)

E-mail: info@scientificpub.com
Website: www.scientificpub.com

Print: 2021

ISBN: 978-81-7233-923-4

Typeset: Rajesh Ojha
Printed in India

About the Editors

Dr Balwinder Singh is Director of Research at Punjab Agricultural University, Ludhiana. He did his Ph.D. in Entomology with specialization in Insect Toxicology in 1984 from PAU, Ludhiana. He was Commonwealth Fellow at Rothamsted Research, Harpenden, U.K. for six months. He has more than 28 years experience in teaching and research in the field of pesticide residues. He has handled many research projects funded by various international and national agencies. He has more than 200 publications to his credit. He has also edited three books and contributed 11 chapters in different books. He is acting as a reviewer for number of International Journals related to field pesticide residues. He is associated with a number of professional societies at the National Level and is the President of the Indian Society for Advancement of Insect Science.

Dr Ramesh Arora is Senior Entomologist in the Department of Entomology, Punjab Agricultural University, Ludhiana. He obtained his M.Sc. and Ph.D. in Entomology from the same institute in 1978 and 1990, respectively. He has been working on development and dissemination of IPM technologies for the last more than 35 years. His current research interests include climate resilient IPM and management of insect pests in vegetable crops. He has more than 250 publications including 80 research articles in refereed journals, 15 books and 40 book chapters to his credit. He is associated with a member of professional societies and served as Editor-in-Chief of Indian Journal of Ecology (1996-2003). He is Vice President, Society of Pesticide Science India, General Secretary, Indian Society for the Advancement of Insect Science, and Editor, Entomology Reporter.

Dr S S Gosal served as Director of Research (2010-1014) at Punjab Agricultural University, Ludhiana. He obtained his M.Sc. and Ph.D. degrees in the discipline of Plant Breeding from the same university. He joined PAU as Assistant Breeder in 1984 and contributed towards strengthening the education and research in the area of plant biotechnology. During 1993, he was awarded a unique Biotechnology Career Fellowship by the Rockefeller Foundation, USA for 6 years, to conduct research in the area of transgenesis at University of Nottingham, England and John Innes Centre Norwich, UK for three months each year. He is elected Member/Fellow of three professional societies and member of several other scientific societies. Dr Gosal is founder Director of School of Agricultural Biotechnology at PAU, Ludhiana. To put his ideas into motion he has competitive research grants for more than 20 adhoc research projects. Besides, he has served as a member of Board of Assessors, Australian Research Council, Canberra. He has undergone a rigorous training on environment risk assessment of GM crops at Danforth Centre for Plant Science Research, St. Louis; APHIS, EPA (USDA) and USTDA, Washington DC, USA. His contributions in the field of plant biotechnology can be gauged from the volume of over 155 original research papers in refereed Indian and foreign journals. He has been the examiner, member of the advisory board and selection committees in more than 15 Universities and Institutes. Dr Gosal has co-authored a text book, two dozen chapters in books, many Practical Manuals and has supervised thesis research as major/co-major advisor/member of advisory committee of more than 70 post graduate students.

Preface

Despite great advances in agricultural productivity and economic well-being in much of the world over the past 50 years, food insecurity and poverty continue to be serious issues in many regions of the world. Since most of the cultivable land is already under cultivation, future increases in food, feed and fibre production have to be achieved with increased productivity and improved crop protection. Ironically, more than a third of the global agricultural production is lost to the activity of animal pests and diseases. Further, the losses are significantly higher in the tropical areas where the food shortages are already serious. The strategy of exclusive reliance on chemical pesticides for minimizing crop losses caused by pests has led to human health and safety hazards and caused a number of ecological and economic problems. The concept of integrated pest management was propounded to minimize these side-effects. But inspite of its inspirational value, most IPM programmes still include economic threshold level based application of chemical insecticides as a major input. In this context, the wide range of biological and molecular approaches discussed in 15 Chapters in this book offer a plethora of environmentally benign alternatives to these chemical insecticides.

The introductory Chapter outlines the role of molecular techniques in improving the efficacy of a diverse range of pest management options including genetically engineered plants, insects and microorganisms as well as host plant resistance and chemical insecticides. The second Chapter discusses the role the biotechnological approaches have played in development of rice genotypes resistant to planthoppers. The third Chapter highlights the importance of molecular taxonomy in studies on fruit flies. The RNAi has emerged as a powerful gene-silencing technique and its potential for utilization in pest management is revealed in Chapter 4. The persistence of pesticide residues in the environment poses significant ecological risks. Bioremediation is emerging as the method-of-choice for dealing with such residues and its potential is explored in Chapter 5.

Insects, like other organisms, are prone to diseases caused by a diverse range of microorganisms. The exploitation of these microbes as a tool in pest management offers immense possibilities which are highlighted in Chapter 6. The gall midge, a major pest of rice, is notorious for evolving new biotypes to overcome plant resistance genes. Chapter 7 explains how an understanding of the molecular basis of these interactions can help in development of durable gall midge resistant genotypes. The wide range of molecular markers and their applications in entomological research are concised in Chapter 8. The success story of *Bt* cotton which has transformed India from a net importer to a major exporter of 'white gold' is highlighted in Chapter 9. The potential of biotechnological techniques in developing pest-resistant fruit plants is outlined is Chapter 10.

Insects being highly versatile organisms have the ability to develop resistance to any pest management tactic which exerts sufficient selection pressure on their populations. Chapter 11 focuses on the strategies for management of resistance in insects to microbial control agents. Chapters 12 and 13 highlight the role of biological and molecular approaches in management of mites and *Conogethes* spp., respectively. Bee diseases and pests present a serious challenge to commercial apiculture. The utilization of molecular techniques for precise identification of these organisms is explained in Chapter 14. There is now a wide array of techniques available to replace the use of conventional insecticides in IPM programmes. But it is essential that all the non-chemical approaches are combined within the framework of IPM. The emergence of biointensive IPM as the preferred alternative to conventional IPM is explained in the concluding Chapter of the book.

It is hoped that the book will fill the wide gap in literature on utilization of biotechnological approaches in biointensive IPM as an alternative to chemical insecticide based IPM for sustainable insect pest management in future. We trust that you, the reader, will find the subject matter interesting and informative. We hope that this compilation answers questions you might have, and serves to stimulate further development of this fascinating science of pest management.

Balwinder Singh
Ramesh Arora
S S Gosal

Contributors

P K Arora
PAU Regional Research Station
Abohar-152 116, India

Ramesh Arora
Department of Entomology
Punjab Agricultural University
Ludhiana 141 004, India

J S Bentur
Directorate of Rice Research
Rajendranagar,
Hyderabad 500 030, India

Manmeet B Bhullar
Department of Entomology
Punjab Agricultural University
Ludhiana – 141 001, India

D S Brar
School of Agricultural Biotechnology
Punjab Agricultural University
Ludhiana – 141 004, India
(Former Head, Plant Breeding,
Genetics & Biotechnology Division
International Rice Research Institute,
Manila, Philippines)

P S Burange
Department of Entomology
Punjab Agricultural University
Ludhiana 141 004, India

A K Chakravarthy
Department of Entomology
University of Agricultural Sciences
Bangalore-560 065, India

Chandrashekharaiah
Department of Entomology
University of Agricultural Sciences
Bangalore-560 065, India

Pardeep K Chhuneja
Department of Entomology
Punjab Agricultural University
Ludhiana 141 004, India

A K Dhawan
Department of Entomology
Punjab Agricultural University
Ludhiana-141 004, India

B Doddabasappa
Department of Entomology
University of Agricultural Sciences
Bangalore-560 065, India

D Fujita
Plant Breeding, Genetics &
Biotechnology Division
International Rice Research Institute
Manila, Philippines

Arshdeep K Gill
Department of Entomology
Punjab Agricultural University
Ludhiana – 141 001, India

P D Kamala Jayanthi
Division of Entomology and Nematology
Indian Institute of Horticultural
Research
Bangalore -560 089, India

K K Jena
Plant Breeding, Genetics &
Biotechnology Division
International Rice Research Institute
Manila, Philippines

S B Kandakoor
Department of Entomology
University of Agricultural Sciences
Bangalore-560 065, India

Uma Kanta
Department of Entomology
Punjab Agricultural University
Ludhiana 141 004, India

Paramjit Kaur
Department of Entomology
Punjab Agricultural University
Ludhiana 141 004, India

Krishan Kumar
PAU Regional Research Station
Abohar-152 116, India

Vijay Kumar
Department of Entomology
Punjab Agricultural University
Ludhiana 141 004, India

Sujan Mamidi
Genomics and Bioinformatics Program
North Dakota State University
Fargo, ND 58102, U.S.A.

Kousik Mandal
Department of Entomology
Punjab Agricultural University
Ludhiana-141 004, India

Prashant Mohanpuria
School of Agricultural Biotechnology
Department of Entomology
Punjab Agricultural University
Ludhiana-141 004, India

G Nagalakshmi
Division of Entomology and Nematology
Indian Institute of Horticultural Research
Bangalore -560 089, India

P C Pathania
Department of Entomology
Punjab Agricultural University
Ludhiana 141 004, India

M A Rashmi
Department of Entomology
University of Agricultural Sciences
GKVK, Bangalore-560 065, India

Surinder K Sandhu
Department of Plant Breeding & Genetics
Punjab Agricultural University
Ludhiana-141 004, India

P S Sarao
Department of Plant Breeding and Genetics
Punjab Agricultural University
Ludhiana – 141 004, India

H C Sharma
International Crop Research Institute for the Semi-Arid Tropics (ICRISAT)
Patancheru - 502 324, India

P R Shashank
Department of Entomology
University of Agricultural Sciences
Bangalore-560 065, India

P S Shera
Department of Entomology
Punjab Agricultural University
Ludhiana 141 004, India

Balwinder Singh
Department of Entomology
Punjab Agricultural University
Ludhiana-141 004, India

Kuldeep Singh
School of Agricultural Biotechnology
Punjab Agricultural University
Ludhiana – 141 004, India

Abraham Verghese
National Bureau of Agriculturally Important Insects
Hebbal, Bangalore – 560 024, India

Contents

Biological and Molecular Approaches in Pest Management
Balwinder Singh, Ramesh Arora & S S Gosal (Eds)
Scientific Publishers (India), Jodhpur, 2015, pp 1-12.

Chapter - 1

Application of Modern Tools of Biotechnology for Pest Management - Prospects and Limitations

H C Sharma
International Crop Research Institute for the Semi-Arid Tropics (ICRISAT), Patancheru 502 324, India
Email: H.Sharma@cgiar.org

1.1. INTRODUCTION

Nearly 30 to 50 per cent of the crop yields are lost due to the ravages of insect pests, and several insect species have the potential to cause 50 to 100 per cent loss during outbreaks (Oerke *et al.*, 1994). Reduction in pest associated losses is one of the potential areas for increasing food production, and it is in this context that we can exploit the tools of biotechnology to minimize the extent of losses due to insect pests. Molecular biology has provided several unique opportunities in crop improvement that include access to molecules novel to crop species, production of transgenic crops expressing insecticidal genes, ability to change the level of gene expression, and the capability to change the spatial and temporal pattern of gene expression (Sharma *et al.*, 2002). Development of more effective insect-resistant varieties and bio-control agents will lead to a reduced reliance on synthetic pesticides, and thereby reduce farmers' crop protection costs, while benefiting both the environment and public health. The promise of biotechnology can be realized by utilizing the information and products generated through research on genomics and transgenics to increase the productivity of crops through enhanced resistance to insect pests.

1.2. GENETIC TRANSFORMATION OF CROP PLANTS FOR INSECT RESISTANCE

Development and deployment of transgenic plants in an effective manner is an important pre-requisite for sustainable and economic use of biotechnology for crop improvement. As a result of advances in genetic transformation and gene expression over the past three decades, there has been rapid progress in using the tools of biotechnology for developing crops for resistance to insects (Hilder and Boulter, 1999; Sharma *et al.*, 2004). While most of the insect-resistant transgenic plants have been developed by using *Bt* δ-endotoxin genes, many studies are underway to use non-*Bt* genes, which interfere with development and the nutritional requirements of insects including:

- Cry toxins *Bt*: Cry1Ab, Cry1Ac, Cry 2a, Cry9c, Cry 2B, Vip 1, Vip 2, etc.
- Plant metabolites: Flavonoids, alkaloids, terpenoids.
- Enzyme inhibitors: SBTI, CpTi.
- Enzymes: Chitinase, lipoxigenase.
- Plant lectins: GNA, ACAL, WAA.
- Toxins from predators: Scorpion, spiders.
- Insect hormones: Neuropeptides and peptidic hormones.

Genes conferring resistance to insects have been inserted into maize, cotton, potato, tobacco, rice, broccoli, lettuce, walnuts, apples, alfalfa, and soybean (Sharma *et al.*, 2004). A number of transgenic crops have now been released for on-farm production or cultivation by the farmers (James, 2005). The first transgenic crop with resistance to insects was grown in 1994. Since then, there has been a rapid increase in the area sown to transgenic crops and transgenic crops are now grown in over 12 countries in the world. Cry type toxins from *Bt* are effective against cotton bollworm, corn earworm, the European corn borer, and rice stem borers (Armstrong *et al.*, 1995; Benedict *et al.*, 1996; Nayak *et al.*, 1997). Successful expression of *Bt* genes against the lepidopterous pests has also been achieved in tomato (Delannay *et al.*, 1989), potato (Jansens *et al.*, 1995), brinjal (Kumar *et al.*, 1998), groundnut (Singsit *et al.*, 1997), pigeonpea (Surekha *et al.*, 2005; Sharma *et al.*, 2006), and chickpea (Kar *et al.*, 1997; Sanyal *et al.*, 2005).

Development and deployment of transgenic plants is carried out under strict biosafety regulations as defined by biotechnology related legislation in each country. The final approval for large scale cultivation of a transgenic crop is given after ascertaining agronomic performance (efficacy and stability of transgene and development of resistance and resistance management strategies), bio-safety of the transgenic to the environment (effects on non-target organisms, gene flow, and effects on biodiversity), biosafety of food, allergenicity, and the consequences of deployment of transgenic crops on quarantine and trade.

Deployment of insect-resistant transgenic plants should be based on the overall philosophy of integrated pest management, and take into account alternate mortality factors, and reduction of selection pressure, and monitor populations for resistance development to design more effective management strategies. Transgenic crops are compatible with other methods of pest control, and yield up to 50 per cent higher than the non-transgenic cultivars even under insecticide protection (Fig. 1) (Sharma and Pampapathy, 2006). Insects such as *Heliothis virescens, Helicoverpa zea, Trichoplusia ni,* and *Spodoptera exigua* are many times more sensitive to insecticide sprays when they have a prior exposure to *Bacillus thuringiensis* (Harris *et al.*, 1998). Thus, transgenic crops can be used in conjunction with other methods of pest control without any detrimental or antagonistic effects. To increase the effectiveness and usefulness of transgenic plants, it is important to develop a strategy to minimize the rate of development of resistance in insect populations through:

- Control of secondary pests
- Resistance management
- Gene pyramiding and gene deployment
- Regulation of gene expression
- Planting refugia and destruction of carryover population
- Control of alternate hosts and use of planting window
- Following integrated pest management from the very beginning.

1.3. GENETIC IMPROVEMENT OF NATURAL ENEMIES

Some of the major problems in using natural enemies in pest control are the difficulties involved in mass rearing and their inability to withstand adverse conditions in the field. Genetic improvement can be useful when the natural enemy is known to be a potentially effective biological control agent, except for one limiting factor. Some of the desirable characteristics for transgenic insects include resistance to pathogens, adaptation to different environmental conditions, high fecundity, and improved host-seeking ability (Atkinson and O'Brochta, 1999). Biotechnological interventions can also be used to broaden the host range of natural enemies or enable their production on artificial diet or non-host insect species that are easy to multiply under laboratory conditions. In addition, there is a tremendous scope for developing natural enemies with genes for resistance to pesticides (Hoy, 2000). Genetic transformation can also be used to improve commercial production of materials such as silk, honey, and lac (Mori and Tsukada, 2000), or production of pharmaceuticals and biomolecules (Yang *et al.*, 2002). However, release of genetically modified insects might have a potential risk to the environment. This is of particular concern when the same vector transmits several disease causing pathogens, as it might be difficult to develop transgenic individuals incapable of transmitting different pathogens.

1.4. GENETIC IMPROVEMENT OF BIOPESTICIDES

Genetic engineering can be used to improve the efficacy of entomopathogenic microorganisms. Efforts to improve *Bt* have largely been focused on increasing its host range and stability. Work on *baculoviruses* is largely focused on incorporation of genes that produce the proteins/toxins, which kill the insects at a faster rate (Bonning and Hammock, 1996), and removal of polyhedrin gene, which produces the protective viral-coat protein, and its persistence in the field (Corey, 1991). Neurotoxins produced by spiders and scorpions have also been expressed in transgenic organisms (Barton and Miller, 1991). Incorporation of benomyl resistance into *Metarhizium anisopliae* and other entomopathogenic fungi could make entomopathogenic fungi more useful for use in integrated pest management (Goettel *et al.*, 1989). The role of neurotoxins from insects and spiders need to be studied in greater detail before they are deployed in other organisms because of their possible toxicity to mammals and other non-target organisms.

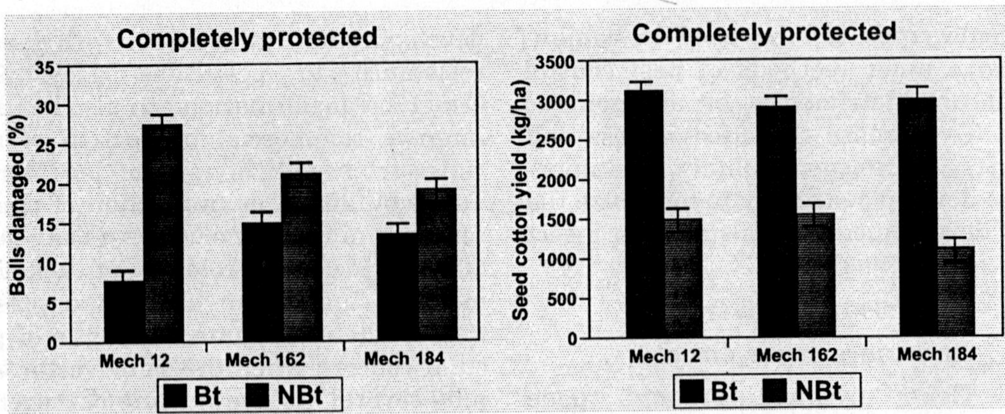

Fig. 1. Bollworm damage (left) and seedcotton yield (right) of transgenic and non-tansgenic cotton hybrids under insecticide protection (Sharma and Pampapathy, 2006).

1.5. MOLECULAR MARKER-ASSISTED SELECTION FOR INSECT RESISTANCE

Recombinant DNA technologies, besides generating information on gene sequences and function, allow the identification of specific chromosomal regions carrying genes contributing to traits of economic interest (Karp *et al.*, 1997). Once genomic regions contributing to the trait of interest have been assigned and the alleles at each locus designated, they can be transferred into locally adapted high-yielding cultivars by making requisite crosses. The offspring with a desired combination of alleles can then be selected for further evaluation using marker assisted selection (MAS). The quality of a marker assisted selection program can only be as good as the quality of the phenotypic data on which the development of that marker was based. It is therefore, essential to use large mapping populations, which are precisely and accurately characterized across seasons and locations. Marker-assisted selection (MAS) can be used to accelerate the pace and accuracy of transferring insect resistance genes into improved cultivars. Several markers have been used to identify QTLs for insect resistance in different crops (Smith, 2005). In contrast to the markers linked to resistance genes inherited as simple dominant traits, the improvement of polygenic traits (QTLs) through MAS is difficult due to involvement of a number of genes, and their interactions (epistatic effects). Several studies on QTLs linked to stem borer resistance in maize underscore the problems involved in using QTLs in MAS. The relative efficiency of phenotypic selection and MAS has been found to be similar (Groh *et al.*, 1998; Willcox *et al.*, 2002). However, phenotypic selection was more favorable due to lower costs. Maximum progress has been made in breeding for insect resistance in common bean by using a combination of phenotypic performance and QTL-based index, followed by QTL based index, and conventional selection (Tar'an *et al.*, 2003).

1.6. UNDERSTANDING GENE SEQUENCE AND FUNCTION

Genes can be discovered using a variety of approaches (Shoemaker *et al.*, 2001), but a routine large-scale approach can commonly be followed by generating and sequencing a library of expressed genes. A large number of ESTs are now available in the public databases for several crops such as *Zea mays*, *Arabidopsis thaliana*, *Oryza sativa*, *Sorghum bicolor*, and *Glycine max*. A comparison of the EST databases from different plants can reveal the diversity in coding sequences between closely and distantly related species, while mapping of ESTs may elucidate the synteny between those species. For understanding gene functions of a whole organism, functional genomics technology is now focused on high throughput methods using insertion mutant isolation, gene chips or micro-arrays, and proteomics. These and other high throughput techniques offer powerful new uses for the genes discovered through sequencing (Hunt and Livesey, 2000).

1.7. METABOLIC PATHWAYS

Genetic engineering can be used to change the metabolic pathways to increase the amounts of secondary metabolites, which play an important role in host-plant resistance to insect pests e.g., medicarpin and sativan in alfalfa, cajanol and stilbene in pigeonpea, deoxyanthocyanidin flavonoids (luteolinidin, apigenindin, etc.) in sorghum, and stilbene in chickpea (Sharma *et al.*, 2002). The expression of phytoalexins in transgenic plants may be difficult due to complexities involved in their biosynthesis. Expression of a bacterial cytokinin biosynthesis gene (*PI-II-ipt*) in *Nicotiana plumbaginifolia* plants has been correlated with enhanced resistance to *M. sexta* and *M. persicae* (Smigocki *et al.*, 2000). Molecular mechanisms underlying the activation of defense genes implicated in phytoalexin biosynthesis are quite common in a large number of plant species. Biotechnology offers the promise to increase the production of secondary metabolites in plants to increase the levels of resistance to insect pests.

1.8. INDUCIBLE RESISTANCE

Induced resistance results in changes in a plant that produce a negative effect on herbivores (Karban and Baldwin, 1997). Chemically induced expression systems or "gene switches" enable temporal, spatial, and quantitative control of genes introduced into plants or those that are already present in the plants to impart resistance to insects. A number of inducible genes have been identified in plants based on endogenous chemical signals such as phytohormones, responses to insect and pathogen attack, or wounding. Effectiveness of the chemical injury inducer, Actigard™ in providing resistance to various insect pests and pathogens in the tomato has been demonstrated by Inbar *et al.* (1998). Proteinase inhibitors and oxidative enzymes such as polyphenol oxidase, peroxidase, and lipoxygenase persist for at least 21 days after induction in damaged tomato leaflets (Stout *et al.*, 1996). Exogenous application of jasmonic acid and salicylic acid has also been shown to induce resistance to several insect pests. The best-studied system utilizes the *PR1-a* promoter from tobacco, which is induced during systemic resistance response following pathogen infection (Uknes *et al.*, 1993). Induced resistance, both in transgenic and non-transgenic plants, offers considerable promise to increase the levels of resistance to insect pests.

1.9. USE OF MOLECULAR MARKERS FOR DIAGNOSIS OF INSECT PESTS AND THEIR NATURAL ENEMIES

For developing appropriate strategies for managing insect pests, it is important to have a correct identification of the pest species. Correct taxonomic identification is also important for import and export of plant material/food grains to implement quarantine procedures. Identification of insect pests has primarily relied on morphological characters of adult life stages. However, intercepted specimens often are not in the adult stage and may be damaged, which seriously handicaps correct identification. The molecular tools now enable precise and rapid identification of insect pests, irrespective of the developmental stage and condition of the samples. The modern tools of biotechnology can be used for detection and identification of insect pests, insect biotypes, and understand genetic diversity, population structure, tri-trophic interactions, and insect plant relationships (Caterino *et al.*, 2000; Heckel, 2003). Molecular markers can also be used to gain a basic understanding of their interaction with environment, and develop sound strategies for pest management.

1.10. DEVELOPMENT OF NEW INSECTICIDE MOLECULES

Crop protection is still dominated by chemical control, and this approach will continue to be important in crop protection in future. Traditionally, the discovery of new agrochemicals has used *in vivo* screens to identify new compounds. Functional genomics offers the opportunity to acquire in-depth knowledge of the genetic make-up and gene function of insect pests that may lead to the discovery of new processes that could be the targets for novel chemistry (Hess *et al.*, 2001). Combining genomics with high throughput biochemical screening can be used to identify a range of new chemicals for pest control. Genomic technologies are now allowing investigation of some previously intractable mechanisms involved in insect resistance to insecticides. New molecular techniques permit fundamental insights into the nature of mutations and genetic processes such as gene amplification, altered gene transcription, and amino acid substitution to underpin insecticide resistance mechanisms. This in turn will lead to high-resolution diagnostics for resistance alleles in homozygous and heterozygous form, especially for insect pests with multiple resistance mechanisms, or for resistance mechanisms not amenable to biochemical assays.

1.11. DOMINANT REPRESSIBLE LETHAL GENETIC SYSTEM TO PRODUCE STERILE INSECTS

The sterile insect technique (SIT) has been employed to control several insect pests. However, this system depends on large-scale production of the target insect, and use of irradiation or chemical sterilization. Release of insects carrying a dominant lethal (RIDL) gene has been proposed as an alternative to the conventional techniques used for insect sterilization (Alphey and Andreasen 2001). This is based on the use of a dominant, repres-sible, female-specific gene for insect control. A sex-specific promotor or enhancer gene is used to drive the expression of a repressible transcription factor, which in turn controls the production of a toxic gene product. A non-sex specific expression of the repressible transcription factor can also be used to regulate a selectively lethal gene product. Insects produced through genetic transformation using this approach do not require sterilization through irradiation, and could be released in the ecosystem to mate with the wild population to produce sterile insects, which will be self-perpetuating.

1.12. PROSPECTS AND LIMITATIONS

There will be tremendous potential benefits to the environment through the deployment of genetically modified crops and use of other tools of biotechnology for pest management (Sharma, 2009). Deployment of insect-resistant crops has been estimated to result in tremendous reduction of pesticides applied for pest control (NRC, 2000). Transgenic plants with insecticidal genes are set to feature prominently in pest management in both developed and the developing world in future. Such an effort will play a major role in minimizing insect-associated losses, increase crop production, and improve the quality of life for the rural poor. In addition to the reduction in losses due to insect pests, the deployment of transgenic plants with insecticidal genes and other molecular approaches for pest management will also lead to more efficient pest management and sustainable crop production.

The benefits of growing transgenic crops to growers have been higher yield, lower input costs, and easier agronomic management. These factors are likely to have substantial impact on the livelihoods of farmers in both industrial and developing countries. On-farm trials carried out with *Bt* cotton in different states of India have shown that the technology substantially reduces pest damage and increases cottonseed yield (Qaim and Zilberman, 2003). The yield gains are much greater than those reported for other countries where genetically modified crops were used mostly to replace and/or enhance chemical pest control. However, there is considerable public debate on the issue, and several claims to the contrary have also been published in the public media. In many developing countries, small-scale farmers suffer pest-related yield losses because of technical and economic constraints. Pest-resistant genetically modified crops can contribute to increased yields and agricultural growth in such situations. Adopting transgenic crops also offers the additional advantage of controlling insect pests that have become resistant to commonly used insecticides (Sharma *et al.*, 2004).

Biotechnological approaches can be used in conjunction with other methods of pest control without any detrimental or antagonistic effects. For rational pest management, products derived through genetic manipulation have to be deployed in combination with conventional insecticides (Schell, 1997). Transgenic crops can make a critical contribution in reducing the dosage and frequency of insecticide application. Insects such as *H. virescens* and *H. zea* are many times more sensitive to pyrethroid sprays when they have a prior exposure to *Bt* toxins (Harris *et al.*, 1998). Transgenic *Bt* cotton and insecticides are highly effective for bollworm control even at lower rates of insecticide application (Brickle *et al.*, 1999).

Not all problems can be solved through the deployment of transgenic crops and other tools of biotechnology as they too have some limitations. Insensitivity of some insect species to *Bt* toxins, emergence of secondary pest problems, gene escape, and possible effects on non-target insects are some of the issues that need to addressed while deploying genetically modified organisms for pest management. There is a considerable debate about the environmental risks such as development of resistance, harmful effects on beneficial insects, and gene flow to the closely related wild relatives of the crops. The evidence on these issues is still inconclusive and warrants careful monitoring before transgenic crops are deployed on a large scale (Sharma and Ortiz, 2000). The biggest risk of modern biotechnology for developing countries is that technological developments may bypass the poor farmers because of a lack of enlightened adoption. It is not that the current biotechnology research in developed countries is irrelevant, but there is a desperate need for research to focus on the problems of small-scale farmers in developing countries. Private sector research is unlikely to take on such a focus, given the uncertainty of future profits in these areas. Without a stronger public sector role, a form of scientific apartheid may develop, in which cutting edge science becomes oriented exclusively toward industrial countries and large-scale farming systems. Efficacy of transgenic crops for controlling the target and the non-target insect pests, and their effects on the natural enemies need to be determined on a regional basis. The significance of such effects would depend on the importance of the immature stages of the target insect for maintaining the populations of the natural enemies (Sharma and Ortiz, 2000). Formulations based on *Bt* are known to be relatively safer to the environment. However, there are certain subtle differences between the *Bt* toxins applied as sprays, and the toxin proteins produced by the transgenic plants that warrant in-depth studies to assess the effects of transgenic crops in the environment on a long term basis.

Transgenics may reduce the numbers of certain host specific natural enemies in areas planted with transgenic crops, but their populations may be maintained on the other crops that serve as a host to the target pests. A few of the known predators are specialists on one insect, and hence, the populations of the predators and parasites with a wide host range would be maintained on the other insect species. Within field impact may be greater for parasitoids that are monophagous. The populations of such natural enemies can only be maintained on the non-transgenic crops or other hosts of the target pest. Finally, the effect of transgenic crops on the abundance of natural enemies should be compared with the non-transgenic fields of the same crop where the natural enemies may be virtually absent because of heavy pesticide application. *Bt*-transgenic plants appear to have little impact on soil biota, such as earthworms, collembolans, and general soil microflora (O'Callaghan *et al.*, 2005).

One of the concerns of deploying transgenic crops is the possibility of vertical and horizontal gene flow. In case of gene transfer to the wild relatives, will it lead to expansion of the niche of the species and result in suppression of diversity in the surrounding areas. Introduction of transgenic plants may also result in an increase in land use for agriculture, where agriculture could not be practiced earlier, i.e., bringing valuable natural ecosystem under agriculture. The greatest risk of a transgenic plants release into the environment is its potential to spread beyond the areas planted. However, there are no records of a plant becoming a weed as a result of plant breeding (Cook, 2000), although there may be some exceptions, e.g., oilseed rape in the European continent. This may be because of low risk of crop plants to the environment, extensive testing of the crop varieties before release and adequate management practices to mitigate any risks inherent in the crop plants. Plant breeding efforts have tended to decrease rather than increase toxic substances, as a result, making the improved varieties more susceptible to insect pests. However, there is a feeling that genes introduced from outside the range of sexual compatibility might present new risks to the environment and humans. However, these apprehensions are not supported by data. Herbicide tolerance is available in many plant species, but it is more cost effective to introduce this trait through genetic transformation. One of the hazards with transgenic plants is the transfer of genes to wild relatives, but this is only a major concern if the wild relatives are under selection pressure (biological control) from that pest. If the target pest does not play any role in population regulation of the wild hosts, gene transfer is unlikely to constitute any hazard. Furthermore, the build up of resistance in the wild relatives can also act as a component of pest management if it acts as an alternate host of the target pest.

1.13. CONCLUSIONS

Predicted growth in world population and the likely effects of climate change will pose a serious challenge to crop production and food security, particularly in developing countries. The augmentation of conventional breeding with the use of marker-assisted selection and transgenic plants promise to facilitate substantial increases in food production. However, knowledge of the physiology and biochemistry of plants will be extremely important for interpreting the information from molecular markers and deriving new and more effective paradigms in plant breeding. The application of DNA marker technologies in exploiting the vast and largely under-utilized pool of favorable alleles existing in the wild relatives of crops will provide a huge new resource of genetic variation to fuel the next phase of crop improvement. In particular, tremendous benefits will be derived through the transfer of genes important for crop protection and crop quality. Tools of biotechnology can be effectively used for detection and identification of insects, improve the efficacy of biocontrol agents, and discover new insecticide molecules, for effective pest management programs. However, rapid and cost effective development, and adoption of biotechnology-derived products will depend on developing a full understanding of the interaction of genes within their genomic environment, and with the environment in which their conferred phenotype must interact.

REFERENCES

Alphey L and Andreasen M (2001) Dominant lethality and insect population control. *Molecular and Biochemical Parasitology* **121**: 173-178.

Armstrong CL, Parker GB, Pershing JC, Brown SM, Sander PR, Duncan DR, Stone T, Dean DA, DeBoer DL, Hart J, Howe AR, Morrish FM, Pajeau ME, Peterse WL, Reich BJ, Rodriguez R, Santino CG, Sato SJ, Schule W, Sims SR, Stehling S, Tarochione LJ and Fromm ME (1995) Field evaluation of European corn borer control in progeny of 173 transgenic corn events expressing an insecticidal protein from *Bacillus thuringiensis*. *Crop Science* **35**: 550-557.

Atkinson PW and O'Brochta DA (1999) Genetic transformation of non-drosophilid insects by transposable elements. *Annals of the Entomological Society of America* **92**: 930-936.

Barton KA and Miller MJ (1991) Insecticidal toxins in plants. *EPA 0431829 A1 910612.*

Benedict JH, Sachs ES, Altman DW, Deaton DR, Kohel RJ, Ring DR and Berberich BA (1996) Field performance of cotton expressing CryIA insecticidal crystal protein for resistance to *Heliothis virescens* and *Helicoverpa zea* (Lepidoptera: Noctuidae). *Journal of Economic Entomology* **89**: 230-238.

Bonning BC and Hammock BD (1996) Development of recombinant baculovirus for insect control. *Annual Review of Entomology* **41**: 191-210.

Brickle DS, Turnipseed SG, Sullivan MJ and Dugger P (1999) The efficacy of different insecticides and rates against bollworms (Lepidoptera: Noctuidae) in B.T. and conventional cotton. In: Richter D (ed.), *Proceedings, Beltwide Cotton Production and Research Conference*, 3-7 January, 1999. Orlando, Florida, USA. Volume 2. Memphis, USA: National Cotton Council. pp. 934 – 936.

Caterino MS, Cho S and Sperling, FA (2000) The current state of insect molecular systematics: a thriving Tower of Babel. *Annual Review of Entomology* **45**: 1-54.

Corey JS (1991) Releases of genetically modified viruses. *Medical Virology* **1**: 79-88.

Delannay X, LaVallee BJ, Proksch RK, Fuchs RL, Sims SK, Greenplate JT, Marrone PG, Dodson RB, Augustine JJ, Layton JG and Fischhoff DA (1989) Field performance of transgenic tomato plants expressing *Bacillus thuringiensis* var *kurstaki* insect control protein. *BioTechnology* **7**: 1265-1269.

Department of Biotechnology (DBT) (1998) *Revised Guidelines for Research in Transgenic Plants*. Department of Biotechnology, Ministry of Science and Technology, Government of India, New Delhi, India. 62 pp.

Goettel MS, St Leger RJ, Bhairi S, Jung MK, Oakle, BR and Staples RC (1989) Transformation of the entomopathogenic fungus, *Metarhizium anisopliae*, using the *Aspergillus nidulans* ben A3 gene. *Current Genetics* **17**: 129-132.

Groh S, Khairallah MM, Gonzalez de Leon D, Willcox M., Jiang C, Hoisington DA and Melchinger AE (1998) Comparison of QTLs mapped in RILs and their test-cross progenies of tropical maize for insect resistance and agronomic traits. *Plant Breeding* **117**: 193-202.

Harris JG, Hershey CN, Watkins MJ and Dugger P (1998) The usage of Karate (lambda-cyhalothrin) oversprays in combination with refugia, as a viable and sustainable resistance management strategy for B.T. cotton. In: *Proceedings, Beltwide Cotton Conference,* 5-9 January 1998, San Diego, California, USA. Volume 2. Memphis, USA: National Cotton Council. pp. 1217-1220.

Heckel DG (2003) Genomics in pure and applied entomology. *Annual Review of Entomology* **48**: 235-260.

Hess FD, Anderson RJ and Reagan JD (2001) High throughput synthesis and screening: the partner of genomics for discovering of new chemicals for agriculture. *Weed Science* **49**: 249-256.

Hilder VA and Boulter D (1999) Genetic engineering of crop plants for insect resistance – a critical review. *Crop Protection* **18**: 199-191.

Hoy, MA (2000) Transgenic arthropods for pest management programs: risks and realities. *Experimental and Applied Acarology* **24**: 463-495.

Hunt SP and Livesey FJ (2000) *Functional Genomics: A Practical Approach*. Oxford University Press, Oxford. 253 pp.

Inbar M, Doodstar H, Sonoda RM, Leibee GL and Mayer RT (1998) Elicitors of plant defensive systems reduce insect densities and disease incidence. *Journal of Chemical Ecology* **24**: 135-149.

James C (2005) *Global Status of Commercialized Biotech / GM Crops: 2003*. ISAAA

Briefs no. 34. International Service for Acquisition on Agri-Biotech Applications (ISAAA) Ithaca, New York, USA. http://www.isaaa.org/Publications/briefs_27.htm.

Jansens S, Cornelissen M, Clercq R de, Reynaerts A and Peferoen M (1995) *Phthorimaea opercullella* (Lepidoptera: Gelechiidae) resistance in potato by expression of *Bacillus thuringiensis* Cry IA(b) insecticidal crystal protein. *Journal of Economic Entomology* **88**: 1469-1476.

Kar S, Basu D, Das S, Ramkrishnan NA, Mukherjee P, Nayak P and Sen SK (1997) Expression of cryIA(c) gene of *Bacillus thuringiensis* in transgenic chickpea plants inhibits development of pod borer (*Heliothis armigera*) larvae. *Transgenic Research* **6**: 177-185.

Karban R and Baldwin IT (1997) *Induced Responses to Herbivory.* Chicago, Illinois, USA: The University of Chicago Press.

Karp A, Edwards KJ, Bruford M, Funk S, Vosman B, Morgante M, Seberg O, Kremer A, Boursot P, Arctander P, Tautz D and Hewitt GM (1997) Molecular technologies for biodiversity evaluation: opportunities and challenges. *Nature Biotechnology* **15**: 625-628.

Kumar PA, Mandaokar A, Sreenivasu K, Chakrabarti SK, Bisaria S, Sharma SR, Kaur S and Sharma RP (1998) Insect-resistant transgenic brinjal plants. *Molecular Breeding* **4**: 33-37

Mori H and Tsukada M (2000) New silk protein: modification of silk protein by gene engineering for production of biomaterials. *Reviews in Molecular Biotechnology* **74**: 95-103.

Nayak P, Basu D, Das S, Basu A, Ghosh D, Ramakrishnan NA, Ghosh M and Sen SK (1997) Transgenic elite *indica* rice plants expressing CryIAc delta-endotoxin of *Bacillus thuringiensis* are resistant against yellow stem borer (*Scirpophaga incertulas*). *Proceedings, National Academy of Sciences, USA* **94**: 2111-2116.

O'Callaghan M, Glare TR, Burgess EPJ and Malone LA (2005) Effects of plants genetically modified for insect resistance on nontarget organisms. *Annual Review of Entomology* **50**: 271-292.

Oerke EC, Dehne HW, Schonbeck F and Weber A (1994) *Crop Production and Crop Protection: Estimated Losses in Major Food and Cash Crops.* Amsterdam, The Netherlands: Elsevier.

Qaim M and Zilberman D (2003) Yield effects of genetically modified crops in developing countries. *Science* **299**: 900-902.

Sanyal I, Singh AK, Meetu K and Devindra AV (2005) *Agrobacterium*-mediated transformation of chickpea (*Cicer arietinum* L.) with *Bacillus thuringiensis cry1Ac* gene for resistance against pod borer insect *Helicoverpa armigera. Plant Science* **168**: 1135-1146.

Schell J (1997) Cotton carrying the recombinant insect poison Bt toxin: no case to doubt the benefits of plant biotechnology. *Current Opinion in Biotechnology* **8**: 235-236.

Sharma HC (2009) *Biotechnological Approaches for Pest Management and Ecological Sustainability.* CRC Press, Boca Raton, Florida, USA. 526 pp.

Sharma HC and Ortiz R (2000) Transgenics, pest management, and the environment. *Current Science* **79**: 421-437.

Sharma HC and Pampapathy G (2006) Influence of transgenic cotton on the relative abundance and damage by target and non-target insect pests under different protection regimes in India. *Crop Protection* **25**: 800-813.

Sharma HC, Crouch JH, Sharma KK, Seetharama N and Hash CT (2002) Applications of biotechnology for crop improvement: prospects and constraints. *Plant Science* **163**: 381-395.

Sharma HC, Sharma KK and Crouch JH (2004) Genetic engineering of crops for insect control: Effectiveness and strategies for gene deployment. *CRC Critical Reviews in Plant Sciences* **23**: 47-72.

Sharma KK, Lavanya M and Anjaiah V (2006) *Agrobacterium*-mediated production of transgenic pigeonpea (*Cajanus cajan* L. Millsp.) expressing the synt! Bt cry1Ab gene. *In Vitro Cellular & Developmental Biology – Plant* **42**: 165-173.

Shoemaker DD, Schadt EE, Armour YD, He P, Garrett-Engel PD and McDonayl PM (2001) Experimental annotation of the human genome using microarray technology. *Nature* **409**: 922-927

Singsit C, Adang MJ, Lynch RE, Anderson WF, Wang A, Cardineau G and Ozias-Akins P (1997) Expression of a *Bacillus thuringiensis* cryIA(c) gene in transgenic peanut plants and its efficacy against lesser cornstalk borer. *Transgenic Research* **6**: 169-176.

Smigocki A, Heu S and Buta G (2000) Analysis of insecticidal activity in transgenic plants carrying the ipt plant growth hormone gene. EUCARPIA TOMATO 2000. XIV meeting of the EUCARPIA Tomato Working Group, Warsaw, Poland, 20-24 August 2000. *Acta Physiologia Plantarum* **22**: 295-299.

Smith CM (2005) *Plant Resistance to Arthropods – Molecular and Conventional Approaches*. Springer Verlag, Dordrecht, The Netherlands. 423 pp.

Stout MJ, Workman J and Duffey SS (1996) Differential induction of tomato foliar proteins by arthropod herbivores. *Journal of Chemical Ecology* **20**: 2575-2594.

Surekha Ch, Beena MR, Arundhati A, Singh PK, Tuli R, Dutta-Gupta A and Kirti PB (2005) *Agrobacterium*-mediated transformation of pigeonpea (*Cajanus cajan* (L.) Millsp.) using embryonal segments and development of transgenic plants for resistance against *Spodoptera*. *Plant Science* **169**: 1074-1080.

Tar'an B, Thomas E, Michaels TE and Pauls KP (2003) Marker assisted selection for complex trait in common bean (*Phaseolus vulgaris* L.) using QTL-based index. *Euphytica* **130**: 423-433.

Uknes S, Dincher S, Friedrich L, Negrotto D, Williams S, Thompson Taylor H, Potter S, Ward E and Ryals J (1993) Regulation of pathogenesis-related Protein-1a gene expression in tobacco. *Plant Cell* **5**: 159-169.

Willcox MC, Khairallah MM, Bergvinson D, Crossa J, Deutsch JA, Edmeades GO, Gonzalez-de-Leon D, Jiang C, Jewell DC, Mihm JA, Williams WP and Hoisington D (2002) Selection for resistance to southwestern corn borer using marker-assisted and conventional backcrossing. *Crop Science* **42**: 1516-1528.

Yang G, Chen Z, Cui D, Li B and Wu X (2002) Production of recombinant human calcitonin from silkworm (*B. mori*) larvae infected by baculovirus. *Current Pharmaceutical Design* **9**: 323-329.

Biological and Molecular Approaches in Pest Management
Balwinder Singh, Ramesh Arora & S S Gosal (Eds)
Scientific Publishers (India), Jodhpur, 2015, pp 13-38.

Chapter - 2

Biotechnological Approaches for Enhancing Resistance to Hoppers in Rice

D S Brar[2,1], P S Sarao[1], Kuldeep Singh[1], K K Jena[2] and D Fujita[2]
[1]*Punjab Agricultural University, Ludhiana, India*
[2]*International Rice Research Institute, Manila, Philippines*
E-mail: preetento@gmail.com

2.1. INTRODUCTION

Rice, *Oryza sativa* L. is the staple food for over half of the world's population. It provides 27 per cent of dietary energy supply and 20 per cent dietary protein intake in the developing countries. More than 90 per cent rice is produced and consumed in Asia. The world rice production has more than doubled from 256 to 650 m tons. This has been possible through the application of the principles of Mendelian genetics and conventional plant breeding methods and with the adoption of improved production technologies. Global rice demand is estimated to rise from 676 m tons to 852 m tons in 2035. Furthermore, rice productivity is continuously threatened by major biotic and abiotic stresses which need to be controlled if the required productivity level is to be achieved. Among the insect pests, plant hoppers cause significant loss in yield. Various strategies available to control pests are 1) chemical control, 2) biological control, 3) crop management and insect pest management (IPM) and 4) host-plant resistance. Of the various strategies, host–plant resistance is the most effective and economical approach to control insect pests. Rice is attacked by six kind of hoppers, brown planthopper (BPH), whitebacked planthopper (WBPH), green leafhopper (GLH), zigzag leafhopper (ZLH), small brown planthopper (SBPH), and green rice leafhopper (GRH). All of them cause yield losses in rice to a variable extent and at various growth stages. These hoppers are also vectors of major viral diseases, such as grassy stunt, ragged stunt, rice stripe virus, black streak, and tungro disease. A number of donors for resistance have been identified and used in breeding varieties resistant to hoppers. However, conventional breeding approaches pose several limitations in transferring insect resistance genes. As a result unlike disease resistance majority of the varieties do not possess insect resistance genes. As many as 27 resistance genes have been identified for BPH, eight for WBPH, 13 for GLH, six for GRH, three for ZLH, and one gene and several QTL for SBPH. A number of genes for BPH resistance have been introgressed across crossability barriers from wild species such as *Oryza officinalis, O. minuta, O. australiensis.* Future research should focus on establishing high-throughput screening protocols for field resistance, identifying new genes for resistance from diverse sources, and developing varieties with durable resistance using MAS through pyramiding of major genes and QTLs.

Recent advances in biotechnology, particularly in molecular biology and genomics offer new opportunities to overcome some of the constraints limiting rice production and to enhance resistance to plant hoppers. Major developments in rice biotechnology include 1) development of a dense molecular map consisting of thousands of SSR and SNP markers, 2) tagging of major genes/QTLs for several agronomic traits for practising marker-assisted selection (MAS), 3) map-based cloning of genes, 4) high throughput *Agrobacterium* mediated transformation of both indica and japonica rice, 5) large assembly of genetic resources for genomics research such as BAC, YAC, T-DNA insertion lines and mutant stocks and 6) huge data on genome sequences of indica and japonica cultivars- a valuable resource for functional genomics and reverse genetics research.

2.2. SOURCES OF RESISTANCE TO HOPPERS IN RICE

Resistant and moderately resistant varieties keep the pest densities to remain below economic threshold levels and best suited to be used in pest management in combination with natural enemies. These varieties also increase the effectiveness of the natural enemies because of a favourable ratio between the densities of the target pest and its natural enemies (Gurr *et al.,* 2011). Painter, (1951) and others demonstrated that clear-cut cases of host resistance existed in crop species of importance to agriculture. This can be grouped into three mechanisms viz., nonpreference/ antixenosis, antibiosis and tolerance (Painter, 1951; Kogan & Ortman, 1978). These three types of resistance mechanisms has been reported in rice by many scientists in different genotypes carrying BPH resistance genes (Khush & Brar, 1991; Liu *et al.,* 2008; Oiu *et al.,* 2010). Li *et al.* (2010) reported that the genes *Bph 14* and *Bph 15* conferred antixenosis and antibiosis effects against BPH. Similarly, Du *et al.* (2009) also reported that *Bph 14* reduces the feeding and growth rate of BPH as an antibiosis mechanism of resistance. In addition resistance to insect pests in plants is manifested due to several mechanisms. Restless behavior of insects on the resistant varieties also increases their vulnerability to the natural enemies. Prolonged developmental period of immature stages increases the susceptibility period of the target pest to the natural enemies or results in the synchronization of insect developmental stages with the peak activity and abundance of the natural enemies. The use of these combinations brings the unrelated mortality factors and provides the density-independent mortality of insect-pests (Gould *et al.,* 1991). Several donors/genes for resistance have been identified and used to develop hopper-resistant varieties (Table 1).

Table 1. Genes and sources of resistance to hoppers in rice

Gene (1)	Genotype(s) (2)
Brown planthopper (BPH)	
Bph1	Mudgo, CO22 (IT 000588), IR747B2-6 (from TKM6), Milyang30, Milyang34 (IT 006216), Nampungbyeo, Chilseongbyeo, Andabyeo, Kanto PL4 (IT173362), Cheongcheongbyeo, Changsongbyeo, Baekunchalbyeo, IR26 (IT001886), IR28 (IT001892), IR29 (IT001893), IR30 (IT001899), Hangangchalbyeo, Yeongpungbyeo, Namyeongbyeo, Gayabyeo, Samgangbyeo, Namcheonbyeo, MTU15, IR26, IR28, IR29, IR30, IR34, IR44, IR45, IR46, IR64 and MGL2
bph2	ASD7 (acc. no. 6303), ASD9, IR 1154-243, Norin-PL4 (line from IR1154-243), Hwacheongbyeo, PTB18, PTB33, H105, Palasithari
Bph3	Rathu Heenati (acc. no. 11730), PTB19, Gangala, Horana Mawee, Muthumanikam, Kuruhondarawala, Mudu Kiriyal, PTB33, IR56, IR58, IR60, IR62, IR68, IR70, IR72, IR74
bph4	Babawee (acc. no. 8978), Gambada Samba, Hotel Samba, Kahata Samba, Thirissa, Sulai, Vellai Illankali, Heenhoranamawee, Kulu Kuruwee, Lekam Samba, Senawee and IR66
bph5	ARC10550 (acc. no. 12507)
Bph6	IR54741-3-21-22, Swarnalata (acc. no. 33964), *O. officinalis* (acc.00896)
bph7	T12 (acc. no. 59689)
bph8	Chin Saba (acc. no. 33016), Col. 5 Thailand and Col. 11 Thailand
Bph9	Pokkali, Balamee and Kaharamana
Bph10	IR65482-4-136-2-2 (*O. australiensis* acc. no. 100882)
bph11	IR 54751-2-44-15-24-3, IR742-23-19-12-3-54 (*O. officinalis*), DV85
Bph12	GSK185-2 (*O. officinalis*), B14 (*O. latifolia*), *O. latifolia* and IR54751-2-34-10-6-2
bph12	*O. officinalis*
Bph13	IR54745-2-21-12-17-6 (*O. officinalis*), *O. eichingeri* (acc. no.105159), *O. officinalis* (acc. no. 00896)
Bph14	B5 (*O. officinalis*), RI35, *O. officinalis*
Bph15	B5 (*O. officinalis*) and *O. officinalis*
Bph16 (t)	M1635-7 (*O. minuta*)
Bph17	Rathu Heenati
Bph18	IR65482-7-216-1-2 (*O. australiensis* acc. no. 100882), BPH2183
bph19	AS20-1 (*O. rufipogon*)

Contd...

(1)	(2)
Bph20	IR71033-121-15 (*O. minuta* acc. no. 101141) and ADR 52
bph20(t)	RBPH54 (*O. rufipogon*)
Bph21	IR71033-121-15 (*O. minuta* acc. no. 101141) and ADR52
bph21(t)	RBPH54 (*O. rufipogon*)
Bph22	IR75870-5-8-5-B-2-B and IR75870-5-8-5-B-1-B, IR71033-62-24 (*O. minuta*), IR75870-5-8-5-B-1-B (*O. glaberrima*)
Bph23	IR71033-121-15 (*O. minuta*)
bph24	IR73678-6-9-B (*O. rufipogon*)
Bph25	ADR52
Bph26	ADR52
Bph27	GX2183
Bph(t)	Sanguizhan
Bph(t)	IR54741-3-21-22 (*O. officinalis*)
Qbph1	B5 (*O. officinalis*)
Qbph2	Col.5 Tailand, B5 (*O. officinalis*), *O. eichingeri* (acc. no. 105159), Yagyaw
Qbph4	Rathu Heenati (acc. no. 11730), Yagyaw
Qbph6	Col. 5 Tailnad
qBPH6(t)	IR71033-121-15 (*O. minuta*)
Qbph7	Yagyaw
Qbph9	Yag[illegible]
Qbph10	Rathu Heenati (acc. no. 11730)
Qbph11	DV85
Small brown planthopper (SBPH)	
Sbph(t)	IR50
Whitebacked planthopper (WBPH)	
Wbph1	N22
Wbph2	ARC1010239
Wbph3	ADR52
wbph4	Podiwi-A8
Wbph5	N'Diang Marie
Wbph6	Guiyigu
Wbph7(t)	B5 (*O. officinalis*)
Wbph8(t)	B5 (*O. officinalis*)
Ovc	Asominori

Contd...

(1)	(2)
Green leafhopper (GLH)	
Glh1	Pankhari 203
Glh2	ASD7
Glh3	IR8
glh4	Ptb8
Glh5	ASD8
Glh6	TAPL796
Glh7	Maddai Karuppan
glh8	DV85
Glh9	IR28
glh10	IR36
Glh11	IR20965-11-3-3 (*O. glaberrima*)
Glh12	ARC10313
Glh13	Asmaita
Glh14	ARC11554
Zigzag leafhopper (ZLH)	
Zlh1	Rathu Heenati
Zlh2	Ptb21
Zlh3	Ptb33

Source: Modified after Brar *et al.* (2009), Fujita *et al.* (2013)

2.3. BREEDING STRATEGY TO ENHANCE HOPPER RESISTANCE

2.3.1. Phenotyping for Hopper Resistance

Screening against hoppers has good possibility of success in selecting resistance/ moderate resistant gene sources and transferring these gene sources into elite agronomic background. This selection requires availability of insects of synchronized stage, diverse germplasm source and good working conditions. For greenhouse screening an efficient rearing programme is required. If field screening is to be conducted then a high natural population is required (hot spots).

Developing an efficient rating method is very important and it depends mostly on degree of plant damage and insect numbers. This is an initial step in the screening technology which is needed to eliminate the majority of susceptible segregants and select the resistant ones. Such large-scale evaluation where insects are offered a free choice of plant materials can be accomplished in the greenhouse, screenhouse or small field plots. The approach for screening and evaluating resistance will depend on the insect and the crop under study, the required insect numbers, as well as the availability of research facilities. If insect damage occurs at more than one stage of plant growth, it is important to evaluate resistance at each of those stages. The conventional seedbox screening or standard seedbox screening test (SSST) is a common method used for greenhouse screening. It is a rapid method of screening large volumes of material for qualitative resistance. In this seeds of test material are sown in a single row of 3.5 cm apart in a seedbox of about 60 × 40 × 10 cm size. Suitable susceptible and resistant checks are sown in similar rows in the same box. Susceptible check (TN1) is sown as outer row which also act as spreader row, in the center of the box half susceptible and half resistant material row is also sown. These boxes are placed in the water through or galvanized iron trays containing water. Seven to ten days after sowing, seedlings are thinned to about 20 plants per row and infested with about 8-10 (BPH & WBPH) and 3 (GLH) 2nd – 3rd instar nymphs per seedling. When susceptible checks are killed (usually after about 1 week), damage score on each line was recorded on 0-9 scale, when 90 to 100 per cent plants of TN 1 were dead following Standard Evaluation System (SES) of rice (Anonymous, 1996). To compare entries a numerical rating system is used to score seedling damage: 0 = no damage; 1 = very slight damage; 3 = first and second leaves of most plants are partially yellow; 5 = pronounced yellowing and stunting or about half of the plants wilting or dead; 7 = more than half of the plants wilting or dead; 9 = all plants dead. Mean score of each line was classified as resistant (0-3.49), moderately resistant (3.50-5.49) and susceptible (5.50-9.00) following Shukla *et al.* (1993).

A second method, the modified seedbox screening test (MSST), is used to overcome some limitations of SSST and for better understanding of "field resistance," that is, when resistance is maintained or increases as plant ages. In this method other practices are similar to that of SSST, but plants are infested at 20 days after sowing with 3 to 5, 2nd – 3rd instar nymphs per plant. The whole seedbox of infested seedlings is covered with a screen cage (65 × 45 × 90 cm) to prevent the insects from escaping from the tray. In this method mortality of the susceptible check is caused by F_1 population of planthoppers, that is, the original nymphs mature and reproduces in the seedbox, and their offspring kill the plants (Velusamy *et al.*, 1986). These two methods have been extremely useful for rapid, high-throughput, and inexpensive screening of the necessarily large amounts of material required to find and isolate resistance genes. Furthermore, they incorporate "free choice," that is, the target insects can choose between the different varieties under test before initiating feeding (SSST and MSST) or oviposition (MSST) behaviour.

For field screening planthopper populations are often too low or too unevenly distributed in the field. This can be overcome by causing BPH population increases through the application of resurgence-inducing insecticides (spray of 0.002% deltamethrin or 0.02% methyl parathion) (Chelliah & Heinrichs, 1980; Kalode *et a1.*, 1982; Heinrichs *et al.*, 1982; Reissig *et al.*, 1982; Panda & Khush, 1995).

2.3.2. Genetics of Hopper Resistance

The earliest reported information on the genetics of resistance in BPH was in 1970 (Athwal *et al.,* 1971). More than 25 genes for resistance to BPH, 8 for WBPH, 13 for GLH and 3 for ZLH have been identified. More recently several QTLs have also been reported. However, due to development of DNA based markers and QTL analysis in 1970-1980s, clear linkage could be drawn between specific region of rice genome and resistance to hoppers. It is useful to breeders in deciding on a breeding methodology and breeding strategies to be adopted for deployment of genes. Entomologists and breeders have investigated the inheritance of resistance to identify diverse genes for resistance to aid in marker-assisted selection. Studies that used adequate methods to generate good quality DNA markers (tight flanking) to indicate clear chromosome regions associated with resistance lead to identification of major QTL for resistance (Hirabayashi *et al.*, 2004; Rahman *et al.*, 2009; Jairin *et al.*, 2010).

2.3.2.1. Brown planthopper (BPH), *Nilaparvata lugens* (Stål)

The brown planthopper (BPH), *Nilaparvata lugens* (Stål) is a typical phloem sap feeder, r-strategist pest which increases exponentially and affects the growth of rice plant. The plants suffer 40 to 70 per cent yield loss if attacked by 100-200 first instar nymphs of BPH at 25 days after transplanting (Bae & Pathak, 1979). It is one of the most serious and destructive pests of rice throughout Asian countries and causes significant yield losses in susceptible cultivars every year (Normile, 2008; Heong & Hardy, 2009). The recent International conferences in 2008 and 2010 exclusively on rice planthoppers analyzed the causes and consequences of BPH outbreak in many Asian countries (Heong & Hardy, 2009). Both nymphs and adults suck sap from the leaves and leaf sheaths, which results in yellowing of leaves, reduced tillering, reduced plant height and increase in unfilled grains. BPH also causes the reduction in chlorophyll, protein content of leaves and photosynthetic rate whereas, severe attack of BPH causes "hopper burn" symptoms (Horgan, 2009; Vanitha *et al.*, 2011). It also transmits virus diseases like grassy stunt, ragged stunt (Ling *et al.*, 1978) and wilted stunt (Chen *et al.*, 1978). Though many chemicals were recommended for the control of this pest (Sarao and Shera, 2013) but due to its feeding behaviour at the base of the plant the farmers are not able to control this pest effectively since it is very difficult to spray the chemical at the base of the plants. Besides, blanket application of insecticides often disrupts the ecological balance of rice ecosystem due to which this pest has already developed resistance against many insecticides in different Asian countries (Gorman *et al.*, 2008; Matsumura & Morimura, 2010).

The sources of resistance to BPH were identified in 1967 (Pathak *et al.*, 1969) and first two genes for resistance, *Bph1* and *bph2*, were identified in 1970 (Athwal *et al.*, 1971). The first resistant variety IR26 with *Bph1* was released in 1973 (Khush, 1971) and it was widely adopted in the Philippines, Indonesia and Vietnam but became susceptible in 1976-77 because of the development of biotype 2 of BPH. Two biotypes of BPH existed before the large-scale introduction of BPH-resistant varieties. By that time, varieties IR36, IR38 and IR 42 with the *bph2* gene had been developed and released (Khush, 1977). IR36 soon replaced IR26 and became the dominant rice variety. Its resistance to BPH lasted for 14 years, until 1991. These were widely adopted in the Philippines, Indonesia, and Vietnam but were found to be susceptible to a South Asian biotype (biotype 4). Several other varieties (IR60, IR62, IR68, IR72, and IR74) were released and were resistant to biotype 3. Some varieties are resistant on the Indian subcontinent but in Southeast and East Asia have *bph5, Bph6* or *bph7* genes for resistance. According to Jairin *et al.*, (2007), *Bph1, bph2, Bph3,* and *bph4* have been used extensively in Thai breeding programs. Improved rice cultivars carrying *Bph1, bph2, Bph3* and *bph4,* however, lost their ability against BPH, although cultivars with *Bph3* have shown a higher degree and broader spectrum of resistance against BPH.

Some of the donors are Mudgo, ASD7, Rathu Heenati, Babawee, ARC10550, Swamalata, Tl2, Chin Saba, Balamawee, *O. officinalis, O. australiensis* and *O. minuta* from cultivated and wild species of rice (Table 1). More than 100 resistant cultivars have been analyzed genetically. Segregation analysis was used to identify the nine earliest BPH- resistance loci (*Bph1*, *bph2*, *Bph3*, *bph4*, *bph5*, *Bph6*, *bph7*, *bph8*, and *Bph9*; Table 2). These loci all contained native *O. sativa* genes and were mainly derived from varieties originating in the Indian sub-continent (India, Bangladesh and Sri Lanka). During the 1990s and 2000s, a further 27 BPH-resistance loci were identified (*Bph10*, *bph11*, *bph12*, *Bph12*, *Bph13*, *Bph13*, *Bph14*, *Bph15*, *Bph16*, *Bph17*, *Bph18*, *bph18*, *bph19*, *Bph20*, *bph20*, *Bph21*, *bph21*, *Bph22*, *Bph22*(6S), *Bph23*, *Bph24*, *BPH25*, *BPH26*, *Bph*(9L), *Bph*(11L), *Qbph11*, and *qBPH6*. These were primarily derived from lines introgressed with wild rice species including *O. minuta* (6 loci), *O. officinalis* (6 loci), *O. rufipogon* (4 loci), *O. australiensis* (1 locus), *O. eichingeri* (1 locus), *O. glaberrima* (1 locus), and *O. latifolia* (1 locus) (Table 2). One locus, *Bph10*, is thought to occur in both *O. australiensis* and *O. officinalis* (based on allelism tests; Ishii *et al.*, 1994). Of the 36 BPH-resistance loci identified so far, PCR-based markers are available for 23. RFLP markers are available for 7 BPH-resistance loci, including *Bph12* (4S). Genetic analyses have not been conducted for six (*bph5*, *bph7*, *bph8*, *Bph22*, *Bph23*, *Bph24*) of the 36 loci and consequently markers are not available.

Mapping of BPH resistance genes

A number of genes for resistance to BPH have been mapped using RFLP, RAPD, and SSR markers. Some QTLs have also been mapped (Table 2). Six of these genes, *Bph1, bph2, Bph9, Bph10, Bph18,* and *Bph21* are located on chromosome 12. *Bph12, Bph15, Bph17* and *Bph20* are located on chromosome 4 (Rahman *et al.,* 2009). *Bph11, Bph13, Bph14* and *Bph19* are located on chromosome 3, *Bph6* on chromosome 11, *Bph3* and *bph4* on chromosome 6, and *Bphl3* on chromosome 2. There is some inconsistency in assigning gene number and its mapping (Table 2). Huang *et al.* (1997) used a doubled-haploid (DH) population (IR64 x Azucena) and located a BPH resistance gene from IR64 on chromosome 12. Three RFLP markers (RG493, RG901, and CD0344) and *sdhl* showed linkage with the BPH resistance gene. *Bph10* from IR65482-4-136-2-2 and *Bph1* from Mudgo for resistance to biotypes 1 and 3 were located near XNpb248. Although the pattern of resistance is different among these varieties, the genes are mapped in a similar position. Renganayaki *et al.* (2002) mapped the *Bph13* gene derived from *O. officinalis* on chromosome 3 using a RAPD marker. The most closely linked marker was converted into an STS marker and is mapped 1.3 cM from the resistance gene. An introgression line derived from *O. sativa* and *O. officinalis,* IR54741-3-21-22, was found to be resistant to an Indian biotype of BPH and resistance was controlled by a single dominant gene. Kim and Sohn (2005) used bulk segregant analysis with 520 RAPD markers for analysis of BPH resistance. One of these primers, OPE18, which amplified a 923-bp band, was tightly linked to BPH resistance. The *Bph1* gene was mapped at a distance of 3.8 cM from the STS marker BpE 18-3.

Table 2. Some examples on molecular mapping of genes/QTLs for BPH resistance in rice

Gene	Chromosome	Donor	Marker	Refernces
(1)	(2)	(3)	(4)	(5)
Bph1	12	Mudgo	G148	Hirabayashi and Ogawa (1995), Sun *et al.* (2006)
	12L	Mudgo	em5814N	Sharma *et al.* (2002)
	12	Samgangbyeo	BpE18-3	Kim and Sohn (2005)
	12L	IR28	XNpb248, XNpb336 AFLPem5814N	Hirabayashi and Ogawa (1995)
	12L	NorinPL3	OPD-7RD7, RG869	Sharma *et al.* (2004)
	12	Gayabyeo	RG457, RM247	Jeon *et al.* (1999)
	12	Samgangbyeo	Os Bphi252	Park *et al.* (2008)
bph2	12	NorinPL4	G2140	Murata *et al.* (1998)
	2	ASD7 (acc. no. 6303)	RM463, RM7102	Sun *et al.* (2006)
	2	NorinPL4 (IR1154-243)	KAM4	Murai *et al.* (2001), Sharma *et al.* (2004)
	12	NorinPL4 (IR1154-243)	G2140	Sun *et al.* (2006)
	2	Yagyaw	5529-1358	Liu *et al.* (2009)
Bph3	6S	Rathu Heenati (acc. no. 6730)	RM589	Jairin *et al.* (2007a)
bph4	6S	Babawee (acc. no. 8978)	RM190, C76A	Kawaguchi *et al.* (2001)
	6S	Babawee	C891, C531	Murata (1998)
	6	Babawee	RM 586-RM589	Jairin *et al.* (2010)
Bph6	4	Swarnalata (acc. no.33964)	RM6997-RM5742	Qiu *et al.* (2010)
Bph9	12L	Kaharamana	RM463, RM5341	Su *et al.* (2006)
	12L	Pokkali	OPR04, S2545	Murata *et al.* (2001)

Contd...

(1)	(2)	(3)	(4)	(5)
Bph10	12L	IR65482-4-136-2-2 (*O. australiensis*)	RG457	Ishii *et al.* (1994)
bph11	3L	IR742-23-19-12-3-54 *(O. officinalis)*	G1318	Hirabayashi *et al.* (1998)
Bph12	4S	B14 *(O. latifolia)*	RM261	Yang *et al.* (2002)
	4S	GSK185-2 *(O. officinalis)*	G271R93	Hirabayashi *et al.* (1998)
	4	*O. latifolia*	RLPP	He (2007)
Bph13	2L	*O. eichingeri* (acc. no. 105159)	RM250,	Liu *et al.* (2001)
	3S	IR54745-2-21-12-17-6 (*O. officinalis)*	AJ09b230, AJ09c	Renganayaki *et al.* (2002)
Bph14 (Qbph1 renamed)	3L	B5 (*O. officinalis)*	R1925-G1318	Yang *et al.* (2004)
Bph15 (*Qbph2* renamed)	4S	B5 (*O. officinalis)*	C820, S11182	Yang *et al.* (2004)
Bph17	4S	Rathu Heenati	RM8213-RM5953	Sun *et al.* (2005)
Bph18	12L	IR65482-7-216-1-2 (*O. australiensis.* acc. no. 100882)	RM1022	Jena *et al.* (2006)
bph19	3S	AS20-1	RM6308-RM3134	Chen *et al.* (2006)
Bph20	4	IR71033-121-15 (*O. minuta* acc. no 101141)	MS10-RM5953	Rahman *et al.* (2009)
Bph21	12	IR71033-121-15 (*O. minuta* acc. no. 101141)	RM3726-RM5479	Rahman *et al.* (2009)
BPH*	12	IR64	RG463, RG901, CDO344	Huang *et al.* (1997)
Bph6	11	IR54741-3-21-22	OPA16 938	Jena *et al.* (2002)
Bph25	6	ADR52	S00310	Myint *et al.* (2012)
Bph26	12L	ADR52	RM5479	Myint *et al.* (2012)
Bph27	4L	*O. rufipogon*	RM16888-RM16846	Huang *et al.* (2013)
Qbph2	2L	Col.5T	RM6843, RM3355	Sun *et al.* (2007)
Qbph3	3	Rathu Heenati	RM313, RM7	Sun *et al.* (2005)
Qbph4	4	Rathu Heenati (acc. no. 11730)	RM8213, RM5953	Sun *et al.* (2005)

Contd...

(1)	(2)	(3)	(4)	(5)
	4S	Yagyaw	RM401, RM335	Liu *et al.* (2009), Sun *et al.* (2005), Yang *et al.* (2004)
Qbph6	6S	Col.5T	RM510	Sun *et al.* (2007)
Qbph7	7	Yagyaw	RM542, RM500	Liu *et al.* (2009)
Qbph9		Yagyaw	RM3533, RM242	Liu *et al.* (2009)
Qbph10	10	Rathu Heenati (acc. no.11730)	RM484, RM496	Sun *et al.* (2005)

Source: Jena (2010), Modified after Brar *et al.* (2009)

Yang *et al.* (2004) developed a high-resolution genetic map of *Bph15* by positioning 21 DNA markers in the target chromosomal region. An assay of the recombinants using subclones in combination with sequence analysis delimited the *Bph15* gene to a genomic segment of approx. 47 kb. Chen *et al.* (2006) fine-mapped *bphl9* to a region of about 1 cM on the short arm of chromosome 3 flanked by RM6308 and RM3134. Sequence information of clones was used to construct a physical map of *bphl9* and the locus was physically defined to an interval of about 60 kb. Sun *et al.* (2005) analyzed Sri Lankan indica rice cultivar Rathu Heenati and found it to be resistant to all four biotypes of BPH. Three loci detected by QTL analysis were assigned to chromosomes 3, 4, and 10. The phenotypic variance of the three QTLs indicated that the QTL on chromosome 4 is a major BPH resistance gene in Rathu Heenati. Linkage analysis indicated that this BPH resistance gene was located between SSR markers, RM8213 and RM5953, on the short arm of chromosome 4, with a map distance of 3.6 cM and 3.2 cM, respectively. This gene was tentatively designated as *Bph17*.

Su *et al.* (2005) located *qbph11,* which explained 68.4% of the phenotypic variation of BPH resistance (biotype 2) in the progenies derived from the cross between DV85 and Kaimaze. An introgression line IR54741-3-21-22, derived from *O. sativa* x *O. officinalis,* was found to be resistant to an Indian biotype of brown planthopper. Genetic analysis of 95 F_3 progeny rows of a cross between the resistant line IR54741-3-21-22 and a BPH-susceptible variety revealed that resistance was controlled by a single dominant gene. RAPD analysis showed that one primer, OPA16938, amplified a resistant parental band in the resistant bulk and a susceptible parental band in the susceptible bulk by bulked segregant analysis. RAPD analysis of individual F_2 plants with primer OPA 16938 showed marker phenotype co-segregation for all, but only one recombinant was identified. The linkage between RAPD marker OPA16938 and the BPH resistance gene was 0.52 cM in the coupling phase. The 938-bp RAPD amplicon was cloned and used as a probe on 122 Clal-digested DH plants from an IR64 x Azucena mapping population for RFLP analysis and was mapped onto rice chromosome 11 (Jena *et al.,* 2002). RAPD marker OPA16938 could be used in a cost-effective way for marker-assisted selection for BPH resistance. Jairin *et al.* (2007a,b) mapped *Bph3* on the short arm of chromosome 6 between two flanking markers, RM589 and RM586, within 0.9 cM and 1.4 cM, respectively. Molecular analysis of three mapping populations derived from Ptb33 x RD6, Rathu Heenati x KDML105, and IR71033-121-15 x KDML105 showed BPH resistance genes located in the same genomic region on the short arm of chromosome 6. A novel resistance gene, *Bph18,* has been identified in breeding line IR65482-7 -216-1-2 that has inherited the gene from the EE genome wild species *O. australiensis* (Jena *et al.,* 2006). The *Bph18* gene has been located on the long arm of chromosome 12 in a 0.843-Mb genomic region flanked by markers RM6869 and Rl0289S. An STS marker, 7312.T4A, deri-ved from the BAC clone OSJNBa0028L05 that encodes a resistance protein sequence present in the flanking region is tightly linked to the *Bph18* gene for BPH resistance. The marker 7312.T4A can be efficiently used for MAS breeding of BPH resistance in rice. Alam and Cohen (1998) identified seven QTLs for BPH resistance on chromosomes 1, 2, 3, 4, 6, and 8 using 175 markers. Ramalingam *et al.* (2003) reported four additional QTLs associated with BPH resistance in the same population using an additional 105 candidate gene markers. Huang *et al.* (2001) reported QTLs in wild species for BPH resistance.

Xu *et al.* (2002) reported seven main-effect QTLs and many epistatic QTLs associated with quantitative resistance to BPH using a recombinant inbred line derived from Tequing x Lemont populations. Su *et al.* (2002) detected QTLs on chromosomes 2, 10, and 12 using a Nipponbare x Kasalath backcross population. Soundararajan *et al.* (2004) used a DH population (IR64 x Azucena) and detected six QTLs on chromosomes 1, 2, 6, and 7 for BPH resistance. Of these, QTLs on chromosome 7 showed association with seedling resistance and QTLs on chromosome 2 with antibiosis, whereas QTLs on chromosomes 1, 6, and 7 were associated with tolerance (Xu *et al.,* 2002). There is a need to extend QTL analysis at different growth stages of the plant and over different environments, including the candidate gene approach. Sun *et al.* (2007) analyzed 147 F3 families derived from BPH-resistant cultivar Col. 5 from Thailand and susceptible cultivar 02428. The BPH population used for infestation was a mixture of biotypes 1 and 2. Two QTLs were identified on chromosome 2 (29.4% phenotypic variation) and chromosome 6 (46.2% phenotypic variation). Comparison of chromosomal locations and reactions to BPH biotypes indicated that the gene on chromosome 6 is different from the previously identified genes. Liu *et al.* (2009) made crosses between BPH-resistant landrace Yagyaw and susceptible cultivar Cpslol7. Four QTLs *(Qbph2, Qbph4, Qbph*7, and *Qbph9)* accounting for 5.64% to 12.77% of the phenotypic variation for BPH resistance were located. The two QTLs showed a significant additive effect. One resistant allele was harbored by the susceptible parent.

Myint *et al.* (2013) reported that *Bph25* co-segregated with marker S00310 on the distal end of the short arm of chromosome 6, and *Bph26* co-segregated with marker RM5479 on the long arm of chromosome 12. Huang *et al.* (2013) reported *Bph27* derived from an accession of Guangxi wild rice, *O. rufipogon* (acc. no. GX2183), was primarily mapped to a 17-cM region on the long arm of the chromosome 4. Fine mapping of *Bph27* was conducted using two BC_1F_2 populations derived from introgression lines of GX2183. Insect resistance was evaluated in the BC_1F_2 populations with 6,010 individual offsprings, and 346 resistance extremes were obtained and employed for fine mapping of the gene.

Marker assisted selection and pyramiding of genes for resistance

Identification of a tightly linked DNA marker is a prerequisite for marker assisted selection (MAS) application (Table 2). The various institutes are directed towards marker assisted back-crossing to introgress the favourable alleles for BPH resistance into elite rice lines and to date many resistant genes have been tested for being linked to markers (Jena *et al.,* 2006; Sun *et al.,* 2005). Classical breeding has supported the development of improved BPH resistant genotypes up to a certain degree. To further improve resistance of rice varieties, it could be emphasized the importance of combining all favourable and complementary physiological traits in a variety, rather than considering BPH resistance as a single trait. Ten BPH-resistance loci (*Bph3*, *Bph6*, *Bph9*, *Bph17*, *Bph18*, *Bph20*, *bph20*, *Bph21*, *bph21*, *BPH25*, *BPH26*, and *Bph*(t)) are suitable for MAS for BPH resistance, albeit with varying levels of BPH virulence in different parts of Asia. Among these genes, *Bph3* and *Bph6*, are the most suitable BPH-resistant genes for MAS because of their detailed resolution and wide public availability as well as an absence of reported virulent BPH populations. Li *et al.* (2006) incorporated *Bph14* and *Bph15* through MAS into a number of parental lines used in hybrid rice breeding in China. Six genes are suitable for MAS for GRH resistance; but leafhopper virulence against two of these (*Grh1* and *Grh2*) has been reported. The development of resistance for all other planthopper and leaf-hopper species using molecular breeding approaches is still severely limited by a scarcity of genetic information and of suitable markers.

Without an understanding of gene function and resistance mechanisms it is difficult to predict the outcome, vis-a-vis resistance strength and spectrum of pyramiding resistance genes or loci. Sharma *et al.* (2004) used MAS for pyramiding *Bph1* and *bph2* into a japonica line. The pyramided line (PYL) showed higher level of resistance than *bph2* alone but was equivalent to *Bph1*. Li *et al.* (2006) incorporated *Bph14* and *Bph15* through MAS into a number of parental lines used in hybrid rice breeding in China and observed that *Bph14* single introgression lines had moderate resistance to BPH whereas, *Bph14*/*Bph15* pyramided lines had higher resistance than the single-gene introgression lines. Park *et al.* (2008) used representational difference analysis (RDA) and found that OsBphi252 is tightly linked to BPH resistance and could be used in MAS. Fujita *et al.* (2009) have evaluated the resistance of NILs and PYLs with *Bph25* and *Bph26* against BPH strains from East Asia. Their results indicate that a PYL containing both genes is resistance against several East Asian BPH strains. Furthermore, Myint *et al.* (2012) demonstrated that a PYL containing both genes could be effective despite the apparent low effectiveness of each gene alone in *Bph25* and *Bph26* monogenic NILs. However, even a *Bph25*-*Bph26* PYL showed low resistance against South Vietnamese strains of BPH (Fujita *et al.,* 2009). Hu *et al.* (2012) reported a pyramided line carrying two resistance genes *Bph14* and *Bph15*. It was evaluated for seedling damage, antixenosis, and honeydew production and found to be more resistant than either the *Bph14*-NIL or the *Bph15*-NIL alone. Likewise PYL for *Bph12* and *Bph6* gene had lower nymphal settling and survival, slower population growth, and caused less damage compared to the monogenic lines (Qiu *et al.,* 2012).

Although PYLs can enhance resistance to planthoppers, care should be taken in the deployment of pyramided varieties as it is still unknown whether PYLs could lead to a more rapid adaptation of BPH to the resistance genes they contain than if the genes were sequentially deployed in monolocus varieties. Tests of the comparative durability of monolocus and pyramided hopper-resistant rice lines have not been conducted, and there are many cases of naturally pyramided rice varieties against which planthopper populations have already adapted. More research is required to better understand the merits and draw-backs of resistance-gene pyramiding for durability and to determine optimal deployment methods for resistant varieties to maximally delay hopper adaptation.

BPH genomics for studying variability and population structure

Four BPH biotypes are known. Biotypes 1 and 2 are widely distributed in South- east Asia, biotype 3 is a laboratory biotype produced in the Philippines, and biotype 4 occurs on the Indian subcontinent. The *Bph1* group confers resistance to biotypes 1 and 3 but is susceptible to biotype 2, the *bph2* group confers resistance to biotypes 1 and 2 but is susceptible to biotype 3, and the *Bph3* group, *bph4, bph8,* and *Bph9* confer resistance to all four biotypes. Genes such as *bph5, Bph6,* and *bph7* confers resistance to biotype 4 only (Khush & Brar, 1991). Many workers have studied the relationship between BPH biotypes and genes for resistance. IR varieties carry *Bph1, bph2,* and *Bph3* genes and *Glh3, glh4, Glh9,* and *glh10.* Wei *et al.* (2009) used a proteomic approach and analyzed the interaction between rice and BPH. Proteins involved in multiple pathways showed significant changes in expression in response to BPH feeding, including jasmonic acid synthesis proteins and oxidative stress-response proteins. Wang *et al.* (2008) used a cDNA microarray containing 1,920 suppression subtractive hybridization clones to detect transcript profile differences in resistant and susceptible cultivars under a control and BPH feeding. In total, 160 unique genes were detected as being significantly affected by BPH feeding. Shi *et al.* (2003) constructed a genomic library from BPH-resistant B5. The library contained 36,864 clones with an average insert size of 60 kb. Eleven clones were identified covering the *QbphI* locus (the locus between markers R1925 and G1318 on chromosome 3). Myint *et al.* (2012) conducted linkage analysis using backcross populations resulted in the identification of BPH-resistance (antibiosis) gene from ADR52. Preliminary near-isogenic lines (pre-NILs) and a preliminary pyramided line (pre-PYL) carrying *Bph25* and *Bph26* were evaluated to characterize the virulence of the Japan migrated BPH population. Although both pre-NILs were susceptible to the virulent BPH strain, the pre-PYL exhibited a high level of resistance.

Mun *et al.* (1999) studied the origin of BPH immigrations as well as the geographic structure in Asian rice growing regions and intraspecific variation in BPH and WBPH. An 850 base pair region of mitochondrial DNA cytochrome oxidase-I (COI) was sequenced from a total of 71 individuals collected from 11 localities in seven countries namely, Korea, Philippines, China, Bangladesh, Malaysia, Vietnam and Thailand. In BPH three haplotypes were found and all populations sampled shared a dominant haplotype. Localities in Korea contained two haplotypes and localities in China and the Philippines contained three. However, in samples from the Indochina peninsula no variation was detected either within or between populations, consistent with a hypothesis of regular migration and gene flow. These populations did not contain some haplotypes found in Korea, suggesting they were not the source of yearly immigration into Korea and further to Japan. Populations from China did share haplotypes with Korea, which was consistent with the hypothesis that China was the source for yearly immigration into Korea. There was insufficient resolution to distinguish among populations in China. For BPH the data suggested that populations south of the Red River Valley in Vietnam experienced regular mixing and were distinct from populations to the north which contributed to yearly immigrations. In WBPH, there was less differentiation among populations. Two haplotypes were found in all populations except Malaysia.

Jing *et al.* (2011) assessed the genetic diversity in populations of the BPH through microsatellite, or simple sequence repeat (SSR), markers from expressed sequence tags (ESTs). They developed 351 EST SSR markers and found clear bands from samples of four BPH populations. They reported high cross-species transferability of markers in the closely related planthopper *Nilaparvata muiri*. The newly developed EST-SSR markers provided sufficient resolution to distinguish within and among biotypes. Analyses based on SSR data revealed host resistance-based genetic differentiation among different BPH populations. The genetic diversity of populations feeding on susceptible rice varieties was lower than that of populations feeding on resistant rice varieties. Similarly, Jairin *et al.* (2013) developed the first genetic linkage map for brown planthopper. This linkage map and the newly developed markers used to create an essential resource and a useful framework for future genetic analyses in BPH and to better understand the plant-insect interactions. It was constructed by integrating linkage data from two backcross populations derived from three inbred BPH strains namely Hadano-66, Chikugo-89, and Koshi-10 to generate the inbred BPH strains. The consensus map consists of 474 SSR, 43 single-nucleotide polymorphisms (SNP), and 1 sequence-tagged site (STS), for a total of 518 markers at 472 unique positions in 17 linkage groups. The linkage groups cover 1093.9 cM, with an average distance of 2.3 cM between loci. The average number of marker loci per linkage group was 27.8. The sex-linkage group was identified by exploiting X-linked and Y-specific markers. Likewise, Sun *et al.* (2012) studied nine microsatellite loci to study the population genetic structure and genetic diversity from the genome of small BPH by constructing (TC)6(AC)5 and (AG)6(AC)5 compound SSR-enriched libraries using suppression-PCR procedures. These loci were found to be highly polymorphic, with 13 to 30 alleles per locus in the three populations namely Jiangsu, Shandong and Zhejiang. The observed and expected heterozygosities ranged from 0.255 to 0.833 and 0.392 to 0.929, respectively.

2.3.2.2. Whitebacked planthopper (WBPH), *Sogatella furcifera* (Horvath)

More than 300 cultivars resistant to the WBPH have been identified and 80 of them have been analyzed genetically. Resistant donors (N22, ARC10239, ADR 52, Podiwi-A8) have been identified (Table 1). Eight genes for resistance *(Wbphl, Wbph2, Wbph3, wbph4, Wbph5, wbph6*

Wbph7, Wbph8) have been identified (Khush, 1984; Tan *et al.*, 2004). Sidhu *et al.* (1979) reported that a single dominant gene (*Wbph1*) was found to be resistant to WBPH in variety N22. Likewise Angeles *et al.* (1981) reported that resistance in ARC 10239 is governed by a single dominant gene (*Wbph2*). Nair *et al.* (1982) investigated 21 additional varieties; 19 had *Wbph1* and two had *Wbph2* + recessive gene. The resistance of two of the 14 varieties analyzed by Hernandez and Khush (1981) was governed by *Wbph2:* 11 varieties each had a single dominant gene that segregated independently of *Wbph1* and *Wbph2.* The dominant gene of one such variety (ADR52) was designated *Wbph3.* Only one variety, Podiwi A8, had a recessive gene, which was designated *wbph4.* Saini *et al.* (1982) analyzed 13 additional varieties. Resistance was governed by *Wbph1* in four varieties, by *Wbph2* in six, by *Wbph1* and *Wbph2* in two, and by a single dominant gene in Hornamawee segregated independently of *Wbph1* and *Wbph2.* Wu and Khush (1985) investigated the inheritance of resistance in 15 varieties. They found that resistance in nine varieties was controlled by *Wbph1* and resistance in four was conferred by two genes. The remaining two varieties had single dominant genes for resistance, which segregated independently of *Wbph1, Wbph2* and *Wbph3.* The dominant gene of N' Diang Marie was designated *Wbph5.* Jayaraj and Murty (1983) studied the inheritance of resistance in nine varieties. Resistance was controlled by a single dominant gene in three varieties and by a recessive gene in the other six varieties. Kadrivel *et al.* (1999) mapped QTLs for resistance to WBPH. QTLs for ovicidal response have also been reported (Yamasaki *et al.*, 1999; Sogawata *et al.*, 2001). Inheritance of resistance in 10 cultivars was investigated by Singh *et al.* (1990).

Eight cultivars-ARC5838, ARC6579, ARC6624, ARC10464, ARC11321, ARC11320, Balamawee, and IR2425-90-4-3-were found to have a single recessive gene for resistance. The recessive genes of IR2425-90-4-3, ARC5838, and ARC11321 were found to be allelic to each other. Resistance in Pbtl9 and IET6288 was found to be under dominant gene control. Sidhu *et al.* (2005) studied the inheritance of resistance in five cultivars. A recessive gene conferred resistance in MR1523 and ARC11367, whereas resistance in NCS2041 was conditioned by a dominant gene. The resistance in Mudgo and MO1 was governed by two independently inherited dominant genes. Padmarathi *et al.* (2007) reported the recessive gene in ARC5984 and ARC6650 as allelic to Podiwi *(wbph4).* The dominant gene in ADR52 was different from that in cultivar Velluthecharra. He (2007) mapped *Wbph7* and *Wbph8* on chromosome 4.

For WBPH resistance, PYLs carrying *Ovc* and other QTLs for ovicidal response (*qOVA-1-3, qOVA-4, qOVA-5-1, and qOVA-5-2*) showed higher egg mortality than the monolocus NIL with the *Ovc* gene alone (Yamasaki *et al.*, 2003). Parihar *et al.* (2010) reported that of the 50 random primers for WBPH resistance, a single linked primer, OPA08, was identified. This primer generated 8-bands, one of which, OPA08-7, was putatively linked to resistant gene as was evident by its presence in almost all the resistant bulks.

2.3.2.3. Green leafhopper (GLH), *Nephotettix* sp.

Green leafhopper (GLH) has capacity to transmit tungro viruses. Till date 14GLH resistance genes have been identified, of which one *Glh14* has RFLP markers (Sebastian *et al.*, 1996).

2.3.2.4. Green rice leafhopper (GRH), *Nephotettix cincticeps* Uhler

Green rice leafhopper (GRH) is mostly found in the temperate regions of East Asia. At least six genes been identified and mapped on chromosomes 3, 4, 5, 6, 8 and 11, respectively (Yasui *et al.*, 2007). Near-isogenic lines (NILs) carrying *Grhl* , *Grh2, Grh4, Grh5,* and *Grh6* in the background of japonica cultivar Taichung 65 (T65) have been developed using MAS. W 1962 (*O. rufipogon)* had *Grh5* and a minor gene on chromosome 4 (Fujita *et al.*, 2006). The nymph mortality of pyramided lines *(Grh2 + Grh4)* and *Grh2 + Grh6* and *Grh4 + Grh6* was higher than that of the NILs each carrying a single resistance gene (Yasui *et al.*, 2007).

2.3.2.5. Zigzag leafhopper (ZLH), *Recilia dorsalis* Motschulsky

The zigzag leafhopper (ZLH) occurs in the tropics and subtropics of Asia. Atleast 3 genes been identified, some donors (Rathu Heenati, Ptb21, Ptb33) for resistance have been identified (Heinrichs *et al.* 1985) (Table 1). The genetics of resistance to the ZLH, WBHP, BPH, and GLH in cultivars Rathu Heenati, Ptb21, and Ptb33 was investigated by Angeles *et al.* (1986). Single dominant genes that segregate independently of each other and that conveyed resistance to ZLH were designated *Zlh1* (Rathu Heenati), *Zlh2* (Ptb21), and *Zlh3* (Ptb33). Tests for the independence of the various genes for resistance to leafhoppers and planthoppers revealed that *Zlh1, Zlh2,* and *Zlh3* are independent of *Wbph3. Zlh2,* and *Zlh3* and also segregated independently of *bph2* and *Bph3*.

2.4. ENHANCEMENT OF RESISTANCE IN RICE TO HOPPERS THROUGH GENETIC ENGINEERING

Plant lectins have severe effects on fecundity, growth and development of insects. However, certain lectins, like from plants of family Amaryllidaceae, show low or no toxicity toward higher animals, but are toxic to insects (Table 3). Among the Amaryllidaceae lectins, the lectin from snowdrop, *Galanthus nivalis* L. agglutinin (GNA) has proved to be non-toxic to mammals and toxic to insects. The lectins are probably involved in the binding to receptors present on the midgut epithelial cells thereby contributing to the insecticidal effect (Powell *et al.* 1998). The bound lectins inhibit abosortion of nutrients or disrupt midgut cells endocytosis of lectins and other toxic metabolites (Eisemann *et al.* 1994).

Rao *et al.* (1998) reported snowdrop lectin (*Galanthus nivalis agglutinin* - GNA) to be toxic toward BPH when administered in an artificial diet. Transgenic rice containing the *gna* gene in constructs in which its expression was driven by a phloem-specific promoter (from the rice sucrose synthase *RSs1* gene) and by a constitutive promoter (from the maize ubiquitin *ubi1* gene) conferred resistance to BPH. PCR and Southern analyses to confirm that the transgenes were transmitted to progeny. Western blot analyses revealed expression of GNA at up to 2.0% of total protein in some of the transgenic plants. GNA expression driven by the *RSs1* promoter was tissue-specific as shown by immune histochemical localization of the protein in the non-lignified vascular tissue of transgenic plants. Insect bioassays and feeding studies showed that GNA expressed in transgenic rice plants decreased survival and overall fecundity (production of off-spring) of the insects, retarded insect development, and had a deterrent effect on BPH feeding. *gna* is the first transgene to exhibit insecticidal activity toward sap-sucking insects in an important cereal crop plant. When fed on GNA expressing transgenic lines the survival of BPH and GLH nymphs decreased by 36-74% and 49-53%, respectively (Rao *et al.,* 1998; Sudhakar *et al.,* 1998; Foissac *et al.,* 2000; Tang *et al.,* 2001; Sun *et al.,* 2002; Nagadhara *et al.,* 2003).

2.4.1. RNAi-Mediated Gene Silencing

RNA interference (RNAi) has been a reverse genetic tool to study gene function in various organisms. RNAi is a potent approach for silencing genes of interest through introducing either double-stranded RNA (dsRNA) or small interfering RNA (siRNA) of the target gene sequence in a host organism. When three dsRNA targeting different sites within a gene encoding vascuolar ATP synthase subunit E (*V-ATPase-E*), were orally delivered into BPH, it resulted in decreased expre-ssion of the target gene (Li *et al.,* 2011). Zha *et al.* (2011) reported a significant reduction in the development of BPH when the insects were fed on transgenic plants expressing dsRNA specific to hexose transporter (*NlHT1*), carboxy peptidase (*Nlcar*) and trypsin-like serine protease (*Nltry*) gene of BPH.

Table 3. Some examples on transgenes used for conferring hopper resistance in rice

Variety	Transformation method	Gene	Insect	Reference
ASD16, M5, FX92, M12	Particle bombardment	*gna*	BPH	Rao *et al.* (1998); Sudhakar *et al.* (1998)
Radon, Nortal	Electroporation	*gna*	BPH	Rao *et al.* (1998)
Eyi105, Ewan5	Particle bombardment	*gna*	BPH	Tang *et al.* (1999)
ASD16, M12	-do-	*gna*	BPH, GLH	Foissac *et al.*(2000)
Eyi105, Ewan5	*Agrobacterium* mediated	*gna*	BPH	Tang *et al.* (2001)
Basmati370, M7	Particle bombardment	*gna*	BPH	Maqbool *et al.* (2001)
Eyi105, Ewan5	-do-	*gna*	SBPH	Sun *et al.* (2002)
Chaitanya, Phalguna	*Agrobacterium* mediated	*gna*	BPH, WBPH, GLH	Nagadhara *et al.* (2003, 2004)
IR64	-do-	*asal*	BPH, GLH	Saha *et al.* (2006); Sengupta *et al.* (2010)
BPT5204, Chaitanya	-do-	*asal*	BPH, WBPH, GLH	Yarasi *et al.* (2008)
Nagdongbyeo	PEG direct gene transfer	*SKT1*	BPH	Lee *et al.* (1999)

2.5. CONCLUSION

Hoppers have been a major biotic stress in rice and efforts have been made to control these through both chemical sprays and resistance breeding. Still hoppers cause significant crop losses throughout the world. With climate change, hoppers are expected to cause more severe losses to rice and chemical control is becoming more and more expensive and one of the environmental hazard. To manage rice hoppers below economic threshold, research emphasis needs to be prioritised.

- Discovering new genes, mapping and cloning of resistance genes and understanding their mode of action.
- Emphasis should be given to MAS for pyramiding of two or more resistance genes.
- Molecular analysis of hopper population using genomic tools to understand their population structure and migration patterns.
- Understanding hopper-rice interaction in detail for devising new strategies to control hoppers.
- Developing transgenics with genes that block biochemical pathways in hoppers or blocking the functioning of developmentally important genes in plants through RNAi.
- Gene deployment should be based on variability in the past populations over smaller geographical regions.

REFERENCES

Alam SN and Cohen MB (1998) Detection and analysis of QTLs for resistance to the brown plant-hopper, *Nilaparvata lugens,* in a doubled haploid rice population. *Theoretical and Applied Genetics* **97:** 1370-1379.

Angeles ER, Khush GS and Heinrichs EA (1981) New genes for resistance to white-backed plant-hopper in rice. *Crop Science* **21:** 47-50.

Angeles ER, Khush GS and Heinrichs EA (1986) Inheritance of resistance to plant hoppers and leaf hoppers in rice. In: *Rice Genetics.* International Rice Research Institute, Manila Philippines. pp 537-549.

Annonymous (1996) *Standard Evaluation System for Rice.* International Rice Research Institute, Los Banos, Philippines.

Athwal DS, Pathak MD, Bacalangco EH and Pura CD (1971) Genetics of resistance to brown planthopper and green leaf hoppers in *Oryza sativa* L. *Crop Science* **11:** 747-750.

Bae SH and Pathak MD (1979) Life history of *Nilaparvata lugens* (Homoptera: Delphacidae) and susceptibility of rice varieties to its attacks. *Annals of the Entomological* Society *of America* **63:** 149-155.

Brar DS, Virk PS, Jena KK and Khush GS (2009) Breeding for resistance Planthoppers in rice. In: (eds.) KL Heong and B Hardy, *Planthoppers: New Threats to the Sustainability of Intensive Rice Production Systems in Asia.* International Rice Research Institute, Los Banos, Philippines, pp. 401-409.

Chelliah S and Heinrichs EA (1980) Factors affecting insecticide-induced resurgence of the brown planthopper *Nilaparvata lugens in* rice. *Environmental Entomology* **9**: 773-777.

Chen CC, Ko VH and Chiu RJ (1978) Rice wilted stunt its transmission by the brown planthopper, *Nilaparvata lugens* (Stål). *Plant Protection Bulletin Taiwan* **20:** 374-376.

Chen JW, Wang L, Pang XF and Pan H (2006) Genetic analysis and fine mapping of a rice brown planthopper (*Nilaparvata lugens* Stal) resistance gene *bphl9(t)*. *Molecular Genetics and Genomics* **275**: 321-329.

Eisemann CH, Donaldson RA, Pearson RD, Cadagon LC, Vuocolo T and Pellam RL (1994) Larvicidal action of lectins on *Lucilia cuprina*; mechanism of action. *Entomologia Experimentalis et Applicata* **72**: 1-11

Foissac X, Loc NT, Christou P, Gatehouse AMR and Gatehouse JA (2000) Resistance to green leafhopper *Nephotettix virescens* and brown planthopper *Nilaparvata lugens* in transgenic rice expressing snow-drop lectin *Galanthus nivalis* agglutinin. *Journal of Insect Physiology* **46**: 573-583.

Fujita D, Doi K, Yoshimura A and Yasui H (2006) Molecular mapping of a novel gene, *Grh5,* conferring resistance to green rice leaf bopper *(Nephotettix cincticeps* Uhler) in rice, *Oryza sativa* L. *Theoretical and Applied Genetics* **113**: 567-573.

Fujita D, Kohli A and Horgan FG (2013) Rice resistance to planthoppers and leafhoppers. *Critical Reviews in Plant Sciences.* 32: 162-191.

Fujita D, Myint KKM, Matsumura M and Yasui H (2009) The genetics of host-plant resistance to rice planthopper and leafhopper. In: (eds.) K.L. Heong and B. Hardy). *Planthoppers: New Threats to the Sustainability of intensive rice Production Systems in Asia*. International Rice Research Institute, Los Banos, Philippines, pp. 389-400.

Gorman K, Liu Z, Denholm I, Bruggen KU and Nauen R (2008) Neonicotinoid resistance in rice brown planthopper, *Nilaparvata lugens. Pest Management Science* **64**: 1122-1125

Gould F, Kennedy GG and Johnson MT (1991) Effects of natural enemies on the rate of herbivore adaptation to resistant host plants. *Entomologia Experimentalis et Applicata* **58**: 1-14.

Gurr GM, Liuz J, Read DMY, Catindig JLA, Chengs JA, Lan LP and Heong KL (2011) Parasitoids of Asian rice planthopper (Hemiptera: Delphacidae) pests and prospects for enhancing biological control by ecological engineering. *Annals of Applied Biology* **158**: 149-176.

He GC (2007) Brown planthopper resistance genes in rice: from germplasm to breeding. *Molecular Plant Breeding* **5**: 175-170.

Heinrichs EA, Medrano FG and Rapusas HR (1985) *Genetic Evaluation for Insect Resistance in Rice.* International Rice Research Institute, Los Banos, Philippines. 356 pp.

Heinrichs EA, Reissig WH, Valencia S and Chelliah S (l982) Rates and effect of resurgence-inducing insecticides on populations of *Nilaparvata lugens* (Hemi-ptera: Delphacidae) and its predators. *Environmental Entomology* **11**: 1269-l273.

Heong KL and Hardy B (2009). *Planthoppers: New Threats to the Sustainability of Intensive Rice Production Systems in Asia.* International Rice Research Institute, Los Baños, Philippines. 460 pp.

Hernandez JE and Khush GS. 1981. Genetics of resistance to white backed planthopper in some rice *(Oryza sativa* L.) varieties. *Oryza* **18**: 44-50.

Hirabayashi H, Angeles ER, Kaji R, Ogawa T, Brar DS, and Khush GS (1998) Identification of brown planthopper resistance gene derived from *O. officinalis* using molecular markers in rice. *Breeding Science (Supplement)* **48**: 82

Hirabayashi H and Ogawa T (1995) RFLP mapping of *Bph-1* (brown planthopper resistance gene) in rice. *Breeding Science* **45**: 369-371.

Hirabayashi H, Ideta O, Sato H, Takeuchi Y, Ando I, Nemoto H, Imbe T, Brar DS and Ogawa T (2004) Identification of a resistance gene to brown planthopper derived from *Oryzaminuta* in rice. *Breeding Research* **6:** 285.

Horgan F (2009) Mechanisms of resistance: a major gap in understanding planthopper-rice interactions, pp. 281-302. In: K L Heong and B Hardy (eds.) *Planthoppers: New Threats to the sustainability of Intensive Rice Production Systems in Asia*. International Rice Research Institute, Los Baños, Philippines.

Hu J, Li X, Wu C, Yang C, Hua H, Gao G, Xiao J and He Y (2012) Pyramiding and evaluation of the brown planthopper resistance genes *Bph14* and *Bph15* in hybrid rice. *Molecular Breeding* **1:** 61-69.

Huang D, Qiu Y, Zhang Y, Huang F, Meng J, Wei S, Li R and Chen B (2013) Fine mapping and characterization of *Bph27*, a brown planthopper resistance gene from wild rice (*Oryza rufipogon* Griff.). *Theoretical and Applied Genetics* **126**: 219–229.

Huang N, He G, Shu L, Li X and Zhang Q. 2001. Identification and mapping of two brown planthopper genes in rice. *Theoretical and Applied Genetics* **102**: 929-934.

Huang NA, Parco A, Mew T, Magpantay G, Mc Couch S, Guiderdoni E, Xu J, Subudhi P, Angeles ER and Khush GS (1997) RFLP mapping of isozymes, RAPD and QTLs for grain shape, brown planthopper resistance in a doubled haploid rice population. *Molecular Plant Breeding* **3**: 105-113.

Ishii T, Brar DS, Multani DS and Khush GS (1994) Molecular tagging of genes for brown planthopper resistance and earliness introgressed from *Oryza australiensis* in to cultivated rice *Oryza sativa. Genome* **37**: 217-221.

Jairin J, Kobayashi T and Yamagata Y (2013) A simple sequence repeat- and single-nucleotide polymorphism-based genetic linkage map of the brown planthopper, *Nilaparvata lugens. DNA Research* **20**: 17–30.

Jairin J, Phengrat K, Teangdeerith S, Vanavichit A and Toojinda T (2007a) Mapping of a broad-spectrum brown plant hopper resistance *gene, Bph3,* on rice chromosome 6. *Molecular Plant Breeding* **19**: 35-44.

Jairin J, Teangdeerith S, Leelagud P, Phengrat K, Vanavichi A and Toojinda T. 2007b. Detection of brown planthopper resistance genes from different rice mapping populations in the same genomic location. *Science Asia* **33**: 347-352.

Jairin J, Sansen K, Wongboon W and Kothcharerk J (2010) Detection of a brown planthopper resistance gene *bph4* at the same chromosome al position of *Bph3* using two different Genetic backgrounds of rice. *Breeding Science* **60:** 71-75.

Jayaraj D and Murty VVS (1983) In: *Proceedings of the 15th International Genetics Congress,* New Delhi, India, 12-21 December 1983. New Delhi (India): Oxford and IBH Publishing Co. p724. (Abstr.)

Jena KK (2010) The species of the genus *Oryza* and transfer of useful genes from wild species into cultivated rice, *O. sativa*. *Breeding Science* **60:** 518-523.

Jena KK, Jeung M, Lee JH, Choi HC and Brar DS (2006) High-resolution mapping of a new brown planthopper (BPH) resistance gene, *Bphl8(t),* and marker-assisted selection for BPH resistance in rice *(Oryza sativa* L.). *Theoretical and Applied Genetics* **112**: 288-297.

Jena KK, Pasalu IC, Rao YK, Varalaxmi Y, Krishnaiah K, Khush GS and Kochert G (2002) Molecular tagging of a gene for resistance to brown planthopper in rice *(Oryza sativa* L.). *Euphytica* **129**: 81-88.

Jeon YH, Ahn SN, Choi HC, Hahn TR and Moon HP (1999) Identification of a RAPD marker linked to a brown planthopper resistance gene in rice. *Euphytica* **107**: 23-28.

Jing S, Liu B, Peng L, Peng X, Zhu L, Fu Q and He G (2012) Development and use of EST-SSR markers for assessing genetic diversity in the brown planthopper (*Nilaparvata lugens* (Stal)). *Bulletin Entomological Research* **102:** 113-122.

Kadrivel P, Maheswaran M and Gunathilagaraj K (1999) Molecular mapping of quantitative trait loci (QTLs) associated with white backed planthopper in rice. *International Rice Research Notes* **24**: 12-14

Kalode MB, Bentur JS, Sain M, Rao UP and Srinivasan TE (1982) An improved method for screening cultures resistant to brown planthopper. *International Rice Research Notes* **7**: 6-7.

Kawaguchi M, Murata K, Ishii T, Takumi S, Mori N, Nakamura C (2001) Assignment of a brown planthopper *(Nilaparvata lugens* Stal) resistance gene *bph4* to the rice chromosome 6. *Breeding Science* **51:** 13-18.

Khush GS and Brar DS (1991) Genetics of resistance to insects in crop plants. *Advances in Agronomy* **45**: 223-274.

Khush GS (1971) Rice breeding for disease and insect resistance at IRRI. *Oryza* **8**: 111-119.

Khush GS (1977) Disease and insect resistance in rice. *Advances in Agronomy* **29**: 265-341.

Khush GS (1984) Breeding for resistance to insects. *Protection Ecology* **7**: 147-165.

Kim SM and Sohn JK (2005) Identification of rice gene *(Bph*1) conferring resistance to brown planthopper *(Nilaparvata lugens* Stal) using SIS markers. *Molecules and Cells* **20**: 30-34.

Kogan M and Ortman EF (1978) Antixenosis-A new term proposed to define Painter's 'Non-preference' modality of resistance. *Annals of the Entomological* Society *of America* **24:** 175-176.

Lee SI, Lee SH, Koo JC, Chun HJ, Lim CO, Mun JH, Song YH and Cho MJ (1999) Soybean Kunitz trypsin inhibitor SKTI confers resistance to the brown planthopper *Nilaparvata lugens* Stal in transgenic rice. *Molecular Plant Breeding* **5**:1-9.

Li J, Chen Q, Lin Y, Jiang T, Wu G and Hua H (2011) RNA interference in *Nilaparvata lugens* (Homoptera: Delphacidae) based on ds RNA ingestion. *Pest Management Science* **67**: 852-859.

Li JB, Xia MY, QHX, He GC, Wan BL and Zha ZP (2006) The evaluation and utilization of new genes for brown planthopper resistance in common wild rice (*Oryza rufipogon* Griff.). *Molecular Plant Breeding* **4**: 365–371.

Li J, Chen Q, Wang L, Liu J, Shang K and Hua H (2010) Biological effects of rice harbouring *Bph 14* and *Bph 15* on brown planthopper, *Nilaparvata lugens. Pest Management Science* **67**: 528-534.

Ling KC, Tiongco ER and Aguiero VM (1978) Rice ragged stunt, a new virus disease. *Plant Disease Reporter* **62:** 701-705.

Liu GQ, Yan HH, Fu Q, Qian Q, Zhang ZT, Zhai WX and Zhu LH (2001) Mapping of a new gene for brown planthopper resistance in cultivated rice introgressed from *Oryza eichingeri. Chinese Science Bulletin* **46**: 1459-1462.

Liu Y, Su C, Jiang L, he J, Wu H, Peng C and Wan J. 2009. The distribution and identification of brown planthopper resistance genes in rice. *Hereditas* **146**: 67-73.

Liu, JL, Yu, JF, Wu, JC & Yin, JL (2008) Physiological responses to *Nilaparvata lugens* in susceptible and resistant rice varieties: allocation of assimilates between shoots and roots. *Journal of Economic Entomology* **101**: 384-390.

Maqbool SB, Riazuddin S, Loc NT, Gatehouse AMR, Gatehouse JA and Christou P (2001) Expression of multiple insecticidal genes confers broad resistance against a range of different rice pests. *Molecular Breeding* **7**: 85-93.

Matsumura M and Morimura SS (2010) Recent status of insecticide resistance in Asian planthoppers. *Japan Agricultural Research Quarterly* **41**: 225-230.

Mun JH, Song YH, Heong KL and Roderick GK (1999) Genetic variation among Asian populations of rice planthoppers, *Nilaparvata lugens* and *Sogatella furcifera* (Hemiptera: Delphacidae): mitochondrial DNA sequences. *Bulletin Entomological Research* **89:** 245-253.

Murai H, Hashimoto Z, Sharma P, Shimizu T, Murata K, Takumi S, Muri N, Kawasaki S and Nakamura C (2001) Construction of a high resolution linkage map of a rice brown planthopper *(Nilaparvata lugens* Stal) resistance gene *bph2. Theoretical and Applied Genetics* **103:** 526-532.

Murata K, Fujiwara M, Kaneda C, Takumi S, Mori N and Nakamura C (1998) RFLP mapping of a brown planthopper *(Nilaparvata lugens* Stal) resistance gene *bph2* of indica rice introgressed into a japonica breeding line "Norin-PL4." *Genes & Genetic System* **73**: 359-364.

Murata K, Fujiwara M, Murai H, Takumi S, Mori N and Nakamura C (2001) Mapping of a brown planthopper *(Nilaparvata lugens* Stal) resistance gene *Bph9* on the long arm of chromosome 12. Cereal *Research Communication* **29**: 245-250.

Myint KKM, Fujita D, Matsumura M, Sonoda T, Yoshimura A and Yasui H (2012) Mapping and pyramiding of two major genes for resistance to the brown planthopper (*Nilaparvata lugens* (Stål) in the rice cultivar ADR52. *Theoretical and Applied Genetics* **124**: 495–504.

Nagadhara D, Ramesh S, Pasalu IC, Rao YK, Krishnaiah NV, Sarma NP, Bown DP, Gatehouse JA, Reddy VD and Rao KV (2003) Transgenic *indica* rice resistant to sap sucking insects. *Plant Biotechnology Journal* **1**: 231-240.

Nagadhara D, Ramesh S, Pasalu IC, Rao YK, Sarma NP, Reddy VD and Rao KV (2004) Transgenic rice plants expressing the snowdrop lectin gene *gna* exhibit high level resistance to the white backed planthopper. *Theoretical and Applied Genetics* **109**: 1399-1405.

Nair RV, Masajo TM and Khush GS (1982) Genetic analysis of resistance to whitebacked plant hopper in twentyone varieties of rice *(Oryza sativa* L.). *Theoretical and Applied Genetics* **61**: 19-22.

Normile D (2008) Reinventing rice to feed the world. *Science* **321:** 330-333.

Padmarathi G, Ram T, Ramesh K, Kondala Rao Y, Pasalu IC and Viraktamath BC (2007) Genetics of white backed planthopper, *Sogatella furcifera* (Horvath) resistance in rice. *SABRAO Journal of Breeding and Genetics* **39**: 99-105.

Painter RH (1951) *Insect Resistance in Crop Plants.* The Mac Millan Co, New York. 520 pp.

Panda N and Khush GS (1995) *Host Plant Resistance to Insects.* CABI, Wallingford (UK), 431 pp.

Parihar A, Pathak AR and Parihar P (2010) Identification of RAPD marker for the whitebacked planthopper (WBPH) resistance gene in rice. *African Journal of Biotechnology* **9**: 1423-1426.

Park DS, Song MY, Park SK, Lee SK, Lee JH, Song SY, Eun MY, Hahn TR, Sohn JK, Yi G, Nam MH and Jeon JS (2008) Molecular tagging of the *Bph1 locus* for resistance to brown planthopper *(Nilaparvata lugens* Stal) through representational difference analysis. *Molecular Genetics and Genomics* **280**: 163-172.

Pathak MD Cheng CH and Furtono ME (1969) Resistance to *Nephotettix cincticeps* and *Nilaparvata lugens* in varieties of rice. *Nature* **223**: 502-504.

Powell KS, Spence J, Bharathi M, Gatehouse JA and Gatehouse AMR (1998) Immunohistochemical and development studies to elucidate the mechanism of action of the snowdrop lectin on the rice brown planthopper. *Journal of Insect Physiology* **44**: 529-539.

Qiu YF, Guo JP, Jing SL, Zhu LL and He GC (2010) High-resolution mapping of the brown planthopper resistance gene *Bph 6* in rice and characterizing its resistance in the 9311 and Nipponbare near isogenic backgrounds. *Theoretical and Applied Genetics* **121**: 1601-1611.

Rahman ML, Jiang W, Chu SH, Qiao Y, Ham TH, Woo MK, Lee J, Khanam MS, Chin JH, Jeung RU, Brar DS, Jena KK and Koh HJ (2009) High-resolution mapping of two brown planthopper resistance genes, *Bph20(t)* and *Bph2l(t)* originating from *Oryza minuta. Theoretical and Applied Genetics* **119**: 1237-1246.

Ramalingam J, Vera Cruz C, Kukreja K, Chittoorr JM, Wu JL, Lee SW, Baraoidan M, George ML, Cohen MB, Hulbert SH, Leach JE and Leung H (2003) Candidate defense genes from rice, barley, and maize and their association with qualitative and quantitative resistance in rice. *Molecular Plant-Microbe Interaction* **16**: 14-24.

Rao KV, Rathore KS, Hodges TK, Fu X, Stoger E, Sudhakar D, Williams S, Christou P, Bharathi M, Bown DP, Spence J, Gatehouse AMR and Gatehouse JA (1998) Expression of snow drop lectin (GNA) in transgenic rice plants confers resistance to rice brown planthopper. *Plant Journal* **15**: 469-477.

Reissig WH, Heinrichs EA and Valencia SL (1982) Insecticide-induced resurgence of the brown planthopper, *Nilaparvata lugenson* rice varieties with different levels of resistance. *Environmental Entomology* **11**: 165-168.

Rengawanayaki K, Fritz AK, Sadasivam S, Pammi S, Harrington SE, Mc Couch SR, Kumar SM and Reddy AS (2002) Mapping and progress toward map-based cloning of brown planthopper biotype-4 resistance gene introgressed from *Oryza officinalis* into cultivated rice, *O. sativa. Crop Science* **42**: 2112-2117.

Saha P, Majumder P, Datta I, Ray T, Roy SC and Das S (2006)Transgenic rice expressing *Allium sativum* leaf lectin with enhanced resisatnce against sap sucking insect-pests. *Planta* **223**: 1329-1343.

Saini RS, Khush GS and Heinrichs EA (1982) Genetic analysis of resistance to white-backed plant-hopper, *Sogatella furcifera* (Horvath), in some rice varieties. *Crop Protection* **1**: 289-297.

Sarao PS and Shera PS (2013) Planthopper management in rice. *Indian Farming* **62**: 19-22.

Sebastian LS, Ikeda R, Huang N, Imbe T, Coffman WR and McCouch SR (1996) Molecular mapping of resistance to rice spherical virus and green leafhopper. *Phytopathology* **86:** 25-30.

Sengupta S, Chakraborti D, Mondal HA and Das S (2010) Selectable antibiotic resistance marker gene free transgenic rice harbouring the garlic leaf lectin gene exhibits resistance to sap sucking planthoppers. *Plant Cell Reports* **29**: 261-271.

Sharma PN, Ketipearachchi Y, Murata K, Torii A, Takumi S, Mori N and Nakamura C (2002) RFLP/AFLP mapping of a brown planthopper *(Nilaparvata lugens* Stal) resistance gene *Bphl* in rice. *Euphytica* **129**: 109-117.

Sharma PN, Torii A, Takumi S, Mori N and Nakamura C (2004) Marker-assisted pyramiding of brown planthopper *(Nilaparvata lugens* Stal) resistance genes *Bph1* and *Bph2* on rice chromosome 12. *Hereditas* **136**: 39-43.

Shi Z, Ren X, Weng Q, Li X and He G (2003) Construction of a genomic library from brown planthopper resistant line using a transformation-competent vector and identification of clones spanning the *Qbph1ocus. Plant Science* **165**: 879-885.

Shukla KK, Gupta AK and Saini RS (1993) Reaction of some rice varieties in the 8th International Bacterial Blight Nursery to White backed planthopper, *Sogatella furcifera* (Horvath). *Journal of Insect Science* **6:** 113-114.

Sidhu GS, Khush GS and Medrano FG (1979) A dominant gene in rice for resistance to white-backed planthopper and its relationship to other plant characteristics. *Euphytica* **28**: 227-232.

Sidhu N, Basal UK, Shukla KK and Saini RG (2005) Genetics of resistance to white-backed planthopper in five rice stocks. *SABRAO Journal of Breeding and Genetics* **37**: 43-49.

Singh JP, Panday MP and Pathak RK (1990) Inheritance of resistance to white-backed planthopper, *Sogatella furcifera* (Horvath) in some rice cultivars. *Euphytica* **46**: 95-100.

Sogawa K, Fujimoto H, Qian Q, Teng S, Liu G and Zhu L (2001) QTLs for ovicidal response to white backed planthopper in rice. *Chinese Rice Research Newsletter* **9**: 5.

Soundararajan RP, Kadirve lP, Gunathilagara JK and Maheswaran M (2004) Mapping of quantitative trait loci associated with resistance to brown planthopper in rice by means of a doubled-haploid population. *Crop Science* **44**: 2214-2220.

Su CC, Cheng XN, Zhai HQ and Wan JM (2002) Detection and analysis QTL for resistance to brown planthopper, *Nilaparvata lugnes* (Stal), in rice (*Oryza sativa* L.) using backcross inbred lines. *Acta Genetica Sinica* **29:** 332-338.

Su CC, Wan J, Zhai HQ, Wang CM, Sun LH, Yasui H and Yoshimura A (2005) A new locus for resistance to brown planthopper identified in indica rice variety DV85. *Plant Breeding* **124**: 93-95.

Su CC, Zhai HQ, Wang CM, Sun LH and Wan JM (2006) SSR mapping of brown plant hopper resistance gene *Bph9* in Kaharamana, an indica rice (*Oryza sativa* L.). *Acta Genetica Sinica* **33**: 262-268.

Sudhakar D, Fu X, Stoger E, Williams S, Barathi M, Gatehouse J and Christou P (1998) High level of expression and immunolocalization of the snowdrop lectin insecticidal protein GNA in transgenic rice plants. *Transgenic Research* **7**: 371-378.

Sun L, Liu Y, Jiang L, Su C, Wang G, Zhai H and Wan J (2007) Identification of quantitative trait loci associated with resistance to brown planthopper in the indica rice cultivar Col. 5 Thailand. *Hereditas* **144**: 48-52.

Sun L, Su C, Wang C, Zai H and Wan J (2005) Mapping of a major resistance gene to brown plant-hopper in the rice cultivar Rathu Heenati. *Breeding Science* **55**: 391-396.

Sun LH, Wang CM, Su CC, Liu YQ, Zhai HQ and Wan J (2006) Mapping and marker-assisted selection of a brown planthopper resistance gene *bph2* in rice (*Oryza sativa* L.). *Acta Genetica Sinica* **33**: 717-723.

Sun X, Wu A and Tang K (2002) Transgenic rice lines with enhanced resistance to the small brown planthopper. *Crop Protection* **21**: 511-514.

Tan GX, Weng Q, Ren X, Huang Z, Zhu LL and He GC (2004) Two white backed plant-hopper resistance genes in rice share the same loci with those for brown plant-hopper resistance. *Heredity* **92**: 212-217.

Tang K, Tinjuangjun P, Xu Y, Sun X, Gatehouse JA, Ronald PC, Qi H, Lu X, Christou P and Kholi A (1999) Particle bombardment-mediated co-transformation of elite Chinese rice cultivars with genes conferring resistance to bacterial blight and sap sucking insect pests. *Planta* **208**: 552-563.

Tang K, Hu Q, Sun X, Wan B, Qi H and Lu X (2001) Development of transgenic rice homozygous lines with enhanced resistance to rice brown planthopper. *In vitro Cellular & Development Biology* **37**: 334-340.

Vanitha K, Suresh S, and Gunathilagaraj K (2011) Influence of brown planthopper *Nilaparvata lugens* (Stal.) feeding on nutritional biochemistry of rice plant. *Oryza* **48:** 142-146.

Velusamy R, Heinrichs EA and Medrano FG (1986) Greenhouse techniques to identify field resistance to the brown planthopper, *Nilaparvata lugens* (Stål) (Homoptera: Delphacidae), in rice cultivars. *Crop Protection* **5**: 328-333.

Wang Y, Wang X, Yuan H, Chen R, Zhu L, He R and He G (2008) Responses of two contrasting genotypes of rice to brown planthopper. *Molecular Plant-Microbe Interaction* **21**: 122-132.

Wei J, Hu W, Liu Q, Cheng X, Tong M, Zhu L, Chen R and He G (2009) Understanding rice plant resistance to the brown planthopper *(Nilaparvata lugens):* a proteomic approach. *Proteomics* **9**: 1-11.

Wu CF and Khush GS (1985) A new dominant gene for resistance to white backed planthopper in rice. *Crop Science* **25**: 505-509.

Xu XF, Mei HW, Luo LJ, Cheng XN and Li ZK (2002) RFLP-facilitated investigation of the quantitative resistance of rice to brown planthopper *(Nilaparvata lugens). Theoretical and Applied Genetics* **104**: 248-253.

Yamasaki M, Tsunematu H, Yoshimura A, Iwata N and Yasui H (1999) Quantitative trait locus mapping of ovicidal response in rice *Oryza sativa* L. against the white-backed planthopper, *Sogatella furcifera* (Horvath). *Crop Science* **39**: 1178-1183.

Yamasaki M, Yoshimura A and Yasui H (2003) Genetic basis of ovicidal response to White backed planthopper (*Sogatella furcifera* Horvath) in rice (*Oryza sativa* L.). *Molecular Breeding.* **12:** 133-143.

Yang H, You A, Yang Z, Zhang F, He R, Zu L and He G (2004) High resolution genetic mapping at the *Bph15* locus for brown plant hopper resistance in rice *(Oryza sativa* L.). *Theoretical and Applied Genetics* **110**: 182-191.

Yang HY, Ren X, Weng QM, Zhu LL and He GC (2002) Molecular mapping and genetic analysis of a rice brown planthopper *(Nilarparvata lugens* Stat) resistance gene. *Hereditas* **136** :39-43.

Yarasi B, Vijaya KS, Pasalu IC, Reddy VD and Rao KV (2008) Transgenic rice expressing *Allium sativum* leaf agglutinin ASAL exhibits high level resistance against major sap sucking pests. *BMC Plant Biology* **8**: 102.

Yasui H, Myint KKM, Fujita D and Matsumura M (2007) Genetics of host plant resistance to planthopper and leaf hoppers species in rice. In: *Proceedings of the JSPS International Seminar*, 22-25 November 2007, Hanoi University of Agriculture, Vietnam. pp 37-44.

Zha W, Peng X, Chen R, Du B, Zhu L and He G (2011) Knockdown of midgut genes by dsRNA-transgenic plant-mediated RNA interference in the hemipteran insect *Nilaparvata lugens*. *PloS One* **65**: e20504.

Biological and Molecular Approaches in Pest Management
Balwinder Singh, Ramesh Arora & S S Gosal (Eds)
Scientific Publishers (India), Jodhpur, 2015, pp 39-51.

Chapter - 3

Molecular Studies in Biosystematics, Phylogeny and Distribution of Tephritid Fruit Flies: A Global Perspective

Abraham Verghese*, M A Rashmi, A K Chakravarthy**, P D Kamala Jayanthi*** and G Nagalakshmi*****

**National Bureau of Agriculturally Important Insects, Hebbal, Bangalore – 560 024, India*
***Department of Entomology, University of Agricultural Sciences, GKVK, Bangalore-560065, India*
****Division of Entomology and Nematology, Indian Institute of Horticultural Research, Hesseraghatta Lake PO, Bangalore -560089, India*
E-mail: directornbaii@gmail.com

3.1. INTRODUCTION

Tephritidae (true fruit flies - due to their close association with fruits and vegetables), is the largest and most diversified family in the order Diptera, comprising about 4,000 species distributed all over the world. It includes many serious pests which pose potential threat to horticulture due to their pestilence, polyphagy as well as quarantine problems due to their limited geographical distribution. The New World genus *Anastrepha*, the primarily African genus *Ceratitis*, and the genus *Bactrocera* contain world's most serious fruit fly pests. Genus *Bactrocera* alone comprises of over 300 species which are well-documented invaders and rank high on quarantine lists worldwide (Clarke *et al.*, 2005). As fruit flies have complexes and immature taxonomy being under-developed, molecular taxonomy is the answer to resolving species identification and resolving cryptic species.

It is of paramount interest to know the origin and spread of pestilent fruit flies, as origin is key to obtain native natural enemies for effective biocontrol. Molecular tools are by far the best means to understand geographical spread from places of origin. In the present chapter we review recent research focusing on the use of molecular strategies for fruit fly identification and management. Molecular techniques have been used to resolve the place of origin of the species and also to track the subsequent spread. It is also being used to resolve the species complexes, a common feature in tephritids. Compared to Drosophilidae one of the fly families very commonly referred to as 'fruit flies' which includes biological model organism, *Drosophila melanogaster* Meigen, far less molecular research has been done in Tephritidae.

3.2. MOLECULAR TAXONOMY

Fruit fly systematics is one of the important fields of applied entomology. Despite the importance of the family the higher classification of the Tephritidae is still in a fluid state (Freidberg 1984; Hancock 1986; Foote *et al.*, 1993; Norrbom *et al.*, 1999). Years of work by many dipterists using morphological characters have yielded inadequate resolution of higher classification within the family Tephritidae because of homoplasy of most morphological characters and difficulty involved in identification of immature stages (egg and maggot). The first and foremost important step in pest management system is correct identification of the pest. Quarantine restrictions further demand quick and accurate identification of fruit fly species including their immature stages. Morphological keys are available for identification of the adults but with some limitations as the presence of homoplasy in tephritids render the identification ambiguous. Therefore, molecular taxonomy comes in as handy tool for tephritid systematics in particular.

Development of molecular systematics using nucleotide sequences along with the conventional taxonomy will enhance the species identification which is not limited by species polymorphism, sex and stage. In molecular identification of fruit flies several molecular markers e.g. allozymes (Berlocher & Smith, 1983), Random Amplified Polymorphic DNA-RAPD (Malviya *et al.*, 2012), Restriction Fragment Length Polymorphism-RFLP (Abd-el-samie & El-fiky, 2011), Amplified fragment length polymorphism-AFLP (Kakouli-Duarte *et al.*, 2001), mitochondrial cytochrome oxidase II (mt COII) + NADH1 dehydrogenase + NADH1 dehydrogenase + tRNALys + tRNAAsp (Smith *et al.*, 2003 a, b, 2005), Single Stranded Conformation Polymorphism - SSCP (Hiss *et al.,* 1994), Single Nucleotide Polymorphism - SNP (Chua *et al.*, 2010), microsatellites (Yu *et al.*, 2001), internal transcribed spacer 1 -ITS1 (Douglas and Haymer, 2001), and nuclear genes viz., period (Barr, *et al.*, 2005), white eye, Carbomyl phosphate synthese - CAD (Jayanthi & Kambhampati, 2005), 16S rRNA (Han & McPheron, 1997), 12S rRNA (Barr *et al.*, 2006) have been extensively used to unravel intra and interspecific genetic variations, population structure, migration pattern and phylogenetic relationships among fruit flies especially in *Rhagoletis, Anastrepha, Ceratitis* and *Bactrocera (=Dacus).*

Presently, the Consortium for the Barcode of Life (COBL) advocates the use of mitochondrial cytochrome oxidase I (mt COI) for species identification, as it exhibits reliable inter-species variation. Tephritid Barcode Initiative (TBI) is a two-year "demonstration project" under COBL that will create an operational system for identifying fruit flies around the world. TBI will barcode at least five representatives of all tephritid fruit flies that are either (1) agricultural pests, (2) beneficial species used for biological control of other pests, (3) closely related to pests or beneficial species; and (4) representative species from other families of fruit flies. TBI plans to obtain barcodes from approximately 2000 species of the estimated 4500 known tephritid species, but it has presently 3882 specimens with barcodes and 729 species with barcodes in COBL database as on 2010.

Microsatellite markers were screened for medfly, *Ceratitis capitata* Wiedemann to understand cross-species amplification in *Ceratitis rosa* Karsch, *Ceratitis fasciventris* (Bezzi) and *Ceratitis cosyra* (Walker) that aid in species diagnosis and population genetics. Through locus homology it is possible to distinguish the different species using a combination of the medfly-based primers. Of the 24 loci screened, 17 were common to all, five were absent in *C. cosyra,* three were absent in *C.rosa,* two were absent in *C. fasciventris,* while one locus was unique to *C. capitata.* Similar results were reported by Douglas and Haymer (2001) and Kakuoil-Duarte *et al.* (2001). Douglas and Haymer (2001) showed that *C.capitata and C.rosa* can be differentiated by the sequence polymorphism of ribosomal ITS1. Kakuoil-Duarte *et al.* (2001), found a repetitive sequence from the genome of *C. capitata* that is differentially specific to *C.capitata* and *C. rosa,* but absent in closely related species, *C. cosyra* using the AFLP technique. In addition, sequencing results indicated that most of the medfly-based microsatellite markers could be useful for population genetic studies an aspect which would facilitate the assessment of their invasive potential, tracing of the geographical origin of colonist populations and risk assessment (Baliraine *et al.,* 2002).

Generally microsatellite markers have high levels of polymorphism and presumptive selective neutrality that would make them the best tool for investigation of population structure and dynamics. In recent years, several dipteran insects including *Bactrocera tryoni* Froggatt (Kinnear *et al.* 1998; Yu *et al.,* 2001), *Glossina palpalis* Wiedemann, (fam: Glossinidae) (Solano *et al.,* 2000; Luna *et al.,* 2001), *C. capitata* (Bonizzoni *et al.* 2000), *Aedes aegypti* (Linnaeus), (fam: Culicidae) (Ravel *et al.,* 2002) have been studied using microsatellites. Dai *et al.,* 2004, isolated six polymorphic microsatellite loci from the Oriental fruit fly *Bactrocera dorsalis* (Hendel). Analyses of 71 individual flies using these markers from six different areas of Taiwan, exhibited a range of allele numbers from five to 25 for each locus. The observed heterozygosity ranged between 0.268 and 0.737 among these loci. No linkage disequilibrium was found and these microsatellite markers proved to have potential utility for population structure and gene flow studies of fruit flies.

The South American fruit fly *Anastrepha fraterculus* (Wiedemann), is a serious pest in Brazilian fruit orchards, commonly referred as *'fraterculus complex'* due to the presence of cryptic species. Rocha and Sevilon (2002) performed a detailed analysis of repetitive DNA in species of the *fraterculus* complex (-as sequences of highly repetitive DNA are usually subjected to rapid evolutionary alterations and even in sibling species, sequences of highly repetitive DNA tend to be unique) (Hankeln *et al.*, 1989). Genomic DNA of *Anastrepha* sp. 1 aff *fraterculus* and *Anastrepha* sp. 2. *fraterculus* (from Brazil) and *Anastrepha* sp.4 aff *fraterculus* (from Ecuador) were cleaved with restriction endonucleases viz., EcoRI and *ClaI,* and all cleavages yielded bands of repetitive DNA sequences. Restriction patterns were similar in sp.1 and sp.2, while sp.4 showed a distinct restriction pattern. DNA fragments from these bands were eluted, labeled and used as probes (S1C-2.1 from a fragment of 2.1 kb yielded by ClaI in sp.1, Southern blot with probe S1C-2.1) produced different hybridization patterns to sp.1 and sp.2 hybridization. In dot blot experiments, probes S1C-0.32 and S2C-0.32 hybridized to DNA from several *Anastrepha* species (sp.1, sp.2, sp.3, sp.4, *A. obliqua, A. amita, A. sorocula, A. grandis, and A. serpentina*). Only probe S2C-0.32 hybridized to DNA from *C. capitata*. These results showed that sp.1, sp.2 and sp.4 from *fraterculus* complex can be distinguished by means of restriction endonuclease patterns of total genomic DNA, besides hybridization patterns of S1C-2.1, S1C-0.32 and S2C-0.32 probes. These results also showed that some of these repetitive DNA sequences are broadly spread among the taxa, since they are shared by representatives of the genus *Anastrepha* and *C. capitata.* This analysis revealed the presence of a typical repetitive DNA, and it is not yet clear whether or not they are coding sequences. Preliminary data on sequencing fragments of 500bp suggest that this fragment really is representative of junk DNA. Additional studies are necessary to elucidate the nature of these sequences. In spite of its unknown nature, they may be employed as taxonomic markers for cryptic species of the *fraterculus* complex due to their differential restriction/hybridization patterns.

The variable region of mitochondrial *cytochrome oxidase subunit I* (COI) gene yielded better estimates of divergence among the eight members of the *Bactrocera tau* (Walker) complex (Wanwisa *et al.*, 2003).

Smith *et al.* (2003 a) inferred the phylogenetic relationships among 24 *Bactrocera* species belonging to nine subgenera from portions of the mitochondrial 16S rRNA, cytochrome oxidase II, tRNALys, and tRNAAsp genes. Two morphological characters that traditionally have been used to define the four groups within the subgenus *Bactrocera* were evaluated in a phylogenetic context by mapping the character states onto the parsimony tree. The mapping analyses suggested that the morphological characters exhibit a simple evolutionary transition from one character state to another. The implications of these results were used for devising a coherent, consolidated classification for *Bactrocera*.

Achievements in Tephritid taxonomy have greatly contributed to both basic research and pest management programmes. A DNA region of the mitochondrial cytochrome oxidase II gene was compared in species representing six genera *viz. Ceratitis, Rhagoletis, Dacus, Bactrocera, Anastrepha, Toxotrypana* and a dendrogram was constructed using the neighbour –joining method with *Liriomyza huidobrensis* (Blanchard) and *Drosophila yakuba* as outgroups. Two main clusters were obtained in the tree, the first grouping being the *Ceratitis* species viz., *C. capitata, C. rosa*, and *C. cosyra,* and the second showing two main branches, one for *Dacus, Bactrocera* and *Rhagoletis,* and the other for *Anastrepha* and *Taxotrypana* (Ferandez *et al*, 2004).

A 1356 base pairs of DNA fragment from the mitochondrial cytochrome oxidase I (CO I) of *Bactrocera philippinensis* Drew & Hancock and *Bactrocera occipitalis* Bezzi sequenced and analysed by Daojian *et al.* (2005) showed the similarity of CO I gene between these species to an extent of 96.8%. A combination of technique with real-time qualitative PCR using two sets of species specific primers along with SYBR Green assay with melting curve analysis is developed to detect and identify these *Bactrocera* sympatric species.

Combining the nuclear (CAD and white eye) and mitochondrial (16s, 12s, NADH1, COII, tRNA) gene data sets from forty seven species of genus *Bactrocera* representing six sub genera through parsimony and Bayesian methods established the potential utility of these genes in tracking phylogenic history (Jayanthi and Kambhampati, 2005).

Morphological identifications of *B. dorsalis*, *Bactrocera correcta* (Bezzi) and *Bactrocera zonata* Saunders using M13 universal primers revealed 92% similarity between species (Asokan *et al.*, 2007).

Phylogenetic relationships among five subgenera, viz., *Austrodacus, Bactrocera, Daculus, Notodacus* and *Zeugodacus* employing the 5′ region of COX-I (1490-2198); showed that the subgenus *Bactrocera* is monophyletic (Asokan *et al.*, 2011). This study further studied the utility of COX-I for the quick and accurate species diagnosis of eggs, larvae, pupae and adults of *B. zonata*, *B. tau*, *and B. dorsalis* and species-specific markers for differentiating *B. zonata* (500bp) and *B. tau* (220bp).

The genetic structure of the Egyptian peach fruit fly *B. zonata* population was analyzed through a portion of mitochondrial cytochrome oxidase I (COI, 369 bp) using RT-PCR. Phylogenetic analyses were performed in order to examine the position of the Egyptian population. The *B. zonata* from Egypt and New Zealand showed a close relationship in neighbour–joining analysis using the seven accession numbers of *B. zonata*. In addition, a theoretical restriction map of the homology portion of the COI gene was constructed using 212 restriction enzymes obtained from the restriction enzyme database to identify the Egyptian and New Zealand *B. zonata* (Abd-El-Samie and Fiky, 2011).

A general working strategy to deal with incomplete reference libraries in the DNA barcoding identification of species was proposed considering that queries with a large genetic distance with their best DNA barcode match are more likely to be misidentified and imposing a distance threshold may profitably reduce the identification errors (Massimiliano *et al.*, 2012). They further modelled relationships between identification performances and distance thresholds in four DNA barcode libraries of Diptera (n = 4270), Lepidoptera (n = 7577), Hymenoptera (n = 2067) and Tephritidae (n = 602 DNA barcodes). In all cases, more restrictive distance thresholds produced a gradual increase in the proportion of true negatives, a gradual decrease of false positives and more abrupt variations in the proportions of true positives and false negatives.

The phylogeny studies of host-plant specialization within the genus *Ceratitis* unraveled the link with host-plant toxicity as a possible mechanism that underlies the evolutionary radiation of stenophagous clades within this genus (Erbout *et al.*, 2010). Using the data from three protein encoding genes and 49 species, the phylogeny of the genus *Ceratitis* was reconstructed. Reconstruction of ancestral character states for host-plant relationships suggested that stenophagy evolved repeatedly and independently within the genus *Ceratitis*. Further, the adverse effects of host-plant toxicity on larval development in polyphagous fruit flies indicated that high concentrations of host-plant toxins pose a significant constraint on host use in polyphagous *Ceratitis* species and the observed phylogenetic patterns for monophyletic stenophagous clusters are the result of an evolutionary process of ecological specialization to toxic hosts.

Genetic discontinuities among geographical regions and levels of population admixture were studied in *Bactrocera cucurbitae* (Coquillett) from 25 worldwide-distributed localities (n=570) with 13 microsatellite loci specifically developed for this species revealed five main groups of populations viz., the African continent La Réunion, Central Asia, East Asia and Hawaii. The proportions of inter-regional assignments and the higher values of genetic diversity in populations from Pakistan, India and Bangladesh suggest that *B. cucurbitae* originated in Central Asia and expanded its range to East Asia and Hawaii on one hand and to Africa and the islands of the Indian Ocean on the other. A number of outliers (10-19 specimens according to different clustering algorithms) showed high levels of admixture (Q>0.70) with populations from different regions and reveal complex patterns of inter-regional gene flow (Massimiliano *et al.,* 2010).

3.3. MOLECULAR TOOLS IN ANALYZING INVASION HISTORY

The introduction of species into a new area can alter successional patterns, mutualistic relationships, community dynamics, ecosystem function, and resource distribution. The molecular methodologies are being proven as useful tools in detecting/analyzing alien introductions. Tephritid fruit flies are very susceptible to such human induced introductions and have greater impact on international marketing and world trade of agricultural produce. Being important pest species of several fruits and vegetables, migration and transport of horticultural produce has lead to (pre-) historic cases of accidental introductions. Although Africa is known to be the origin of several fruit fly introductions and establishments worldwide (the most notorious species being the mediterranean fruit fly, *C. capitata*), with the intensification of fruit trade, several importing countries become highly vulnerable to introduction of alien fruit fly species. Notable invasive species include *B. zonata, Bactrocera invadens* Drew, Tsura & White, *Bactrocera latifrons* (Hendel), *B. cucurbitae* and *C. capitata.*

DNA barcoding, using a short gene sequence taken from standardized portions of the genome, allows recognition of indigenous and alien species even in immature stages that are inseparable on morphological grounds. This allows quarantine officers and national plant protection agencies to identify intercepted material during inspections or surveying activities. Molecular phylogeny indicates the lineages among species, recognizing clusters of closely related taxa. Combined with host-plant data, patterns in host-insect relationships can be revealed. Population genetics can provide an insight in the macrogeographic population structure and interrelationships between regions indicating possible sources of introductions or migrations. Thus molecular phylogeography allows us to reconstruct the pathways of geographic range expansions. The application of molecular tools in these fields is demonstrated by De Meyer *et al.* (2010) with examples of fruit fly pests originating from or introduced to the African mainland.

3.4. MOLECULAR METHODOLOGIES IN DETECTION OF GUT BACTERIA

The bacteria in fruit flies alimentary tract have been known to be advantageous in their biology. The classical tcchniques of gut bacterial studies in fru'.t flies were known with easy culturable gut bacteria but unculturable and not-yet cultured bacteria could not be detected. Molecular techniques now can reveal real time gut bacterial community structure in insect alimentary tract. Classical and molecular cloning techniques were employed to study gut bacterial community structure in *B. cucurbitae* gathered from wild (Thaochan *et al.*, 2010). Genomic DNAs were isolated from alimentary tract and amplified with universal primer of 16S rRNA gene for bacteria. The community diversity and community similarity were analyzed with shannon-weaver index, rarefaction analysis, Morisita's index, multidimensional scaling analysis and phylogenetic tree analysis. Classical technique revealed only one bacterial family viz. *Enterobacteriaceae* that belonged to *Grammaproteobacteria* in *B. cucurbitae* species, whereas the molecular study revealed five bacterial taxa, *Alpha-, Beta-, Grammaproteobacteria, Actinobacteria* and *Firmicutes*. The former three bacterial taxa were classified in phylum *Proteobacteria*. Interestingly, *Wolbachia* sp., known as bacterial arthropod pathogen and an unculturable bacterium, were detected through the molecular study. Therefore molecular cloning techniques will expose the ecology of gut bacterial community structures more precisely than conventional methods.

Characterization of three bacterial symbionts (BC1, BC2 and BC3) of fruit fly *B. tau* including morphological, biochemical and 16S rDNA (rrs gene) analysis was done by Prabhakar *et al.* (2009), to determine their taxonomic position. Morphological and biochemical characterization placed two bacteria (BC1, *Klebsiella oxytoca* and BC2, *Pantoea agglomerans*) into family *Enterobacteriaceae* and the third one (BC3, Staphylococcus sp.) into family *Staphylococcaceae*. 16S rDNA gene sequence comparison with the available NCBI database sequences further confirmed the characterizations of bacterial symbionts. Molecular phylogeny of *K. oxytoca* and *P. agglomerans* closely related to the other free living enterobacterial members with 74 to 93% sequence homology (genetic distance 0.000 to 0.085); however, they showed only 74 to 87% similarity with other insect symbionts (genetic distance 0.090 to 0.121). *Staphylococcus* sp. showed 94% sequence homology with other members of family Staphylococcaceae with the genetic distance of 0.013. Further, they found population of these symbionts in adult fruit flies increased exponentially up to the 10th day of adult emergence and thereafter it became almost constant.

In a study carried out at IIHR, Bangalore gut bacteria was isolated and characterized from males/females of *B. dorsalis* using eubacterial primers, targeting 16S rRNA gene. Based on morphological, physiological, biochemical and molecular characteristics, the bacterial isolates from *B. dorsalis* males/ females were discovered to be gram positive (family *Enterobacteriaceae*). Further simila. work with respect to *B. cucurbitae* is being envisaged (personal communication Nagalakshmi 2012*).

3.5. TRANSCRIPTOME ANALYSIS

Lack of genetic information of any pest is an obstacle in understanding the mechanisms behind its development and its ability to resist insecticides. Analysis of the transcriptome and its expression profile data is essential for extending the genetic information resources thereby providing a shortcut that will support the insecticide resistance studies in fruit flies.

De novo assembly of *B. dorsalis* transcriptome using short read sequencing technology generated 484,628 contigs, 70,640 scaffolds, and 49,804 unigenes (Shen *et al.,* 2011). Of those unigenes, 27,455 (55.13%) matched known proteins in the NCBI database, Digital gene expression (DGE) libraries showed differences in gene expression profiles at different developmental stages (eggs, third-instar larvae, pupae, and adults) and provided comprehensive insight into gene expression profiles at different developmental stages. This facilitates the study of the role of each gene in the developmental process and insecticide resistance in particular for *B. dorsalis*.

3.6. PEST MANAGEMENT

Molecular means have been established as effective tools for pest management and biological control programme as parasitoid detection and identification is a necessary step in the development and implementation of biological control strategies employing augmentive release at present, but the supportive genomic or transcriptomic data for members of tephritidae remains limited. DNA-based methods have been used in recent years to identify natural enemies of pest species where morphological differentiation is problematic. Molecular techniques also offer a considerable advantage over traditional morphological methods for fruit fly and parasitoid discrimination as well as within-host parasitoid identification, which currently relies on dissection of immature parasitoids from the host, or lengthy and labour-intensive rearing methods. Here we review recent research focusing on the use of molecular strategies for fruit fly management.

An extensive transcriptome study for the oriental fruit fly *B. dorsalis* (Peng *et al.,* 2012) revealed over 350 million bases of cDNA derived from the whole body of *B. dorsalis* at different developmental stages. In a single run, 747,206 sequencing reads with a mean read length of 382 bp were obtained. These reads were assembled into 28,782 contigs and 169,966 singletons. The mean contig size was 750 bp and many nearly full-length transcripts were assembled. Additionally, they identified a great number of genes that are involved in reproduction and development as well as genes that represent nearly all major conserved metazoan signal transduction pathways, such as insulin signal transduction. Furthermore, transcriptome changes during development were analyzed. A total of 2,977 differentially expressed genes (DEGs) were detected between larvae and pupae libraries, while there were 1,621 DEGs between adults and larvae and 2,002 between adults and pupae. These DEGs were functionally annotated with KEGG pathway annotation and nine genes were validated by qRT-PCR. Data obtained by them represent the extensive sequence resources available for *B. dorsalis* and provide for the first time access to the genetic architecture of reproduction and development as well as major signal transduction pathways in this tephritid fruit fly, allowing us to elucidate the molecular mechanisms underlying courtship, oviposition, development and detailed analyses of the signal transduction pathways.

Over 100 species of Hymenoptera (Braconidae) have been reared from fruit-infesting tephritids with the majority from the subfamily Opiinae (White, and Elson Harris, 1992; Wharton and Marsh, 1978; Carmichael *et al.*, 2005). Parasitoids from several genera within this subfamily including *Diachasmimorpha*, *Fopius*, *Aganaspis* and *Psyttalia* (Hymenoptera: Braconidae) are thought to be the most promising or are currently employed in the biological control of fruit fly pests (Clausen, 1978; Wharton, 1989). The correct identification of a natural enemy is therefore fundamental to its effective use as a biological control agent. The genus *Psyttalia* (Braconidae: Opiinae) contains several species being used or considered for use in the biological control of various fruit-infesting tephritid pests.

Rugman-Jones *et al.* (2009) studied *Psyttalia concolor* (Szepligeti) species complex of subsaharan Africa which are effective tephritid parasitoids. However, there is considerable confusion in the identity of members of this complex because they are difficult to differentiate morphologically. They provided information on nuclear and mitochondrial DNA markers that may be used for separation of various populations. Phylogenetic analyses using sequence data from 28sD2 and COI gene regions illustrated relationships among 10 identifiable groups of *Psyttalia* populations, and they subsequently discussed (and assigned) available species names for these populations. The name *Psyttalia humilis* (Silvestri) is available of sub-Saharan populations that are morphologically similar to *P. concolor*, the name applicable to populations from the Mediterranean region and their findings have a bearing on the potential use of several populations in biological control efforts against olive fly in California.

3.7. CONCLUSION

Molecular studies help in species identification irrespective of the insect stage and are thus, very useful in quarantine and early detection and prevention of invasive fruit flies. Such studies can play a role in improving/ identifying strains of male fruit flies that have a mating edge over wild males in sterile insect technique. Identifying maggots if accidently introduced from imported fruits is crucial which is possible only with molecular methods. In the years to come, endosymbionts will have an important say in IPM and these are best studied through molecular tools. At IIHR, efforts are on to have a barcode library of all Indian fruit flies to facilitate identification services. Needless to say, fruit fly management will receive a boost with appropriate dependence on biotechnological tools.

REFERENCES

ABD-El-Samie EM and EL-Fiky ZA (2011) Molecular phylogeny and identification of the peach fruit fly, *Bactrocera zonata*, established in Egypt. *Journal of Insect Science* **11**: 177.

Asokan R, Krishnakumar NK and Verghese A (2007) Molecular identification of fruit flies, *Bactrocera* spp. (Diptera: Tephritidae) using mitochondrial cytochrome oxidase I. *Current Science* **93** (12): 25.

Asokan R, Rebijith KB, Singh SK, Sidhu AS, Siddharthan S, Raveen P, Karanth K, Llango E and Ramamurthy VV (2011) Molecular identification and phylogeny of *Bactrocera* species (Diptera: Tephritidae) *Florida Entomologist* **94**(4):1026-1035.

Baliraine FN, Bnizzoni M, Osir EO, Lux SA, Mulaa FJ, Quilici S, Gomulski LM, Gasperi G and Malacrida AR (2002) Medfly microsatellite markers for species diagnosis and population genetic analysis in three other fruit fly (Diptera: Tephritidae) species, *Procedings of VI International Fruit fly Symposium 6-10 May, stellenbosch, South Africa*, pp. 411-414.

Barr NB, Cui L and Mcpheron BA (2005) Molecular Systematics of Nuclear Gene *period* in Genus *Anastrepha* (Tephritidae). *Annals of Entomological Society of America* **98**: 173-180.

Barr NB, Copeland RS, De Meyer M, Masiga D, Kibogo HG, Billah MK, Osir E, Wharton RA and Mcpheron BA (2006) Molecular diagnostics of economically important *Ceratitis* fruit fly species (Diptera: Tephritidae) in Africa using PCR and RFLP analyses. *Bulletin of Entomological Research* **96**: 505-521.

Berlocher SH and Smith DC (1983) Segregation and mapping of allozymes of the apple maggot fly. *Journal of Heredity* **74**: 337-340.

Bonizzoni M, Malacrida AR, Guglielmino CR, Gomulski LM, Gasperi G and Zheng L (2000) Microsatellite polymorphism in the Mediterranean fruit fly, *Ceratitis capitata*. *Insect Molecular Biology* **9:** 251–261.

Carmichael AE, Wharton RA and Clarke AR (2005) Opiine parasitoids (hymenoptera: Braconidae) of tropical fruit flies (Diptera: Tephritidae) of the Australian and south pacific region. *Bulletin of Entomological Research* **95**: 545- 569.

Chua TH, Song BK and Chong YV (2010) Development of allele-specific single-nucleotide polymorphism-based polymerase chain reaction markers in cytochrome oxidase I for the differentiation of *Bactrocera papayae* and *Bactrocera carambolae* (Diptera: Tephritidae). *Journal of Economic Entomology* **103**(6): 1994-1999.

Clarke AR, Armstrong KF, Carmichael AE, Milne JR, Raghu S, Roderick GK and Yeates DK (2005) Invasive phytophagous pests arising through a recent tropical evolutionary radiation: the *Bactrocera dorsalis* complex of fruit flies. *Annual Review of Entomology* **50**: 293–319.

Clausen CP (1978) Tephritidae (trypetidae, trupaneidae) In: Clausen, CP (ed.) *Introduced Parasites and Predators of Arthropod Pests and Weeds: A World Review*. Agricultural Handbook No. 480, USDA Agricultural Research Service, Washington, DC, USA.

Dai SM, Lin CC and Chang C (2004) Polymorphic microsatellite DNA markers from the Oriental Fruit Fly *Bactrocera dorsalis* (Hendel) *Molecular Ecology Notes* **4**: 629–631.

Daojian Y, Zhilin C, Runjie Z and Wenying Y (2005) Real-time qualitative PCR for the inspection and identification of *Bactrocera philippinensis* and *Bactrocera occipitalis* (Diptera: Tephritidae) using SYBR Green Assay. *The Rafeles Bulletin of Zoology* **53**(1): 73-78.

De Meyer M, Sunday E, Massimiliano V, Fathiya K, Maulid M and Anna M (2010) In and out of Africa: invasion history of fruit flies on the African Mainland (diptera, tephritidae). *VIII International Symposium on Fruit Flies of Economic Importance.* Valencia.

Douglas LJ and Haymer DS (2001) Ribosomal ITS1 Polymorphisms in *Ceratitis capitata* and *Ceratitis rosa* (Diptera: Tephritidae). *Annals of Entomological Society of America* **94**: 726-731

Erbout N, Lens L & Massimiliano V and De Meyer M (2010) Host plant toxicity, stenophagy and evolutionary radiation in the genus ceratitis (diptera, tephritidae). *VIII International Symposium on Fruit Flies of Economic Importance,* Valencia.

Ferandez P, Segura D, Callejas C and Ochando MD (2004) A phylogenetically study of the family Tephritidae (Insecta: Diptera) using a mitochondrial DNA sequence. *Procedings of VI International Fruit fly Symposium 6-10 May, stellenbosch, South Africa,* pp.439-443.

Foote RH, Blanc FL and Norrbom AL (1993) *Handbook of the Fruit Flies (Diptera: Tephritidae) of America North of Mexico.* Comstock Publishing Associates, Ithaca. 571pp.

Freidberg A (1984) Gall Tephritidae (Diptera). In: Ananthakrishnan, TN (ed.) *Biology of Gall Insects.* Oxford & IBH publishing Co., New Delhi, pp. 129-167.

Han HY and Mcpheron BA (1997) Molecular phylogenetic study of Tephritidae (Insecta: Diptera) using partial sequences of the mitochondrial 16S ribosomal DNA. *Molecular Phylogeneties & Evolution* **7**(1):17-32.

Hancock DL (1986) Classification of the trypetinae (Diptera: Tephritidae) with a discussion of the Afro tropical fauna. *Journal of Entomological Society of South Africa* **49**: 275-305.

Hankeln T, Keyl HG and Schmidt ER (1989) DNA-probes for the investigation of chromosome evolution in chironomus. Part II: Repetitive sequences. *Acta Biolica Debr Oecologica Hungarica* **2**: 2019-227.

Hiss RH, Norris DE, Dietrich CH, Whitcomb RF, West DF, Bosio CF, Kambhampati S, Piesman J, Antolin MF and Black W C (1994) Molecular taxonomy using single strand conformational polymorphism (SSCP) analysis of mitochondrial ribosomal DNA genes. *Insect Molecular Biology* **3**: 171–182.

Jayanthi PDK and Kambhampati S (2005) Phylogeny of *Bactrocera* species inferred from mitochondrial and nuclear genes. National Meet of Entomological Society of America 13-17th November, 2004, Salt Lake City, Utah, USA.

Kakouli-Duarte T, Casey DG and Burnell AM (2001) Development of a Diagnostic DNA Probe for the Fruit Flies *Ceratitis capitata* and *Ceratitis rosa* (Diptera: Tephritidae) Using Amplified Fragment-Length Polymorphism. *Journal of Economic Entomology* **94**(4): 989-997.

Kinnear MW, Bariana HS, Sved JA and Frommer M (1998) Polymorphic micro-satellite markers for population analysis of a tephritid pest species, *Bactrocera tryoni. Molecular Ecology* **7**: 1489–1495

Luna C, Bonizzoni M, Cheng Q, Robinson AS, Aksoy S and Zheng L (2001) Microsatellite polymorphism in tsetse flies (Diptera: Glossinidae). *Journal of Medical Entomology* **38**: 376–381.

Malviya S, Bajpai N and Tewari RR (2012) RAPD-PCR Based Genetic Relationship of Muscid Flies (Diptera: Musci-Dae). *International Journal of Pharmacology & Biological Science* **3**(3) (B): 1018 – 1024.

Massimiliano V, Delatte H, Backeljau T and De Meyer M (2010) Macrogeographic population structuring in the cosmo-politan agricultural pest *Bactrocera cucurbitae* (Diptera: Tephritidae). *VIII International Symposium on Fruit Flies of Economic Importance,* Valencia.

Massimiliano V, Jordaens K, Breman FC, Backeljau T and De Meyer M (2012) Identifying Insects with Incomplete DNA Barcode Libraries, African Fruit Flies (Diptera:Tephritidae) as a Test Case. *PLoS ONE* **7**(2): e31581. doi:10.1371/ journal.pone.0031581

Norrbom AL, Carroll LE and Freidberg A (1999) Status of knowledge. In: Thompson, FC (ed.) *Fruit Fly Expert Identification System and Systematic Information Database.* North American Dipterists Society, Backhays Publishers, Leiden, *Myia* **9:** 9-47.

Peng T, Zheng W, He W and Zhang H (2012) High-Throughput Sequencing to Reveal Genes Involved in Reproduction and

Development in *Bactrocera dorsalis* (Diptera: Tephritidae). *PLoS ONE* **7**(5): e36463. doi:10.1371/journal.pone.0036463

Prabhakar C, Sood P, Kapoor V, Kanwar S, Mehta P and Sharma P (2009) Molecular and biochemical characterization of three bacterial symbionts of fruit fly, *Bactrocera tau* (Tephritidae: Diptera). *Journal of General and Applied Microbiology* **55**: 479-487

Ravel S, Herve JP, Diarrassouba S, Kone A and Cuny G (2002) Microsatellite markers for population genetic studies in *Aedes aegyptii* (Diptera: Culicidae) from Cote d'Ivoire: evidence for a microgeographic genetic differentiation of mosquitoes from Bouake. *Acta Tropica* **82**:39–49.

Rocha LS and Selivon D (2002) Studies on highly repetitive DNS in cryptic species of the *Anastrepha frateculus* complex (Diptera: Tephritidae), *Procedings of VI International Fruit fly Symposium,* 6-10 May, Stellenbosch, South Africa, pp.415-418.

Rugman-Jones A, Paul F, Robert Wharton B, Tom Van Noort A and Richard S (2009) Molecular differentiation of the *Psyttalia concolor* (Szépligeti) species complex (Hymenoptera: Braconidae) associated with olive fly, *Bactrocera oleae* (Rossi) (Diptera: Tephritidae), in Africa. *Biological Control* **49**: 17–26

Shen GM, Dou W, Niu JZ, Jiang HB and Yang WJ (2011) Transcriptome Analysis of the Oriental Fruit Fly (*Bactrocera dorsalis*). *PLoS ONE* **6**(12): e29127. doi:10.1371/journal.pone.0029127

Smith PT, Kambhampati S and Armstrong KA (2003a) Phylogenetic relationships, character evolution, and biogeography of *Bactrocera* (Diptera: Tephritidae) inferred from mitochondrial DNA sequences. *Molecular Phylogenetics and Evolution* **26**: 8-17.

Smith PT, Kambhampati S and Armstrong KA (2005) Phylogenetic relationships, character evolution, and biogeography of *Bactrocera* (Diptera: Tephritidae) inferred from a combined analysis of mitochondrial DNA. *Insect Systematics and Evolution* **36**: 343-360.

Smith PT, Mc Pheron BA and Kambhampati S (2003b) Phylogenetic analysis of mitochondrial DNA supports the monophyly of *Dacini* fruit flies (Diptera: Tephritidae) inferred from mitochondrial DNA sequences. *Annals of the Entomological Society of America* **95**: 658-664.

Solano P, De LA, Rocque S, De Meeus T, Cuny G, Duvallet G and Cuisance D (2000) Microsatellite DNA markers reveal genetic differentiation among populations of *Glossina palpalis gambiensis* collected in the agro-pastoral zone of Sideradougou, Burkina Faso. *Insect Molecular Biology* **4**: 433–439.

Thaochan N, Anuchit C and Rai D (2010) Classical and molecular study of gut bacterial community structure in alimentary tract of *Bactrocera cucurbitae* (coquillett). *VIII International Symposium on Fruit Flies of Economic Importance,* Valencia.

Wanwisa J, Visut B and Pattamaporn K (2003) Molecular phylogeny of tephritid fruit flies in the *Bactrocera tau* complex using the mitochondrial *COI* sequences. *Genome* **46**: 112–118.

Wharton RA and Marsh PM (1978) New world opiinae (Hymenoptera: Braconidae) parasitic on Tephritidae (Diptera). *Journal of Washington Academic of Sciences* **68**: 147-167.

Wharton RA (1989) Classical Biological Control of Fruit-Infesting Tephritidae. In: Robinson, AS, Hooper, G (eds.) *World Crop Pests: Fruit Flies, Their Biology, Natural Enemies and Control*; Elsevier: Amsterdam, the Netherlands, pp. 303 - 313.

White IM and Elson-Harris MM. (1992) *Fruit Flies of Economic Significance: Their*

Identification and Binomics; CAB International: Wallingford, UK.

Yu H, Frommer M, Robson MK, Meats AW, Shearman DC and Sved JA (2001) Microsatellite analysis of the Queensland fruit fly *Bactrocera tryoni* (Diptera: Tephritidae) indicates spatial structuring: implications for population control. *Bulletin of Entomological Research* **91**: 139–147.

Biological and Molecular Approaches in Pest Management
Balwinder Singh, Ramesh Arora & S S Gosal (Eds)
Scientific Publishers (India), Jodhpur, 2015, pp 52-72.

Chapter - 4

RNA Interference Research: Current Status and Future Outlook for Utilization in Insect Pest Management

Prashant Mohanpuria*, S K Sandhu and Ramesh Arora*****

**School of Agricultural Biotechnology*
***Department of Plant Breeding & Genetics*
****Department of Entomology, Punjab Agricultural University Ludhiana-141004, India*
E-mail: pmohanpuria@pau.edu

4.1. INTRODUCTION

RNA interference (RNAi) or gene silencing is a post-transcriptionally, RNA dependent, sequence-specific process of turned down or shut off expression of certain genes, which suppresses the production of a specific protein in an organism. Historically, gene silencing is known by different names such as cosuppression, post-transcriptional gene silencing (PTGS) in plants, quelling in fungi and RNAi in animals but now a days, RNAi term has been frequently used by researchers for gene silencing in all eukaryotes. Its mechanism has been reported to be conserved in a wide range of eukaryotes including plants, nematodes, fungi, protozoa, insects, and all vertebrates. From previous science history, RNAi is known to protect the organisms from invading infectious nucleic acid containing pathogens like bacteria, fungi and viruses etc. and thus, it must be the part of defense system to protect the genome of a species. It is also a natural process of regulating gene expression in plants, animals and other organisms. Gene expression can be regulated by way of silencing it at transcriptional level *i.e.* transcriptional gene silencing (TGS) (Miglani 2013) and at post-transcriptional level *i.e.* (PTGS) (Fagard and Vaucheret 2000). TGS involves gene disruption and heritable changes in chromatin structure etc. In former, the physical structure of a gene is disrupted by the insertion of a piece of DNA of known sequence such as T-DNA or transposon elements, and treatment with mutagens or irradiation. TGS and PTGS both seem to be the example of an RNA-mediated epigenetic modification which reinforces the prevailing concept of 'homology-dependent gene silencing' where DNA–DNA, RNA–DNA and RNA–RNA interactions are considered as potential triggers for gene silencing (Matzke *et al.*, 2005). It requires conventional DNA methyltransferases, histone modifying enzymes and some other specific proteins. The phenomenon of PTGS was first identified as "co-suppression" by Napoli *et al.* (1990) in transgenic petunia flowers where the introduction of chalcone synthase gene (CHS A) of anthocyanin biosynthesis pathway resulted in white or chimeric flowers instead of expected dark purple coloration. It was due to the silencing of endogenous homologous gene of petunias. In 1998, Fire and Mello reported the dsRNA as the real inducer of some gene suppressor in the worm *Caenorhabditis elegans* and the term 'RNAi' came into existence (Hannon, 2002). Later on, a number of gene silencing phenomena that inactivate genes at the post transcriptional level were identified (Miglani, 2013). Antisense RNA suppression is an early form of PTGS employed by scientists and it involves the base pairing of antisense RNA strand with its complementary sense RNA strand (mRNA) of the corresponding target gene intended to silence (Brantl, 2002). This complementary base pairing prevents the translation of the target mRNA and also makes mRNA susceptible to endonucleases (Kimball, 2002). But due to the lesser specificity and stability as compared to RNAi, antisense RNA suppression was not preferred by the researchers. In RNAi, RNA itself is responsible for the sequence specific degradation of a target mRNA. It requires the introduction of homologous dsRNA to specifically target a gene product, resulting in complete loss or reduction in appreciable amount of expression of the target gene. The complex mechanism of RNAi involves two highly conserved enzyme systems; one is called Dicer (in case of fungi, insects and other animals) and Dicer-like (DCL) elements (only in plants), and second one is called RNA-induced silencing complex (RISC) or RNA-induced transcriptional silencing (RITS) complex. Cytoplasmic localized Dicer or nuclear localized DCL family members are large, conserved, multidomain proteins that contain one or two putative RNA helicase domains (at N-terminal), a PAZ (Piwi/Argonaute/Zwille) domain, tandem ribonuclease III (RNase III) domains and one dsRNA-binding domain (at C-terminal) (Yu and Kumar, 2003; Tang *et al.*, 2003).

In RNAi small functional RNA molecules which are either produced endogenously or exogenously by different ways play major role. Based upon the origin of small RNAs, they are classified into small interfering RNA (siRNA) and micro RNA (miRNA). The miRNAs are chemically and functionally similar to siRNAs, but are evolutionarily conserved and produced endogenously from miRNA genes present in the genome. These are produced from larger precursors that are transcribed from non-protein-coding genes in most of the plants and animals (Zhou *et al.*, 2008). Majority of the plant miRNA genes exist as independent transcriptional units, which are transcribed by RNA polymerase II into primary transcripts (pri-miRNAs) of 1-2kb or longer. The pri-miRNAs are processed into precursor miRNAs (pre-miRNAs) with a specific hairpin shaped stem-loop structure. The processing of these precursor molecules is accomplished by nuclear-localized RNase III enzyme DICER-LIKE 1 (DCL1) in case of plants

(Kurihara and Watanabe 2004). Further, the processed molecule is exported from nucleus to cytoplasm by a miRNA transporter, HASTY (HST). After methylation of mature miRNA, it joins with Argonaute to form RISC that can subsequently target specific mRNAs in the same way as siRNAs do (Jones-Rhoades *et al.*, 2006).

Long exogenous dsRNA initiates RNAi by activating Dicer which converts it into 20-26nt small interfering RNAs (siRNA), molecules with two base overhangs at 3' end. The endonuclease activity of Dicer cuts off the dsRNA. These siRNAs are then incorporated into RISC or RITS complex, where one strand *i.e.* sense strand (the "passenger" strand) is eliminated and the other *i. e.* antisense strand "guide" strand is retained. In case of plants, *C. elegans* and fungi, the siRNA produced are used as primers for the generation of new dsRNA by RNA-dependent RNA polymerase (RdRP). The new dsRNA also subsequently serves as target for the DCL and is processed into secondary siRNA for the amplification of silencing signals. The core catalytic protein components of RISC or RITS are Argonaute proteins which are responsible for siRNA binding as well as ribonuclease activity to cut the target mRNA. While single-stranded siRNAs (guide) loaded into active RISC to search for its homology with target mRNA and then direct PTGS through mRNA cleavage/translational repression (Bartel 2004) or nonsense-mediated mRNA decay (NMD) (removal of 5'capping and/or deadenylation) (Sen and Blau 2005; Shi *et al.*, 2012). PTGS is mainly based on RNA-RNA interactions. Argonaute proteins are localized to specific regions in the cytoplasm called P-bodies (cytoplasmic bodies or GW bodies), which are regions with high rates of mRNA decay (Sen and Blau 2005). Disruption of P-bodies decreases the efficiency of RNAi, suggesting that they are the site of a critical step in the RNAi process (Jakymiw *et al.*, 2005). Sometimes, the mature single-stranded siRNAs loaded into active RITS complex to direct TGS through chromatin modifications/remodelling or RNA-directed DNA methylation (RdDM) (Mohanpuria *et al.*, 2010; Miglani 2013). TGS is mainly based on RNA-DNA interactions (Fig. 1). In addition to this, introduction of long dsRNA or siRNA has also been adopted as a tool to inactivate gene expression *i.e.* TGS, both in cultured cells and in living organisms (Meister and Tuschl 2004). Till date dsRNAs resulting in an RNAi response have been used to study functional genomics in more than 30 insect species from a variety of different orders, including Diptera, Coleoptera, Lepidoptera, Isoptera, Orthoptera, Hymenoptera and Hemiptera. This approach has been extremely successful in the functional characterization of genes involved in variety of physiological processes such as development, reproduction, behavior and immunity in insects (Belles, 2010). Using this approach successfully knock down of gene expression and altered phenotype of developing progeny have been shown in the grasshopper, *Schistocerca americana* (Kennerdell and Carthew 1998), cockroach, *Blattella germanica* (Martin *et al.*, 2006), rust-red flour beetle, *Tribolium castaneum* (Bucher *et al.*, 2002), milkweed bug, *Oncopeltus faciatus* (Liu and Kaufman 2004a and Liu and Kaufman 2004b) and cricket, *Gryllus bimaculatus* (Ronco *et al.*, 2008 and Mito *et al.*, 2008). But the introduction of dsRNA into the insect body cavity results in knock down of the target gene and if it influences the vitality of the pests, then this could be a very useful approach for integrated pest management for future.

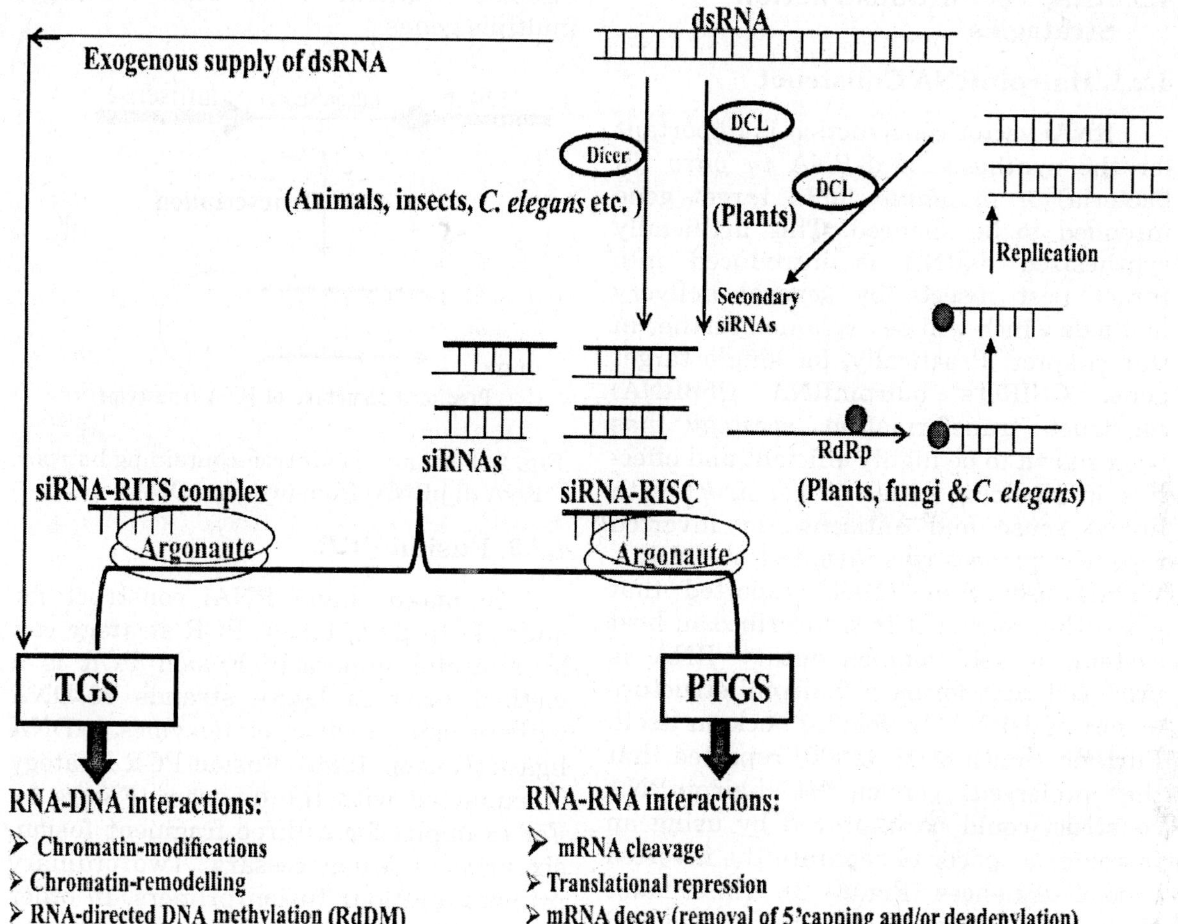

Fig 1. Representative model of gene silencing mediated by RNA interference (RNAi) in plants, animals, nematode worms and fungi. RNAi is triggered by long dsRNA molecules of the target gene(s). The dsRNA is processed into 21-26 nt siRNAs through cytoplasmic localized Dicer (in fungi, insects, nematodes like *C. elegans* and other animals) or nuclear localized Dicer-like (DCL) (in plants) multidomain protein. While siRNAs can also be used as primers for the generation of new dsRNA by RNA-dependent RNA polymerase (RdRP) the new dsRNAs can subsequently serve as target for the DCL and are processed into secondary siRNAs for amplification of silencing signals in plants, fungi and *C. elegans*. The mature single-stranded siRNAs loaded into active RNA-induced transcriptional silencing complex (RITS) direct transcriptional gene silencing (TGS) through chromatin modifications/remodelling or RNA-directed DNA methylation (RdDM). Single-stranded siRNAs loaded into active RNA-induced silencing complex (RISC) direct post transcriptional gene silencing (PTGS) through mRNA cleavage or translational repression or mRNA decay. Further, based on RNA-DNA interactions, dsRNA may also direct TGS.

4.2. RNAi Vector Construction Strategies

4.2.1. HairpinRNA Construct

RNAi vector construction is important for the synthesis of dsRNA *in vitro* (in bacteria) or *in planta* for a target gene intended to be silenced. This artificially synthesized dsRNA is introduced into target pest insects by several delivery methods which will be explained further in this chapter. Practically, for single target gene, CSIRO's hairpinRNA (ihpRNA) construct transformation strategy has been shown to be highly efficient and effective in silencing genes. A construct with linked sense and antisense or inverted repeat sequences was invented by CSIRO. Waterhouse *et al.* (1998) reported that when this construct is transcribed in host system, a self complementary RNA is produced that forms a 'hairpin' structure (a sort of dsRNA) by folding back on itself. Further, Smith *et al.* (2000) reported that the preferred version of hairpinRNA construct could be improved by using an intron or a spacer to separate the inverted repeat sequences (Figure 2). This makes the construct more stable in bacteria and increases silencing efficiency also. Even, the ihpRNA strategy increases the gene silencing by 90-100 per cent (Wesley *et al.*, 2001). For making RNAi cassette, a minimum partial cDNA sequence (100-400bp) of a gene is amplified through polymerase chain reaction (PCR) in sense and in antisense direction separately using the primer pairs incorporated with different restriction sites which are also present in RNAi vector cloning site. But these restriction sites must not be present in between the gene fragment(s) amplified for RNAi construct formation. This simple restriction digestion based strategy creates a difficult situation if we want to target multiple genes.

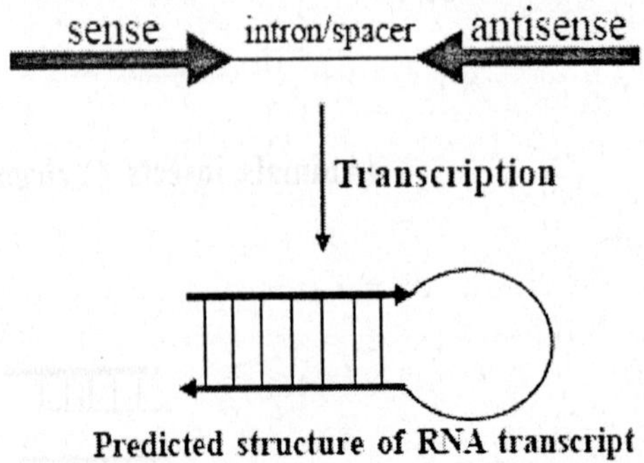

Fig. 2. Formation of intron-containing hairpin RNA (ihpRNA) from inverted DNA repeat.

4.2.2. Fusion PCR

To make single RNAi construct for multiple targets, fusion PCR strategy can be a useful approach. Fusion PCR is a method used to ligate strands of DNA without using restriction enzymes or DNA ligase (Horton 1989). Fusion PCR strategy is explained with the diagram (Figure 3). For example, for a three fragment fusion, six primers are necessary: Two ordinary primers and four fusion primers. In other aspects than the fusion primer design and the number of steps to produce a final product, fusion PCR follows ordinary PCR protocol. Fragments from the genes that are to be recombined are amplified in separate PCRs. The primers are designed so that the ends of the products contain complementary sequences. When these PCR products are mixed, denatured, and reannealed, the strands having the matching sequences at their 3' ends overlap and act as primers for each other. Extension of this overlap by DNA polymerase produces a recombinant DNA molecule. This fusion PCR product (containing three target gene fragments) can be amplified in sense direction and also in antisense directions and then ihpRNA construction strategy can be employed further. Induction of RNAi in a system requires the introduction of homologous dsRNA or intron-containing hairpin RNA (ihpRNA) specific to a target gene. But the RNAi core machinery (DICER, DCL and RISC, RITS) is conserved in all eukaryotic systems. But as far as systemic RNAi or spread of RNAi signals is concerned only plants and nematode worms have this feature. Working on gene silencing in plants and insect systems is in progress and the above mentioned ihpRNA strategy has been employed successfully in plants (Mohanpuria *et al.*, 2011). These constructs are delivered to plants using *Agrobacterium* or biolistic mediated approaches for the expression of RNAi cassette.

4.3. ADVANTAGES OF UTILIZING RNAi IN PEST MANAGEMENT

Studies have demonstrated that administration of *in vitro* synthesized dsRNA via oral route can induce systemic RNAi in a variety of pest species belonging to different insect orders. Second instar nymphs of *Rhodnius prolixus* Stål were fed dsRNAs targeting transcripts encoding the anticoagulant heme-binding protein nitrophorin 2, resulting in significantly reduced levels of anticoagulant activity in salivary glands (Araujo *et al.*, 2006). Third instar larvae of the light brown apple moth *Epiphyas postvittana* Walker when fed dsRNAs targeting the transcripts encoding a larval gut enzyme and a pheromone binding protein (PBP) in adult antennae produced reduced levels of both the transcripts in the tissues in which they are normally expressed. Here, the ingested dsRNA was not only taken up by larval midgut cells, but also was transported to the cells in the eye/antennal disc, where it persisted for at least 18 days from third larval instar through adult eclosion (Turner *et al.*, 2006).

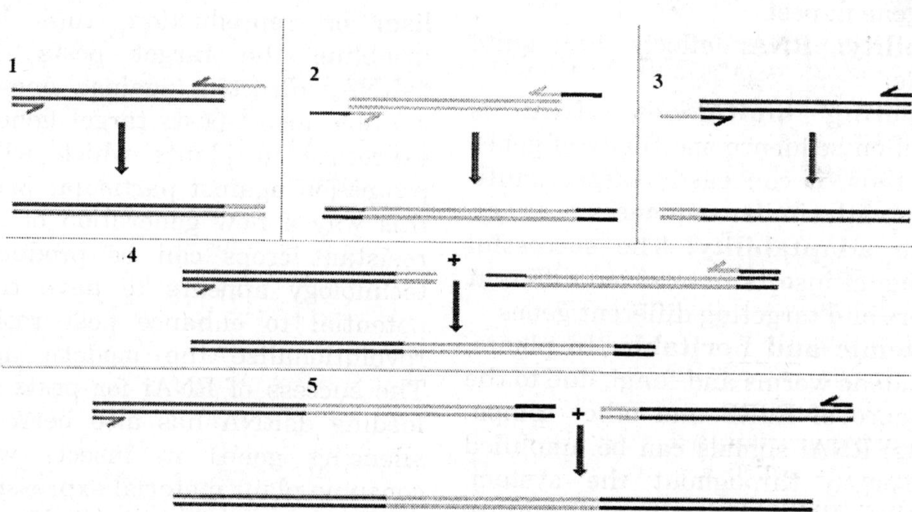

Fig. 3. Essential steps for fusing three cDNA fragments using fusion PCR. 1-3: Each fragment is run with primers that results in regions of homology corresponding to the next piece. 4: Two fragments with homology are run together, resulting in fusion and amplication. 5: The product from step 4 is run with the last homology fragment, resulting in fusion and amplication of the final product.

In addition to oral dsRNA delivery, RNAi was also induced by spraying an aqueous solution of dsRNA directly onto larvae of Asian corn borer *Ostrinia furnacalis* Guenée resulting in developmental stunting or mortality of pest. Further, eggs soaked in dsRNA solutions had significantly decreased rates of hatching relative to control treatments and that fluorescently labeled dsRNA delivered to eggs persisted in larvae to reach gut, hemocytes and silk fiber (Wang *et al.*, 2011). The following characteristics of RNAi make it a preferred technique for utilization in pest management.

- **Highly conserved system:** The RNAi core components like Dicer/DCL and RISC/RITS are well conserved in most of the eukaryotes including plants and insects.
- **Specificity to target the gene:** RNAi is sequence-specific gene silencing technique and thus is much better as compared to others to target the gene in pest.
- **Stability:** RNAi effects are quite stable.
- **Targeting multigenes:** RNAi is based on sequence specificity of genes and thus, it can easily target multigenes with similar sequences.
- **Wide adaptability:** The successful testing of insects from seven different orders and targeting different genes.
- **Systemic and heritable:** In plants, nematode worms and fungi, due to the presence of RdRP and other components, RNAi signals can be amplified or spread throughout the system. Further, RNAi is dominant in nature and its effects are heritable from one generation to other.
- **Alternate approach for producing insect-resistant transgenic plants**: RNAi technology holds potential where *Bt*-based approaches have proved difficult like protection against sap-sucking hemipteran pests such as aphids, leafhoppers and whitefly etc. Moreover it has no side effects known yet.

4.4. REQUIREMENTS FOR UTILIZING RNAi FOR PEST MANAGEMENT

Every year billions of dollars worth of food, feed and fiber crops are lost because of various plant diseases, crop pests or the toxins produced by these which contaminate the food and feed, making it unconsumable and unmarketable. RNAi as a powerful tool for gene silencing can be employed to control these crop pests. Because of stability and specificity, RNAi has been used as an important tool for functional genomics also, and now it is widely used in insect genetic research. RNAi can potentially shut down the formation of proteins related to metabolism or reproduction, thus killing or disabling the target pests. For this, dsRNA, directed against one or more suitable insect-pests target genes is to be expressed in plants which will provide protection against particular pest, and by this way a new generation of insect-pest resistant crops can be produced. RNAi technology appears to have tremendous potential to enhance pest resistance to revolutionalize the modern agriculture. The success of RNAi for pests control by feeding dsRNA has also been shown by silencing genes in insects when they consume plant material expressing dsRNA constructs against well-chosen target genes. Use of RNAi to produce pest resistance plants depends upon two key points: (i) identification of a suitable pest target, and (ii) dsRNA delivery method, which includes *in planta* expression of dsRNA or delivery of sufficient amount of intact dsRNA for uptake by the pest insects.

4.4.1. Identification of Suitable Targets in Pest-Insects

To identify a suitable target, one has to look for the vital genes/proteins essentially required for the survival of particular insect pest. The uniqueness of the target for that particular pest should also be taken care. Thus, the potential targets for implementing RNAi in pest control can be osmoregulation/water balance, reproduction, moulting, detoxification, energy production and neurotran-smission systems etc. As far as vitality of insect pest is concerned, osmoregulation/water balance, reproduction and moulting seem to be indispensible. Calcitonin like receptor gene (CLR) and aquaporin gene (AQP) present in sap-sucking pests could be suitable targets because osmoregulation is an important vital physiological process in insect pests. The steroid hormone regulates molting, metamorphosis, reproduction and various other developmental processes in insect pests. Further, ecdysteroid receptor and its ligands are not detected in vertebrates and therefore, it would be a good candidate for insect-pest management strategies employing RNAi. Injection of dsRNA targeting *Chickadee* gene has been reported to cause severe disruption of normal whitefly oocyte development. In addition to these, detoxification system which includes cytochrome P450 and glut-athione-*S*-transferase, energy production system which includes vacuolar ATPase (vATPase), and neurotransmission system which includes acetylcholines-terase genes, have been targeted in order to enhance the control/ resistance against important agricultural pests.

4.4.2. dsRNA Delivery Methods

Various dsRNA delivery methods namely microinjection into hemocoel/body cavity, feeding/ingestion through artificial diet, dsRNA expression in transgenic plants (Mao *et al.*, 2007; 2011), direct spray of dsRNA (Wang *et al.*, 2011), soaking in dsRNA solution (Sivakumar *et al.*, 2007) and electroporation (Karim *et al.*, 2010) have been discussed with their advantages and limitations for the feasible RNAi research for insect pest management. Microinjection is widely used and very efficient research tool to introduce dsRNA into organisms *in vivo*, not only for nematodes but also for many insect pests. This method is more specific in volume (picolitres) injected and in delivery of dsRNA with immediate access to a tissue of choice or into hemolymph and thus avoiding any possible barriers to get into insect pest (Yu *et al.*, 2013). The concentration of dsRNA injected has been varied in the range of 0.25μg to 100μg in different sets of experiments (Bautista *et al.*, 2009; Terenius, 2011). dsRNA specific to midgut and salivary gland expressed genes have been microinjected into hemocoel of *Bemisia tabaci* and reduced the expression of these genes up to 70 per cent as compared to control. For delivery of dsRNA through ingestion/feeding, the dsRNA can be expressed in bacteria like *E. coli* (DE3) or it can be synthesized *in vitro*, and then fed to insect pests either by mixing with food or by supplying as solution droplets (Turner *et al.*, 2006). The success of RNAi in insect pest management by feeding dsRNA has been shown by silencing genes in insects when they consume plant material expressing ihpRNA or dsRNA constructs against well-chosen target genes. Feeding of dsRNA specific to target gene(s) at different life normal stages of important insect pests have been described in

Table 1. Whitefly nymphs were collected from cotton leaf into bioassay vials and fed with sterilized artificial diet containing dsRNA of vATPase, ATP translocase, actin and tubulin genes, and 29% to 97% mortality of whitefly was achieved after 6 days of feeding. Similarly, different bacterial-expressed dsRNA fragments of hormone receptor gene were fed through artificial diet to larva of cotton bollworm (*Helicoverpa armigera*) which resulted in mortality within 1-7 days of feeding (Xiong *et al.*, 2013). dsRNA expression into transgenic plants such as tobacco, *Arabidopsis* and rice have been utilized for control of lepidopteran, coleopteran and hemipteran insects pests. Using dsRNA fragments specific to Hormone receptor gene were expressed in tobacco transgenic lines and fed to *H. armigera* resulted in inhibition of pupation and deformities of adult emergence within 5 days of feeding. In another example of dsRNA production *in planta*, the model plants *Nicotiana tabacum* and *Arabidopsis thaliana*, that could lead to targeted genes such as Cytochrome P450 monoxygenase and V-ATPase A silencing respectively, in cotton boll worm (Mao *et al.*, 2007; 2011) and western corn rootworm (Baum *et al.*, 2007). In recent years, soaking the organism into a dsRNA solution seems to be a popular method for triggering RNAi response and it has found quite effective with insect cell lines (Sivakumar *et al.*, 2007). But direct spray of dsRNA on newly hatched *Ostrinia furnalalis* larvae has also been tried by Wang *et al.* (2011) which resulted in considerable mortalities ranging from 40% to even 100% after treatment and this effect correlated with the downregulation of the target gene expression. This method can be effectively utilized for field application of dsRNA as well if we select a very unique target for a particular pest.

Table 1. Overview of studies on RNAi in insect parts

Insect pest	Target gene	Insect organ	Insect stage	Application Method	Amount dsRNA	Length (bp)	mRNA Silencing	Recovery	Gene expression evaluation method	Reference
(1)	(2)	(3)	(4)	(5)	(6)	(7)	(8)	(9)	(10)	(11)
Coleoptera										
Diabrotica virgifera virgifera	a	Gut	Neonates	Artificial diet	<5.4 ng/cm^2	-	-	-	LC_{50}	Baum *et al.* (2007)
	Vacuolar ATPase subunit A	Gut	Neonates	Artificial diet	52 ng/cm^2	-	Drastic reduction	-	Northern blot, growth	
	Vacuolar ATPase subunit A	Gut	Neonates	Transgenic plant	-	246	-	-	Damage to the roots of the transgenic corn	
	α-Tubulin	Gut	Neonates	Artificial diet	52 ng/cm^2	-	Gradual decrease	-	Northern blot, growth	
	α-Tubulin	Gut	Neonates	Transgenic plant	-	-	-	-	Growth	
Diabrotica undecimpuncata	Vacuolar ATPase subunit A and E	Gut	Neonates	Artificial diet	780 ng/cm^2	-	-	-	Mortality	
Leptinotarsa decemlineata	Vacuolar ATPase subunit A and E	Gut	Neonates	Artificial diet	52 ng/cm^2	1842, 300, 134	-	-	Mortality	

(1)	(2)	(3)	(4)	(5)	(6)	(7)	(8)	(9)	(10)	(11)
Phyllotreta striolata	Arginine kinase	Gut	Adults	Leaves	0.05–3.2 ng/ml	332	Significant decrease	-	RT-PCR, enzyme assay, mortality, development, fecundity & fertility	Zhao *et al.* (2008)
Diptera										
Glossina morsitans morsitans	Midgut protein Tsetse EP	Midgut	Male adults	Blood meal	10μg	315	30–55%	No	RT-PCR, northern blot, protein assay and mortality	Walshe *et al.* (2009)
	Transferrin	Fat body	Male adults	Blood meal	10μg	-	No effect	-	RT-PCR, northern blot and protein assay	
Hemiptera										
Acyrthosiphon pisum	Water specific aquaporin	Gut	6 day aphids	Artificial diet	1μg	-	50%	5 days	qPCR and change in osmotic pressure	Shakesby *et al.* (2009)
Myzus persicae	Coo2	Salivary glands	Adults	Transgenic plants	-	710	60%	-	RT-PCR, facundity	Pitino *et al.* (2011)
	Rack1	Gut	Adults	Transgenic plants	-	309	60%	-	RT-PCR, qPCR, facundity	
	hunchback (hb)	Gut	2nd instars	Artificial diet	0.75 mg/ml	564, 528	18-54%	-	RT-PCR, qPCR and mortality	Mao and Zeng (2012)
Bemisia tabaci	Ecdysone receptor *(EcR), E75, Cyp315a1, Cyp18a1*	Gut	4th instar	Transgenic plants	-	-	41-48%	-	RT-PCR, qPCR, mortality	Luan *et al.* (2013)
Rhodnius prolixus	Nitroporin 2	Salivary glands	2nd instars 4th instar	Artificial diet Artificial diet	13μg 80μg	502 502	42±10% No effect	7days -	RT-PCR and plasma Coagulation RT-PCR	Araujo *et al.* (2006)
Hymenoptera										
Apis mellifera	Toll-related receptor 18W	Whole organism	Larvae	Soaking	1.26μg	852, 330, 234	60%	-	qPCR, mortality and morphologicala bnormalities	Aronstein *et al.* (2006)
	Vitellogenin	Fat body carcass	2nd instar	Natural diet	0.5 or 3μg	504	90%	-	qPCR, weight and development	Nunes & Simões (2009)
Isoptera										
Reticulitermes flavipes	Cellulose enzyme	Salivary glands	Workers	Artificial diet	13μg	500	60%	8days	qPCR, cellulase assay, mortality, fitness, feeding and weight	Zhou *et al.* (2008)
	Caste regulatory hexamerin storage protein	Fat body	Workers	Artificial diet	5.5μg	500	50-70%	No	qPCR and protein assay, mortality, fitness, feeding and weight	

Contd...

(1)	(2)	(3)	(4)	(5)	(6)	(7)	(8)	(9)	(10)	(11)
Lepidoptera										
Epiphyas postvittana	Larval gut carboxylas terase	Gut	2rd instars	Droplet	1μg	370	<50%	No	qPCR	Turner *et al.* (2006)
	Pheromo-ne binding protein	Antennae	3rd instars	Droplet	1.3 μg	-	60%	4 days	qPCR	
Helicoverpa armigera	Cytochro-me P450 (CYP6AE14)	Midgut	3rd instars	Transgenic plant	-	469	Decrease	-	qPCR, Northern blot and gossypol tolerance	Mao *et al.* (2007; 2011)
	Glutathi-one-S-tra-nsferase	Midgut	3rd instars	Transgenic plant	-	Diced	Decrease	-	Northern blot and growth	
	Acetyl-choline-sterase	Whole organism	Neo-nates	Artificial diet	0.7–1.4μg	21	Drastic reduction	Yes	RT PCR, AChE activity, mor-tality, weight, growth	Kumar *et al.* (2009)
		Whole organism	Neo-nates	Leaves	0.7–1.4μg	21	-	-	Mortality	
Plutella xylostella	Cytochr-ome P450 (CYP6BG1)	Midgut + carcass	4th instars	Droplet	0.250μg	345	98%	-	qPCR and permethrin resistance	Bautista *et al.* (2009)
Spodoptera litura	Amino-peptidase N	Gut	Neo-nates	Artificial diet + soaking	-	756	No effect	-	qPCR	Rajago-pal *et al.* (2002)
Spodoptera frugiperda	Allato-statin C	Brain	5th instar	Droplet	3μg	331	80±10%	-	qPCR	Griebler *et al.* (2008)
	Alloto-tropin 2	Brain	5th instar	Droplet	3μg	195	97±2%	-	qPCR	
	Cytochro-me P450 (CYP6BF1v4)	Midgut + carcass	4th instars	Droplet	0.4μg	381	85%	-	qPCR and permethrin resistance	
Orthoptera										
Gryllus bimaculatus	Sulfaki-nins	Brain	Adults	Droplet	10–60μg	342	-	-	Food uptake	Meye-ring-Vos and Muller (2007)

- Not determined

[a] Putative COPI coatomer, ESCRTIII, ribosomal protein S4V-ATPase subunits, mov34, ribosomal protein rps-14, actin, apple ATPase, ESCRTI-Vps28 and ribosomal protein L9 and L19.

Source: Modified after Huvenne and Smagghe (2010)

Recently, a new delivery method *i.e.* chitosan/dsRNA or siRNA nanoparticles (dsRNA entrapped by polymer chitosan) - mediated delivery method seemed quite efficient as nanoparticles helped in maintaining stability of dsRNA during delivery and accumulation in insect pest body. The dsRNA forms stable 100-400 nm particles in association with chitosan, through the electrostatic forces between the positive charges of the amino group in chitosan and the negatively-charged RNA and thus nanoparticles facilitated the delivery of intact dsRNA (McCarroll and Kavallaris 2012; Zhang *et al.*, 2010).

From the different delivery methods used so far, the feeding of dsRNA seems to the most appealing method because of its ease, being less invasive, cost effective, and time saving and, above all, being the natural route of delivery. Moreover, this method opens the possibility of developing new methods of pest control through the production of species-specific hairpin RNA against pests in transgenic plants (Price and Gatehouse 2008). Finally, the RNAi based control can be facilitated by high throughput dsRNA delivery method. Despite these useful researches, the mechanisms for uptake of dsRNA or siRNA in insect-pests are yet to be explored. Various researchers have enlisted various RNAi-mediated genes knockdown in insect pests. If we compare the conventional insecticidal discoveries which took around 8-10 years for their optimization with spent of $ 80-100 million, RNAi-based insecticidal discovery took hardly $ 2-4 million in 2-4 years of span, and then the later seems quite economic in money as well as in time (Gong *et al.*, 2013).

4.4.3. Mechanisms of Uptake of dsRNA/siRNA by Cells and Spread of Silencing Signals in Plants and Insects

Several proteins such as RNA dependant RNA polymerase (RdRP), systemic RNAi defective mutants (sid) or *sid-1*-like (sil), dsRNA binding motif-containing proteins (dsRBM), Dicer or DCL and Argonautes play several important roles in coordination and thus are important for uptake of dsRNA and spread/amplification/maintenance of RNAi responses in different organisms especially in plants and insects.

4.4.3.1. Based on RdRPs proteins

Systemic RNAi in plants is based on the RNA dependant RNA polymerase (RdRP) which facilitates the within cell amplification of silencing. The spread/ movement of the dsRNA/siRNAs generated from dsRNA between plant cells can take place through the plasmodesmata and/or via the vascular system (phloem). Instead of plants, in the worm *C. elegans* and the fungus *Neurospora crassa*, amplification of silencing signal and cell-to-cell RNAi spreading has been observed. RdRP have also been identified in animal species such as *Branchiostoma floridae* (Vienne *et al.*, 2003). RdRP-related acivity was described in *Drosophila* embryo extracts (Lipardi *et al.*, 2005) but the presence of RdRP was never confirmed in insects (Tomoyasu *et al.*, 2008). A strong RNAi response in *Tribolium castaneum* instead of absence of any RdRP shows possibility of critical involvement of other RNAi core machinery and non-conservation of RdRP in insects and animals. But based on few findings, there could be possibility of presence of RdRP gene orthologs in insects and animals. The spread of dsRNA/siRNA in insects and animals is based on any other mechanism than in the plants or in *C. elegans*.

4.4.3.2. Based on Sid/Sil proteins

Mechanism of uptake of RNAi trigger and further spread of silencing signals has been explained with systemic RNAi defective mutants (sid) which involve three proteins. Relation/cooperation/co-ordination between these proteins make possible the uptake and/or spread of dsRNA molecules into *C. elegans*. SID-1 is an essential multimeric, transmembrane protein contains a long amino-terminal extracellular domain followed by an array of transmembrane domains which are inferred to form a channel for transport of dsRNA molecules passively into the *C. elegans* cells (Winston *et al.*, 2002). However, SID-2 protein either binds to or delivers the dsRNA from environment or it modifies the SID-1 molecule to activate the transport and thus, it induces the delivery/uptake of dsRNA to the gut cells through the endocytosis pathway (Winston *et al.*, 2007; Whangbo and Hunter 2008). Instead of SID-1 and 2, SID-5 has been implicated for dsRNA export from cells (Hinas *et al.*, 2012).

Although, *Drosophila* do not have any *sid-1* but overexpression of *sid-1* in *Drosophila* culture cells enhances the ability of the cells to uptake the dsRNA from the culture media, suggesting an important conserved role for Sid-1 in dsRNA uptake (Feinberg and Hunter 2003). Moreover, several *sid-1*-like (sil) genes have been identified in *Bombyx mori*, *Tribolium castaneum, Apis mellifera* and *Aphis gossypii* etc. but these lack a robust systemic RNAi response except in *Tribolium castaneum* and this could be suggested that Sil proteins might not be involved in systemic RNAi or a involved in a different function than dsRNA uptake in these insect pests. Although, these *sil* genes donot have sequence identity with *sid-1* genes while they have a greater sequence identity with the *C. elegans* gene *chup-1* (tag-130). Further, chup-1 is a cholesterol transporter and no involvement of it has been reported in RNAi (Valdes *et al.*, 2012). But certainly these SID/SIL proteins have role in transfer/ spread of dsRNA. The use of an alternate mechanism for systemic RNAi could be possible in several insects (Tomoyasu *et al.*, 2008). Further, lack of sil genes suggests the early loss of these genes during the path of evolution in different insect orders.

4.4.3.3. Based on dsRBM proteins

dsRNA binding motif-containing proteins (dsRBM) act with Dicer and determine the specificity to incorporate small RNA molecules into a silencing complex. Thus, it affects the efficiency of RNAi in different organisms. In *D. melanogaster* Dicer-2 (DCR-2), which predominantly processes siRNAs, requires R2D2 protein (that contains two dsRBMs) for the incorporation of siRNA into RISC (Liu *et al.*, 2003). Similarly, *C. elegans* contains RDE-4 protein which has been shown to bind with RDE-1 Ago protein (Tabara *et al.*, 2002; Parker *et al.*, 2006), and therefore may have a similar function to R2D2 in forming active RISC. These dsRBM-containing proteins (effector proteins) allow Dicer to recognize different sources of exogenous dsRNA and stimulate the dicer activity (Parker *et al.*, 2006). These RNA-binding proteins then facilitate the transfer of cleaved siRNAs to the RISC complex (Liu *et al.*, 2003). The *Tribolium* genome contains two R2D2-like proteins while *Drosophila* has only one. An additional R2D2 with Dcr-1 might contribute to the robust RNAi response in *Tribolium*. Further, dsRBM proteins are known to bind only to long dsRNA and thus, they might be involved in the maintenance of dsRNA in cells and in producing a longer-lasting RNAi effect (Tomoyasu *et al.*, 2008). This fact was also supported by the presence of ATP-dependent mechanism in retaining dsRNA inside the cells of *C. elegans* (Feinberg and Hunter 2003). Further, if we see the system of insect cells, there the presence of cytoplasmic Dicer shows any new maintenance/spreading mechanism in them. Thus, there might be existence of any dsRNA maintenance mechanism in insect cells (Fig. 4).

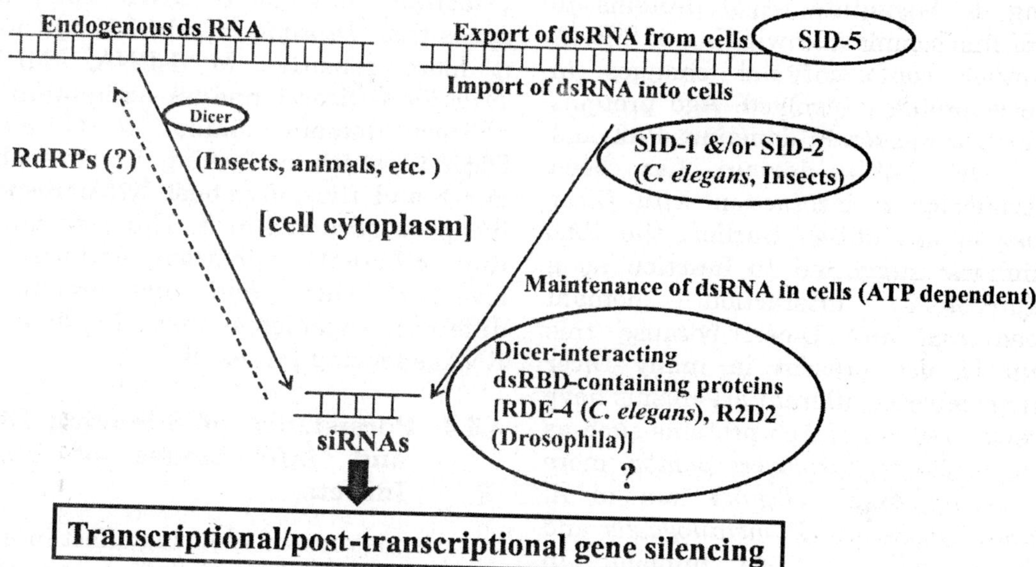

Fig 4. Mechanisms of dsRNA uptake and spreading of RNAi signals in insects, nematode worms and other animals, gene silencing. Import of dsRNA into cells is facilitated by systemic RNAi defective mutants (SID) proteins such as SID-1 for making a channel for transport of dsRNA molecules, SID-2 either binds to or delivers the dsRNA from environment or it modifies the SID-1 molecule to activate the transport and thus, it induces the delivery/uptake of dsRNA to the gut cells through the endocytosis pathway, and SID-5 has been implicated for dsRNA export from cells. The dsRNA-binding motif (dsRBD)-containing proteins (effector proteins) such as RDE-4 and R2D2 allow Dicer to recognize different sources of exogenous dsRNA and stimulate dicer activity. These RNA-binding proteins then facilitate the transfer of cleaved siRNAs to load into silencing complex for gene silencing. dsRBM proteins are known to bind only to long dsRNA and thus, they might be involved in the maintenance of dsRNA in cells and in producing a longer-lasting RNAi effect. Further, the presence of ATP-dependent mechanism is required to retain dsRNA inside the cells of *C. elegans* to show the possibility of existence of any dsRNA maintenance mechanism in cells of living organisms. There could be possibility of presence of RdRP gene orthologs or the spread of dsRNA/siRNA in insects and other animals is based on any other mechanism than in the plants or in *C. elegans*.

4.4.3.4. Based on Dicer and Argonaute proteins

Dicer family proteins have several conserved motifs. Dicer is required for the processing of dsRNA that comes from specific sources. In Drosphila, Dicer-2 is required for long dsRNA processing and is stably associated with dsRBM-containing proteins (Liu *et al.*, 2003). Four Dicer-like (DCL) proteins (DCL1 to DCL4) identified in *Arabidopsis*, three of which have been involved in processing of dsRNA come from different sources (Xie *et al.*, 2004). In *C. elegans*, only one Dicer (DCR-1) has been identified which cooperates with the dsRBM protein RDE-4 during RNAi.

Argonaute proteins are core components of RISC and are involved in siRNA-based silencing and even some Argonaute proteins are also involved in TGS (Fig. 4) as a component of RNA induced transcriptional silencing complex (RITS) (Lippman and Martienssen 2004; Verdel *et al.*, 2004). Activation of RISC is a ATP-dependent process, which involves binding of Argondute (Ago) proteins to RNA of this complex, unwinding of SiRNA and other conformational changes in ribonucleoproteins involved. Ago proteins contain two conserved domains: PAZ and PIWI. The PIWI domain has been implicated in an interaction with Dicer (Tahbaz *et al.*, 2004). Earlier, the PAZ domain was suggested to function as a protein-protein interaction domain between Ago and Dicer, because this domain is also present in many Dicer family members. Different organisms have different numbers of Ago proteins such as one in *Schizosaccharomyces pombe*, more than twenty in *C. elegans*, ten in *A. thaliana*, five each in *D. melanogaster* and *Tribolium* and eight in humans. In Drosophila, AGO2 is required for siRNA-triggered mRNA degradation (Okamura *et al.*, 2004). Based on the phylogenetic analysis of Argonaute proteins, they have been categorized into different groups which are important for RNAi pathway and transcriptional silencing. Out of which, secondary Argonautes are associated with amplification of RNAi effects in *C. elegans* but are not present in *Drosophila* and *Tribolium* and thus are not conserved in other insects (Tomoyasu *et al.*, 2008). Ago proteins may have evolved an intrinsic sequence specificity that allows them to bind preferentially to small RNAs of specific sequence. Also, they may utilize specific adaptor proteins which are associated with the different cellular sources of dsRNA (Farazi *et al.*, 2008) and thus, can be related with the uptake of dsRNA into the cells.

In animals, the best studied dsRNA uptake/spread mechanism is that of *C. elegans* where, ≥50 bp long dsRNA has been reported for its systemic spread from one cell to other and not the amplified siRNA molecules (Scott *et al.*, 2013). *Tribolium* has more RNAi component genes than *Drosophila*, which might make it more sensitive to dsRNA. Further, *Tribolium*-Dicer-1 has an amino-terminal helicase domain, similar to *C. elegan* Dicer-1, which is lacking in Drosophila-Dcr-1 and thus, a robust RNAi response has been observed in it. Thus, we can say that different expression patterns and levels of Dicer and Ago proteins in different organisms may regulate the RNAi silencing processes.

4.4.4. Persistence of Silencing Effect and Life Stages of Target Insects

Persistence of silencing effect in insect body can be determined in terms of the spread of RNAi signal molecules *i. e.* systemic RNAi from one cell to other. In plants, nematodes and fungi systemic RNAi have been observed because of the presence of RdRP genes which are discussed earlier in this chapter. In absence of RdRP, dsRNAs in insects are likely not to be amplified further and there is need to deliver these directly to the cells and tissues where suppression of target gene expression is to take place. In insects, RNAi is not always systemic e.g. fly cells can take up dsRNA, which cannot spread throughout the entire body (Saleh *et al.*, 2006). Concept of dsRNA uptake and their spread in terms of cell-autonomous and non-cell autonomous RNAi have been described where level of spread would be greater in systemic non-cell autonomous RNAi than in the cell autonomous RNAi (Whangbo and Hunter 2008; Huvenne and Smagghe 2010). But the nematode *C. elegans* (Huvenne and Smagghe 2010), beetle *Tribolium* (Tomoyasu *et al.*, 2008) and grasshopper *Schistocerca americana* (Dong and Friedrich, 2005) have strong systemic RNAi reactions based on the presence of sid/sil proteins which are responsible for import and export of dsRNA molecules from insect cells. In most of the insect pest RNAi studies, first instar to fourth instar nymphal stages have been used for injection/feeding of dsRNA molecules. However, feeding of transgenic plants expressing dsRNA to several insect pests has explored the persistence of silencing effects from one life stage to other stages in targeted insect pests (Table 1). Loss-of-function of gap gene hunchback (*hb*), a key regulator in the anteroposterior patterning of *Acyrthosiphon pisum* has resulted in segmentation defects in the next generation. *Aphb* ingested RNAi was performed at the second instar stage to knockdown the *Aphb* expression and continuous feeding of *Aphb* double-stranded RNA mixed in artificial diet led to the reduction of *Aphb* transcripts up to adult stage and rised the insect lethality (Mao and Zeng, 2012). Recently, leaf-mediated dsRNA feeding method showed that silencing of genes in the ecdysone pathway reduced the survival and delayed the development of the whitefly during nymphal stages while a limited effect was observed on survival and fecundity of adult whiteflies (Luan *et al.*, 2013). Further more studies are required to understand the spread of silencing RNA within the insect body and the genes involved in this process. Understanding the molecular mechanisms of systemic spread of RNA signals inside the insect body will facilitate the application of RNAi technology in pest management.

4.5. CONCLUSIONS

RNAi technology is now becoming a new approach for the control of pests in agriculturally important crop plants. To achieve an effective RNAi response for pest control, suitable vital targets and dsRNA delivery methods need to be selected. Microinjection is an important method for accurate amounts of dsRNA delivery to the exact place of choice. The dsRNA feeding or dsRNA spray can be utilized for large scale gene screening or for field application. As far as field applicability of RNAi is concerned, the target selection must be carefully selected to avoid from any off-target effects or responses. Utilization of multiple gene targets could further strengthen the results of this technology in control of pests. RNAi technology holds potential where *Bt*-based approaches have proved difficult like protection against sap-sucking hemipteran pests such as aphids, leafhoppers and whitefly. Recently, increasing number of insect genomes sequenced will help in refining the applicability of RNAi in selection of suitable target genes. Once the mechanism of dsRNA uptake or RNAi signals spread and amplification is fully understood, applicability of RNAi would improve in different agriculturally important pests. Recent advances in RNAi technology reveal that it could serve as a novel and important tool in pest management directly as well as help in developments of a new generation of insect-resistant transgenics.

REFERENCES

Araujo RN, Santos A, Pinto FS, Gontijo NF, Lehane MJ and Pereira MH (2006) RNA interference of the salivary gland nitrophorin 2 in the triatomine bug *Rhodnius prolixus* (Hemiptera: Reduviidae) by dsRNA ingestion or injection. *Insect Biochemistry and Molecular Biology* **36**: 683–693.

Aronstein K, Pankiw T and Saldivar E (2006) SID-1 is implicated in systemic gene silencing in the honey bee. *Journal of Apicultural Research* **45:** 20-24.

Bartel DP (2004) MicroRNAs: genomics, biogenesis, mechanism and function. *Cell* **116:** 281–297.

Baum JA, Bogaert T, Clinton W, Heck GR, Feldmann P, Ilagan O, Johnson S, Plaetinck G, Munyikwa T, Pleau M, Vaughn T and Roberts J (2007) Control of coleopteran insect pests through RNA interference. *Nature Biotechnology* **25**: 1322–1326.

Bautista MAM, Miyata T, Miura K and Tanaka T (2009) RNA interference-mediated knockdown of a cytochrome P450, CYP6BG1, from the diamondback moth, *Plutella xylostella*, reduces larval resistance to permethrin. *Insect Biochemistry and Molecular Biology* **39:** 38–46.

Belles X (2010) Beyond *Drosophila*: RNAi In vivo and functional genomics in insects. *Annual Review of Entomology* **55**: 111-128.

Brantl S (2002) Antisense-RNA regulation and RNA interference. *Biochim Biophys Acta* **1575:** 15-25.

Bucher G, Scholten J and Klingler M (2002) Parental RNAi in *Tribolium* (Coleoptera) *Current Biology* **12**: R85–R86.

Dong Y and Friedrich M (2005) Nymphal RNAi: systemic RNAi mediated gene knockdown in juvenile grasshopper. *BMC Biotechnology* **5**: 25.

Fagard M and Vaucheret H (2000) (TRANS) GENE SILENCING IN PLANTS: How Many Mechanisms? *Annual Review Plant Physiology Plant Molecular Biology* **51:** 167-194.

Farazi TA, Juranek SA and Tuschl T (2008) The growing catalog of small RNAs and their association with distinct Argonaute/Piwi family members. *Development* **135**: 1201-1214.

Feinberg EH and Hunter CP (2003) Transport of dsRNA into cells by the transmembrane protein SID-1. *Science* **301**: 1545-1547.

Gong L, Chem Y, Hu Z and Hu M (2013) Testing insecticidal activity of novel chemically synthesized siRNA against *Plutella xylostella* under laboratory and field conditions. *PLoS One* **8**(5): e62990.

Griebler M, Westerlund SA, Hoffmann KH and Meyering-Vos M (2008) RNA interference with the allatoregulating neuropeptide genes from the fall armyworm *Spodoptera frugiperda* and its effects on the JH titer in the hemolymph. *Journal of Insect Physiology* **54:** 997-1007.

Hannon GJ (2002) RNA interference. *Nature* **418:** 244-251.

Hinas A, Wright AJ and Hunter CP (2012) SID-5 is an endosome-associated protein required for efficient systemic RNAi in *C. elegans*. *Current Biology* **22**: 1938-1943.

Horton RM, Hunt HD, Ho SN, Pullen JK and Pease LR (1989) Engineering hybrid genes without the use of restriction enzymes:gene splicing by overlap extension. *Gene* **77**: 61-68.

Huvenne H and Smagghe G (2010) Mechanisms of dsRNA uptake in insects and potential of RNAi for pest control. A review. *Journal of Insect Physiology* **56**: 227-235.

Jakymiw A, Lian S, Eystathioy T, Li S, Satoh M, Hamel J, Fritzler M and Chan E (2005) Disruption of P bodies impairs mammalian RNA interference. *Nature Cell Biology* **7** (12): 1267–1274.

Jones-Rhoades MW, Bartel DP and Bartel B (2006) MicroRNAs and their regulatory roles in plants. *Annual Review of Plant Biology* **57**: 19-53.

Karim S, Troiano E and Mather TN (2010) Functional genomics tool: gene silencing in *Ixodes scapularis* eggs and nymphs by electroporated dsRNA. *BMC Biotechnology* **10**: 1.

Kennerdell JR and Carthew RW (1998) Use of dsRNA-mediated genetic interference to demonstrate that *frizzled* and *frizzled 2* act in the wingless pathway *Cell* **95:** 1017–1026.

Kimball SR (2002). Regulation of global and specific mRNA translation by amino acids. *Journal of Nutrition* **132:** 883-886.

Kumar M, Gupta GP and Rajam MV (2009) Silencing of acetylcholinesterase gene of *Helicoverpa armigera* by siRNA affects larval growth and its life cycle. *Journal of Insect Physiology* **55:** 273-278.

Kurihara Y and Watanabe Y (2004) *Arabidopsis* micro-RNA biogenesis through Dicer-like 1 protein functions. *Proceedings National Academy Science USA* **101**: 12753-12758.

Lipardi C, Baek HJ, Wei Q and Paterson BM (2005) Analysis of short interfering RNA function in RNA interference by using *Drosophila* embryo extracts and schneider cells. *Methods in Enzymology* **392**: 351-371.

Lippman Z and Martienssen R (2004) The role of RNA interference in heterochromatic silencing. *Nature*: **431**: 364-370.

Liu P and Kaufman T (2004a) *Hunchback* is required for suppression of abdominal identity, and for proper germband growth and segmentation in the intermediate germ band insect *Oncopeltus fasciatus. Development* **131**: 1515–1527.

Liu P and Kaufman T (2004b) *Kruppel* is a gap gene in the intermediate germband insect *Oncopeltus fasciatus* and is required for development of both blastoderm and germband-derived segments. *Development* **131**: 4567–4579.

Liu Q, Rand TA, Kalidas S, Du F, Kim HE, Smith DP and Wang X (2003) R2D2, a bridge between the initiation and effector steps of the *Drosophila* RNAi pathway. *Science* **301:** 1921–1925.

Luan JB, Ghanim M, Liu SS and Czosnek H (2013) Silencing the ecdysone synthesis and signaling pathway genes disrupts nymphal development in the whitefly. *Insect Biochemistry and Molecular Biology* **43**(8): 740-746.

Mao J and Zeng F (2012) Feeding-based RNA interference of a gap gene is lethal to the pea aphid, *Acyrthosiphon pisum. PLoS One* **7**: e48718.

Mao YB, Cai WJ, Wang JW, Hong GJ, Tao XY, Wang LJ, Huang YP and Chen XY (2007) Silencing a cotton bollworm P450mono-oxygenase gene by plant-mediated RNAi impairs larval tolerance of gossypol. *Nature Biotechnology* **25:** 1307–1313.

Mao YB, Tao XY, Xue XY, Wang LJ and Chen XY (2011) Cotton plants expressing CYP6AE14 double-stranded RNA show enhanced resistance to bollworms. *Transgenic Research* **20:** 665–673.

Martin D, Maestro O, Cruz J, Mané Padrós D and Bellés X (2006) RNAi studies reveal a conserved role for RXR in molting in the cockroach *Blattella germanica. Journal of Insect Physiology* **52**: 410–416.

Matzke M, Kanno T, Huettel B, Jaligot E, Mette F, Kreil P, Daxinger L, Rovina P, Aufsatz W and Matzke M (2005) RNA-directed DNA methylation, In: P. Meyer (Ed.), *Plant Epigenetics, Annual Plant Reviews,* Vol. 19, Blackwell Publishing, Oxford, pp. 69-105.

McCarroll J and Kavallaris M (2012) Nanoparticle delivery of siRNAs as a novel therapeutic for human disease. *Australian Biochemist* **43**: 9-20.

Meister G and Tuschl T (2004) Mechanisms of gene silencing by double stranded RNA. *Nature* **431**: 343-349.

Meyering-Vos, M and Muller, A (2007) RNA interference suggests sulfakinins as saliety effectors in the cricket *Grylus bimaculatus. Journal of Insect Physiology.* **53:** 840-848.

Miglani GS (2013) *Gene Regulation.* Narosa Publishing House Pvt. Ltd., New Delhi & Alpha Science International, Oxford, UK, pp. 8.3-8.18.

Mito T, Ronco M, Uda T, Nakamura T, Ohuchi H and Noji S (2008) Divergent and conserved roles of extradenticle in body segmentation and appendage formation, respectively, in the cricket *Gryllus bimaculatus*. *Developmental Biology* **313**: 67-79.

Mohanpuria P, Kumar V, Ahuja PS and Yadav SK (2011) Producing low-caffeine tea through post-transcriptional silencing of caffeine synthase mRNA. *Plant Molecular Biology* **76**(6): 523-534.

Mohanpuria P, Kumar V, Mahajan M, Mohammad H and Yadav SK (2010) *Gene Silencing: theory, techniques and applications,* In: Anthony J. Catalano (Ed.) *Genetics- Research and Issues,* Nova Science Publishers, Inc. New York, pp. 321-334.

Napoli C, Lemieux C and Jorgensen R (1990) Introduction of a chimeric chalcone synthase gene into petunia results in reversible co-suppression of homologous genes in trans. *Plant Cell* **2**: 279-289.

Nunes FMF and Simóes ZLP (2009) A non-invasive method for silencing gene transcription in honeybees maintained under natural conditions. *Insect Biochemistry and Molecular Biology* **39:** 157-160.

Okamura K, Ishizuka A, Siomi H and Siomi MC (2004) Distinct roles for Argonaute proteins in small RNA-directed RNA cleavage pathways. *Genes & Development* **18:** 1655–1666.

Parker G, Eckert D and Bass B (2006) RDE-4 preferentially binds long dsRNA and its dimerization is necessary for cleavage of dsRNA to siRNA. *RNA* **12**(5): 807–818.

Pitino M, Coleman AD, Maffei ME, Ridout CJ and Hogenhout SA (2011) Silencing of aphid genes by dsRNA feeding from plants. *PLoS One* **6**: 25709.

Price DRG and Gatehouse JA (2008) RNAi-mediated crop protection against insects. *Trends in Biotechnology* **26:** 393-400.

Rajagopal R, Sivakumar S, Agrawal N, Malhotra P and Bhatnagar RK (2002) Silencing of midgut aminopeptidase N of *Spodoptera litura* by double-stranded RNA establishes its role as *Bacillus thuringiensis* toxin receptor. *Journal of Biological Chemistry* **277:** 46849-46851.

Ronco M, Uda T, Mito T, Minelli A, Noji S and Klingler M (2008) Antenna and all gnathal appendages are similarly transformed by homothorax knock-down in the cricket *Gryllus bimaculatus. Development Biology* **313**: 80-92.

Saleh MC, Rij RPV, Hekele A, Gillis A, Foley E, O'Farrell PH and Andino R (2006) The endocytic pathway mediates cell entry of dsRNA to induce RNA silencing. *Nature Cell Biology* **8:** 793–802.

Scott JG, Michel K, Bartholomay L, Siegfried BD, Hunter WB, Smagghe G, Zhu KY and Douglas AE (2013) Towards the elements of successful insect RNAi. *Journal of Insect Physiology* **59**: 1212-1221.

Sen G and Blau H (2005) Argonaute 2/RISC resides in sites of mammalian mRNA decay known as cytoplasmic bodies. *Nature Cell biology* **7**(6): 633–636.

Shakesby AJ, Wallace IS, Isaacs HV, Pritchard J, Roberts DM and Doublas AE (2009) A water-specific aquaporin involved in aphid osmoregulation. *Insect Biochemistry and Molecular Biology* **39:** 1-10.

Shi C, Baldwin IT and Wu J (2012) *Arabidopsis* plants having defects in nonsense-mediated mRNA decay factors UPF1, UPF2, and UPF3 show photoperiod-dependent phenotypes in development and stress responses. *Journal of Integrative Plant Biology* **54**(2): 99-114.

Sivakumar S, Rajagopal R, Venkatesh GR, Srivastava A and Bhatnagar RK (2007) Knockdown of aminopeptidase-N from *Helicoverpa armigera* larvae and in transfected Sf21 cells by RNA interference reveals its functional interaction with *Bacillus thuringiensis* insecticidal protein Cry1Ac. *Journal of Biological Chemistry* **282:** 7312–7319.

Biological and Molecular Approaches in Pest Management
Balwinder Singh, Ramesh Arora & S S Gosal (Eds)
Scientific Publishers (India), Jodhpur, 2015, pp 73-96.

Chapter - 5

Bioremediation of Pesticides in the Environment

Balwinder Singh and Kousik Mandal

Department of Entomology, Punjab Agricultural University, Ludhiana-141004, India
E-mail: kousik11@gmail.com

5.1. INTRODUCTION

Pesticides are essential tools of modern agriculture and form an integral part of overall strategy because production has hardly any meaning without protection. Pesticides have made a great impact on human health, production and preservation of food, fibre and other cash crops by controlling disease vectors and by keeping in check many species of unwanted insects and plants. Although, various methods are used to control pests in different pest management systems, pesticides continue to be the major component of most of the pest control programmes and will probably remain so during the near future (Singh and Dhaliwal, 2001).

It has been observed that less than 1 per cent of pesticides applied reach the target pests, leaving 99 per cent pesticide to affect the environment (Pimental, 1995). It was reported that widespread use of pesticides has resulted in global contamination of the ecosystem. No matter, how and where the pesticides are applied, they finally reach to the soil, which acts as a reservoir for all toxic chemicals in the environment. These pesticides in the soil may get attached to its different components, where they may undergo many intra- and inter- compartmental transport mechanisms as well as transformation processes. The persistence of a pesticides depend on how deeply they are mixed into the soil; even the most persistent compounds disappear relatively faster when on the soil surface, yet when incorporated into the soil they are very persistent. Once a pesticide enters the soil, it undergoes transformation caused by microbial, chemical, and photochemical processes.

Pesticides in soil undergo a variety of degradative, transport, and adsorption/

desorption processes depending on the chemical nature of the pesticide and soil properties. Pesticides interact with soil organisms and their metabolic activities (Singh and Walker, 2006) and may alter the physiological and biochemical behaviour of soil microbes. Microbial biomass is an important indicator of microbial activities and provides direct assessment of the linkage between microbial activities and the nutrient transformations and other ecological processes (Schultz and Urban, 2008). Many recent studies reveal the adverse impacts of pesticides on soil microbial biomass and soil respiration (Zhou *et al.*, 2006). Some microbial groups are capa-ble of using applied pesticide as a source of energy and nutrients to multiply, whereas the pesticide may be toxic to other organis- ms (Johnsen *et al.*, 2001). Likewise some-times, application of pesticides reduces microbial diversity and may also inhibit or kill certain group of micro-organisms.

In soil, microorganisms (bacteria, fungi, etc.) are primarily responsible for pesticide degradation. Thus, the microbial breakdown of pesticides is considered to be the most important catabolic reaction in soil. It is reported that microbes present in environment degrade these xenobiotic compounds and use them for their metabolic requirements. This process is known as biodegradation/biotransformation. Microbial degradation is an important mechanism to control the fate of pesticides in soils and is generally considered to be desirable both from an environmental perspective as well as an agricultural point of view. Biodegradation of pesticides in soil is controlled by the number and activity of microbes in the soil and environmental conditions. Degradation studies in soils are essential for evaluation of persistence of pesticides and their breakdown products.

Pesticides can be degraded by microbes, either by using them as a substrate for growth and energy or by cometabolism without supplying energy for the microorganisms. Whether microorganisms can utilize pesticide molecules as the sole carbon source has been the subject of investigation. This biodegradation/biotransformation of pesticides by soil microbes may be utilized for remediation of contaminated sites (Singh, 2002).

Microbes can transform a variety of pesticides and utilize them as a source of energy. Organochlorine pesticides have a considerably high persistence in the environment owing to their stable structure and very less biodegradation. Organophosphates and carbamates are rapidly degraded chemically as well as by microbes. Synthetic pyrethroids are not recommended for outdoor purposes, as they are highly photolabile and breakdown easily in sunlight (Singh, 2002).

Current legislation and recent waste management strategies have placed significant emphases on waste minimization, recycling and remediation rather than disposal, which is now perceived as being the least desirable option. The persistence of organo-xenobiotics in the environment is a matter of significant public, scientific and regulatory concern because of the potential toxicity, mutagenicity, carcinogenicity and ability to bioconcentrate up the trophic ladder. These concerns continue to drive the need for the development and application of remediation techniques (Colleran, 1997).

Bioremediation is the use of biological treatments, for the cleanup of hazardous chemicals in the environment. At present, employing the biochemical abilities of microorganisms is the most popular strategy for the biological treatment of contaminated soils and waters. Microorganisms, have a unique ability to interact both chemically and physically with a huge range of both man-made and naturally occurring compounds leading to a structural change to, or the complete degradation of, the target molecule (Head, 1998). The relatively recent development of bioremediation has added to existing cleanup strategies currently available for the restoration and rehabilitation of contaminated sites and can be conducted either in situ or ex situ. This biological strategy is dependent on the catabolic activities of the indigenous microflora, optimizing the conditions in situ for growth and biodegradation.

In contrast, bioremediation can be regarded as a more effective and environmentally friendly strategy since it results in the partial or complete biotransformation of organo-xenobiotics to microbial biomass and stable, innocuous end products (Colleran, 1997). The acceptance of bioremediation as a viable cleanup strategy, however, in many cases also depends on cost i.e. cannot be more expensive than existing chemical and physical treatments. It indicates that the bioremediation strategies listed are competitive in terms of cost as well as in terms of the impact on the contaminated matrices. It is only over the last 3-5 years that the use of composting strategies in biodegradation/bioremediation of organic pollutants has been seriously adopted; as a result there is a lack of general information as well as a limited number of pollutant/ pollutant mixtures treated.

For bioremediation to be successful, the bioremediation methods depend on having the right microbes in the right place with the right environmental factors for degradation to occur. The right microbes are bacteria or fungi, which have the physiological and metabolic capabilities to degrade the pollutants. Bioremediation offers several advantages over conventional techniques such as landfilling or incineration. Bioremediation can be done on site, is often less expensive and site disruption is minimal, it eliminates waste permanently, eliminates long-term liability, and has greater public acceptance, with regulatory encouragement, and it can be coupled with other physical or chemical treatment methods. Bioremediation has also its limitations. Some chemicals are not amenable to biodegradation, for instance, heavy metals, radionuclides and some chlorinated compounds. In some cases, microbial metabolism of contaminants may produce toxic metabolites. Bioremediation is a scientifically intensive procedure, which must be tailored to the site-specific conditions, which means one has to do treatability studies on a small scale before the actual clean up of the sites.

5.2. BIODEGRADATION AND BIOREMEDIATION

Biodegradation/biotransformation is a naturally occurring bioprocess, using innumerable strains of microbes that can effectively degrade/transform the toxic contaminants to non-toxic/harmless byproducts. Bioremediation development has made it possible to greatly accelerate natural bioprocesses by selecting, concentrating and acclimating microbes to degrade or transform many xenobiotic contaminants in a few week rather than months or years. Biostimulation and bioaugmentation are important in development of bioprocess and bioremediation of xenobiotic contaminants (Singh, 2002).

By definition, bioremediation is the use of living organisms, primarily microorganisms, to degrade the environmental contaminants into less toxic forms. It uses naturally occurring bacteria and fungi or plants to degrade or detoxify substances hazardous to human health and/or the environment. The microorganisms may be indigenous to a contaminated area or they may be isolated from elsewhere and brought to the contaminated site. Contaminant compounds are transformed by living organisms through reactions that take place as a part of their metabolic processes. Biodegradation of a compound is often a result of the actions of multiple organisms. When microorganisms are imported to a contaminated site to enhance degradation we have a process known as bioaugmentation (Vidali, 2001).

Biodegradation and bioremediation are matching processes to an extent that both of these are based on the conversion or metabolism of pesticides by microorganisms. The difference between these two is that, the biodegradation is a natural process whereas bioremediation is a technology. In bioremediation, we use microbes to degrade the pesticides *in situ*. A successful bioremediation technique requires an efficient bacterial strain that can degrade largest pollutant to minimum level. Adequate rate of biodegradation is required to attain the acceptable level of pesticide residues or its metabolites at contaminated site in a limited time frame. The rate of biodegradation in soil depends on four variables i.e. (i) availability of pesticide or metabolite to the microorganisms, (ii) physiological status of the micrrorganisms, (iii) survival and/or proliferation of pesticide degrading microorganisms at contaminated site and (iv) sustainable population of these microorganisms. Therefore, to attain an achievable bioremediation, it requires the creation of unique niche/or microhabitats for desired microbes, so they can be successfully exploited.

So, bioremediation is a technology of cleaning and remedying the soil through biological methods by means of non-pathogenic microorganisms that feed on the contaminating substances. The microorganisms are used to reduce the complexity of organic molecules (biotransformation), or for degradation to complete mineralization (biodegradation). Some defined bacterial species are able to degrade, to a limited extent, all hydrocarbons present in heavy fuel oil or oil sludge. Some of the polluting components may be dissolved only by the coupled metabolic activity of multiple genera of microorganisms. A consortium (mixed culture) of microorganisms can conduct these complex processes of degradation, while at the same time, being more resistant, on average, to changes in the ecosystem than just a single microbial species (Brenner *et al.*, 2008).

5.3. STRATEGIES FOR BIOREMEDIATION

Different techniques are employed depending on the degree of saturation and aeration of an area. *In situ* techniques are defined as those that are applied to soil and groundwater at the site with minimal disturbance. So, *in situ* techniques involve treatment of the contaminated material in place. *Ex situ* techniques are those that are applied to soil and groundwater at the site which has been removed from the site via excavation (soil) or pumping (water). So, *ex situ* technologies are those treatments which involve the physical removal of the contaminated material for treatment process. *Bioaugmentation* techniques involve the addition of microorganisms with the ability to degrade pollutants (Boopathy, 2000; Vidali, 2001).

5.3.1. *In situ* Bioremediation

These techniques are generally the most desirable options due to lower cost and less disturbance since they provide the treatment in place avoiding excavation and transport of contaminants. *In situ* treatment is limited by the depth of the soil that can be effectively treated. In many soils effective oxygen diffusion for desirable rates of bioremediation extend to a range of only a few centimeters to about 30 cm into the soil, although depths of 60 cm and greater have been effectively treated in some cases (Vidali, 2001).

The most important land treatments are:

a) Bioventing: Bioventing is the most common *in situ* treatment and involves supplying air and nutrients through wells to contaminated soil to stimulate the indigenous bacteria. Nitrogen and phosphorous are two commonly added nutrients (Rockne and Reddy, 2003). Bioventing employs low air flow rates and provides only the amount of oxygen necessary for the biodegradation while minimizing volatilization and release of contaminants to the atmosphere. It works for simple hydrocarbons and can be used where the contamination is deep under the surface (Vidali, 2001). A basic bioventing system includes a well and a blower, which pumps air through the well and into the soil (Lee *et al.*, 2006). It can be used where the contamination is deep under the surface, being also a long term cleanup option lasting from several months to years (Gavrilescu, 2005).

b) Biosparging: Biosparging involves the injection of air under pressure below the water table to increase groundwater oxygen concentrations and enhance the rate of biological degradation of contaminants by naturally occurring bacteria. Biosparging increases the mixing in the saturated zone and there by increases the contact between soil and groundwater. The ease and low cost of installing small-diameter air injection points allows considerable flexibility in the design and construction of the system (Vidali, 2001). Neither of these processes are suitable for compounds which may volatilize too quickly. These procedures have not been used frequently with pesticide-contaminated sites (Chawla *et al.*, 2013).

c) Bioaugmentation. Bioremediation frequently involves the addition of microorganisms indigenous or exogenous to the contaminated sites. Two factors limit the use of added microbial cultures in a land treatment unit: 1) non indigenous cultures rarely compete well enough with an indigenous population to develop and sustain useful population levels and 2) most soils with long-term exposure to biodegradable waste have indigenous microorganisms that are effective degrades if the land treatment unit is well managed (Vidali, 2001).

Bioaugmentation is usually paired with biostimulation in which adequate amounts of water, nutrients and oxygen are also introduced into the contaminated site in order to enhance the activity of introduced microbial degraders (Couto *et al.*, 2010) or to promote cometabolism (Lorenzo, 2008) .

In-situ, or in place, treatment can be conducted through several different mediums. One of these mediums is fungi as a degrader of pesticides. The application of white rot fungi for use in pesticide bioremediation has been extensively studied in recent years. In its natural environment, white rot fungi lives on woody tissues that contain lignin, a recalcitrant compound. The white rot fungi contain enzymes necessary to degrade lignin, as well as other toxic and recalcitrant compounds. In lab testing, it has been found that this fungi is capable of degrading certain pesticides 45-75% more effectively than control samples (Fragoeiro, 2005). However, in situ bioremediation has some limitations: 1) it is not suitable for all soils, 2) complete degradation is difficult to achieve and 3) natural conditions (e.g. temperature) are difficult to control for optimal biodegradation (Chawla *et al.,* 2013).

5.3.2 *Ex situ* Bioremediation

These techniques involve the excavation or removal of contaminated soil from ground.

a) Landfarming: Landfarming is a simple technique in which contaminated soil is excavated and spread over a prepared bed and periodically tilled until pollutants are degraded. The goal is to stimulate indigenous biodegradative microorganisms and facilitate their aerobic degradation of contaminants. In general, the practice is limited to the treatment of superficial 10–35 cm of soil. Since landfarming has the potential to reduce monitoring and maintenance costs, as well as clean-up liabilities, it has received much attention as a disposal alternative (Vidali, 2001). Landfarming is particularly suited for pesticide-contaminated soils (Felsot *et al.*, 1992). It is a full-scale bioremediation technology in which the size and location of the spreading operation is decided based upon the application rate of pesticides to avoid concentrations that would be unsafe in soil, groundwater or crops. Generally, the rate is similar to the label rate on the pesticide label, which is the suggested rate of application of pesticide per unit of land or soil (Chawla *et al.*, 2013). The soil is tilled periodically to aerate the waste and remix the active microbes. In this process, several soil conditions are monitored to ensure a high rate of degradation. These criteria include, but are not limited to, moisture content, aeration and pH. Throughout the bioremediation process, moisture and nutrients may be added to the soil as needed, and the pH can be controlled by adding crushed limestone or lime (Federal Remediation Technology Roundtable, 2006).

b) Composting: Composting is a technique that involves combining contaminated soil with nonhazardous organic amendants such as manure or agricultural wastes. The presence of these organic materials supports the development of a rich microbial population and elevated temperature characteristic of composting (Vidali, 2001). Typical compost temperatures are in the range of 55° to 65° C. The increased temperatures result from heat produced by microorganisms during the degradation of the organic material in the waste (Blanca *et al.,* 2007). Composting can be performed using windrows, aerated static piles, or specially designed composting vessels. The contained systems typically allow treatment to be completed in less time than the windrow or aerated pile by providing better control of composting conditions. Rapid treatment time is offset by the high initial cost of the composting reactor (Gavrilescu, 2005). Moisture, pH, temperature, and explosives concentration are monitored. At the completion of the composting period, the windrows would be disassembled and the compost is taken to the final disposal area (Shukla *et al.*, 2010).

c) Biopiles: Biopiles are a hybrid of landfarming and composting (Fahnestock *et al.*, 1998). It is designed to provide optimum temperature, moisture content, aeration and nutrient conditions to promote rapid degradation (Gavrilescu, 2005). Essentially, engineered cells are constructed as aerated composted piles. Typically used for treatment of surface contamination with petroleum hydrocarbons they are a refined version of landfarming that tend to control physical losses of the contaminants by leaching and volatilization. Biopiles provide a favourable environment for indigenous aerobic and anaerobic microorganisms (Vidali, 2001).

d) Bioreactors: Slurry reactors or aqueous reactors are used for ex situ treatment of contaminated soil and water pumped up from a contaminated plume. Bioremediation in reactors involves the processing of contaminated solid material (soil, sediment, sludge) or water through an engineered containment system. A slurry bioreactor may be defined as a containment vessel and apparatus used to create a three-phase (solid, liquid, and gas) mixing condition to increase the bioremediation rate of soil bound and water-soluble pollutants as a water slurry of the contaminated soil and biomass (usually indigenous microorganisms) capable of degrading target contaminants. In general, the rate and extent of biodegradation are greater in a bioreactor system than in situ or in solid-phase systems because the contained environment is more manageable and hence more controllable and predictable (Vidali, 2001).

Fulekar, (2009) carried out bioremediation of fenvalerate by *Pseudomonas aeruginosa* in a scale up bioreactor and concluded that this technology would be beneficial to pesticides industry for bioremediation of pesticide fenvalerate.

e) Precipitation or flocculation: Precipitation has long been the primary method of treating metal-laden industrial wastewaters. As a result of the success of metals precipitation in such applications, technology is being considered and selected for use in remediating ground water containing heavy metals including their radioactive isotopes. This method involves the conversion of soluble heavy metal salts to insoluble salts that will precipitate. The precipitate can then be removed from the treated water by physical methods such as clarification (settling) and/or filtration (Chawla *et al.*, 2013).

f) Microfiltration: Membranes with a pore size of 0.1–10μm perform micro filtration. Microfiltration membranes are used at a constant pressure to remove dissolved solids in waste water. More than 90% of original waste water can be recovered and reused (Chawla *et al.*, 2013).

g) Electro dialysis: Electro dialysis is a technique that employs electrical current and special membranes which are semi permeable to ions based on their charge. Membranes that permeate cations and membranes that permeate anions are placed alternately with flow channels between them and electrodes are placed on each side of the membranes. The electrodes draw their counter ions through the membranes so that these are removed from the water (Chawla *et al.*, 2013). Banasiak *et al.*, (2011) used this method for the removal of endosulfan from polluted water.

Ex-situ bioremediation generally requires a shorter amount of time to be completed than in-situ treatment (Federal Remediation Technology Roundtable,

2006). Furthermore, ex-situ treatment is easier to control due to the possibility of continuous mixing during treatment. On the other hand, the ex-situ method also requires excavation and a possible increase in labour costs.

Some limitations of ex-situ bioremediation drastically affect the feasibility of using this technology. In all ex-situ treatments discussed, a large amount of space is required for treatment. In urban areas, this is a viable concern. Volatile contaminants must be either contained or pre-treated, as they may volatilize into the air if not monitored, and may cause air pollution (Federal Remediation Technology Roundtable, 2006).

5.4. ROLE OF MICROBES

A major part of the pesticides are acted upon by the microorganisms present in the soil and water. Microorganisms usually occupy volume of less than 0.1 per cent of the soil, but are responsible for numerous transformation that cycle elements energy in nature. Transformations of pesticides do not always occur by only one type of organisms or by a few selected organisms, but are effected by actinomycetes, fungi and bacteria (Table 1).

Table 1. Microorganisms involved in biodegradation of some important pesticides

Pesticides	Microorganisms	References
(1)	(2)	(3)
Organochlorines insecticides		
DDT	*Fusarium solani*	Mitra *et al.*, 2001
HCH	*Sphingobium indicum* B90A	Raina *et al.*, 2008
Endosulfan	*Aspergillus niger*	Mukherjee and Gopal, 1994
	Mucor thermo-hyalospora MTCC 1384	Shetty *et al.*, 2000
	Fusarium ventricosum F1, *Fusarium ventricosum* F4, *Pandoraea* sp. B4, *Pandoraea* sp. Lin-3	Siddique *et al.*, 2003
	Klebsiella oxytoca KE-8	Kwon *et al.*, 2005
	Stenotrophomonas maltophilia, *Rhodococcus erythropolis*	Kumar *et al.*, 2007
	Bordetella sp. B9	Goswami and Singh, 2009
	Aspergillus sydoni	Goswami *et al.*, 2009
	Achromobacter xylosoxidans CS5	Li *et al.*, 2009
Organophosphates insecticides		
Acephate	*Exiguobacterium* sp. BCH4 and *Rhodococcus* sp. BCH2	Phugare *et al.*, 2012
	Pseudomonas aeruginosa Is-6	Ramu and Seetharaman, 2014
Chlorpyriphos	*Pseudomonas aeruginosa*, *Bacillus cereus*, *Klebsiella* sp., *Serratia marcecens*	Lakshmi *et al.*, 2009

Contd...

(1)	(2)	(3)
Diazinon	*Serratia* sp. and *Pseudomonas* sp	Cycoń *et al.*, 2009
	Serratia marcescens DI101	Abo-Amer, 2011
Dichlorvos	*Rhodobacter sphaeroides* EBL0706	Zhao *et al.*, 2009
Dimethoate	*Raoultella* sp. X1	Liang *et al.*, 2009
Malathion	*Pseudomonas aeruginosa*, *Bacillus pseudomycoides*, *Bacillus licheniformis*	Thabit and EL-Naggar, 2013
	Acinetobacter johnsonii MA19	Xi *et al.*, 2009
Methyl parathion	*Acinetobacter radioresistens* USTB-04	Liu *et al.*, 2007
	Bacillus pumilus T1	Ali *et al.*, 2011
	Serratia sp. DS001	Pakala *et al.*, 2007
Monocrotophos	*Aspergillus Niger* MCP1	Jain *et al.*, 2012
	Paracoccus sp.M-1	Jia *et al.*, 2006
	Aspergillus oryzae ARIFCC 1054	Bhalerao and Puranik, 2009
	Pseudomonas aeruginosa F10B, *Clavibacter michiganense* subsp. *insidiosum* SBL 11	Singh and Singh, 2003
Phorate	*Ralstonia eutropha* AAJ1	Rani *et al.*, 2009
	Azotobater chroococcum	Kadam and Gangawane, 2005
Profenophos	*Pseudomonas putida*, *Burkholderia gladioli*	Malghani *et al.*, 2009
Triazophos	*Diaphorobacter* sp. TPD-1	Yang *et al.*, 2011
	Bacillus sp. TAP-1	Tang and You, 2012
Carbamates insecticides		
Carbaryl	*Micrococcus* sp.	Doddamani and Ninnekar, 2001
	Pseudomonas cepacia	Venkateswarlu *et al.*, 1980
Carbofuran	*Paracoccus* sp. YM3.	Peng *et al.*, 2008
	Pseudomonas sp. NJ-101	Bano and Musarrat, 2004
Synthetic pyrethroids insecticides		
β-cyfluthrin	*Pseudomonas stutzeri* S1	Saikia *et al.*, 2005
λ-cyhalothrin	*Pseudomonas fluorescens*, *Trichoderma viride*	Tamilselvan *et al.*, 2008
Cypermethrin	*Micrococcus* sp.	Tallur *et al.*, 2008
New chemistry insecticides		
Imidacloprid	*Leifsonia* sp. PC 21	Anhalt *et al.*, 2007
Acetamiprid	*Pigmentiphaga* sp. D2	Yang *et al.*, 2013
Fipronil	*Paracoccus* sp.,	Kumar *et al.*, 2012
	Bacillus thuringiensis	Mandal *et al.*, 2013

Contd...

(1)	(2)	(3)
Fungicides		
Carbendazim	*Rhodococcus jialingiae* djl-6-2	Wang *et al.*, 2010
Bromuconazol	*Aeromonas hydrophila*	Bordjib *et al.*, 2012
Tebuconazole	*Enterobacter sakazakii, Serratia* sp.	Sehnem *et al.*, 2010
Hebicides		
Alachlor	*Streptomyces* sp. LS166, *Streptomyces* sp. LS177, *Streptomyces* sp. LS182	Sette *et al.*, 2004
Atrazine	*Arthrobacter* sp. DAT1	Wang *et al.*, 2013
Butachlor	*Rhodococcus sp.* B1	Liu *et al.*, 2012

Bacteria as a group are quite versatile. They have been shown to play a major role in the degradation of pesticides. Research into pesticide degradation by microorganisms was focussed primarily on bacteria, few studies were performed with fungi, actinomycetes etc. Generally, bacteria are easier to culture and grow more quickly than fungi; they are more amenable to genetic manipulation techniques and the chances of mutants reverting back are fewer.

Organochlorines such as DDT can be degraded by *Fusarium solani* (Mitra *et al.*, 2001). Endosulfan can be degraded by *Aspergillus niger* (Mukherjee and Gopal, 1994), *Klebsiella oxytoca* KE-8 (Kwon *et al.*, 2005) and *Bordetella* sp. B9 (Goswami and Singh, 2009). Organophosphates such as acephate, chlorpyriphos, diazinon, dichlorvos, dimethoate, malathion and phorate can be degraded by different bacteria. *Diaphorobacter* sp. TPD-1 (Yang *et al.*, 2011) and *Bacillus* sp. TAP-1 (Tang and You, 2012) degrade triazophos. Pesticides belonging to carbamate can be degraded by *Micrococcus* sp. (Doddamani and Ninnekar, 2001) and *Pseudomonas cepacia* (Venkateswarlu *et al.*, 1980), *Paracoccus* sp. YM3 and *Pseudomonas* sp. NJ-101 (Bano and Musarrat, 2004).

5.5. FACTORS AFFECTING BIOREMEDIATION OF PESTICIDES

The ultimate sink of the pesticides applied in agriculture as well as public health care is soil. Soil acts as storehouse of various kinds of microbes. It receives the chemicals in various forms and acts as a scavenger of harmful substances. Various factors like moisture, pH, temperature, presence of appropriate bacteria, physical and chemical properties of pesticides, concentration of pesticides, biological availability, nutrient availability affect the binding of individual and mixtures of pesticides in soil. This in turn influences the effectiveness of bioremediation strategies.

a) **Moisture content:** Microorganisms carrying out a metabolic transformation require adequate moisture for their growth and activity. Moisture obviously is not a limiting factor in oceans, in fresh waters, or in subterranean aquifers. However, an inadequate supply of water can severely restrict biodegradation in surface soil, in which drying to suboptimal water levels is common. In our study, for example, the optimum moisture level for the biodegradation of oily sludge was found to be at 30 to 90% of the soil's water-holding capacity (Dibble and Bartha, 1979). The optimum moisture level will depend on the properties of soil, the compound in question, and whether the transformation is aerobic and anaerobic. The last factor is of particular significance because excess water displaces air from the pores in soil, and a waterlogged soil soon becomes anaerobic and unfavourable for aerobic processes. Decreasing the moisture content of soil diminishes rates of degradation, a result of an inadequate supply of water to sustain proliferation, metabolism, or both.

b) pH: Soil pH is important because it can affect availability of nutrients. Most microbial species can survive only within a certain pH range. At extremes of acidity or alkalinity, activity declines. At more moderate pH values, biodegradation trends to be faster. If a compound in a particular environment can be metabolized by a diverse group of organisms, the range of pH values at which degradation occurs frequently is broader than if only one species can bring about the transformation. Apart from pesticides, however, the effect of pH on biodegradation of polluting chemicals has received scant attention, although it is common practice to add lime to bioremediate acid soils or subsoil materials containing harmful organic compounds (Alexander, 1999). Many investigators observed the important role of pH in the adsorption of compounds with acidic functional groups on activated carbon and soil because neutral and ionic forms display very different adsorption properties.

c) Temperature: Temperature influences the rate of biodegradation by controlling rate of enzymatic reactions within micro-organisms. Generally, speed of enzymatic reactions in the cell approximately gets doubled for each 10°C rise in temperature. In general, the optimum temperature for pesticide biodegradation ranges from 25 to 30°C in soil, soil slurry, and bacterial cultures. The optimum temperature for degradation may vary in field environments depending on the bacterial population. An increase in temperature might enhance pesticide removal simply by increasing biological activity (within limits), or might also affect biodegradation by enhancing bioavailability through reduced sorption. Bacteria capable of degrading pesticide at extreme temperatures (<5°C or >40°C) have not been reported. Some studies suggest that the influence of temperature is not as significant as that of other soil conditions and characteristics (i.e., redox conditions, moisture content) that affect volatility, sorption and bioavailability.

d) Presence of appropriate bacteria: In situ degradation is dependent on the presence, numbers and activity level of bacteria which possess the appropriate enzymatic capability. In addition, environmental conditions must be suitable such that degradation can proceed. Degradation rates are strongly influenced by a wide variety of environmental factors, including soil type, moisture content, temperature, and pH, all of which affect microbial activity. For degradation to proceed, bacteria, or their enzymes, must be in contact with the pesticide, and the pesticide or its metabolites must be transported into the cell. Some bacteria display chemotactic responses to substrates, which may play an important role in the environment;

others grow in a filamentous form towards potential substrates (Fewson, 1988a).

To date, most studies of microbial pesticide degradation have focused on surface soils. Biodegradation of pesticides in the rhizosphere of plants and in the soil subsurface has not been addressed in any detail, nor has the influence of soil heterogeneity and spatial variability. Although it is becoming increasingly apparent that substantial numbers of metabolically diverse bacteria exist in subsurface environments, their ability to metabolize pesticides is virtually unknown. Metabolism of atrazine has been detected in subsoils but the microbes involved were not isolated (McMahon *et al.*, 1992).

e) Physical and chemical properties of pesticides: The volatility of a particular pesticide in soil is dependent on a variety of physical and chemical properties of that pesticide. Vapour pressure, solubility in water, nucleus type and number, kind(s), and position(s) of functional groups all play important roles and can greatly modify the kinetics of volatilization and enable volatility of a compound to be predicted. Generally, volatilization from soil is only considered a factor in the fate of pesticides if the vapour pressure of the compound is greater than 10^4 mmHg (Racke, 1990).

For example, the highly toxic phosphates do not persist for more than three months while chlorinated hydrocarbon insecticides like chlordane are known to persist at least for 4-5 years and some times more than 15 years. Persistence of pesticides poses a threat to the livestocks and human health. Longer persistence leads to the accumulation of residues in soil which may result into the increased bioaccumulation by plants to the level at which the consumption of plant products may prove deleterious to human beings as well as livestocks (Chawla *et al.*, 2013).

f) Concentration of pesticide: A significant influence on biodegradation of pesticide could be the concentration at which they are present in contaminated soils. There is evidence that pesticide biodegradation rates in soil follow first-order kinetics and are concentration dependent. Pesticide may affect soil microbial populations, stimulating growth of certain microorganisms and exerting toxic effects and inhibiting growth of others. In soil and culture media containing higher pesticide concentrations, that are more representative of what might be found at post-production sites and waste disposal sites, the numbers of some microbial species or their activities may be affected. Bioremediation of pesticide contaminated soil is usually only necessary on industrial post-production, or waste dumping sites, as opposed to crop fields where application rates were lower and the pesticide are expected to be degraded by naturally occurring micro-organisms.

g) Biological availability: In soils, biodegradation of pesticides can be limited by the poor availability of the chemical. A number of factors influence availability, though effects are primarily due to either low solubility and/or strong adsorption to the soil matrix (Linn *et al.*, 1993). The chemical structure of a pesticide, its size and functional groups all determine the water solubility of the compound. Some pesticides are highly soluble, and others are applied as their water-soluble salts or amides to enhance dispersal. Less soluble pesticides are applied as an emulsion or suspension of fine solids in an aqueous carrier (O'Conner, 1994). In soils, pesticide molecules may bind tightly to soil organic matter, therefore reducing biological availability. Pesticide sorption can either enhance or decrease microbial degradation rates in soil. The concentration of pesticide applied to the soil may be toxic, but once bound to soil an apparent reduction in toxicity is seen allowing biodegradation to proceed.

h) Nutrient availability: Availability of essential nutrients such as carbon, nitrogen, phosphorus and oxygen may also limit the rate at which bacteria degrade pesticides. The concentration of the pesticide must not be so high as to be toxic, nor so low such that biodegradation does not proceed due to a lack of induction of appropriate degradative enzymes and uptake mechanisms, failure to induce sufficient enzyme activities, or problems in providing sufficient energy even for cell maintenance (Fewson, 1988b; Pahm and Alexander, 1993). Some pesticides, such as 2,4-D and carbofuran, serve as a carbon source for bacterial growth, however, in the presence of alternative readily degradable carbon sources, bacteria are less likely to degrade the pesticide.

5.6. MICROORGANISMS IN BIOREMEDIATION

Microorganisms excel at using organic substances, natural or synthetic, as sources of nutrients and energy. The explanation for their remarkable range of degradative abilities is that, by the time human beings came on the scene, microorganisms had already coexisted for billions of years with an immense variety of organic compounds. The vast diversity of potential substrates for growth led to the evolution of enzymes capable of transforming many unrelated natural organic compounds by many different catalytic mechanisms. The resulting giant "library" of microbial enzymes serves as raw material for further evolution whenever a new chemical becomes available (Butler and Mason, 1997; Ellis, 2000).

Depending on their behavior in the environment, organic compounds are often classified as biodegradable, persistent or recalcitrant. A biodegradable organic compound is one that undergoes a biological transformation (Blackburn and Hafker, 1993; Liu and Suflita, 1993). A persistent organic compound does not undergo biodegradation in certain environments; and a recalcitrant compound resists biodegradation in a wide variety of environments. While partial biodegradation is usually an alteration by a single reaction, primary biodegradation involves a more extensive chemical change. Mineralization is a parallel term to biodegradation, referring to complete degradation to the end products of CO_2, water and other inorganic compounds.

Microorganisms with the ability to degrade a wide variety of compounds, like benzene, phenol, naphthalene, atrazine, nitroaromatics, biphenyls, polychlorinated biphenyls (PCBs) and chlorobenzoates, have been isolated and characterized (Sangodkar *et al.*, 1989; Dickel *et al.*, 1993; Faison, 2001). Although simple aromatic compounds are biodegradable by a variety of degradative pathways, their halogenated counterparts are more resistant to bacterial attacks and often necessitate the evolution of novel pathways (Chakrabarty, 1982; Engasser *et al.*, 1990). Many of the environmentally important chemicals introduced for industrial uses are halogenated and halogenation, which is often implicated as a reason for persistence. Among the halogenated compounds, the chlorinated compounds are the most extensively studied (Chaudhary and Chapalamadugu, 1991; Cork and Krueger *et al.*, 1992). The presence of chlorine atoms on the aromatic nucleus is known to greatly retard the rate of degradation.

5.7. BIOTECHNOLOGY IN BIOREMEDIATION

Biotechnological processes for the bioremediation of chemical pollutants offer the possibility of in situ treatments and are mostly based on the natural activities of microorganisms. Biotechnological processes to destroy hazardous wastes offer many advantages over physicochemical processes. When successfully operated, biotechnological processes may achieve complete destruction of organic wastes. However, an important factor limiting the bioremediation of sites contaminated with certain hazardous compounds is the slow rate of degradation (Iwamoto and Nasu, 2001). This slow degradation rate often limits the practicality of using microorganisms in remediating contaminated sites. This is an area where genetic engineering can make a marked improvement. Molecular techniques can be used to increase the level of a particular protein or enzyme or series of enzymes in bacteria with an increase in the reaction rate (Chakrabarty, 1986). The easiest way to create an appropriate genetically engineered strain is to begin with an organism that already possesses much of the necessary degradative enzymatic machinery.

Microbial consortia involved in environmental biotechnology, e.g. activated sludge and soil/sediment consortia, are very complex, enabling them to act on a variety of pollutants. In the early 1990s, when good molecular biological techniques were developed to study microbial ecology researchers started to analyze microbial populations relevant to pollutant degradation in the environment (environmentally relevant microorganisms - ERMs) (Watanabe and Baker, 2000). Thus began studies of microbial consortia using molecular techniques, such as direct DNA isolation from environmental samples and after amplification determining the sequences of specific genes. After identification, the sequence can be compared to a large database comprising 16S rRNA, sequences of previously isolated organisms.

a) DNA extraction: Direct soil DNA extraction techniques are now an important part of microbial ecology investigations. They are very useful for determining the presence of native bacteria or bacteria introduced into environments, GMOs or GEMs – genetically modified (engineered) organisms (Torsvik, 1980; Sommerville *et al.*, 1989; Saano and Lindstrom, 1995). A large fraction, often 90-99% of microbial cells present in environmental samples are not culturable on microbiological media. Viable microbial cells can be only partially recovered from complex environmental samples by traditional plating methods. Their detection via molecular techniques is often required to have a true understanding of the microbial ecology of any environment.

b) PCR (Polymeras chain reaction): The development of this technique was a major methodological discovery in molecular biology. The method is being used in every type of laboratory interested in molecular biology from diagnostics to pure research. The high specificity, sensitivity and reproducible consistency of PCR detection of specific DNA sequences in complex environmental samples has contributed significantly to the advancement of knowledge in environmental microbiology diagnostics, detection of mutation, generation of DNA probes by PCR, and the cloning of PCR products (Pillai *et al.*, 1991; Toze, 2000; Watson and Blackwell, 2000). The most obvious application of this technique is to enhance gene probe detection of specific gene sequences. By amplification of a target sequence, PCR enhances the detection of rare sequences in complex mixtures of DNA isolated from environmental samples. The PCR is a powerful DNA amplification technique with the potential to provide a sensitive and specific method for monitoring microorganisms in the environment (Steffan and Atlas, 1991).

c) 16S rRNA gene sequencing: This identification method can be used with *Archaea* as well as *Eukarya* and *Bacteria*. It is based on determining the phylogenetic position of the known microbe among known microorganisms. This determination is based upon a particular DNA strand – its 16S rRNA gene sequence. This sequence is considered the best for these evolutionary measurements because it is highly conserved among all species (Head *et al.*, 1998). Obtaining the 16S rRNA sequence is accomplished in a variety of ways. One of the most common and effective is PCR, which replicates the 16S rRNA strand. This amplified material is then sequenced. Next, the sequenced 16S rRNA is compared to the sequences of other microorganisms that have been placed in a database (the Ribosome Database).

5.8. PHYTOREMEDIATION

Phytoremediation is defined as the use of green plants and their associated microorganisms, soil amendments and agronomic techniques to remove content or render harmless environmental contaminants. Vegetation based remediation shows potential for accumulating, immobilizing, and transforming a low level of persistent contaminants. In natural ecosystems, plants act as filters and metabolize substances generated by nature. Phytoremediation is an emerging technology that uses plants to remove contaminants from soil and water. The term "phytoremediation" is relatively new, coined in 1991. Its potential for encouraging the biodegradation of organic contaminants requires further research, although it may be a promising area for the future. Phytoremediation is well suited for use at very large field sites where other methods of remediation are not cost effective or practicable; at sites with a low concentration of contaminants where only polish treatment is required over long periods of time; and in conjunction with other technologies where vegetation is used as a final cap and closure of the site.

Phytoremediation techniques have five types, classified based on the contaminant fate: phytoextraction, phytotransformation, phytostabilization, phytodegradation, Rhizofiltration (Vidali, 2001).

a) Phytoextraction or phytoaccumulation: Phytoextraction or phytoaccumulation is the process used by the plants to accumulate contaminants into the roots and aboveground shoots or leaves. This technique saves tremendous remediation cost by accumulating low levels of contaminants from a widespread area. Unlike the degradation mechanisms, this process produces a mass of plants and contaminants (usually metals) that can be transported for disposal or recycling.

b) Phytotransformation or phytodegradation: Phytotransformation or phytodegradation refers to the uptake of organic contaminants from soil, sediments, or water and, subsequently, their transformation to more stable, less toxic, or less mobile form. It is the degradation of contaminants into smaller units by the plant enzymes or via release of exudates that help to degrade pollutants via co-metabolism in the rhizophere. The term "Green Liver Model" is used to describe phytotransformation as plants behave analogously to the human liver when dealing with these xenobiotic compounds.

c) Phytostabilization: Phytostabilization is a technique in which plants reduce the mobility and migration of contaminated soil. It refers to the holding of contaminated soils and sediments in place by vegetation and to immobilizing toxic contaminants in soils. Leachable constituents are adsorbed and bound into the plant structure so that they form a stable mass of plant from which the contaminants will not reenter the environment.

d) Phytodegradation or rhizodegradation: Phytodegradation or rhizodegradation is the breakdown of contaminants through the activity existing in the rhizosphere. This activity is due to the presence of proteins and enzymes produced by the plants or by soil organisms such as bacteria, yeast, and fungi. Rhizodegradation is a symbiotic relationship that has evolved between plants and microbes. Plants provide nutrients necessary for the microbes to thrive, while microbes provide a healthier soil environment.

e) Rhizofiltration: Rhizofiltration is a water remediation technique that involves the uptake of contaminants by plant roots. Rhizofiltration is used to reduce contamination in natural wetlands and estuary areas.

5.9. ADVANTAGES OF BIOREMEDIATION

Bioremediation is a natural process and is therefore. perceived by the public as an acceptable waste treatment process for contaminated material such as soil (Gavrilescu, 2005). It is also less disturbing to the environment. Bioremediation uses the resources available in nature to clean up contamination. (Chawla *et al.*, 2013). Bioremediation offers several advantages over conventional techniques such as land filling or incineration. Bioremediation can be done on site, is often less expensive and site disruption is minimal, it eliminates waste permanently, eliminates long-term liability, and has greater public acceptance, with regulatory encouragement, and it can be coupled with other physical or chemical treatment methods (Boopathy, 2000).

Bioremediation of pesticides have following advantages.

- Bioremediation is a natural process and is therefore perceived by the public as an acceptable waste treatment process for contaminated material such as soil. Microbes able to degrade the contaminant increase in numbers when the contaminant is present; when the contaminant is degraded, the biodegradative population declines. The residues for the treatment are usually harmless products and include carbon dioxide, water, and cell biomass.
- Theoretically, bioremediation is useful for the complete destruction of a wide variety of contaminants. Many compounds that are legally considered to be hazardous can be transformed to harmless products. This eliminates the

chance of future liability associated with treatment and disposal of contaminated material.

- Instead of transferring contaminants from one environmental medium to another, for example, from land to water or air, the complete destruction of target pollutants is possible.
- Bioremediation can often be carried out on site, often without causing a major disruption of normal activities. This also eliminates the need to transport quantities of waste off site and the potential threats to human health and the environment that can arise during transportation.
- Bioremediation can prove less expensive than other technologies that are used for clean-up of hazardous waste.

5.10. DISADVANTAGES OF BIOREMEDIATION

Bioremediation can also be considered problematic because sometimes it may lead to partial degradation of the contaminant which may result in toxic and potentially volatile products. It is generally very costly, labor intensive and can take several months for the remediation to achieve acceptable levels. Especially, bioremediation of pesticides requires a longer treatment time than other possible detoxifying processes (Rockne and Reddy, 2003). Biological processes are often highly specific. Important site factors required for success include the presence of metabolically capable microbial populations, suitable environmental growth conditions and appropriate levels of nutrients and contaminants. Bioremediation is limited to those compounds that are biodegradable. Not all compounds are susceptible to rapid and complete degradation. Biological processes are often highly specific. Important site factors required for success include the presence of metabolically capable microbial populations, suitable environmental growth conditions, and appropriate levels of nutrients and contaminants. Research is needed to develop and engineer bioremediation technologies that are appropriate for sites with complex mixtures of contaminants that are not evenly dispersed in the environment. Contaminants may be present as solids, liquids, and gases. Bioremediation often takes longer than other treatment options, such as excavation and removal of soil or incineration (Chawla *et al.*, 2013).

5.11. CONCLUSIONS

Bioremediation of hazardous wastes is a relatively new technology that has undergone more intense investigation as of recent decades. This process is focused on destroying or immobilizing toxic waste materials. Pesticides are a common hazard around the world, as these chemicals are leaching into soils, groundwater and surface water and creating health concerns in many communities. The persistence of pesticides makes their removal and detoxification a more urgent undertaking. The bioremediation of pesticides can be divided into two broad categories: in-situ and ex-situ treatment. Both methods have significant advantages and disadvantages.

The choice of the bioremediation strategy should be made on the basis of type of pesticide, environmental matrix and the organisms present in the ecosystem. Since, the organism is the only eligible factor, the knowledge about features, advantages or disadvantages of organisms could be a decisive factor on bioremediation proficiency. Some parameters have to be addressed to assure bioremediation. In bacteria and fungi, pH, temperature, cell count, biomass growth rate, substrate bioavailability, and moisture are some of them. Plants require less supervision, but finding the best phytoremediator could be a hard and time consuming task. Derivative bioremediation is a promising strategy. To get all the benefits from this strategy is necessary to carefully select the most adequate enzyme, and to have it well-characterized. Regardless, further research on biodegradation or biotrans-formation mechanisms in plants, bacteria, fungi or algae is imperative if bioremediation strategies are to be implemented or improved.

The main aim of this chapter is to provide the scientific understanding needed to harness natural processes and to develop methods to accelerate these processes for the bioremediation of contaminated environments. Except few limiting factors, this technology has the ability to rejuvenate the contaminated environments effectively. However, rapid advances in the last few years have helped us in the understanding of process of bioremediation. The use of culture independent molecular techniques has definitely helped us to understand the microbial community dynamics, structure and assisted in providing the insight in to details of bioremediation which has surely facilitated to make the technology safer and reliable. In this context, bioremediation in relation to process optimization, validation and its impact on the ecosystem can be performed and by judicious use of the models that can predict the activity of microorganisms that are involved in bioremediation with existing geochemical and hydrological models, transformation of bioremediation from a mere practice into a science is now a reality. With the exciting new development in this field and focus on interdisciplinary research and using it on gaining the fundamental knowledge necessary to overcome the obstacles facing current technologies and also with respect to ethical, legal, and social issues involved this technology will go a long way in cleaning the environment in near future.

REFERENCES

Abo-Amer A (2011) Biodegradation of diazinon by *Serratia marcescens* DI101 and its use in bioremediation of contaminated environment. *Journal of Microbiology and Biotechnology* **21**:71-80.

Alexander M (1999) *Biodegradation and Bioremediation*. Academic press, California, USA.

Ali M, Ahmed KN, Hameed A and Ahmed S (2011) Biodegradation of methyl parathion by newly isolated *Bacillus pumilus* T1. *Minerva Biotecnologica* **23**:39-43.

Anhalt JC, Moorman TB and Koskinen WC (2007) Biodegradation of imidacloprid by an isolated soil microorganism. *Journal of Environmental Science and Health Part B* **42**: 509-514.

Banasiak L, Schafer A and Van der Bruggen B (2011) Sorption of pesticide endosulfan by electrodialysis membranes, *Chemical Engineering Journal* **166**: 233-239.

Bano N and Musarrat J (2004) Characterization of a novel carbofuran degrading *Pseudomonas* sp. with collateral biocontrol and plant growth promoting potential. *FEMS Microbiology Letters* **231**:13-17.

Bhalerao TS and Puranik PR (2009) Microbial degradation of monocrotophos by *Aspergillus oryzae. International Journal of Biodeterioration & Biodegradation* **63**: 503-508.

Blackburn JW and Hafker WR (1993) The impact of biochemistry, bioavailability and bioactivity on the selection of bioremediation techniques. *Trends in Biotechnology* **11**: 328-333.

Blanca AL, Angus JB, Spanova K, Lopez-Real J and Russell NJ (2007) The influence of different temperature progıammes on the bioremediation of polycyclic aromatic hydrocarbons (PAHs) in a coal-tar contaminated soil by in-vessel composting. *Journal of Hazardous Materials* **14**: 340-347.

Boopathy R (2000) Factors limiting bioremediation technologies. *Bioresource Technology* **74**: 63-67.

Bordjiba O, Ouartsi N and Bjeribi R (2012) Biodegradation of pesticide: bromuconazol by microbial consortium in biphasic system. *The Online Journal of Science and Technology* **2**: 20-25.

Brenner K, You L and Arnold FH (2008) Engineering microbial consortia: a new frontier in synthetic biology. *Trends in Biotechnology* **26**: 483-489.

Butler CS and Mason J R (1997) Structure, function analysis of the bacterial aromatic ring hydroxylating dioxygenases. *Advances in Microbial Physiology* **38**: 47–84.

Chakrabarty AM (1982) Genetic mechanisms in the dissimilation of chlorinated compounds. In: Chakrabarty A M (ed.) *Biodegradation and Detoxification of Environmental Pollutants*. CRC, Boca Raton. pp 127–139.

Chakrabarty AM (1986) Genetic engineering and problems of environmental pollution. *Biotechnology* **8**: 515–530.

Chaudhary GR and Chapalamadugu S (1991) Biodegradation of halogenated organic compounds. *Microbiological Reviews* **55**: 59–78.

Chawla N, Suneja S, Kamlesh K and Kumar R (2013) Bioremediation: An emerging technology for remediation of pesticides. *Research Journal of Chemistry and Environment* **17**: 88-105.

Colleran E (1997) Uses of bacteria in bioremediation. In: Sheehan D (ed.), *Methods in Biotechnology: Bioremediation Protocols.* pp. 3-22.Humana Press, New Jersey.

Cork DT and Krueger JP (1992) Pesticide biodegradation. *Encyclopedia of Microbiology* **3**: 357–361.

Couto MNPFS, Monteiro E and Vasconcelos MTSD (2010) Mesocosm trials of bioremediation of contaminated soil of a petroleum refinery: comparison of natural attenuation, biostimulation and bioaugmentation, *Environmental Science and Pollution Research* **17**: 1339–1346.

Cycoń M, Wójcik M and Piotrowska-Seget Z (2009) Biodegradation of the organophosphorus insecticide diazinon by *Serratia* sp. and *Pseudomonas* sp. and their use in bioremediation of contaminated soil. *Chemosphere* **76**: 494-501.

Dibble JT and Bartha R (1979) The effect of environmental parameters on the biodegradation of oily sludge. *Applied and Environmental Microbiology* **37**: 729-739.

Dickel O, Haug W and Knackmus H J (1993) Biodegradation of nitrobenzene by a sequential anaerobic–aerobic process. *Biodegradation* **4**: 187–194.

Doddamani HP and Ninnekar HZ (2001) Biodegradation of carbaryl by a *Micrococcus* species. *Current Microbiology* **43**: 69-73.

Ellis BML (2000) Environmental biotechnology informatics. *Current Opinion in Biotechnology* **11**: 232–235.

Engasser KH, Auling G, Busse J and Knackmus HJ (1990) 3-Fluorobenzoate enriched bacterial strain FLB 300 degrades benzoate and all three isomeric monofluoro-benzoates. *Archives of Microbiology* **153**: 193–199.

Fahnestock FM, Wickramanayake GB, Kratzke KJ and Major WR (1998) *Biopile Design, Operation and Maintenance*

Handbook for Treating Hydrocarbon Contaminated Soil, Battelle Press, Columbus, OH.

Faison BD (2001) Hazardous waste treatment. *SIM News* **51**: 193–208.

Federal Remediation Technology Roundtable (2006) *Remediation Technologies Screening Matrix and Reference Guide,* Version 4.0. (http://www.frtr.gov//matrix2/section1/toc.html#Pref)

Felsot AS, Mitchell JK, Bicki TJ and Frank JF (1992) Experimental design for testing landfarming of pesticide-contaminated soil excavated from agrochemical facilities. *Pesticide Waste Management* **22**: 244–261.

Fewson CA (1988a) Microbial metabolism of mandelate: a microcosm of diversity. *FEMS Microbiology Letters* **54**: 85-110.

Fewson CA (1988b) Biodegradation of xenobiotic and other persistent compounds: the cause of recalcitrance. *Trends in Biotechnology* **6**: 148-153.

Fragoeiro SIS (2005) Use of Fungi in Bioremediation of Pesticides. Cranfield University. Ph. D. Thesis. (https://dspace.lib.cranfield.ac.uk/bitstream/1826/906/2/Fragoeiro+ thesis.pdf# search=%22bioremediation%20of%20pesticides%20and%20herbicides%22)

Fulekar MH (2009) Bioremediation of fenvalerate by *Pseudomonas aeruginosa* in a scale up bioreactor. *Romanian Biotechnological Letters* **14**: 4900-4905.

Gavrilescu M (2005) Fate of pesticides in the environment and its bioremediation. *Engineering in Life Sciences* **5**: 497-526.

Goswami S and Singh DK (2009) Biodegradation of α and β endosulfan by in broth medium and soil microcosm by bacterial strain *Bordetella* sp. B9. *Biodegradation* **20**: 199-207.

Goswami S, Vig K and Singh DK (2009) Biodegradation of α and β endosulfan by *Aspergillus sydoni. Chemosphere* **75**: 883-888.

Head IM (1998) Bioremediation: towards a credible technology. *Microbiology* **144**: 599-608.

Head LH, Saunders JR and Pickup RW (1998) Microbial evaluation, diversity, and ecology: A decade of ribosomal RNA analysis of uncultivated microorganisms. *Microbial Ecology* **35**: 1-21.

Iwamoto T and Nasu M (2001) Current bioremediation practice and perspective. *Journal of Bioscience and Bioengineering* **92**: 1–8.

Jain R, Garg V, Singh KP and Gupta S (2012) Isolation and characterization of monocrotophos degrading activity of soil fungal isolate *Aspergillus niger* MCP1 (ITCC7782.10). *International Journal of Environmental Sciences* **3**: 1841-1850.

Kadam TA and Gangawane LV (2005) Degradation of phorate by Azotobacter isolates. *Indian Journal of Biotechnology* **4**: 153-155.

Jia K, Cui Z, He J, Guo P and Li S (2006) Isolation and characterization of a denitrifying Monocrotophos degrading *Paracoccus* sp. M-1. *FEMS Microbiology Letters* **263**: 155–162.

Johnsen K, Jacobsen CS and Torsvik V (2001) Pesticides effects on bacterial diversity in agricultural soils - A review. *Biology and Fertility Soils* **33**: 443-453.

Kumar K, Devi SS, Krishnamurthi K, Kanade GS and Chakrabarti T (2007) Enrichment and isolation of endosulfan degrading and detoxifying bacteria. *Chemosphere* **68**: 317-322.

Kumar R, Singh B and Gupta VK (2012) Biodegradation of fipronil by *Paracoccus* sp. in different types of soil. *Bulletin of Environmental Contamination and Toxicology* **88**: 781–787.

Kwon GS, Sohn HY, Shin KS, Kim E and Seo BI (2005) Biodegradation of the organochlorine insecticide, endosulfan and the toxic metabolite, endosulfan sulfate, by *Klebsiella oxytoca* KE-8. *Applied Microbiology Biotech*nology **67**: 845-850.

Lakshmi CV, Kumar M and Khanna S (2009) Biodegradation of chlorpyriphos in soil by enriched cultures. *Current Microbiology* **58**: 35-38.

Lee TH, Byun IG, Kim YO, Hwang IS and Park TJ (2006) Monitoring biodegradation of diesel fuel in bioventing processes using in situ respiration rate. *Water Science and Technology* **53**: 263-272.

Li W, Dai Y, Xue B, Li Y, Peng X, Zhang J and Yan Y (2009) Biodegradation and detoxification of endosulfan in aqueous medium and soil by *Achromobacter xylosoxidans* strain CS5. *Journal of Hazardous Mater*ials **167**: 209-216.

Liang Y, Zeng F, Qiu G, Lu X, Liu X and Gao H (2009) Co-metabolic degradation of dimethoate by *Raoultella* sp. X1. *Biodegradation* **20**: 363-373.

Linn DM, Carski TH, Brusseau ML and Chang FH (1993) Sorption and degradation of pesticides and organic chemicals in soils. SSSA Special Publication no. 32. Soil Science Society of America.

Liu H M, Cao L, Lu P, Ni H, Li Y X, Yan X, Hong Q and Li SP (2012) Biodegradation of Butachlor by *Rhodococcus* sp. Strain B1 and Purification of Its Hydrolase (ChlH) Responsible for N-Dealkylation of Chloroacetamide Herbicides. *Journal of Agricultural and Food Chemistry* **60**: 12238–12244.

Liu F, Hong M, Liu D, Li Y, Shou P, Yan H and Shi G (2007) Biodegradation of methyl parathion by *Acinetobacter radioresistens* USTB-04. *Journal of Environmental Science*s **19**: 1257-1260.

Liu S and Suflita JM (1993) Ecology and evolution of microbial populations for bioremediation. *Trends in Biotechnology* **11**: 344–352.

Lorenzo DV (2008) Systems biology approaches to bioremediation, *Current Opinion in Biotechnology* **19**: 579-589.

Malghani S, Chatterjee N, Yu HX and Luo Z (2009) Isolation and identification of profenophos degrading bacteria. *Brazilian Journal of Microbiology* **40**: 893-900.

Mandal K, Singh B, Jariyal M and Gupta VK (2013) Microbial degradation of fipronil by *Bacillus thuringiensis*. *Ecotoxicology and Environmental Safety* **93**: 87-92.

McMahon PB, Chapelle FH, Falls WF and Bradley PM (1992) Role of microbial processes in linking sandstone diagenesis with organic-rich clays. *Journal of Sedimentary Petrology* **62**: 1-10.

Mitra J, Mukherjee PK, Kale SP and Murthy NB (2001) Bioremediation of DDT in soil by genetically improved strains of soil fungus *Fusarium solani*. *Biodegradation* **12**: 235-245.

Mukherjee I and Gopal M (1994) Degradation of β-endosulfan by *Aspergillus niger*. *Toxicology & Environmental Chem*istry **46**: 217-221.

O'Conner B (1994) Selecting inocula for the biodegradation of organic compounds at low concentration. *Microbial Ecology* **25**: 275-286.

Pahm MA and Alexander M (1993) Selecting inocula for the biodegradation of organic compounds at low concentrations. *Microbial Ecology* **25**: 275-286.

Pakala SB, Gorla P, Pinjari AB, Krovidi RKK, Baru R, Yanamandra M, Merrick M and Siddavattam D (2007) Biodegradation of methyl parathion and p-nitrophenol: evidence for the presence of a p-nitrophenol 2-hydroxylase in a Gram-negative *Serratia* sp. strain DS001. *Applied Microbiology Biotechnology* **73**: 1452-1462.

Peng X, Zhang JS, Li YY, Li W, Xu GM, Yan YC (2008) Biodegradation of insecticide carbofuran by *Paracoccus* sp. YM3.

Journal of Environmental Science and Health Part B **43**: 588-594.

Phugare SS, Gaikwad YB and Jadhav JP (2012) Biodegradation of acephate using a developed bacterial consortium and toxicological analysis using earthworms (*Lumbricus terrestris*) as a model animal. *International Biodeterioration & Biodegradation* **69**: 1-9.

Pillai SD, Josephson KL, Bailey RL, Gerba CP and Pepper II (1991) Rapid method for processing soil samples for polymerase chain reaction amplification of specific gene sequences. *Applied and Environmental and Microbiology* **57**: 2283-2286.

Pimental D (1995) Amount of pesticides reaching target pests: environmental impacts and ethics. *Journal of Agricultural and Environmental Ethics* **8**: 17-29.

Racke KD (1990) Pesticides in the soil microbial ecosystems. In: Racke K D and Coats J R (eds.) Enhanced Biodegradation of Pesticides in the Environment. pp. 1-12. ACS Symposium Seies 426, American Chemical Society, Washington, DC, USA.

Raina V, Suar M, Singh A, Parkash O, Dadhwal M, Gupta SK, Dogra C, Lawlor K, Lal S, Meer JR, Holliger C and Lal R (2008) Enhanced biodegradation of hexachlorocyclohexane (HCH) in contaminated soils via inoculation with *Sphingobium inidicum* B90A. *Biodegradation* **19**: 27-40.

Ramu S and Seetharaman B (2014) Biodegradation of acephate and methamidophos by a soil bacterium *Pseudomonas aeruginosa* strain Is-6. *Journal of Environmental Science and Health Part B* **49**: 23–34.

Rani R, Lal R, Kanade G S and Juwarkar A (2009) Isolation and characterization of a phorate degrading bacterium. Letters in Applied Microbiology **49**:112-116.

Rockne K. and Reddy K (2003) *Bioremediation of Contaminated Sites.* University of Illinois, Chicago.

Saano A and Lindstrom K (1995) Isolation and identification of DNA from soil, In: Alef K and Nannipieri P (eds), *Methods in Applied Soil Microbiology and Biochemistry*. Academic Press, pp. 440-451.

Saikia N, Das SK, Patel BKC, Niwas R, Singh A and Gopal M (2005) Biodegradation of beta-cyfluthrin by *Pseudononas stutzeri* strain S1. *Biodegradation* **16**: 581-589.

Sangodkar UMX, Aldrich TL, Haugland RA, Jhonson J, Rothmel RK, Chapman PJ and Chakrabarty AM (1989) Molecular basis of biodegradation of chloroaromatic compounds. *Acta Biotechnology* **9**: 301–316.

Schultz P and Urban NR (2008) Effects of bacterial dynamics on organic matter decomposition and nutrient release from sediments: A modeling study. *Ecological Modelling* **210**: 1–14.

Sehnem NT, Souza-cruz P and Peralba MDC (2010) Biodegradation of tebuconazole by bacteria isolated from contaminated soils. *Journal of Environmental Science and Health Part B* **45**: 67-72.

Sette LD, Mendonça LA, Costa A, Marsaioli AJ, and Manfio GP (2004) Biodegradation of alachlor by soil streptomycetes. *Applied Microbiology and Biotechnology* **64**: 712–717.

Shetty PK, Mitra J, Murthy NBK, Namitha KK, Savitha KN and Raghu K (2000) Biodegradation of cyclodiene insecticide endosulfan by *Mucor thermohyalospora* MTCC 1384. *Current Science* **79**: 1381-1383.

Shukla KP, Singh NK and Sharma S (2010) Bioremediation: Developments, current practices and perspectives. *Genetic Engineering and Biotechnology Journal* **3**: 1-20.

Siddique T, Okeke BC, Arshad M and Frankenberger WT Jr (2003) Biodegradation Kinetics of Endosulfan by *Fusarium ventricosum* and a *Pandoraea* Species. *Journal of Agricultural and Food Chemistry* **51**: 8015-8019.

Singh BK and Walker A (2006) Microbial degradation of organophosphorus compounds. *FEMS Microbiology Reviews* **30**: 428–471.

Singh B and Dhaliwal GS (2001) Pesticides, Environment and Society: Issues and options. In: *Punjab Society, Perspectives & Challenges.* Manohar Publications, New Delhi, pp. 324-332.

Singh DK (2002) Microbial degradation of insecticides: an assessment for its use in bioremediation. In: Singh V P and Stapleton R D Jr. (eds.) *Biotransformations: Bioremediation Technology for Health and Environmental Protection*. pp. 175-188. Elsevier Science B.V.

Singh S and Singh DK (2003) Utilization of monocrotophos as phosphorus source by *Pseudomonas aeruginosa* F10B and *Clavibacter michiganense* subsp. *insidiosum* SBL11. *Canadian Journal of Microbiology* **49**: 101-109.

Sommerville CC, Knight IT, Stanb WL and Colwell RR (1989) Simple, rapid method for direct isolation of nucleic acids from aquatic environments. *Applied and Environmental Microbiology* **55**: 548-554.

Steffan RJ and Atlas RM (1991) Polymerase chain reaction. Applications in environmental microbiology. *Annual Review of Microbiology* **45**: 137-161.

Tallur PN, Megadi VB and Ninnekar HZ (2008) Biodegradation of cypermethrin by *Micrococcus* sp. Strain CPN1. *Biodegradation* **19**: 77-82.

Tamilselvan C, Pramila B, Rajaguru K, Raja K and Bairas B (2008) In-vitro degradation of the insecticide lambda cyhalothirn by the antagonistic organisms, *Pseudomonas fluorescens* and *Trichoderma viride* under aerobic condition. *Pestology* **32**: 33-36.

Tang M and You M (2012) Isolation, identification and characterization of a novel triazophos-degrading *Bacillus* sp. (TAP-1). *Microbiological Research* **167**: 199-205.

Thabit TMA and EL-Naggar MAH (2013) Malathion degradation by soil isolated bacteria and detection of degradation products by GC-MS. *International Journal of Environmental Sciences* **3**: 1467-1476.

Torsvik VL (1980) Isolation of bacterial DNA from soil. *Soil Biology & Biochemistry* **12**: 15-21.

Toze S (2000) PCR and the detection of microbial pathogens in water and wastewater. *Water Research* **33**: 3545-3556.

Venkateswarlu K, Chendrayan K and Sethunathan N (1980) Persistence and biodegradation of carbaryl in soils. *Journal of Environmental Science and Health Part B* **15**: 421-429.

Vidali M (2001) Bioremediation. An overview. *Pure and Applied Chemistry* **73**: 1163-1172.

Wang Q, Xie S and Hu R (2013) Bioaugmentation with *Arthrobacter* sp. strain DAT1 for remediation of heavily atrazine–contaminated soil. *International Biodeterioration & Biodegradation* **77**: 63-67.

Wang Z, Wang Y, Gong F, Zhang J, Hong Q and Li S (2010) Biodegradation of carbendazim by a novel actinobacterium *Rhodococcus jialingiae* djl-6-2. *Chemosphere* **81**: 639-644.

Watanabe K and Baker PW (2000) Environmentally relevant microorganisms. *Journal of Bioscience and Bioengineering* **89**: 1-11.

Watson RJ and Blackwell B (2000) Purification and characterization of a common soil component which inhibits the polymerase chain reaction. *Canadian Journal of Microbiology* **46**: 633-642.

Xi S, Liu J, Li L and Qiao C (2009) Biodegradation of malathion by *Acinetobacter johnsonii* MA19 and optimization of cometabolism substrates. *Journal of Environmental Sciences* **21**: 76-82.

Yang C, Li R, Song Y, Chen K, Li S and Jiang J (2011) Identification of the biochemical degradation pathway of triazophos and its intermediate in *Diaphorobacter* sp. TPD-1. *Current Microbiology* **62**: 1294-1301.

Yang H, Wang X, Zheng J, Wang G, Hong Q, Li S, Li R and Jiang J (2013) Biodegradation of acetamiprid by *Pigmentiphaga* sp. D-2 and the degradation pathway. *International Biodeterioration & Biodegradation* **85**: 95-102.

Zhao K, Yu Y, Jiang D, Wang D, Li Z M, Huang GZ and Bai ZH (2009) Degradation of dichlorvos by *Rhodobacter sphaeroides*. *Huan Jing Ke Xue* **30**: 1199-1204.

Zhou Y, Liu W and Ye H (2006) Effects of pesticides metolachlor and S-metolachlor on soil microorganisms in aquisols. II. Soil respiration. *Ying Yong Sheng Tai Xue Bao* **17**: 1305–1309.

Biological and Molecular Approaches in Pest Management
Balwinder Singh, Ramesh Arora & S S Gosal (Eds)
Scientific Publishers (India), Jodhpur, 2015, pp 97-152.

Chapter - 6

Microbial Control in Insect Pest Management: Achievements and Challenges

Ramesh Arora

Department of Entomology, Punjab Agricultural University, Ludhiana-141 004, India
E-mail:rarora@pau.edu

6.1. INTRODUCTION

Impressive food production increases in large parts of the world during the past 40 years have been based on massive increases (15-20 times) in the use of synthetic chemical pesticides (Oerke, 2006), which is not sustainable. There is ample evidence of the adverse effects on the environment and human health associated with the use of synthetic pesticides. This has led to a strong consumer push for environmentally benign pest control methods, coupled with regulatory actions to reduce risk, such as the withdrawal of many synthetic pesticides. Retailers and regulators are setting much lower maximum residue limits for synthetic pesticides, requirements from which biopesticides have been exempt, and governments are moving to favor explicitly nonsynthetic pesticide based control options in regulations. The high cost of discovering, developing and registering new synthetic pesticides and the rapid emergence of pest resistance have also contributed to an increased interest in biopesticides (Glare *et al.,* 2012). This chapter discusses the role of microbial control agents (MCAs) including bacteria, viruses, fungi, nematodes or bioactive metabolites produced directly by these organisms for management of insect pests (Tanada and Kaya, 1993; Arora *et al.,* 2000b; Vega and Kaya, 2012). The classification of several groups of entomopathogens has undergone drastic changes in the last decade. Unless otherwise stated, the classification scheme followed by various authors in the recent monumental book on insect pathology edited by Vega and Kaya (2012) has been followed here.

6.2. VIRAL ENTOMOPATHOGENS

Viral diseases are one of the most investigated infections in insects. According to International Committee on Taxonomy of viruses (www.ictvonline.org), those pathogenic to insects belong to at least 13 families plus a number of yet unclassified viruses.

A feature of many insect viruses, which does not occur in viruses infecting plants or vertebrates, is that they are occluded i.e. the virions are embedded within a proteinaceous body called occlusion bodies (OBs). The occluded viruses belong to three virus families. Each of these families appears to have independently evolved the ability to produce OBs consisting largely of a single protein. The OBs allow these viruses to persist for long periods in the environment and nearly all of these viruses replicate only in the larval stages of holometabolous insects. In some instances occluded virus is known to have survived for decades in the absence of any host insects and still be infectious (Hunter-Fujita *et al.*, 1998). The important groups of viral entomopathogens are briefly introduced hereunder, while the baculoviruses are discussed in some detail.

A. Occluded viruses

Baculoviruses (BVs)

The baculoviruses make up a large (>700 baculoviruses) family (Baculoviridae) of viruses, with more isolates recorded than for any other group of insect viruses. They have a large circular double stranded (ds) DNA genome contained within the enveloped nucleoprotein core of a rod-shaped virion. Baculoviruses have been isolated exclusively from insects and have been widely exploited for pest management. These are discussed in detail in the succeeding section.

Cypoviruses

These viruses belong to genus *Cypovirus* within family *Reoviridae*, a RNA family of viruses that infect a wide range of vertebrates, invertebrates and plants. The cypoviruses earlier known as cytoplasmic polyhedrosis viruses (CPVs), are characterized by the occurrence of occlusions only in the cytoplasm of infected midgut cells and have been isolated from more than 250 species of insects, from the orders Lepidoptera, Diptera and Hymenoptera. The genus Cypovirus consists of 16 different species designated as CPV-1 to CPV-16, differentiated on the basis of electrophoretic mobilities of the 10 double-stranded RNA molecules that comprise their genomes, as well as nucleotide sequence and antigenic variations. Some additional species have also been proposed recently (Harrison and Hoover, 2012).

Cypovirus occlusions vary in size from 0.1 to 10 μg and are composed of a polyhedrin protein distinct from the polyhedrins and granulins of baculoviruses. These occlusions contain icosahedral non-enveloped virions of 50-70nm diameter having only a single shell surrounding the virus core.

Entomopoxviruses

The entomopoxviruses form a distinct subfamily the *Entomopoxvirinae*, within the family *Poxviridae* and have been recorded from about 60 insect species. This subfamily has three insect related genera: *Alphaentomopoxvirus*, found in Coleoptera; *Betaentomopoxvirus*, isolated from Orthoptera and Lepidoptera; and *Gammaentomopoxvirus*, isolated from Diptera.

The viral occlusions of EPVs, referred to as spheroids, are oval shaped and range in size from 5 to 20 μm. The major protein comprising the matrix of the spheroid is called spheroidin. The virions are either ovoid (alpha and betaentomopoxviruses), or brick shaped (gammaentomopoxviruses).

Nudiviruses

This group includes viruses resembling baculoviruses in terms of replicating in the cell nucleus of their host insects and producing rod-shaped enveloped virions, but generally do not produce proteinaceous occlusions. These viruses were earlier included in family baculoviridae but are now treated as separate group.

The group includes the OrNV from the rhinoceros beetle, *Oryctes rhinoceros*, the HzNV-1 infecting *Helicoverpa zea*, HzNV-2 also called the gonad-specific virus also infecting *H. zea* and GbNV infecting crickets, *Gryllus bimaculatus*. The nudiviruses differ considerably from each other in their structure and biology.

B. Non-Occluded Viruses

Majority of the non-occluded viruses attack other groups of organisms besides insects, including vertebrates and even plants. The only groups which are restricted to insects are the ascoviruses, po y-dna viruses, nodaviruses and tetraviruses.

Ascoviruses

Ascoviruses (Genus-*Ascovirus*, Family: *Ascoviridae*) have been recorded so far, only from noctuids. They cause a chronic, fatal disease of larvae with an unusual cytopathology in which the infected cell nucleus enlarges and then disintegrates in to fragments. The cell size is enlarged and large number of virion containing vesicles form. Ascoviruses are very poorly infectious *per os* and the main route of infection appears to be by injection when parasitic wasps that have previously been contaminated from an infected larva lay their eggs in a host larva. Ascoviruses form large enveloped virions with complex symmetry containing a linear double stranded DNA genome.

Polydnaviruses (PDVs)

The polydnaviruses (Family: *Polydnaviridae*) are unusual and complex viruses with a genome consisting of multiple supercoiled DNAs of variable size from 2 to more than 28 kbp. They replicate only in the calyx (part of ovaries) of parasitic Hymenoptera. These viruses appear to have an obligate symbiotic relationship with parasitoid wasps in the families Braconidae and Ichneumonidae and are classified (in to genera) as *Bracovirus* and *Ichnovirus*, respectively, according to the host. PDV genes have proved useful as promoters in recombinant baculoviruses.

Nodaviruses

The nodaviruses (Family: *Nodaviridae*) are named after nodamura virus, which was isolated from mosquitoes in Japan. They infect a range of insects in Lepidoptera and Coleoptera. Because of the simplicity of these viruses, their molecular biology is relatively well characterized and black beetle virus was the first insect virus for which complete genome sequence was determined.

Tetraviruses

The tetraviruses (Family: Tetraviridae) is a small family of insect viruses all isolated from Lepidoptera. Virus particle are non-enveloped with diameters 35-

40nm. They were formerly named after the type species Nudaurelia-β-virus, one of a complex of viruses (α-eta) isolated from the pine tree emperor moth *Nudaurelia cytherea*. *H. armigera* stunt virus is also included in this family. The family includes two genera: *Betatetravirus* (type species: *Nudaurelia capensis betavirus*) and *Omegatetravirus* (type species: *Nudaurelia capensis omegavirus*).

6.2.1. Baculoviruses

Viruses in the family Baculoviridae are the best known of all the insect viruses. This is because they are very common in some of our most important insect pests, disease symptoms are easily recognized and they have the most potential for development as biopesticides. The major reason for interest in baculoviruses is their environmental safety. They are often, genus or species specific. Consequently, they are safe to all non-target organisms including humans, animals, plants and even beneficial insects. Baculoviruses are infections *per os* (by mouth) and exhibit efficient horizontal transmission. They replicate rapidly and cause extensive cell and tissue destruction in the host body and thus are virulent mortality agents. They are capable of causing natural epizootics and infecting a larger proportion of insects in a population. The occlusion bodies show great persistence in water and other carriers, as well as in soil. This enables formulation and application of virus-based pesticides just like chemical pesticides. On the other hand, the cost of development and registration is low as compared to chemical pesticides. Baculoviruses can be mass produced by simple cottage industries in contrast to chemical pesticides which require sophisticated industrial plants for manufacturing and formulation (Miller, 1998; Arora *et al.*, 2000b; Battu *et al.*, 2002; Harrison and Hoover, 2012).

6.2.1.1 Classification of Baculoviridae

The baculoviruses have undergone a series of reorganizations and changes in status and nomenclature. As per a prior classification scheme, the family *Baculoviridae* comprised of two genera: *Nucleopolyhedrovirus* containing all the NPVs and *Granulovirus* containing all the GVs. But as per the current classification (Jehle *et al.*, 2006) accepted by the International Committee on Taxonomy of Viruses (ICTV), the *Baculoviridae* are classified into following four genera (http://www.ictvonline.org/ index/asp).

Alphabaculovirus: This genus contains NPVs that infect Lepidoptera and the type species is *Autographa californica multiple nucleopolyhedrovirus* (AcMNPV).

Betabaulovirus: The genus consists of GVs, all of which have been isolated from Lepidoptera and the type species in *Cydia pomonella granulovirus* (CpGV).

Gammabaculovirus: This genus consists of NPVs from *Neodiprion* spp. of sawflies and the type species is *Neodiprion lecontei nucleopolyhedrovirus* (Nele NPV). The genomes of gammabaculoviruses that have been sequenced to date tend to be significantly smaller than those of other baculoviruses and do not contain any genes encoding a BV envelope fusion protein, something found in baculoviruses from the other genera.

Deltabaculovirus: The genus includes the NPV isolated from the mosquito *Culex nigripalpus*. The viral occlusion protein matrix of this virus is significantly larger than those of the viruses from the other three genera, and appears to share no significant amino acid sequence identity with other polyhedrins and granulins. The type species for the family is *Culex nigripalpus nucleopolyhedrovirus* (Cuni NPV).

6.2.1.2. Life Cycle of Baculoviruses

Initial infection by baculoviruses generally occurs when host larva picks up viral occlusions from the environment while feeding. The occlusion bodies dissolve rapidly in the alkaline digestive juices of the midgut. It has been suggested that OB dissolution is mediated by alkali-accelerated breaking of disulphide bonds of polyhedrin trimers and the ionization of tyrosine side chains at an alkaline pH (Ji *et al.*, 2010). The released occlusion derived virions (ODVs) together with the ingested food are contained within the peritrophic matrix (PM) which protects the midgut epithelium from abrasion by the food particles and also restricts access of ingested pathogens and toxins to the epithelium.

The first barrier for the virions to cross in order to reach the midgut epithelium is the PM. It has been demonstrated that many baculoviruses secrete a protein termed enhancin to faciliate crossing the PM. The enhancins act by degrading the insect intestinal mucin, a major glycoprotein of the PM, and also facilitate viral entry into midgut cells (Wang and Granados, 1997). After passing through the PM, the released ODVs fuse with the microvilli forming the brush border of the midgut epithelium and pass through into the columnar cells (Horton and Burand, 1993) and eventually into the nucleus through the nuclear pores (Summers, 1971; Ohkawa *et al.*, 2010). The virus undergoes several rounds of replication in the nucleus. After assembly in the nucleus, nucleocapsids exit the nucleus, travel to plasma membrane, thus producing the budded virion phenotype. For alphabaculoviruses and most betabaculoviruses, budded virus serves as a vehicle to spread infection to other tissues in the host. In the remaining baculoviruses, the infection does not spread beyond the midgut (Harrison and Hoover, 2012).

The secondary spread of the virus beyond the midgut epithelial layer is not easy to explain because separating the midgut epithelial layer from the rest of internal organs is the basal lamina, a fibrous matrix of glycoproteins secreted by the epithelial cells, which might serve as an effective barrier against microbial infections. This protective basal lamina surrounds all the internal tissues except the haemocytes. As the budded virus is synthesized and passed out of the midgut epithelial layer, it collects along the basal lamina. There is evidence that the passage of budded virus to other tissues occurs via the tracheal matrix. The tracheal system consists of a continuous series of tubes that act to deliver the oxygen to the organism's internal organs. These tubes penetrate the basal lamina of the organs and tissues where they supply oxygen. The terminal tracheloar cells termed tracheoblasts that service larval midgut cells have been reported to be the secondary targets of infection in baculovirus-infected hosts, and systemic infection of larvae spreads from the tracheoblasts to epidermal cells in tracheal trunks and on to other tissues in the haemocoel. Infection of other tissues is associated with the tracheae of those tissues, indicating that baculoviruses use the tracheal system as a conduit to bypass the basement membrane (Engelhard *et al.*, 1994; Harrison and Hoover, 2012).

As infection spreads along the tracheal epithelium from the foci of infection, the virus gains access to various other tissues such as the epidermis and the fat body. The appearance of virus within the fat body and epidermis indicates that *in vivo* spread of the virus is almost complete and the larva will soon succumb to the infection. Ultimately the ODVs are occluded when polyhedrin or granulin condense around mature, enveloped virions to form a crystalline matrix (Arnott and Smith, 1968).

The phenotypic effects of the virus are not manifested in the larvae during the initial few days after infection. However, several days following ingestion of the virus the larvae show symptoms of both behavioral and morphological changes. The infected larvae exhibit feeding inhibition and the larval period is prolonged. In some species, the infected larvae migrate to the apical parts of the plant and die (Hofmann, 1891; Vasconceleos *et al.,* 1996). The dead larvae can be seen hanging upside down by their prolegs. The virus usually undergoes several generations of replication within the host resulting in the production of large quantities within the larvae. As a result, the larva appears swollen, with a glossy cuticle. The haemolymph contains large amounts of virus giving it a cream colour. During the terminal stages the cuticle turns dark. By the time the larva stops feeding completely, other signs such as cessation of frass production, loss of turgor pressure and darkening of the cuticle become apparent indicating infection, which will result in the mortality of the larvae. In many species, when the infection is advanced, the larva takes the shape of a loose sack containing a mass of liquefied infected tissues. Eventually, the host cells lyse releasing the virus and the cuticular wall of the larva becomes fragile. When the thin body wall bursts the progeny OBs are released into the environment. Lepidopteran NPVs and some GVs encode a chitinase and a cathepsin protease that are required for the characteristic wilting and dramatic "melting" of the host larva after death (Hawtin *et al.,* 1997).

6.2.1.3. Host Range

Most baculoviruses infect only a single or a few closely related species. As a result, these are safe to all non-target organisms including natural enemies and pollinators. As a consequence, secondary pest out breaks and pest resurgence are avoided. However, baculoviruses are unsuitable for use in situations where two or more pest species need to be simultaneously controlled. This also limits their commercial exploitation except in case of pest species which are of great economic significance. Some of the baculoviruses developed as microbial pesticides have a broader host range e.g. *Autographa californica* MNPV, *Anagrapha falsifera* NPV and *Mamestra brassicae* NPV. But only a few of the pest species are susceptible at concentrations that translate to economically feasible application rates in the field (Moscardi, 1999).

6.2.1.4. Field Stability and Persistence

Baculoviruses are highly susceptible to damage by exposure to sunlight and/or U.V. light (Young and Yearian, 1974). Crude, unrefined and unformulated baculovirus preparations have been found in some cases to exhibit a degree of persistence and performance in the field equal or superior to refined, formulated baculovirus applications, possibly owing to absorbance of UV light by the insect parts in the crude preparation (Cherry *et al.*, 2000; Tadeu *et al.*, 2002). Compounds that reflect or absorb UV light have been tested and used in baculovirus formulations to prolong persistence. Various additives/ adjuvants like charcoal, India-ink, egg-albumin, mollases, optical brightners, etc have been suggested to be used alongwith formulations of various entomopathogens including baculoviruses (Angus and Luthy, 1971). Certain substrates such as boric acid (Morales *et al.*, 1997), chitinase (Shapiro *et al.*, 1987), extracts of neem tree (Cook *et al.*, 1996) and optical brighteners of the stilbene group have been shown to enhance baculoviruses activity. Mixtures of baculovirus with optical brighteners of the stilbene group seem to have excellent potential for use in formulated products because they can enhance viral activity at concentrations as low as 0.01 per cent, reduce killing time, and provide protection against UV solar radiation (Shapiro, 1992; Shapiro and Argauer, 1997). These substances have also been reported to enhance the infectivity of baculoviruses as evidenced by a significant reduction in LC_{50} in bioassays. The degree of reduction in LC_{50} mediated by optical brightners has depended on the specific virus-host pair, the optical brightners used, the larval instar tested, and the assay procedure, with values ranging from 2.6 to approximately 5731-fold (Lasa *et al.*, 2007; Harrison and Hoover, 2012).

6.2.1.5. Role in Insect Pest Management

The benefits of baculoviruses as means to control pests stem from their host specificity and environmental safety. But the narrow host range also limits their utility. The commercially available viral pesticides are listed in Table 1 and selected examples of utilization of baculoviruses in pest management are discussed hereunder.

a. Management of velvetbean caterpillar with AgNPV

The velvetbean caterpillar, *Anticarsia gemmatalis* is the key pest of soybean in the America, ranging from north-central areas of Argentina to the southeastern USA. In Brazil, it is the most important defoliator in all soybean producing regions, demanding an average of two insecticide applications for its control during the season.

In the 1970s, a nucleopolyhedrovirus AgMNPV, was isolated in different regions of Brazil. Initial field trials with AgMNPV indicated its potential in management of *A. gemmatalis* (Allen and Knell, 1977; Moscardi *et al.*, 1981). In pilot scale field trials with AgNPV at 50 LE/ha, reduction of over 80 per cent in larval populations was obtained and yields were on a par with pesticide treated plots. The development and implementation of AgMNPV biopesticide programme was carried out by *Embrapa* (Brazilian Agricultural Research Corporation). Implementation of the programme began in the 1982-83 season, when *ca.* 2000 ha of soybean crop were treated. Initially, small amounts of AgMNPV were produced in host larvae reared on artificial diet at *Embrapa* soybean facilities. Virus use gained momentum with the development of a WP formulation in 1986 and by 1990 one million hectares were treated every year. From 1990 on Embrapa signed contracts with five companies for commercial production of AgMNV. The covered area increased to approximately 1.5 million ha in 1995, with most of the AgMNPV being produced in the field during each season (Moscardi, 1999). Field production of the virus became a big business, involving different persons and small companies that specialized in selling AgMNPV-killed caterpillars to the private companies that registered the product for commercialization. During 2002-03 season, approximately 45t of AgMNPV-killed caterpillars were collected and sold to private companies (US$ 10-12/kg), representing about 2.0 million ha – equivalents of the biological insecticide. This material was stored at 10ºC for further processing and formulation (Moscardi, 2007).

Despite its value for producing high amounts of AgMNPV at low cost, field production presented some problems that restricted the expansion of its use or affected the quality of the end product. Owing to pressure on virus field producers, the procedure for collection of dead larvae in the field shifted from hand picking to shaking plants over pieces of cloth placed over the ground in between soybean rows. This shift resulted in collection of dead larvae, live host larvae, collection of other insects, and debris, resulting in material with higher amounts of extraneous organic matter other than AgMNPV-killed caterpillars. Therefore, research efforts were focused on improving the commercial laboratory production methodology. Research efforts helped to reduce the diet cost and improve the quality of the product. A pilot laboratory was established in 2003 followed by a commercial laboratory in 2004. The latter had a capacity to inoculate 800,000 to 1,000,000 larvae/day, resulting in AgMNPV to treat 1.8 to 2.0 million ha/yr. Currently, this is the largest laboratory facility worldwide for the production of a viral pesticide (Moscardi, 2007).

The use of AgMNPV in Brazil has generated substantial economic, ecological and social benefits. The current annual savings in terms of reduced cost of pest control amount to US$ 14 million. More importantly since beginning of the programme, 23 million litres of chemical pesticide use was avoided due to use of AgMNPV. The programme has also contributed to socio-economic improvement of poor families in AgMNPV field-production region (Moscardi, 2007).

b. Management of Codling moth with CpGV

The codling moth *Cydia pomonella* is a worldwide pest of a number of crops, including apples, pears and walnuts. Standard commercial management practices consist of several sprays per season of synthetic pesticides. The *C. pomonella* granulovirus (CpGV) was first found in apple and pear orchards growing near Chihuahua, Mexico in 1964. The pathogensis of CpGV is similar to that of other baculoviruses associated with lepidopterans. One outstanding characteristic of CpGV is its extremely high virulence as revealed by the LD_{50} value of 1.2 granules/larva for neonates (Sheppard and Stairs, 1977). The CpGV was commercially produced in 1980 as SAN 406 by Sandoz Inc. and exempted from tolerance by USA EPA in 1981. Several research and development projects were conducted in European countries in the 1980s. Over the years, CpGV has been registered in a number of countries under various trade names and is used to control codling moth on over 100,000 ha in Europe (Asser Kaiser *et al.,* 2007). Populations of codling moth in some areas of Germany, France and other parts have been reported to have developed resistance to CpGV from 2003 onwards. Resistance levels up to

100,000 fold have been reported from Germany. Screening of CpGV isolates from other areas has led to the identification of isolates which overcome the virus resistance in the pest (Berling *et al.*, 2009).

Table 1. Examples of commercially available viral pesticides registered for pest control in different countries.

Host insect	Crop(s)	Commercial name(s)	Country
(1)	(2)	(3)	(4)
Granuloviruses			
Adoxophyes orana (Fisher von Roesberstamm)	Apple	Capex 2	Switzerland
Adoxophyes sp.	Tea	-	Japan
Agrotis segteum (Denis & Schiffermuller)	-	Agrovir	Germany, China
Cydia pomonella (Linnaeus)	Apple, pears	Carpovirusine CYD-X Granusal Virin-GyAp Madex	France USA Germany Russia Czechoslovakia
Erinnyis ello (Linnaeus)	Cassava	-	Brazil
Homona magnanima Diakonoff	Tea	-	Japan
Phthorimaea operculella Boyd	Potatoes	PTM Baculovirus	Peru, Egypt, Tunisia
Pieris rapae (Linnaeus)	Cabbage	-	China
Plodia interpunctella (Hubner)	Stored almonds, raisons	-	USA
Plutella xylostella (Linnaeus)	Cabbage	-	China
Nucleopolyhedroviruses			
Anagrapha falcifera (Kirby)*	Cotton, vegetable	-	USA
Anticarsia gemmatalis (Hubner)	Soybean	Baculoviron, Baculovirus Nitral, Coopervirus, Protege	Brazil
Autographa californca (Speyer)*	Cabbage, cotton, ornamentals	VPN 80	Guatemala
Buzura suppressalis (Guenee)	Tea, tung oil tree	-	China
Helicoverpa armigera (Hubner)	Cotton, tomato	Virin-HS Helicovex VIVUS MAX	Russia Europe Australia

Contd...

(1)	(2)	(3)	(4)
Helicoverpa zea (Boddie)	Cotton	Elcar, Gemstar	USA
Hyphantrea cunea (Drury)	Forest, mulberry	Virin-ABB	Russia
Lymantria dispar(Linnaeus)	Forests	Gypchek	USA
		Disparvirus	Canada
		Virin-ENSH	Russia
		-	China
Mamestra brassicae (Linnaeus)*	Cabbage	Mamestrin	France
		Virin EKS	Russia
Neodiprion abietis	Balsamfir	Abietev	Canada
Neodiprion lecontei (Fitch)	-	Lecont virus	USA, Canada
Neodiprion sertifer (Geoffroy)	-	Monisarmiovirus Virox	Finland, UK
Orgyia pseudotsugata (McDummough)	Forests	TB Biocontrol 1	USA
Spodoptera exigua (Hubner)	Ornamental, vegetables	SPOD-X	USA
	Garden pea, grapes, ornamentals	-	Thailand
Spodoptera frugiperda (J.E. Smith)	Maize	-	Brazil
Spodoptera littoralis (Boisduval)	Cotton	Spodopterin	Africa
Spodoptera litura (Fabricius)	Vegetables, cotton, rice, peanuts	-	China
Spodoptera sunia (Guenee)	Vegetables	VPN 82	Guatemala

* These viruses have been developed primarily against lepidopteran species other than original hosts because of their wide host range.

Source: Modified after Moscardi (1999)

c. Sawflies and other forest pests

The larvae of many species of sawflies feed in pine trees. The NPVs of several species of sawflies in the forest ecosystem seem particularly suitable for long-term control. *Neodiprion sertifer, N. lecontei, N. swainei* and *Gilpinia hercyniae* have been successfully controlled with one or two introductions of their respective NPVs into field populations (Cunningham, 1982). Short term control of other species of sawflies e.g. *N. abieties* and *N. taedai linearis* is also obtained by application of respective NPVs against early instars.

Forest Lepidoptera subjected to the applications of baculoviruses include the Gypsy moth *Lymantria dispar*, the Douglas fir tussock moth, *Orygyia pseudotsugata*, the spruce budworm *Choristoneura fumiferana*, the Western spruce budwom *C. occidentalis*, the jackpine budworm *C. pinus*, the pine beauty moth *Panolis flammea* and the fall webworm *Hyphantra cuncea*. Several of these baculoviruses are now used commercially in USA, Canada and other countries (Huber, 1986).

d. Pests of plantation crops

A non-occluded virus (Or NV) of the rhinoceros beetle *Oryctes rhinoceros*, a major pest of coconut palms, has been successfully used in pest control in South-east Asia and the Pacific since 1967. Adults from the field or those reared in the laboratory are infected by immersing them in a viral suspension and then allowed to crawl through sawdust mixed with virus. These infected adults are then released in the field to disseminate the virus. OrNV has been the most successful classical microbial control agent and became established at every place it was released and where post-release reports were available.

e. The *Helicoverpa / Heliothis* complex

The genera *Helicoverpa* and *Heliothis* contain some of the most destructive pests of agricultural crops including *H. armigera, H. assulta, H. zea* and *Heliothis virescens*. Between 1960 and 1980, over 200 field trials were conducted in various countries for the control of these pests on different crops. Results were best in soybean and chickpea, intermediate in maize and sorghum, but success was less evident in cotton (Burges, 1981).

The first baculovirus to be developed for commercial use was Elcar, the NPV of *H. zea* registered by The United States Environmental Protection Agency (USEPA) in 1975. Elcar was infective against all the major *Helicoverpa / Heliothis* spp. and provided efficient control in soybean, sorghum, maize, tomato, chickpea and navybeans. Due to reduced interest production was stopped in 1982. With renewed interest in baculoviruses, Biosys introduced Gemstar LC, a liquid concentrated formulation of HzSNPV in 1996. *Hear*SNPV isolates produced for *H. armigera* control are marketed under trade names Helicovex and VIVUS MAX in Europe and Australia, respectively (Buerger *et al.,* 2007). Of the 11 baculoviruses registered as commercial insecticides in China, *Hear*SNPV is produced in greatest quantity (Sun and Peng, 2007). *Hear*SNPV is also commercially produced in India for control of *H. armigera* on variety of crops (Cherry *et al.,* 2000).

f. Pests of stored products

The GV of the Indian meal moth, *Plodia interpunctella* has the potential for long-term control in storage bins. A freeze dried, powdered formulation of PiGV with a relatively long shelf life was developed (Cowan *et al.,* 1986). This formulation could be applied as a spray or dust and the cost of application was comparable to conventional fumigant treatment (Vail and Tebbets, 1991).

6.2.1.6. Genetic Improvement of Baculoviruses

To improve the efficiency of viral insecticides, focused on increasing virulence, expanding host range, and enhancing photostability, several strategies comprising both conventional (strain selection, serial passage and mutation) as well as DNA recombinant technology have been employed. The genetic engineering has been used for expressing physiological effectors (hormones, enzymes or antisense), interrupting the normal metabolism of the insect, expressing insect specific toxins such as neurotoxin and Bt toxins or deleting baculoviral genes that may interact with the reaction of the viruses to the insect (Wood and Hughes, 1996; Bonning *et al.,* 2003; Harrison and Hoovar, 2012; Arora and Shera, 2014).

An old fashioned methodology of cross infectivity testing of baculoviruses that includes screening of long recognized strains for possible encroachments on generic, tribal, sub-familial, familial, or even ordereal taxonomic frontiers in their host relationships may prove to be highly rewarding. An isolate of the spruce budworm *Choristoneura fumiferana* NPV was fed to the neonate cabbage looper *Trichoplusia ni* and the wax moth *Galeria melonella* larvae. After passage, the virus was able to infect new hosts (Stairs *et al.,* 1981). Wood *et al.* (1981) could isolate a mutant strain of AcMNPV designated HoB, which produced a large number of occlusion bodies in infected cells and which had higher virulence in insects than the parent strain. Smirnoff (1961) selected a strain of an NPV of *N. sertifer* with increased virulence from the first larvae to die after inoculation. Insect viruses have also been selected for resistance to UV radiation. One such strain of granulovirus of codling moth *Cydia pomonella* (CpGV) was 5.6 times as resistant to UV light as the original isolates and remained infective for twice as long in the field (Brassel and Benz, 1979).

Genetic engineering methods have been widely deployed to produce baculoviruses with improved insecticidal activity since first proposed by Miller *et al.* (1983). Recombinant baculoviruses have been assembled that express neurotoxic peptides from the venoms of scorpions, mites, spiders, sea anemones and ants (Szolajska *et al.,* 2004). When active toxin is produced, infection of larvae with recombinant viruses expressing these toxins resulted in cessation of feeding, paralysis and a reduction in mean or median survival times of larvae by approximately 10 to 60 per cent compared to wild-type virus (Inceoglu *et al.,* 2006). Over expression of genes for diuretic hormones, modified juvenile hormone esterases and a pheromone biosynthesis activating neurohormone also resulted in reduced survival time as compared to the wild-type viruses (Harrison and Hoover, 2012).

Field tests with recombinant HzSNPV and *Hear*SNPV clones expressing genes for insecticidal proteins have demonstrated that these viruses provided greater plant protection as compared to that provided by wild-type viruses (Smith *et al.,* 2000; Sun *et al.,* 2009). However, the recombinant baculoviruses produced reduced quantity of progeny OBs which may create problems in field perpetuation as well as commercial production of the virus.

6.3. BACTERIAL ENTOMOPATHOGENS

As with other living organisms, insects are intimately associated with bacteria at all stages of their lives. Entomopathogenic bacteria consist of three major divisions based on the presence or structure of the cell walls: bacteria with a Gram-negative type cell wall (Gracilicutes), Gram-positive type cell wall (Firmicutes), and Eubacteria lacking a cell wall (Tenericutes). The important groups of bacterial entomopathogens are discussed here in.

6.3.1 Genus *Bacillus* (Family: Bacillaceae, Order: Bacillales, Class: Bacilli, Phylum: Firmicutes)

This genus contains catalase-positive bacteria with peritrichous flagella and rod-shaped cells that may be arranged in chains. Under adverse conditions they undergo sporulation, which is not repressed by air, to form an oval-shaped endospore that in some species may be accompanied by parasporal bodies. Genus *Bacillus* contains some of the most important entomopathogenic species known so far.

***Bacillus Thuringiensis* Berliner**

Bacillus thuringiensis (Bt) is a motile, Gram-positive, spore forming, phyloplane and soil-inhabiting bacterium and in addition to endospores, produces a proteinaceous parasporal crystal in the sporangium at the time of sporulation. The formation of crystal proteins (usually referred to as Cry proteins) is an essential feature that distinguishes Bt from very closely related species *B. cereus*. Bt is widely distributed in nature, and is being discovered repeatedly from different geographical areas and habitats frequented by varied complex of class Insecta. Dead insects, litter of sericultural farms, soil, water, dust, etc., form the so-called resource materials for the world-wide isolation of this bacterium. There are only a few reported cases of natural epizootics of *B. thuringiensis* in lepidoteran pests. Nevertheless, it is the most widely exploited microbial control agent.

B. thuringiensis was first isolated in 1901, from infected silkworms *Bombyx mori* by the Japanese bacteriologist Shigetane Ishiwata. He called the organism "*Sottokin-Bacillus*" which translates into "Sudden death-Bacillus" but this name did not last. The bacterium was rediscovered by the German biologist Ernst Berliner, who isolated it in 1911, from the infected larvae of Mediterranean flour moth *Ephestia kuehniella* (Zeller), collected from a mill in the province of Thuringe and named it *Bacillus thuringiensis* (Berliner, 1915). There have been several excellent reviews on completion of a century of discovery and naming of Bt (Nester *et al.,* 2002; Sanahuja *et al.,* 2011; Sanchis, 2011).

The first attempts to use Bt for insect control took place in 1920s against the Gypsy moth, *Lymantria dispar* (L.) in the United States and against the European corn borer, *Ostrinia nubilalis* (Hubner) in easteran Europe (Sanchis, 2011). The first commercial *B. thuringienis* product, Sporeine was produced in France in 1938, by Laboratoire Libec (Entwistle *et al.,* 1993). However, it did not become very popular. In 1970, Howard Dulmage isolated a potent strain of *B. thuringiensis* var. *kurstaki* from diseased mass reared larvae of pink bollworm, *Pectinophora gossypiella* (Saunders), and designated it the HD-1 strain (Dulmage, 1970).

6.3.1.1. Classification of *B. thuringiensis*

Diverse biochemical, morphological and antigenic methods have been used to describe and classify new Bt isolates. Early efforts concentrated on serological and biochemical methods, including morphological and biochemical typing, esterase pattern of vegetative cells and serotyping of the H flagellar antigen in vegetative cells. (Jurat-Fuentes and Jackson, 2012). Initially, gross correlation between classifications of Bt isolates based on biochemical characters and H serotyping with pathotypes was noted (Dulmage *et al.,* 1981), but this relation became more complicated as the number of Bt isolates with novel activities increased dramatically (de Barjac and Frachon, 1990).

While pathogenicity cannot be considered a reliable taxonomic criterion, the distinction of Bt strains according to their pathogenicity (pathotype) is useful for practical purposes considering that the primary interest in Bt is its ability to kill insects. However, the ability to discriminate among Bt strains based on pathotype is greatly challenged by the production of multiple toxins with diverse specificities by individual isolates and the increasing number of pathotypes described. Thus most Bt serovars include strains that are pathogenic against insects in diverse taxonomic orders. Since toxicity is correlated with production of toxins, identification of toxins produced by a Bt strain, may be used to predict the range of activity more accurately (Jurat-Fuentes and Jackson, 2012).

Early attempts at Cry toxin classification were based on specificity and gene sequence similarities (Hofte and Whitely, 1989). These authors proposed a nomenclature keeping the Cry epithet (for designating crystal), followed by a Roman numeral indicating the insect class against which the encoded toxin is active, and by two alphabetic designations to indicate relative amino acid homology. Under this approach, Cry I toxins were active against lepidopteran larvae, Cry II had dual specificity against lepidopteran and dipteran larvae, Cry III were pathogenic to coleopteran larvae and Cry IV specific for dipteran larvae. Discrepencies between insecticidal specificity and sequence similarity as the number of described Cry toxins increased, led to problems with this nomenclature. To resolve these problems, a Bt Toxin Nomenclature Committee was formed in 1993 and it proposed a revised nomenclature procedure to overcome limitations of the previous method (Crickmore *et. al.*,1998). The revised nomenclature system is based solely on amino acid homology and is currently used to name and classify new crystal toxins. Toxin names are assigned according to degree of evolutionary divergence from other toxins based on phylogenetic trees, which allows for grouping of closely related toxins. Roman numerals have been substituted for Arabic numerals for the first rank, assigned for toxins displaying at least 45 per cent sequence identity. The secondary rank is represented by a upper case English language alphabet which groups toxins displaying up to 75 per cent amino acid sequence identity. The tertiary rank displayed by a lower case English language alphabet identified toxins displaying 75-95 percent sequence identity. The quaternary rank, denoted by an Arabic numeral, distinguishes toxins that are more than 95 per cent identical, is optional and is assigned to independently sequenced genes currently.

This classification also includes families of highly related crystal proteins produced by bacteria other that Bt: Cry 16 and Cry 17 from *Clostridium bifermentans*, Cry 18 from *Paenibacillus popilliae*, Cry 43 from *P. lentimorbus* and *P. popilliae* (Jurat-Fuentes and Jackson, 2012). The updated toxin lists, toxin dendrograms and further details are available at the Bt toxin Nomenclature website (http://www.lifesci.susssex.ac.uk/home/Neil_crickmore/Bt/).

Despite their sequence diversity, all Cry proteins share a similar overall tertiary structure, as exemplified by the six structures solved thus far by X-ray crystallography (Cry1Aa, Cry2Aa, Cry3Aa, Cry3Bb, Cry4Aa and Cry4Ba). The C-terminal portion is involved in crystal formation and is not part of the activated toxin, as it is cleaved off in the insect gut.

The N-terminal portion is the toxin itself, and it comprises three domains. Domain I is a bundle of seven helixes, six of which are amphipathic encircling the seventh hydrophobic helix, and this domain is responsible for membrane insertion and pore formation. Domain II consists of three anti-parallel β-sheets with exposed loop regions, and domain III is a β-sandwich. Both domains II and III confer receptor binding specificity thus helping to define the host range (Boonserm *et al.*, 2006; Sanahuja *et al.*, 2011).

6.3.1.2. Life Cycle and Mode of Action

A generalized Bt infective cycle commences with the ingestion of spores and /or crystalline inclusions by a susceptible host. After ingestion, cessation of feeding and paralysis follows in the host, which can be reversible in some species. Disease becomes irreversible with increased hemolymph pH after toxins increase the permeability of the midgut epithelial barrier.

The insects susceptible to Bt have been classified into 3 types: Both type I and type II insects suffer from midgut paralysis within a few minutes after infesting the crystal. In type I insects this continues to a general paralysis correlating with an increase in pH and potassium concentration in the hemolymph. The examples of type I insects include the lepidopterans, *B. mori*, *Anthraea pernyi*, *Philosamia ricini,* and the dipterans *Simulium vittatum* and *Aedes aegyptii*. Type II insects do not develop general paralysis and die without much hemolymph pH change. The examples include *Pieris rapae, Lymantria dispar, Ephestia Cautella* and *Corcyra Cephalonica*. The type III hosts are not killed by crystals alone and require the presence of spores for mortality. *Galleria mellonella* and *Anagastra kuehniella* are examples of the type III group (Jurat-Fuentes and Jackson, 2012).

The mode of the action of *Bt* is complex and there are some differences in the action of different subspecies. In the most studied case of Bt *kurstaki* the crystals of *delta*-endotoxin are stabilized, by a high number of disulphide bridges between individual proteins, within each crystal. Species highly susceptible to Bt e.g. *Papilio demoleus* have a high midgut pH (9.7-11.0), very large quantity of the reducing substances and high proteolytic activities in the midgut. In contrast resistant species e.g. *Spodoptera litura* have a lower midgut pH (8.2–8.5), small amount of reducing substances and low proteolytic activity. A highly reductive environment in the insect midgut breaks the inter- protein disulphide bonds easily, liberating the full length protoxins of 130 kDa (Adams *et al.*, 1996).

Once the large soluble protoxins are free of crystal structure, the larval midgut proteolytic enzymes begin their action. The primary proteolytic components of most lepidopterans are serine proteases, particularly trypsin and chymotrypsin like. These enzymes cleave about 29 amino acids from the N-terminus and about 500 amino acids from the C-Terminus. The final toxin core often referred to as the trypsin resistant fragment is only half (65-67Kda) the protoxin size (Adams *et al.*, 1996).

The peritrophic membrane acts like a molecular sieve allowing free flow of digestive enzymes and nutrients up to molecular mass of about 100 kDa. Large molecules containing microbial contaminants present on the leaf surface are retained by the peritrophic membrane and are eliminated in larval frass. The activated Bt toxin easily diffuses across the peritrophic membrane to interact with the gut epithelium. These is a ballooning, exfoliation and breakdown of the insect midgut epithelium.The ultrastructural changes in the affected midgut cells are an enlargement of the nuclei, alteration of the endoplasmic reticula to vacuole-like configurations, the disintegration of microvilli and the deformation of the basal infoldings of the epithelium. It has been revealed that the toxin binds to specific receptors on the midgut epithelial cells. The attached toxin generates small pores in the cell membrane. The formation of pores causes the loss of semi-permeability of the plasma membrane, resulting in a net inflow of ions accompanied by the influx of water, cell swelling and cell lysis. The receptor sites for different endotoxins vary in structure and the intensity of binding may also differ for different crystal proteins. This may account for the variation in host specificity of different subspecies/strains of Bt (Adams *et al.*, 1996).

During recent years Broderick *et al* (2006, 2009) have presented evidence that Bt does not kill the larvae of Gypsy moth in the absence of gut bacteria. Elimination of midgut microbial community by oral administration of antibiotics abolished Bt insecticidal activity, and reestablishment of an *Enterobacter* sp. that normally residues in the midgut microbial community restored *Bt*-mediated killing. *Escherichia coli* engineered to produce Bt insecticidal toxin killed Gypsy moth larvae irrespective of the presence of other batceria in the midgut. However when the engineered *E. coli* were heat killed and then fed to the larvae, the larvae did not die in the absence of the indigenous midgut bacteria. *E. coli* and *Enterobacter* sp. achieved high population in the haemolymph in contrast to Bt, which appeared to die in haemolymph. Bt induced mortality, thus, required the presence of enteric bacteria.

However, subsequent investigations questioned the methodology used and verified that undesirable effect of the antibiotic treatment (Raymond *et al.*, 2009) and differences in gut flora composition or bioassay methodology (Van Frankenhuyzen *et al.*, 2010) explained the unexpected results. Dependence of Bt pathogenicity on alternate gut bacteria is also unlikely given the observation of gut flora inhibiting Bt pathogenicity (Hernandez-Martinez *et al.*, 2010) and enhancement of Bt virulence by antimicrobials.

6.3.1.3. Commercialization

The first attempts to use *B. thuringienesis* var. *thuringienesis* for insect control took place in 1920s against the Gypsy moth *Lymantria dispar* in the northeastern United States and against the European corn borer *Ostrinia nubilalis,* in Hungary. The first commercial Bt product 'Sporeine' was produced in France by Laboratoire Libec (Entwistle *et al.*, 1993). However, due to the advent of DDT, further progress in this direction was halted.

Edward A. Steinhus, Insect Pathologist at the University of California, Berkeley, is credited with the resurgence of interest in Bt. In 1956, Steinhaus convinced the President of the Pacific Yeast Products Company of Wasco in California to produce Bt and the product 'Thuricide' was soon available. A number of such products were also produced in USSR and other European countries but all the products had moderate efficacy and were not very successful. In 1970, Dulmage isolated another more potent strain of *B. thuringienesis* var. *kurstaki* and designated it as HD-1 strain. HD-1 proved be 2 to 200 times more toxic against key agricultural pests and became the basis for products that were competitive with chemical insecticides in performance and cost. This strain became commercially available through Abbot laboratories as 'Dipel' in the early 1970s. Expansion in the use of Bt in the 1970s and 1980s, was largely due to the development of methods of large scale fermentation, and increased efficiency in production and quality control, so that formulations with high activity could be developed (Table 2) (Sanchis, 2011). Bt products became the first choice for management of forest pests in USA and Europe where application of conventional chemical insecticides was highly restricted. During the 1970s and 1980s, its use against the spruce budworm *Choristoneura fumiferana* and the Gypsy moth in North American forests accounted for over 60 per cent of the world sales.

An illustrating example of the use of Bt in forestry was the control of an unexpected outbreak of the nun moth, *Lymantria monacha*, which attacked more than 600,000 ha of forests in Poland in 1994. An emergency programme in which several public and private institutions were involved, was put into practice by applying a special forestry formulation of Btk from Novo Nordisk called Foray 48B. The results were excellent with Bt providing about 95 per cent control of the pest while having a minimum impact on non-target and beneficial organisms (Rosell *et al.*, 2008). After forestry, organic farming especially of vegetable crops, was another niche area where Bt was widely used, for management of lepidopteran and coleopteran pests.

6.3.1.4. Genetic Improvement of *Bacillus thuringiensis*

The first step towards improving Bt strains naturally involved the isolation of new strains with new or higher insecticidal activity against targeted insect pests. The ability to identify and clone Bt Cry genes and the characterization of the specific activities of individual Cry proteins as well as the availability of recombinant DNA technology, led to the development of new strategies for improving the exploitation of Bt for increasing its entomopathogenic potential (Arora and Shera, 2014).

Amongst the critical milestones was the discovery that the genes coding for the toxin crystals were located on transmissible plasmids enabling exchange of genetic information between Bt Strains (Gonzales *et al.*, 1982). This opened up the way to manipulation of genes, including transfer between *Bt* strains. Conjugation was used to develop strains with optimized activity against a given insect pest or strains with broadened toxicity spectrum (Sanchis, 2000) Bt toxin genes have been cloned and expressed in wide variety of organisms (baculoviruses to cyanobacteria) as well in plants in attempts to improve their delivery and efficacy against insect pests. Several commercial products and novel strains of Bt have been developed through biotechnological interventions (Table 2).

Examples of combination pesticides produced by conjugation include Foil, a product based on EG2424, which produces Cry1Ac from Bt *kurstaki* (active against European corn borer *Ostrinia nubilalis*) and Cry3A from Bt *tenebrionis* (active against Colorado potato beetle *Leptinotarsa decemlineata*) (Carlton and Gawron-Burke, 1993). Ecogen constructed strain EG 2348, the active ingredient of the Condor® bioinsecticide, that contained a combination of Cry genes encoding crystal proteins particularly active against specific lepidopteran pests of soybean. But it is not always possible to introduce plasmids by conjugation, particularly if this results in the coexistence of two plasmids with similar origins of replication in the same cell, as they would be incompatible. In such cases, the relevant genes can be inserted on to a different plasmid in *Escherichia coli* using standard cloning methods and that plasmid can be introduced into an existing Bt strain by artificial transformation (Arantes and Lereclus, 1991).

Table 2: Natural and genetically modified Bt products registered for agricultural and medical/veterinary use

Bt strain	Company	Product	Target insect	Crop/Ecosystem
(1)	(2)	(3)	(4)	(5)
Bt *kurstaki* HD-1	Abbott Laboratories Chicago, IL USA	Naturally occuring *B. thuringiensis* Biobit XL, Dipel DF, ES 4L, 8L, 10G, SG pluse, Foray 48B, 48F, 76B	Lepidoptera	Field and vegetable crops, greenhouse orchard fruits and nuts, ornamentals, forestry, stored products
Bt *kurstaki* HD-1	Thermo Trilogy Corp. Columbia, MD USA	Steward, Thuricide, Vault	Lepidoptera	Vegetable crops, orchards, ornamentals, tea
Btk Strain SAH	Thermo Trilogy Corp.	Javelin, Delfin, 32LV	Lepidoptera	Field crops, vegetables, fruits, nuts, forestry
Bt *kurstaki*	Abbott	Bactospeine, Futura	Lepidoptera	Field crops, vegetables, fruits, forestry
Bt *kurstaki*	Thermo Trilogy	Able, Costar	Lepidoptera	Field crops, vegetables, fruits, nuts, forestry
Bt *aizawai*	Abbott	Florbac	Lepidoptera	Row crops
Bt *aizawai*	Abbott	Xentari	Lepidoptera	Field crops, vegetables, fruits, ornamentals, greenhouses
Bt *tenebrionis*	Abbott	Novodar	Coleoptera	Potato, tomato, eggplant, ornamentals shade trees
Bt *tenebrionis*	Thermo Trilogy	Trident, Trident II	Coleoptera	Potato, tomato, eggplant
Bt *kurstaki*	BioDalia, Dalia, Israel	Bio-Ti	Lepidoptera	Avocado, tomato, pine forests

Contd...

(1)	(2)	(3)	(4)	(5)
Bt *kurstaki*	Rimi, Tel Aviv, Israel	Bitayon (Granular feeding baits)	*Btrachedra amydraula*	Date palms, vegetable and field crops
Bt *galleriae*	Tutzcorin Alkali Chemicals & Fertilizer Ltd., India	Spicturin	Lepidoptera	Cruciferous crops
Bt YB-1520	Huazhong Agric. University, China	Mainfeng pesticide	Lepidoptera	Cotton, vegetables, maize, fruit trees,
Bt *chinensis* CT-43 (no flagella)	Huazhong Agric. University China	Shuangdu preparation	Lepidoptera, Diptera, Coleoptera	Cotton, forest, maize, gardens
Genetically Modified				
Aizawai recipient *kurstaki* donor, GC-91	Thermo Trilogy Corp.	Agree, Design Turex (transconjugant)	Lepidoptera, (Resistant *P. xyllostella*)	Vegetables
Kurstaki recipient, *aizwai* donor	Ecogen Inc. Langhorne, PA, USA	Condor, Cutlass (transconjugant)	Lepidoptera	Row crops
Bt *kurstaki* ED 7841	Ecogen Inc.	CRYMAX	Lepidoptera	Vegetables, horticultural, ornamentals
Bt *kurstaki* ED7826	Ecogen Inc.	Leptinox (recombinant)	Lepidoptera	Turf, hay, row, crops
Bt *kurstaki* EQ 7673	Ecogen	Raven (recombinant)	Lepidoptera, Coleoptera	Row crops, Potato , tomato, eggplant
Delta endotoxin encapsulated in *Pseudomonas fluorescens*	Mycogen Corp., San Diego, CA, USA	MVP, MATTCH, MTRACK (CellCap)	Lepidoptera, Coleoptera	Row crops, Potato, tomato, eggplant
Bt *israelensis*	Cynamid	Acrobe	Diptera	Medical pests
Bt *israelensis*	Novo-Nordisk	Skeetal	Mosquito	Ditches, swamps, lakes, ponds
Bt *israelensis*	Zoecon	Taknar, Teknar HPD	Black flies, mosquitoes	Ditches, swamps, lakes, ponds
Bt *israelensis*	Abbott	Vectobac G, Gnatrol, Bactimos	Mosquitoes, black flies, fungus gnats	Ditches, Swamps ponds, greenhouse, ornamentals, vegetables
Bt *israelensis* 187 and CS-8	Huazhong Agric. University	Jie Jue Ling Prep.	Diptera	Rice fields, ditches ponds

Source: Modified after Baum *et al.* (1999), Kaur, (2000)

A new type of biocide GCS-*BtA* based on "Germany-China Scientific Cooperation" research was developed by conjugation of *delta*-endotoxin from *B. thuringiensis* with abamectin, a toxin of *Streptomyces avermitilis* (ex Burg *et al.*) The conjugated GSCA-*BtA* had a broader host spectrum and a faster killing speed than either the Bt crystal or abamectin alone for the control of agricultural pests (Liu and Sengonca, 2003). GSCC-*BtA* has been recommended for use in the integrated pest control programmes in the vegetable crops as it proved highly effective not only against *Plutella xylostella* (Linnaeus), *Myzus persicae* (Sulzer) *Phyllotreta vittata* Fabricus, *Tretranychus cinnabarinus* (Boisd.), *Franklniella occidantalis* (Pergande), *Aphis fabae* Scopoli and *Cameraria ohridella* (Deschka &Dimic) but also displayed safety to some predators (Zhu and Sengonka, 2006)

The use of *B. thuringiensis* as insecticide is limited in field application because of the rapid inactivation of toxins and spores after exposure to sunlight (Pusztai *et. al.*, 1991). *B. thuringiensis* mutants producing melanin from *Btk* and other strains obtained by inducing mutagenesis utilizing ethyl methane sulphonate (EMS) were found to be more resistant to UV irradiation (Saxena *et al.*, 2002) and also showed highly insecticidal activities against potato tuber moth, *Phthorimaea operculella* (Zeller) (Mohamed *et al.*, 2010). *Bt* var. *tenebrionis* strain NB-125, a mutant strain NB-176 was developed by gamma irradiation which contained two to three copies of Cry 3A. The commercial product Novodor ® (Abbot) based on this mutant strain was able to produce large thromboid crystals composed primary of Cry 3A toxin thus enhancing the field activity (Gurtler and Petersen 1991).

Bt genes have also been cloned and expressed in endophytic, epidophytic, epiphytic and/or aquatic bacteria as well as eukaryotic plants to improve their delivery and residual activity. The rationale for using live endophytic or epiphytic bacteria as hosts is to prolong the persistence of Cry proteins in the field by using a host that can propagate itself at the site of feeding and continue to produce crystal proteins (Sharma, 2009).

Crop Genetics International (CGI) transferred *Cry1Ac* gene from *Btk* ND-73 into *Clavibacter xyli* var. *cynodontis* (Cxc) an endophytic bacterium that colonizes the vascular system of various plants including maize. The transgenic Incide® Cxc infects the internal tissues of growing plants and gene introduced into this bacterium encoded a protein toxic to the larvae of the European corn borer (Lampel *et al.*, 1994). Endophtic isolates of *B. cereus* have been used as hosts for *cry 2A* genes and *B. megaterium* isolate that persists in phylloshere has been used as host for *cry1A* genes (Mahaffe *et al.*, 1994; Bora *et al.*,1994). Plant colonizing bacteria such as *Pseudomonas fluorescens*, *P. cepaica* Burkholder, *Rhizobium leguminosaraum* Jordan *and Azopirillium* spp have also been used to produce and deliver Bt proteins (Udayasuriyan *et al.*, 1995). Two endophytes *Rhizobium meliloti* Dangeard and *R. leguminosarum* (Frank) were introduced with δ-endotoxin *cry 3A* gene from *Bt tenebrionis* for the control of two coleopteran insects *Sitona hispidulus* (Fabricus) *and S. linatus* (Linnaeus) respectively (Bezdicek *et al.*, 1991). A strain RSI identified as *Bacillus* sp. has been isolated, which is capable of colonizing in cotton leaves. It has been found to be an excellent colonizer of cotton phyllosphere. *Cry 1A* gene of *Bt* was introduced into it by conjugal transfer. It has been shown that the resultant transconjugant colonizes cotton plants for prolonged period and also protects the plant from *H. armigera* attack for more than 30 days (Narayanan, 2002).

A cloned *Bt* endotoxin gene (*cry 1 Ab*) has been expressed in the corn root colonizing Gram-negative bacterium, *P. fluorescens* to deliver lepidopteran active toxin beneath the soil surface for root feeding insects. The recombinant bacteria were subsequently killed by heat and iodine and thus the pseudomonad cell protected the *Bt* protein from environmental degradation, ensuring longer residual activity. The symbiotic relationship between pigeonpea, *Cajanaus indica* Adans and nitrogen fixing bacterium *Bradyrhizobium* has also been utilized for improving delivery system for soil insects. Nambiar *et al.* (1990) expressed diptern active *cry 4D* gene from *Bt israelensis* into *Bradyrhizobium* sp. that nodulates pigeonpea. The transfer of plasmid by conjugative mobilization into this species provided protection against root nodule damage by *Rivellia angulata* (Hendel) on pigenopea. Similarly, cloned gene from coleopteran active *Bt tenebrionis* have been transferred to *Rhizobium leguminosarum* by conjugation. Pea and white clover plants showed reduced root and nodule damage by larvae of *Sitona lepidus* Gyllen when inoculated with toxin genes containing *Rhizobium* strains (Skot *et al.*, 1990).

6.3.2. Genus *Lysinibacillus*

(Family: Bacilliaceae)

Lysinibacillus sphaericus (syn. *Bacillus sphaericus*) is a highly heterogenous species that contains both saprophytic and pathogenic strains (Ahmed *et. al.*, 2007). The bacterium is strictly aerobic and lacks genes encoding enzymes and transport systems to use sugars as a carbon source, so that amino acids and organic acids are used instead. The bacterium is ubiquitous and cosmo-politan in soil and soil-aquatic systems. It produces a round spore located in a terminal or subterminal position in a swollen sporangium. It has attracted particular attention due to its pathogenicity to mosquito larvae and its longer environmental persistence as compared to Bt var. *irraelensis* (Yap, 1990).

Flagellar H-serotyping (de Barjac *et. al.*, 1980) was used to classify *B. sphaericus* strains into 49 serotypes, of which 9 (H1, H2, H3, H5, H6, H9, H26 and H48) contain mosquitocidal strains. *B. sphaericus* strains 1593 and 2297 isolated from diseased *Culex* spp. mosquito larvae and strain 2362 from adult blackfly *Simulium* sp., increased interest in its commercial use. Strains 2362 and C3-41 are the most widely used as active ingredients in commercial *B. sphaericus* products (Jurat-Fuentes and Jackson, 2012).

Susceptibility to *B. sphaericus* is highly variable depending on the strain and the mosquito species and is due to production of toxins. A generalized order of susceptibility is *Culex* > *Anopheles* > *Mansonia* > *Aedes*. In contrast, the order of susceptibility to Bt sub sp. *israelensis* is *Aedes* > *Culex* > *Anopheles* > *Mansonia* (Yap, 1990).

Initial pathology of *B. sphaericus* can be observed in larval midgut. Partial inhibition of feeding in infected *Cx. pipens* larvae was observed within 10 minutes post-ingestion, with feeding stopping between 2 and 5 hours, and death occurring within 2 days (Singer, 1981). The main virulence factors of *B. sphaericus* are toxins that target the midgut epithelium. The mosquitocidal properties of *B. sphaericus* are produced by the action of binary (Bin) toxin and mosquitocidal (Mtx) toxins, although some strains also produce additional Cry toxins. The Bin is observed as a single parasporal crystal during late stages of sporulation. This toxin is composed of equimolar amounts of two proteins of 42 kDa (Bin A) and 51 kDa (Bin B), and both are necessary for maximum toxicity. Both Bin A and Bin B have been shown to be proteolytically processed to smaller proteins. The Bin B protein acts as binding domain for Bin A, which is internalized in the midgut cell and exerts its toxic properties. Several strains of *B. sphaericus* produce the so-called mosquitocidal toxins (Mtx) during the vegetative stage. These toxins are soluble and do not form crystalline inclusions (Jurat-Fuentes and Jackson, 2012).

B. sphaericus products display advantages over Bt subsp. *israelensis*, including longer environmental persistence, recycling in the environment, and activity in polluted waters. *B. sphaericus* spores are less sensitive to inactivation by UV radiation than other *Bacillus* spp. spores owing to high concentrations of small acid-soluble proteins and DNA repair systems. Encapsulation of spores in calcium alginate microcapsules increases resistance to UV radiation and high temperature (Lacey, 1990). The safety of *B. sphaericus* against non-target invertebrates and vertebrates has been will documented. The high specificity of *B. sphaericus* limits potential adverse effects on non-target aquatic species including mosquito predators, other invertebrates and fish (Lacey, 2007).

6.3.3. Genus *Paenibacillus* (Family: Paenibacillaceae, Order: Bacillales, Class: Bacilli, Phylum: Firmicutes). **Milky-disease organisms**

A number of refractile bacilli multiply and sporulate in the haemolymph of scarabaeid beetles. The haemolymph becomes turbid and the posterior portion of larval body turns opaque and milky white; thus the name milky disease. The syndrome was first described in the grubs of the Japanese beetle, *Popillia japonica* Newman in 1933. The bacteria causing type A and type B milky disease were identified as *Paenibacillus* (=*Bacillus*) *popilliae* and *P. lentimorbus*, respectively (Dutky, 1940).

The accidental introduction of the Japanese beetle in 1916 in a nursery in New Jersey, USA led to the isolation of a number of pathogens of the beetle. The beetle became a serious pest and fed on almost 300 species of plants. The US Bureau of Entomology established the Japanese beetle laboratory in 1917 at Rivertone, New Jersey to study the biology, ecology and control measures of the pest. *B. popilliae* is the most important pathogen isolated from Japanese beetle and was the first insect pathogen to be registered in USA for insect control.

B. popilliae is the most important of various Bacilli causing milky disease. Its host range within scarabaeids is fairly broad infecting about 70 insect species. The thick walled sporangium of *B. popilliae* contains a spore and parasporal body, giving the cells a foot-print like appearance. The spores have to be ingested and germinate at high pH and enzyme rich conditions in the scarab digestive tract within 2 days; the rods then penetrate into midgut cells. The bacteria multiply and invade into haemocoel where multiplication continues rapidly filling up much of the cavity. A few spores are also formed at this stage but the main phase of sporulation occurs later and is completed by 14 to 21 days when the larva develops the typical milky appearance. Infected larvae remain active until just prior to death, when they become sluggish and moribund. The infection is generally long drawn, except when the infection occurs in the first instar larva or after a larva receives a massive bacterial dose. When infected in a given instar, the larva generally does not moult and metamorphose. The milky-white haemolymph when exposed to air, does not darken as in the case of a normal, uninfected haemolymph. The cause of death is unknown and may be due to a depletion of essential nutrients and the presence of heat labile toxins or a disturbance of oxidative enzymes (Klein, 1981).

Following death the host disintegrates and spores are released into the soil. Spores are ingested by scarabaeid larvae as they feed on roots and organic matter in the soil. Larvae with obvious symptoms of milky disease are common in the field after peak feeding of third instars. Symptoms develop in 3-4 weeks but underfed larvae succumb to disease without exhibiting external symptoms. Milky disease develops best at 20-25°C and is inhibited at less than 15°C and more than 30°C. Presently, 10 biovars of *B. popilliae*, are known throughout the world including North America, Australia, Europe and Asia. These biovars differ in the morphology of sporangia, spores and parasporal body. All are, however, fastidious in their cultural requirements and only limited production is possible *in vitro* (Milner, 1981). It grows on rich media containing yeast extract, casein hydrolyzate or an equivalent amino acid source and sugar. Thiamine and barbituric acid are also required. The spores are mass produced by injection of spores into the haemocoel of grubs. The *in vivo*, production technique makes the product very expensive. In India, natural infection of *B. popilliae* has been isolated from white grubs and the commercial product has been found promising for the control of white grubs (Jayaraj *et al.*, 1994).

The advantages of using *B. popilliae* include narrow host range, complete safety to man, other vertebrates and all non-target organisms, compatibility with other control agents including insecticides and EPNs and high persistence in the soil. It's use is however limited by high cost of *in vivo* production, slow rate of action, lack of effect on adults which also cause obvious damage and need for large areas to be treated for effect. Narrow host range is also a limitation because other species of grubs if present are not infected.

B. popilliae has been successful as a classical microbial control agent against the Japanese beetle in turfs and pastures. The bacterial spores once established in the soil either through natural dispersal or application persist in the soil through recycling for years.

6.3.4. Genus *Brevibacillus*
(Family: Paenibacillaceae)

Brevibacillus laterosporus is an aerobic spore-forming bacterium characterized by its ability to produce a canoe-shaped lamellar parasporal inclusion adjacent to the spore. Some strains produce crystalline inclusions of various shapes and sizes, which are released separately from spores during lysis of the sporangium (Ruiu *et al.* 2013). It's a natural inhabitant of water (Laubach, 1916), soil (Oliveira *et al.*, 2004; Ruiu *et al.*, 2006), and has also been isolated from insects (McCray, 1917). It's biopesticidal potential has been reported against insects in different orders, such as Coleoptera (Boets *et al.*, 2004), Lepidoptera (Oliveira *et al.*, 2004), mosquitoes and black flies (Favret and Yousten, 1985; Rivers *et al.*, 1991), house flies (Ruiu *et al.*, 2006), against nematodes (Bone and Singer, 1991; Singer, 1996) and against phytopathogenic fungi (Saikia *et al.*, 2011).

Different insecticidal toxins produced by diverse *B. laterosporus* strains have been identified and characterized against various targets. Boets *et al.* (2004) highlighted the insecticidal role of secreted binary toxins with Bt Cry proteins acting against different coleopteran species (*Diabrotrica* spp., *Leptinotarsa* spp. and *Anthonomus* spp.). Recently, certain mosquitocidal *B. laterosporus* strains were discovered to produce insecticidal crystals similar to those typical of *Bt*. Proteins contained in these parasporal bodies were isolated and correlated to the insecticidal action of these strains (Zubasheva *et al.*, 2010). Enhancement of this larvicidal activity was obtained by bioencapsulation in Protozoa (Zubasheva *et al.*, 2011). On the other side, the toxicity against flies is in relation to spores of strains lacking parasporal crystals (Ruiu *et al.*, 2007).

As a result of recent studies, the ultrastructural effects of this entomopathogenic species was shown against house fly larvae exposed to a *B. laterosporus* treated diet. The progressive symptomatology and midgut changes induced in intoxicated larvae are similar to those caused by *Bt*, thus suggesting a toxin mediated process involving the alteration of midgut epithelial membrane permeability leading to osmotic disorders, cell alteration and disruption (Ruiu *et al.*, 2012). Promising field applications with experimental formulations in dairy farms against flies have been reported (Ruiu *et al.*, 2008 and 2011).

6.3.5. Gram –Negative Bacteria

(Phylum : Protobacteria)

Gram-negative bacteria historically have been separated from Gram-positive bacteria by their inability to absorb crystal violet dye. Gram-negative bacteria are characterized by the absence of spores and can occur in rods and cocciform shapes as aerobic or facultatively unaerobic species. These bacteria are widely distributed and abundant in the environment.

These bacteria have low pathogenicity and low invasive capacity; but once they enter the insect's haemocoel, they multiply rapidly and may cause death within a brief period. The digestive tracts of most insects contain many non-spore forming bacteria, many of which are potential pathogens. As long as these bacteria remain in the midgut, they are relatively harmless. Temperature extremes, other pathogens, parasites, and poor nutrition are some of the stress factors that can cause these bacteria to gain entrance into haemocoel. Non-spore forming bacteria are undoubtedly important mortality factors under stress situations, but little effort has been made at utilizing them as microbial control agents

Gram-negative bacteria are well known for their ability to form natural transconjugants and have been widely used for genetic transformation. Genetic research on disparate groups of Enterobacteriaciae has shown the presence of common genes constituting a 'toxin complex' (Tc) found among the genera *Serratia*, *Photorhabdus*, *Xenorhabdus*, *Pseudomonas*, *Yersinia* and *Paenibacillus*. Genes for Tc have been found both on plasmids and within the bacterial chromosome (Jurat-Fuentes and Jackson, 2012).

6.3.5.1. Family Enterobacteriaceae (Order: Enterobacteriales, Class: Gammaproteobacteria)

Many species in the family Enterobacteriaceae live in the digestive tracts of animals, including insects. Members of the genus *Serratia* comprising 10 species are ubiquitous in nature and are commonly found as saprophytes in water, soil, milk and other foods. Some strains cause infections in vertebrates including humans. *Serratia* spp. commonly infect insect species in the orders Orthoptera, Coleoptera Hymenoptera, Lepidoptera and Diptera. Identification of infection in insects was facilitated by the characteristic red pigmentation of several strains of *S. marcescens* and *S. plymutica*. *S. marcescens* is known to associate with and colonize the digestive tract of a broad range of insects and once inside the haemocoel, causes rapid mortality. Some insects, however, are susceptible to *S. marcescens* strains through the oral route, such as tsetse flies, *Glossina* spp., blow fly *Lucilia sericata*, and May beetles *Melolontha melolontha*. However, concerns about relatedness to human pathogens have limited research on *S. marcescens* as a microbial control agent (Jurat-Fuentes and Jackson, 2012).

a. *Serratia entomophila*

The New Zealand grass grub *Costelytra zealandica* (Scarabaeidae: Coleoptera) is an important pest of grasslands and rye grass-clover pastures. The grass grub is an amazing source of pathogens with 30 primary pathogens being isolated from it. Among these amber disease is the most distinct. The disease is caused by ingestion of the bacteria, *Serratia entomophila*, rapidly causing a cessation of feeding and gut clearance, which gives the larvae their amber appearance. Once feeding ceases, the gut becomes a vesicle for bacterial growth. The infected insects remain in a non-feeding state for weeks to months until weakened gut tissues rupture, leading to bacterial invasion and septicemia. Ultimately the cadaver breaks down to release bacteria into the soil. The isolated bacteria can be cultured through *in vitro* fermentation and applied back to the soil causing disease. The impact of natural epizootics of amber disease on grass grub populations stimulated the development of a selected strain as a commercial biocontrol agent (Jackson *et al.*, 1992). In 1980s, Monsanto developed it into its first biopesticide product 'Invade". Invade was highly effective; one litre of fermenter product was enough to treat one hectare of pasture. However, it had a limited shelf life, and, for successful results, had to be placed in the soil in a large volume of water using modified seed drills. New Zealand fertilizer company, Balance Agrinutrients Ltd., developed a granular formulation 'Bioshield' which was easy to apply and it became very popular. Key to the success of the bacterial pathogen is bacterial recycling. Once established, the pathogenic bacteria persist until the grass grub population declines to low levels or is killed by adverse soil conditions (Jackson, 2007).

b. *Yersinia entomophaga*

This recently discovered species was isolated from diseased larvae of the New Zealand grass grub, *Costelytra zealandica* White (Coleoptera Scarabaeidae) (Hurst *et al.*, 2011a). It is a non-spore-forming bacterium that secretes a multisubnit toxin complex (Yen-Te) showing homology with toxin complexes produced by *Photorabdus* spp. Proteins in the complex include three protein families termed A (YenA1, YenA2), B (YenB) and C (YenC1, YenC2) and two chitinases (Chi1 and Chi2) with high endochitinase activity (Hurst *et al.*, 2011b). These toxins have a broad insecticidal range of activity including Coleoptera and Lepidoptera. Histopathological studies on the effects caused by the ingestion of Yen-Tc revealed the progressive disorganization and deterioration of the midgut epithelium of *C. zealandica*. These effects are reminiscent of those previously reported for other Tc producing bacteria, like *P. luminescens* and *X. nematophila* (Marshall *et al.*, 2012).

Recently, the insecticidal activity of formulations containing *Y. entomophaga* against the pasture pest porina (*Wiseana* spp. larvae) has been reported highlighting its persistence and survival in the field, thus suggesting its use as a microbial alternative for the management of porina (Ferguson *et al.*, 2012; Ruiu *et al.*, 2013).

6.3.5.2. Genus *Wolbachia* (Family: Anaplasmataceae, Order: Rickettsiales, Class: Alphaproteobacteria)

Wolbachia are a common and widespread group of obligate endosymbiotic bacteria found in the reproductive tissue of arthropods and filarial nematodes. *W. pipientis* is the sole member of the genus. It has been estimated that up to two thirds of all insect species may be infected by this bacterium.

In their arthropod hosts, *Wolbachia* induce a variety of reproductive manipulations that enhance the fitness of infected females compared to their uninfected counterparts. The commonality among all these methods is that they all increase the number of female offsprings generated by infected females relative to their uninfected counterparts. As *Wolbachia* are maternally inherited, this increases the proportion of infected individuals in the population and ensures the spread of the bacteria. The selective advantage allows *Wolbachia* to sweep rapidly into insect populations. *Wolbachia* has the ability to interfere with pathogen development in the host, providing the insect with a fitness benefit (pathogen protection) or altering its ability to become infected with and transmit vertebrate pathogens (pathogen interference). *Wolbachia* has great potential to be used as an agent for controlling arthropod-borne diseases. Reproductive manipulations of host reproduction may allow the use of *Wolbachia* for population replacement, to drive desirable genes into an insect population, and the development of incompatible insects for population suppression. The use of strains that shorten insect lifespan is an alternative strategy for pathogen control (Hughes and Rasgon, 2012).

The incompatible insect technique uses *Wolbachia*-mediated cytoplasmic incompatibility to suppress insect populations in a similar fashion to the SIT. Matings between infected males released into the population and wild type females are incompatible and do not produce progeny. Life-shortening Wolbachia symbionts have been successfully transferred into dengue causing *Aedes aegypti*. The bacteria are maternally inherited and halve the life span of adult mosquitoes in the laboratory. This approach may significantly reduce the dengue transmission potential of *A. aegypti*.

6.3.5.3. *Chromobacterium subtsugae* (Family: Neisseriaceae, Order: Neisseriales, Class: Betaproteobacteria)

Firstly isolated from forest soil in Maryland, USA, *Chromobacterium subtsugae* is a Gram negative betaproteobacterium developing violet-pigmented colonies, due to the production of violecein, a tryptophan derivative synthesized by different bacterial species (Hoshino, 2011). It was indicated as a novel species, as a result of 165 rRNA gene sequencing and phenotypic characterization, which allowed it to be distinguished from *C. violaceum*, a species that shares the ability to produce violacein (Duran and Menck, 2001).

Its insecticidal potential by ingestion was initially proved against diverse insect species in different orders (i.e. Coleoptera, Lepidoptera and Hemiptera). These included the Colorado potato beetle (*Leptinotarsa decemlineata* Say), the Western corn rootworm (*Diabrotica lvirgifera* Le Conte), the Southern corn rootworm (*Diabrotica undecimpunctata* Mannerheim), the small hive beetle (*Aethina tumida* Murray), the diamondback moth (*Plutella xylostell* L.) the sweet potato whitefly (*Bemisia tabaci* Gennadius) and the Southern green stink bug (*Nezara viridula* L.) (Martin *et al.*, 2007a). In addition to mortality, significant sub-lethal effects, with special regard to feeding inhibition, were observed in most cases. Martin *et al.* (2007b) associated the insecticidal activity to either viable cells or cells killed by autoclaving, thus suggesting a mechanism of action mediated by heat-stable toxins. The stability of toxic factors was proved to be maintained also after freezing of pH changes (Martin *et al.*, 2007c). Successively, analytical studies on a crude extract combined with bioassay on the Beet armyworm (*Spodoptera exigua* Hubner), confirmed the production of bioactive compounds during *C. subtsugae* stationary growth phase (Koivunen *et al.*, 2009).

The mode of action of this new entomopathogenic bacterium is complex and different are the metabolites that have recently been associated to the insecticidal effects (Asolkar *et al.*, 2012). *C. subtsugae* strain PRAA4-IT has now been developed into a commercially available product (Grandevo, Marrone Bio Innovations Inc.) for the management of a broad spectrum of chewing and sucking insect species (Ruiu *et al.*, 2013).

6.4. FUNGAL ENTOMOPATHOGENS

The first pathogens found to cause diseases in insects were fungi because of the conspicuous macroscopic growth on the surfaces of their hosts. Many fungal pathogens however form no superficial growth or produce sparse, inconspicuous or minute external structures that are difficult to detect by the inexperienced investigator.

The diseases caused by fungi are termed as mycoses. Of the estimated 100,000 identified species of fungi (Blackwell, 2011), nearly 1000 fungal entomopathogens placed in over 100 genera are known (Roberts and Humber, 1981; St. Leger and Wang, 2009). However, recent molecular phylogenetic studies have indicated that the number of species of fungal entomopathogens may be much higher (Rehner, 2009). The molecular studies have also resulted in a dramatic change in the classification of fungi. Most fungal entomopathogens occur in the orders Entomophthorales (Class: Zygomycetes) and Hypocreales (Class: Sordariomycetes). The members of Hypocreales are generally considered as opportunistic pathogens infecting many species of several insect orders and host death is commonly associated with toxin production overwhelming host defense mechanisms. In contrast Entomophthorales have evolved into higher parasitic forms and host death occurs by tissue colonization with little or no use of toxins. They are however, difficult to rear on artificial media and mass production is not feasible (Roy *et al.,* 2006). In addition to these two fungal orders, there are several other groups especially within the ascomycete fungi which parasitize insects. Further the microsporidia, which were earlier included among protozoa, are now considered to be fungi.

6.4.1. Structure and Reproduction

The fungi may consist of single cell (e.g. yeast) or more commonly, branched filaments or hyphae collectively known as a mycelium. The hyphae are uninucleate or multinucleate segments or they are coenocytic with numerous nuclei not separated by transverse walls. Many of the entomopathogenic fungi are dimorphic i.e. they exist in both hyphal and yeast like forms. A specialized fungal form, a selerotium, is a hard, frequently spherical structure consisting of a mass of sterile hyphae considered to be resistant to adverse environmental conditions (Alexopoulos *et al.,* 1996). Fungi can reproduce sexually (telemorphic) or asexually (anamorphic). Asexual reproduction occurs through mycelial fragmentation or production of reproductive spores known as conidia. Conidia are the result of mitotic division and are formed directly on hyphal strands or from conidiogenous cells often at the ends of hyphae. The shape and arrangement of conidia and conidiphores (the hyphae where conidiogenous cells are formed) are useful in identification of fungal entomopathogens (Humber, 1997; Vega *et al.,* 2012).

Various methods of sexual reproduction occur in fungi. A thick-walled zygospore is produced in Zygomycota after the fusion between morphologically indistinguishable gametangia (antheridia, oogania). Sexual fusion in some Ascomycota is by the transfer of a nonmotile microconidium or spermatidium to a receptive hypha, but in many others, sexual fusion is between nondifferntiated hyphae. Nuclear fusion is followed by meiosis and usually mitosis to form eight haploid nuclei that develop into spores inside a sac or ascus. Parasexual recombination involves mitotic crossing over, whereby hyphae with two different types of nuclei (i.e. heterokaryons) will fuse and produce a recombinant type without the involvement of sexual reproduction (Tanada and Kaya, 1993).

A matting of reproductive structures produced on the exterior of the insect by many fungal entomopathogens is called muscardine e.g. white muscardine by *Beauveria bassiana*, green muscardine by *Metarhizium anisopliae*, yellow muscardine by *Aspergillus flavus,* and red muscardine by *Sorosporella uvella*.

6.4.2. Host Range

Fungi infect insects in all orders, most commonly Hemiptera, Diptera, Coleoptera, Lepidoptera, Orthoptera and Hymenoptera. In some insect orders, the immature stages are more often infected than the adult stage; in others, reverse may be the case. Host specificity varies considerably; some fungi infect a broad range of hosts and other are restricted to a few or single insect species. Many fungi have different strains which vary considerably in their spectrum of activity. Considering the two most diverse orders of fungal entomopathogens, the entomophthorales primarily consist of fungi with narrow host ranges, while the Hypocrealean fungi exhibit a variety of host ranges from narrow to very broad (Vega *et al.,* 2012).

6.4.3. The Infection Process

The development of mycoses involves a number of steps (Tanada and Kaya, 1993; Vega *et. al.,* 2012).

Adhesion of spore on the insect cuticle. The adhesion of the spores into the insect's cuticle may be a passive mechanism and involve mucilageous material and spore surface structures. Wang and St. Leger (2007) reported that *M. anisopliae* conidia adhesion to insect cuticle is mediated by an adhesion-like protein, known as MAD1.

Germination of the spore. Spore germination largely depends on environmental humidity and temperature, and to a lesser extent light conditions, nutritional environment and physical factors. The germinating spore, attached to the host, forms a germ tube that serves as a penetrant hypha. An appressorium may also be produced in some cases.

Penetration of integument. The mode of penetration mainly depends on the property of the cuticle, its thickness, sclerotization, and the presence of antifungal and nutritional substances. The newly formed larva and newly formed pupa are more susceptible to infection than those in which the cuticle has fully hardened. Medium-chain fatty acids like palmi-toleic, linoleic and linolenic acids in the epicuticle prevent the invasion of fungus through the integument.

The process of penetration through the insect's integument by a germi-nating hypha involves physical as well as chemical (enzymatic) forces. The enzymes involved in the digestion of integument include lipase-protease- chitinase in a sequence. Fungi may also infect insects through the buccal cavity, spiracles, and other external openings (e.g. sense organs) of an insect (St. Leger, 1991).

Multiplication in the haemocoel. After the germinating hypha has penetrated the insect's integument and entered the haemocoel, it produces yeast like hyphal bodies, essentially blastos-pores that multiply by budding, In addition to hyphal bodies, hyphal strands and wall-less protoplasts may develop in the haemocoel.

The death of the insect ends the parasitic development of the fungus. The fungus then grows saprophytically in the haemocoel to form a mycelial mass that turns into a hard or firm sclerotium. Spores are produced within the sclerotium or on sporophores (sporangiophore, conidiophore). The sporophores and sterile hypae emerge from the cadaver under favourable environmental conditions, to form the characteristic mycelial growth on the insect's integument.

Spore dispersal. Spore dispersal can be an active or passive process and depends on the characteristics of the spore and sporophores. Certain fungi form conidia that are enclosed in a mucilaginous slime. Such conidia may attach to a passing insect for dispersal. Spores of many species are forcibly discharged from sporophores and carried by the wind.

6.4.4. Mycotoxins

Many fungi produce toxins during the course of infection in an insect. These toxins called mycotoxins may have a role in the pathogenicity of fungi but it remains unclear (Molnar *et al.*, 2010). *Beauveria* spp. have been reported to produce several different cyclodepsipeptides (peptide lactones) which alter the permeability of cell membranes. The most studied *B. bassiana*-produced cyclodepsipeptide is beauvericin, which is toxic to brine shrimp, the fall armyworm cell lines and mosquito larvae. Another compound, bassianolide, is toxic to silkworm larvae.

Another group of cyclodepsipeptides called destruxins are produced by *Metarhizium anisopliae*. Some of the 38 different destruxins or their analogues have been shown to be insecticidal. *Metarhizium* spp. have also been reported to produce cytochalasins, myroridins, viridoxin, helvolic acid, hydroxyfungarins, fusarin analogues, aurovertins, macrolides and tyrosine betaine (Vega *et al.*, 2012).

6.4.5. Role in Pest Management

Sucking pests usually do not ingest significant quantities of bacteria or baculoviruses and are consequently not amenable to control by these groups of pathogens. Even for lepidopterous insects, eggs, pupae and adults are largely non-feeding stages. Fungal entomopathogens with their contact activity can be used under such situations. Both Bts and baculoviruses are largely active against Lepidoptera; for many major pest species from Coleoptera, Diptera, Orthoptera, Hemiptera and Thysanura, fungi emerge as obvious candidates. Phytophogous mites are also susceptible to fungal pathogens.

Fungi have a comparative advantage of having a multiple site/multiple-step mechanism of action which minimizes the chances of occurrence of resistance in target pests.

There are good opportunities for the isolation of fungal strains in the original habitats of exotic pests, thus making possible to use these pathogens within classical biocontrol programmes. Fungi can be highly virulent against phytophagous insects while still presenting a significantly lesser impact on their natural enemies because of a certain level of physiological specialization on the pest host, as a result of repetitive natural selection. de Faria and Wright (2007) identified 171 mycoinsecticidal products based on 15 fungal entomipathogens used as biopesticides around the world (Table 3).

Table 3. Fungal species developed into mycoinsecticides and their target insect groups.

Species	No. of products	Target insect groups
(1)	(2)	(3)
Fungi: Anamorphic Hypocreales		
Aschersonia aleyrodis Webber	1 (0.6%)	Hem.: Aleyrodidae
Beauveria bassiana (Bals.), Vuill.	58 (33.9%)	Col., Lep., Hem., Thy., Acari., Orth., Hym., Dip., Sip.,
Beauveria brongniartii (Sacc.) Petch	7 (4.1%)	Col. (Scarabaeidae, Cerambycidae), Hem., Lep., Dip.
Hirsutella thompsonii F.E. Fisher	3 (1.8%)	Acari

contd...

(1)	(2)	(3)
Isaria fumosorosea Wize	10 (5.8%)	Hem.: (Aleyrodidae, Aphididae), Acari, Thy., Col., Nematoda
Isaria sp.	1 (0.6%)	Hem.: (Aleyrodidae), Thy., Nematoda
Lecanicillium longisporum (Petch) R. Zare & W. Gams	2 (1.2%)	Hem.: (Aleyrodidae, Aphididae, Ortheziidae)
Lecanicillium muscarium (Petch) R. Zare & W. Gams	3 (1.8%)	Hem.: (Aleyrodidae, Aphididae), Thy., Acari
Lecanicillium sp.	11 (6.4%)	Hem.: (Aleyrodidae, Aphididae), Thy., Dip., Acari, Nematoda
Metarhizium anisopliae (Metschn.) Sorokin	58 (33.9%)	Col. (Curculionidae, Scarabaeidae, Nitidulidae), Hem., Cerocpidae, Aphididae, Miridae, Cicadellidae, Pentatomidae, Delphacidae)., Iso., Orth., Blat., Dip., Thy., Lep. (Crambidae), Hym., Sip., Acari
Metarhizium anisopliae var. *acridum* Driver & Milner	3 (1.8%)	Orth: Acrididae, Pyrgomorphidae
Nomuraea rileyi (Farl) Samson	1 (0.6%)	Lep.
Fungi: Anamorphic species identified as		
Sporothrix insectorum de Hoog & H.C. Evans	3 (1.8%)	Hem.: Tingidae
Fungi: Zygomycota: Zygomycetes: Entomophthorales		
Conidiobolus thromboides Drechsler	2 (1.2%)	Hem.: (Aleyrodidae, , Ortheziidae), Thy.
*Lagenidium giganteum** Couch	1 (0.6%)	Dip.: Culicidae
Mixes (2 or more species)	7 (4.1%)	
Total	**171 (100%)**	

* Recently excluded from Fungi and placed in the Stramenopiles (Oomycetes) (Green, 2011).
Source: Modified after de Faria and Wraight (2007)

The white muscardine fungus, ***Beauveria bassiana*** (Family: Cordicipitaceae, Order: Hypocreales) was the first disease in animals shown to be caused by a fungus or any microorganism. Agostino Bassi demonstrated the contagious and pathogenic nature of the fungus infecting silkworms and also developed measures for controlling the disease. In his honour, the fungus was later on named as *Beauveria bassiana*. It occurs worldwide and attacks a wide range of insects. It has been detected from over 700 species of insects mainly Lepidoptera, Coleoptera and Hemiptera and some species in Diptera and Hymenoptera. The fungus has been found infecting the lungs of wild rodents, nasal passages of humans, horses, and giant tortoises. Some of the major insect pests that are susceptible to this fungus include the European corn borer *Ostrinia nubilalis*; codling moth, *Cydia pomonella*; Japanese beetle, *Popillia japonica*; Colorado potato beetle, *Leptinotarsa decemlineata*; cabbage caterpillar *Pieris brassicae*; cotton whitefly *Bemisia tabaci*; tarnished plant bug *Lygus lineolaris*, brown planthopper, *Nilaparvata lugens*; and leafhoppers *Nephotettix* spp. The application of *B. bassiana* for the control of pine moth *Dendrolimus* sp. in China probably represents one of the most successful examples of microbial control. The program started in 1950 and it is calculated that at least one million hectares of pine forests are now treated.

Strains of *B. bassiana* (Bb) exhibit considerable variations in host range and virulence. Strain Bb-147 is registered on maize in Europe for control of European corn borer *O. nubilalis* and the Asiatic corn borer *O. furnacalis*. Bb strain GHA is registered in the US for control of whitefly, thrips, aphids and mealybugs. Bb strain ATCC74040 is registered against many homopterous, heteropterous and coleopterous insects (Khachatourians *et al.*, 2002).

The green-muscardine fungus ***Metarhizium anisopliae*** (Family: Clavicipitaceae, Order: Hypocreales) is as commonly and widely distributed fungus as *B. bassiana*. It has been detected from over 300 spp of insects. It is mainly effective against Coleoptera, Orthoptera, Homoptera, Lepidoptera and Isoptera. It is especially active against chrysomelid, curculionid and scarabaeid beetles (Zimmerman, 1993). Recent phylogenetic classifications have split the former *M. anisopliae* into several species (Bischoff *et al.*, 2009).

The former species ***Verticillium lecanii*** has been split into several species by recent phylogenetic classification including *L. lecanii* (Zare and Gams, 2001). ***Lecanicillium lecanii*** (Family: Clavicipitaceae, Order: Hypocreales) is a common pathogen of scale insects in tropical and sub-tropical environments. It is known as the 'white-halo' fungus because of the appearance of the white mycelium around the edges of the infected scale.

Nomuraea rileyi (Family: Clavicipitaceae, Order: Hypocreales) is pathogenic to a majority of the noctuids including *H. armigera*, *H. zea*, *S. litura*, *T. ni* and *Plusia* sp. Exposure of the infected larvae to light is essential for maximal conidial production. Epizootics of this fungus often develop late in the growing season of crops. Cost effective technology for its multiplication on artificial media has been developed by the Directorate of Oilseeds Research, Hyderabad.

M. anisopliae* var. *acridum (*Metarhizium* strain IMI330189) recently reclassified as *M. acridum* has been developed as a commericial biopesticide 'Green Muscle®' for the management of grasshoppers and locusts in Africa by LUBILOSA, an international consortium of agencies headed by CABI (Lomer *et al.*, 2001; Moore, 2008).

The host range of *M. acridum* is narrow, and at field application rates, it is safe to most non-target organisms including and to mammals. Dry spores of the fungus are formulated in oil and applied at very low application rates of 25-75 g/ha (1.25- 3.75 x 10^{12} spores/ha) and at a volume of 0.5 to 2 litres (Langewald and Kooyman, 2007). ULV spray application is an excellent way to release a relatively high concentration inoculum of the fungus into the host's environment. The inoculum reaches the insect host's cuticle and the spores germinate, penetrate and the fungus grows inside the host's body. This process takes time and the speed of kill is slow. But infected grasshoppers consume less food and are more prone to predation than healthy individuals. Depending on environmental conditions after death, the cadavers sporulate, producing new fungus inoculum. The long persistence of fungal spores in the field, which under optimal conditions can accrue to more than a year, plays a very important role in the efficacy (Douthwaite *et al.*, 2001; Langewald and Kooyman, 2007).

Entomopohaga maimaiga (Family: Entomophthoraceae, Order: Entomophthorales) commonly known as the 'Gypsy fungus' has been highly successful as a classical microbial control agent against the Gypsy moth *Lymantria dispar*. The Gypsy moth is an introduced pest of hardwood forests in large areas of USA. Among the natural enemies of Gypsy moth introduced in USA, *E maimaiga* from Japan was released in 1910-1911 in the Boston area and in 1985-1986 in New York and Virginia. But the fungus was presumed not to have established in either of these areas. In 1989, Gypsy moths, infected with *E. maimaiga,* were discovered in Connecticut (Andreadis and Weseloh, 1990), the first report of this in North America. By 1992 the fungus had spread across the contiguous distribution of Gypsy moth in the north-eastern USA. Since then the fungus has spread behind the leading edge of Gypsy moth spread.

The fungus spreads on its own through aerial dispersal of actively ejected asexual spores from cadavers of dead Gypsy moth larvae. The fungus persists in the top layer of soil as resting spores, which have been shown to persist for at least 11 or 12 years and probably longer. The only major limitation in its use is that *E. maimaiga* is an obligate pathogen and is not easy grow in the laboratory (Hajek, 2007).

6.4.6. Genetic Improvement

Several studies have reported genetic modifications of fungal entomopathogens. *M. anisopliae* has been genetically modified for (i) resistance to fungicide benomyl and the herbicides bialaphos and glufosinate ammonium (ii) increased speed of kill via protease over-expression (iii) chitinase overproduction, and (iv) expression of a scorpion toxin gene. *B. bassiana* has been modified for chitinase overproduction and for expression of: a chitin-degrading protease from *M. anisopliae*, a scorpion neurotoxin, VIPs from *B. thuringiensis* and a chitinase gene from *O. unilateralis* (Vega *et al.*, 2012).

6.5. MICROSPORIDIAN ENTOMOPATHOGENS

Microsporidians are obligate, intracellular, pathogenic protists related to the Fungi. Most of the more than 1300 described species in approximately 186 genera are pathogens of invertebrate animals, while some species are known to attack vertebrates (Becnel and Andreadis, 1999). Microsporidia are important primary pathogens of both pests and beneficial insects. The first known microsporidian *Nosema bombycis,* was described from the silkworm, *Bombyx mori,* by Nageli in 1857. Another two species *N. apis* and *N. ceranae* are pathogenic on honey bees. Microsporidia are also important regulatory factors in populations of insect pests. Traditionally microsporidia have been considered as protozoans. But recent molecular phylogenetic studies have revealed that microsporidia are related to entomopathogenic zygomycetes (Corradi and Slamovits, 2010; Solter *et al.*, 2012).

Insects are represented as type hosts of 90 microsporidian genera, nearly half of all those described. Most microsporidia must be eaten to infect an insect, but there may also be some natural transmission, for example by parasitoids and predators. The pathogen enters the insect body via the gut wall, spreads to various tissues and organs, and multiplies, sometimes causing tissue breakdown and septicemia. Infected insects may be sluggish and smaller than normal, sometimes with reduced feeding and reproduction. The weakened insects are more likely to be susceptible to adverse weather and other mortality factors.

Paranosema **(syn.** ***Nosema*****)** ***locustae*** (Phylum: Microspora, Class: Microsporea, Order: Microsporida, Family: Nosematidae) is the only commercially available species of microsporidian marketed under several labels (Noloc, Nolo Bait, Hopper Stopper) for the control of locusts, grasshoppers and crickets. *P. locustae* has an unusually broad host range and is known to infect 121 species of orthoptera (Lange, 2005). Other microsporidia with far greater virulence against grasshoppers than that of *P. locustae* have been described, including *N. entomophaga, N. cuneatum* and *Johenrea locustae*. However, their greater virulence hampers development of an effective production system.

Spores of *P. locustae* are impregrated into wheat bran flakes and applied on the field. It takes one to three weeks for the grasshoppers to be infected. Following ingestion, the spore ruptures and activates the disease in the host. *P. locustae* primarily infects the adiptocytes of fat body tissue, resulting in the disruption of the host's metabolism and energy stores (Solter *et al.*, 2012). Infected individuals are lethargic and slow, making them easy prey for birds. *P. locustae* is not toxic to birds, animals or other insects.

Nosema affects all stages of the insect but is most effective against second and third instar nymphs. Like other microsporidians, it is not good as an instant or rescue control but successful as a long term population management tool especially in forage and ranglands. Recent reports have indicated that *P. locustae* infection can disrupt swarm formation in migratory locusts. It has been suggested that *P. locustae* and the native gut bacteria that produce the aggregation pheromone are mutually refractory. The reduction in aggregation pheromone reduces the production of the neurotransmitter serotonin (which initiates gregarization) and dopamine (which induces and maintains gregarization), preventing swarm behavior of migratory locusts (Shi *et al.*, 2014). Other microsporidians, being exploited on a limited scale, include *N. fumiferae* for spruce budworm *C. fumiferana, N. pyrausta* for European corn borer *O. nubilalis* in Illinois, *Vavraia (Pleistophora) culicis*, introduced into culex mosquito populations in Nigeria; *Nosema portugal* and *Endoreticulatus* sp. against *L. dispar* in Maryland.

6.6. ENTOMOPATHOGENIC NEMATODES (EPNs)

Nematodes commonly referred to as roundworms, eelworms, or threadworms are bilaterally symmetrical, colorless, unsegmented, translucent, usually elongate and cylindrical throughout their body length. The body is covered by a non-cellular elastic cuticle, which differs chemically from the chitinous cuticle of arthropods, and lacks appendages.

Nematodes are diverse organisms and, except for arthropods, have habitats more varied than any other group of animals. Nematodes occur as free-living organisms and as facultative or obligate parasites of plants and animals. Those parasitic on insects are known as entomogenous nematodes. Parasitism by entomogenous nematodes can have various deleterious effects on their hosts including sterility, reduced fecundity, reduced longevity, delayed development, or other behavioural, physiological or morphological aberrations and in some cases, rapid mortality. Parasitic associations with insects have been described from 23 nematode families. Seven of these families contain species that have potential for biological control of insects (Koppenhofer and Kaya, 2002).

EPNs are not actually microbial organisms but are always included as part of the microbial control agents. In fact, the non-feeding infective juveniles (IJs) possess attributes of both insect parasitoids or predators and microbial pathogens. Like parasitoids/predators they have chemoreceptors and are motile; like pathogens they are highly virulent, killing their host quickly. Most nematodes have simple life cycles and undergo three main stages of development: egg, juvenile and adult. The mated female deposits her eggs in the environment and the juvenile usually undergoes one moult in the egg and emerges as a second-stage juvenile. The majority of nematode species moult four times before becoming adults. These moults may occur in the egg, free in the environment, or in the insect host. Some EPNs have a resistant stage called the 'dauer juvenile' (DJ). The DJ is the third stage nematode, which is usually ensheathed in the second-stage cuticle and occurs commonly in the rhabditids. Immature nematodes are like the adults in appearance and structure, and therefore their development is analogous to ametabolous insects (Tanada and Kaya, 1993).

6.6.1 Important Entomopathogenic Groups

Nematodes in the families Heterorhabditidae and Steinernematidae, both belonging to Order Rhabditida (Class Secernentea), have the ability to quickly kill (1-4 days) their hosts thanks to their mutualistic association with bacteria in the genus *Photorhabdus* and *Xenorhabdus*, respectively (Table 4). These bacteria are facultative anaerobes in the family Enterobacteriaceae. EPNs have been recovered from soils throughout the world and are very common soil organisms. Each nematode species is specifically associated with one symbiont species, although a symbiont species may be associated with more than one nematode species (Koppenhofer and Kaya, 2002). The number of species, within both the families, is increasing rather rapidly. Adams *et al* (2006) reported 42 and 6 recognized species in Steinernematidae and Heterorhabditidae respectively. Recently Lewis and Clarke (2012) reported that the number of species in the respective families has increased to 67 and 17.

Both the groups are obligate pathogens in nature. The only stage that survives outside of a host is the third stage non-feeding infective juvenile (IJ). The IJs carry cells of their bacterial symbionts in their intestinal tract. After locating a suitable host, the IJs invade it through natural openings (mouth, spiracles, anus) or through thin areas of host cuticle (common only in heterorhabditids) and penetrate into host haemocoel. The IJs recover from their development arrestment, release the symbionts, and bacteria and nematodes cooperate to overcome the host's immune response. The bacteria propagate and produce substances that rapidly kill the host and protect the cadaver from colonization by other microorganisms. The nematodes start developing, feed on the bacteria and host tissues metabolized by the bacteria, and go through 1-3 generations. Depleting food resources in the host cadaver lead to the development of a new generation of IJs that emerge from the cadaver in search of new hosts (Koppenhofer and Kaya, 2002).

The only nematode species outside the above mentioned groups, which has been successful in management of insect pests is the facultative parasite *Beddingia* (*Deladenus*) *siricidicola* (Family: Phaenopsitylenchidae, Order: Tylenchida, Class: Secernentia) which has been released as a classical microbial control agent against the wood-wasp *Sirex noctilio* in several countries in the Southern hemisphere. The wood-wasp is a devastating pest of several North American species of pine which have been introduced into New Zealand, Australia, Brazil, Uruguay, South Africa and Argentina. *B. siricidicola* has been introduced and has established in each of these countries (Bedding and Iede, 2005).

6.6.2. Role in Pest Management

EPNs have been evaluated against a large number of insect pests with results varying from no effect to excellent control. Nematodes have been applied most successfully in habitats that provide protection from environmental extremes, especially in soil, their natural habitat, and in cryptic habitats. Excellent control has been achieved against insects that bore into plants probably because their cryptic habitats contain less limiting factors (NEs of nematodes) than soil. Nematode based biopesticides are currently available for the control of a number of coleopterous (flea beetles, root weevils, rootworms), dipterous (leafminers, sciarid flies) and lepidopterous (cutworms, armyworms, crown borers, stem borers, etc) pests on a variety of crops. Around a dozen species of nematodes are currently produced as biopesticides mostly in North America and Europe (Table 4).

Table 4. Important species of entomopathogenic nematodes commercially available as biopesticides alongwith their respective bacterial symbionts

Sr No	Nematodes species	Symbiont species
1	*Steinernema carpocapasae* (Weiser)	*Xenorhabdus nematophilus*
2	*S. feltiae* (Filipjev)	*X. boveinii*
3	*S. glaseri* (Steiner)	*X. poinarii*
4	*S kraussei* (Steiner)	*X. boveinii*
5	*S. longicaudum* Sherr & Wang	Undercribed
6	*S. riobrave* (=*riobravis*) Cabinillas, Poinar & Raulston	*X. cabanillasii*
7	*S. scapterisci* Nguyen & Smart	*Xenorhabdus* sp.
8	*Steinernema* spp.	*Xenorhabdus* spp.
9	*Heterorhabditis bacteriophora* Poinar	*Photorhabdus luminescens*
10	*H. indica* Poinar, Karunakar & David	*P. luminescens*
11	*H. megidis* Poinar, Jackson & Klein	*P. temperata* sub sp. *cinerea*
12	*Heterorhabditis* spp.	*Photorhabdus* spp.

Source: Modified after Kaya *et al.* (2006)

The major limitations in use of EPNs as biocontrol agents are high cost of production, sensitivity to UV light and desiccation on the plant surface. The survival of IJs on the leaf surface can be increased by development of improved formulations containing antidesicants and surfactants (Shapiro Ilan *et al.*, 2012).

Production approaches for EPNs include *in vivo* and *in vitro* methods (solid and liquid fermentation). For laboratory use and small scale field experiments, *in vivo* production appears to be the appropriate method. *In vivo* production is also appropriate for niche markets and small growers where a lack of capital, scientific expertise or infrastructure cannot justify large investments into *in vitro* technology. *In vitro* production technology is used when large scale production is needed at reasonable quality and cost (Shapiro Ilan *et al.*, 2012).

The *in vitro* solid culture on a three dimensional rearing medium consisting of crumpled polyether polyurethane foam was developed by Bedding (1981). The approach was later expanded to autoclavable bags with filtered air being pumped in (Bedding, 1984). The *in vitro* liquid culture methods were developed and refined during 1990s. Liquid culture offers increased cost efficiency relative to other production methods, but it also demands greater capital investment and a higher level of technical expertise (Shapiro Ilan *et al.*, 2012).

6.7. STATUS OF MICROBIAL CONTROL IN PUNJAB, INDIA

The Punjab state located in the northwest India has a geographical area of 5.03 million hectares out of which 4.23 million hectares is under cultivatgion. The State has emerged as the grain bowl of the country during the post-independence period. With a geographical area of only 1.54 per cent of the country, the state accounts for nearly 21 per cent of wheat, 11 per cent of rice and 11 per cent of cotton produced in the country. The quantum jump in food grain production from 3.1 mt in 1960-61 to 29.2 mt in 2011-12 has been made possible by a combination of technological break throughs and bold policy initiatives. However, the state has paid a rather heavy price in terms of ecological and environmental problems caused by indicriminiate use of chemical pesticides and fertilizers. There is an urgent need to reduce dependence on pesticides and manage pests through ecofriendly approaches. The progress achieved in development and utilization of microbial control agents (MCAs) for management of insect pests in the state is discussed here in.

6.7.1. Indigenous Entomopathogens Detected

In Punjab, first systematic survey for insect pathogens was undertaken in 1969 and it resulted in the identification of a number of bacterial, fungal and viral pathogens (Battu *et al.,* 1971). Regular surveys since then have added to this list and presently thirty species of entomopathogens attacking important crop pests in the state are known (Table 5). More than twenty of these are baculoviruses including the nuclear polyhedrosis viruses (NPVs) and the granulosis viruses (GVs). Natural incidence of many of these viruses among insect pests and their role in regulation of population of these pests has been studied during the last four decades. All these NPVs except the NPV of *Thysanoplusia orichalcea* (Fabricius) (ToNPV) remained sporadic or enzootic in the field. Two epizootics of ToNPV have killed *en mass* host semiloopers infesting Egyptian clover and mint crops during 1983 and 1987, respectively (Battu, 1992). The NPV of *Spilosoma obliqua* Walker was the most important biotic mortality agent acting on the pest throughout Northern India but had not acquired epizootic status in the State (Battu and Ramakrishan, 1989).

Table 5. Entomopathogens detected from diseased insects collected during field surveys in Punjab, India

Entomopathogen	Host Insect	
	Common name	Scientific name
(1)	(2)	(3)
BACULOVIRUSES	Castor semilooper	*Achaea janata* Linnaeus
Nucleopolyhedrovirus (NPV)	Cutworms	*Agrotis* sp.
	Green semilooper	*Anomis flava* (Fabricius)
	Red hairy caterpillar	*Amsacta* sp.
	Cucurbit semilooper	*Chrysodeixis chalcites* Esper.
	Brinjal hadda	*Epilachna vigintioctopunctata* Fabricius
	Castor hairy caterpillar	*Euproctis lunata* Walker
	American bollworm	*Helicoverpa armigera* (Hubner)
	Cauliflower head borer	*Hellula undalis* (Fabricius)
	Armyworm	*Mythimna separata* (Walker)
	Cabbage butterfly	*Pieris brassicae* (Linnaeus)
	Diamondback moth	*Plutella xylostella* (Linnaeus)
	Cabbage semilooper	*Thysanoplusia orichalcea* (Fabricius)
	Bihar hairy caterpillar	*Spilosoma obliqua* Walker
	Tobacco caterpillar	*Spodoptera litura* (Fabricius)
	Lucerne caterpillar	*Spodoptera exigua* (Hubner)
	Leppet moth	*Trabala vishnou* Lefevere
Granulovirus (GV)	Castor semilooper	*A. janata*
	Cabbage butterfly	*P. brassicae*
	Bihar hairy caterpillar	*S. obliqua*
	Tobacco caterpillar	*S. litura*
BACTERIA		
Bacillus thuringiensis Berliner	Armyworm	*M. separata*
	Cabbage semilooper	*T. orichalcea*

contd...

(1)	(2)	(3)
Streptococcus faecalis Andrew and Horder	Bihar hairy caterpillar	*S. obliqua*
	Tobacco caterpillar	*S.litura*
	Cabbage butterfly	*P. brassicae*
Serratia marcessens Bizio	Thistle borer	*Ostrinia kasmirica* Moore
FUNGI		
Nomuraea sp.	Rice leaf folder	*Cnaphalocrocis medinalis* (Guenee)
Beauveria sp.	American bollworm	*H. armigera*
Aspergillus flavus Link	Tobacco caterpillar	*S. litura*
	Bihar hairy caterpillar	*S. obliqua*
	Maize borer	*Chilo partellus* (Swinhoe)
Metarhizium anisopliae (Metchnikoff)	Sugarcane pyrilla	*Pyrilla perpusilla* Walker
Fusarium sp.	Maize border	*C. partellus*
PROTOZOA		
Gregarina minuta Ishii	Rust red flour beetle	*Tribolium castaneum* Herbst

Source: Modified after Arora *et al.* (2002a)

Among other insect pathogens, indigenous bacterium *Bacillus thuringiensis* Berliner var. *thuringiensis* has been recovered from armyworm, *Mythimna separata* (Walker) and cabbage semilooper, *T. orichalcea*; non-spore forming bacterium *Streptococcus faecalis* Andrews and Horder from *S. litura, S. obliqua* and *P. brassicae;* fungal pathogen *Nomuraea* sp. from rice leaf folder, *Cnaphalocrocis medinalis* (Guenee) and *Beauveria* sp. from *Helicoverpa armigera* (Hubner) (Arora *et al.*,2002a).

6.7.2. Laboratory Studies

Basic and applied studies have been conducted on important indigenous pathogens as well as on commercially available microbial products.

6.7.2.1. Propagation of Nucleopolyhedroviruses

Being obligate pathogens, NPVs have to be propagated in their respective healthy host larvae of suitable age by using appropriate viral inoculum levels and maintaining inoculated host insects at appropriate environmental conditions.

In general, lower levels of yield of occlusion bodies (OBs) were obtained from naturally infected host insects of all metamorphic stages. Laboratory studies on yield levels of OBs from experimentally NPV infected *Hellula undalis* (Fabricius) larvae revealed that prolonged survival period (up to 5 days) of the diseased larvae resulted in as much as nearly 6 times more production of OBs than that obtained at the end of 3 days (Battu, 1989). A study of the comparative occlusion yield of NPV of *S. obliqua* revealed that *per os* inoculum administration killed cent per cent treated larvae and resulted in highest yields of OBs. Dipping of host larvae in alkali-liberated viral suspension and in diseased haemolymph suspension resulted in a higher incubation period and lower yields of OBs. The administration of 2 x 10^8 OBs / ml to third-fourth instar mixed larval population of *S. obliqua* by *per os* feeding resulted in a yield of 6.2 x 10^9 OBs / larva at 10 days after treatment (DAT) (Battu and Arora, 1999).

In case of *S. litura* 5-day old larvae, an ideal inoculum of 3.38 x 10^6 OBs / ml resulted in a yield of 8.5 x 10^9 OBs / larva at 8 DAT. In *H. armigera,* a maximum yield of 2 x 10^9 OBs per larva was obtained by feeding 4 x 10^6 OBs / ml, through chickpea foliage to 5-6 days old host larvae at room temperature at 9 DAT. A concentration of 1.23 x 10^4 OBs ml^{-1} was ideal inoculum for medium-sized *H. armigera* larvae using leaf-dip method and yielded 6.8 x 10^9 OBs / larva at 11 DAT (Singh, 1999).

Host rearing of *H. armigera* in the laboratory requires expensive air conditioning and humidity control equipment. An inexpensive oviposition cage made of an earthen pot has been designed and is being successfully used for obtaining eggs from gravid *H. armigera* females in hot and dry weather at room temperature (Arora and Battu, 1996).

6.7.2.2. Bioassay Studies

Bioassay studies have been conducted using commercially produced as well as indigenous isolates of *B. thuringiensis* and NPVs of important pests. It was found that LC_{50} of the NPV against larvae of leppet moth, *Trabala vishnou* (Lefevere), was 1.113 x 10^3 OBs ml^{-1}. The slope value of the dosage mortality curve was 0.3158 which indicated great variability in the population. Dipel 8L with an LC_{50} value of 0.013 per cent was more effective against *P. xylostella* larvae than Biolep (LC_{50} = 0.033%) and Halt (LC_{50} = 0.22%) among the commercially available Bt products. However, local Bt isolates BtgR2 (LC_{50} = 0.0046%), BtgR4 (LC_{50} = 0.0073%) and BtgP1 (LC_{50} = 0.0119%) from soils of Patiala and Ropar districts were more effective than even Dipel 8L (Arora *et al.*, 2002a).

Geographic isolates of NPVs are known to vary in virulence and efficacy, against field populations of the target pests. Three local *Hear* NPV isolates were compared with commercially supplied NPV in bioassay studies against larvae of *H. armigera*. The PAU isolate with an LC_{50} value of 2.1558 x 10^6OBs / ml was the most effective, followed by the Karaiwala (3.0546 x 10^6OBs / ml) and Mudki (3.0804 x 10^6OBs / ml) isolates. The Bangalore (Karnataka) and Godhra (Gujarat) isolates with LC_{50} values of 6.7284 x 10^6OBs / ml and 7.509 x 10^6OBs / ml respectively, were less effective (Arora *et al.*, 1997). Another bioassay study indicated that PAU-I isolate, with an LC_{50}value of 3.006 x 10^6OBs / ml following feeding of II instrar *H. armigera* larvae on NPV-treated cauliflower leaves, was the most virulent. The LC_{50} values for PAU-2, Mansa-1, Mansa-2 and Malerkotla isolates were 5.79 x 10^6, 4.926 x 10^6, 11.256 x 10^6 and 13 x 86 x 10^6OBs / ml, respectively. Time mortality assay also revealed that PAU-1 isolate was the most virulent with lowest median survival time (ST_{50}) values (138.0 – 220.8 hrs) against second instar larvae at all the treated concentration (Padmanabhan, 2001).

6.7.2.3. Food Consumption and Utilization Studies

A major disadvantage of biopesticides is a slowing down of development and the resulting prolongation of larval duration. In castor semilooper, *Achaea janata* (Linnaeus), both NPV and GV caused a retardation of growth 3-4 days after infection, as well as disturbances in the moulting process particularly towards later stages of disease development. Significant reduction in the absolute quantity of food consumed by semiloopers infected by each of the two viruses was recorded. The diseased larvae continued to loose weight compared with control till their death (Battu, 1986).

Studies on food consumption of Bt intoxicated *P. xylostella* larva was 7.45 mg / day in BtgR2 and 8.62 mg / day in Dipel 8L as compared to 12.32 mg / day in control. The indices of growth like approximate digestibility (AD), efficiency of conversion of indigested food to body substance (ECI) and relative growth rate (RGR) were also adversely affected (Arora *et al.*, 2002b). This resulted in a concentration dependent increase in the post treatment larval duration of *P. xylostella* following feeding on *B. thuringiensis* treated food (Singh *et al.*, 2001).

6.7.2.4. Interaction of Microbial and Chemical Pesticides

Slow mortality caused by microbial pesticides is a major limitation which prevents their use on agricultural crops. A combination of microbial pesticides with reduced dosages of chemical pesticides may prove useful under such situations to provide quick mortality. Laboratory studies have been conducted to study the interaction of recommended insecticides with NPVs of some important insect pests. Four recommended insecticides were evaluated for their compatibility with SoNPV for the management of *S. obliqua.* Dichlorvos and HCH at 20 and 50 ppm were found compatible with the virus. Relatively reduced virosis observed in fenitrothion (50 ppm) and endosulfan (20 ppm) indicated an antagonistic effect (Battu *et al.*, 1992).

A mixing of SINPV with dichlorvos (20 and 50 ppm), endosulfan (20 ppm) and fenitrothion (50 ppm) resulted in positive cotoxicity factors indicating synergism against *S. litura* larvae. In contrast, endosulfan (50 ppm) and fenitrothion (20 ppm) showed subadditive antagonistic response. Similar studies with *Hear*NPV against *H. armigera* larvae showed true synergism with endosulfan (20 and 50 ppm) and fenvalerate (20 and 50 ppm) while carbary1 (20 and 50 ppm) showed true antagonism (Singh *et al.*, 2001).

6.7.3. Field Trials

During the last decade, large scale field trials have been conducted on several crops to evaluate the efficacy of microbial pathogens.

6.7.3.1. Vegetable Crops

The diamondback moth, *Plutella xylolstella* (Linnaeus), is the key pest of cauliflower and cabbage crops in the state. *Bacillus thuringiensis* Berliner var. *kurstaki* - based commercially available biopesticides have been found highly effective against the pest. Weekly sprays of Dipel 8L / Biolep / Biobit @ 0.750 1 / kgha-1 effectively controlled the pest. These biopesticides caused > 80, > 95 and 100 per cent mortality after the first, second and third spray, respectively. In contrast, endosulfan and fenvalerate could kill only about 50 per cent of the larval population after each spray. The biopesticides were also moderately effective against other lepidopterous pests including head border, *Hellula undalis* (Fabricius), tobacco caterpillar, *S. litura* and *H. armigera* attacking cabbage and cauliflower crops in the state (Arora *et al.*, 2000a).

Spodoptera litura is a sporadic pest of vegetable crops especially cauliflower in the state. Weekly sprays of SINPV @ 2.25x10^{12} OBs/ha caused 23, 67 and 96 per cent larval mortality 5 days after the first, second, and third spray, respectively. The mean curd yield in the medium and higher dosage of NPV was 157.2 and 181.1 Q / ha, respectively, as compared to 182.5 and 184.3 in the two insecticide standards (Battu *et al.*, 1998). Earlier studies have indicated that mixing of charcoal 1 per cent enhanced the efficacy of evening spray of SINPV on cauliflower NPV could be applied either singly or tank mixed with Bt-based biopesticides (Battu, 1992). Thus, the total pest complex of cruciferous vegetable crops may be managed by use of biopesticides.

On tomato, weekly sprays of *Hear*-NPV @ 2.25x10^{12} OBs/ha during the fruiting period were found as effective as the standard insecticide treatment for the management of tomato fruit borer *H. armigera*. The mean percent fruit borer damage in *Hear*NPV treated plots was 12.1 as compared to 13.5 in insecticide standard (Carbaryl 50 WP 1 kg ai / ha) and 31.6 in untreated control. The mean tomato yield of four pickings during this period was 121.2, 121.5 and 82.1 q / ha in NPV, carbaryl and untreated control, respectively (Padmanabhan and Arora, 2002).

6.7.3.2. Cotton

Field evaluation of commercially available NPV of *H. armigera* as well as Bt based biopesticides particularly Dipel 8L either alone or in alternation with the recommended insecticides indicated low to moderate effectiveness against bollworms. Pooled analysis of data of 3 year's field trials on American cotton with Dipel 8L showed that seed cotton yield was 1426, 1640, 1819 and 1978 kg ha^{-1} in Dipel 8L alone, Dipel 8L alternated with endosulfan, recommended spray schedule and untreated control, respectively. In a large field scale trial, three weekly sprays of *Hear*NPV PAU-1 isolate at 2.7x10^{12} OBs/ha^{-1} in combination with the recommended spray schedule (RSS) were compared with the RSS alone and farmers' spray schedule (FSS) on the farmers' field. The mean per cent bollworm incidence was 22.46, 30.93 and 42.16 on boll basis in *Hear* NPV plus RSS, RSS alone and FSS, respectively (Battu, 2000).

An IPM module involving seed treatment with Gaucho (5 g kg^{-1}) followed by neem (Neem Azal 5 ml 1^{-1} water), endosulfan (875 g a.i. ha^{-1}), Biolep (0.750 kg ha-1), fenvalerate (50 g a.i. ha^{-1}), *Hear* NPV (2.7x10^{12} OBs/ha^{-1}), triazophos (600 g a.i. ha^{-1}), and chlorpyriphos (1 kg ha^{-1}), was evaluated for the management of cotton pests of American cotton genotype F-1378 at Ludhiana. The IPM module registered 42.0 and 27.2 per cent boll and loculi damage, respectively, as compared to corresponding per cent damage value of 32.6 and 20.3 in RSS and 67.9 and 40.6 in untreated check. The seed cotton yield obtained was 908, 728 and 406 kg ha^{-1} in RSS, IPM module and untreated check, respectively (Battu *et al.*, 1998). *Hear*-NPV, applied alone @ 1.5x10^{12} OBs and 3.0x10^{12} OBs/ha, in combination or interspersed between insecticide sprays @ 250 LE / ha alongwith standard insecticide sprays, was evaluated for management of cotton bollworm complex. The yield of seed cotton in plots receiving NPV sprays in combination or interspersed with insecticides was on par with plots receiving insecticide sprays alone (Dhawan *et al.*, 2001).

6.7.3.3. Other crops

Field trials have revealed that *H. armigera* on Egyptian clover and sunflower can also be managed by application of *Hear*NPV. Sunflower head borer, *H. armigera* could also be effectively managed by 2-3 weekly sprays of *Hear*-NPV@ 2.25×10^{12} OBs/ha^{-1}. Six days after two weekly sprays of NPV at 375 LE / ha, number of *H. armigera* larvae per plant was 1.0 as compared to 1.16 in carbaryl 50 WP (1.25 kg a.i./ha) and 3.16 in untreated control. The mean seed yield was 1349.5, 1235.0 and 633.0 kg / ha in NPV, carbaryl and untreated control, respectively (Arora *et al.*, 2002a). On Egyptian clover, *Hear* NPV @ 1.5×10^{12} OBs/ha^{-1} was found effective against small and medium sized larvae of *H. armigera* Nucleopolyhedrovirus has also been found promising for the management of Bihar hairy caterpillar, *S. obliqua*, on several oilseed crops including sunflower, groundnut and sesamum. SoNPV 1.5×10^{12} OBs/ha^{-1} was quite effective against the lower and medium larval instars of *S. obliqua* on these crops (Arora *et al.*, 1996; Battu and Arora, 1996).

6.7.4. Safety Evaluation

Both nucleopolyhedroviruses and granuloviruses have a limited host range and are usually genus or species specific and, therefore, quite safe to non-target organisms including natural enemies, pollinators, domestic animals and man. Extensive safety testing of NPV of *S. obliqua* following FAO / WHO protocols revealed that it was not hazardous to human beings and higher animals (Battu and Ramakrishnan, 1983).

The NPV of *S. litura* did not harm the larval-pupal parasitoid, *Parasarcophaga misera* (Walker), which is one of the important parasitoids of the pest under Punjab conditions. Moreover, the coexistence of braconid, ichneumonid and tachinid parasitoids within NPV-infected larvae and pupae of *S.litura*, *S. obliqua*, *H. armigera*, *E. lunata* and *P. brassicae,* without any ill effects on these natural enemies has been observed quite frequently (Battu *et al.*, 1998). The presence of low levels of virulent viral OBs of host insect NPVs in adult parasitoids bred from infected insects points towards the possible role of such parasitoids as natural reservoirs and transmitters of these viruses in the environment (Battu, 1992).

The effect of predation of the NPV-infected lepidopterans *S. obliqua, M. separata, T. orichalcea, S. litura and T. vishnou* on the survival of one reptilian *Calotes versicolor* Daudin, two avian *Corvus splendens* Viellot and *Passer domesticus* Linnaeus and two mammalian *Mus musculus* Linnaeus and *Rattus* Linnaeus predators was studied. Anticipated ingested levels of 0.29 to 3.5×10^7 POBs predator^{-1} h^{-1} field predation of infected insects as well as single oral administration of 14.90×10^8 and 7.45×10^8 POBs of NPVs of *S. obliqua* and *S. litura* per rat and mouse, respectively in the laboratory, proved innocuous to respective vertebrate hosts (Battu, 1987).

Unlike chemical pesticides, the NPVs do not persist on the crop or in the environment. More than 90 per cent activity of NPVs of *S. obliqua, S. litura* and *H. armigera* on different crops is lost within a week. Therefore, there are no hazards due to residues of these pathogens on the crops. This is particularly useful for export of agricultural commodities especially vegetables, fruits, basmati rice, cotton lint, etc. where presence of pesticide residues is affecting exports of these commodities.

6.8. CONCLUSIONS

The entomopathogens including viruses, bacteria, fungi, microsporidia and nematodes are undoubtedly important natural control agents regulating the populations of insect pests. But so far only transgenic crops (cotton, soybean, corn) containing the delta-endotoxin gene from *B. thuringiensis* have attained a dominant status internationally. Foliar products based on *B. thuringiensis* also occupy a significant share in the global pesticide market, due to their use for pest management in forestry, organic cultivation and speciality high value crops. Other entomopathogens including baculoviruses, fungi and nematodes, etc. have so far been able to capture only niche markets in specialized areas. This is because the microbial control agents have a rather narrow spectrum of activity, are slow acting (especially as compared to chemical pesticides) and are easily inactivated on the foliar surface.

However, recent advances in molecular biology have provided the necessary tools for genetic improvement of entomopathogens to overcome these limitations. There have also been significant improvements in mass production technology for all the major groups of entomopathogens, during the last three decades. However, further refinements in production technology are required to make these products competitive with chemical pesticides.

The environmental and ecological problems caused by extensive and intensive usage of chemical pesticides have forced the society to minimize the usage of these toxicants. Therefore, the chemical insecticide market is slowly shrinking. In contrast, microbial control agents have shown annual growth rate of more than 10 per cent during the last decade due to their safety, specificity and self-perpetuating action (Glare *et al.*, 2012).

There is a need to intensify research on entomopathogens in all the important areas: search for new microbial species/ strains causing epizootics in insect pests in nature; studies on tritrophic/ multitrophic interactions involving host plants-insect-pests (hosts)- pathogens; molecular studies for understanding the mechanism of action and for increasing host range, enhancing the speed of kill and improving photostability; development of economical methods for mass production alongwith improvements in formulation and application technology. With advances in these areas, the microbial control agents appear poised to play a greater role, as environmentally benign alternatives to chemical pesticides for sustainable insect pest management in future.

REFERENCES

Adams BJ, Fodor A, Koppenhofer HS, Stackebrandt E, Stock SP and Klein MG (2006) Biodiversity and systematics of nematode-bacterium entomopathogens. *Biological Control* **37**: 32-49.

Adams LA, Liu CL, McIntosh SC and Starnes RL (1996) Diversity and biological activity of *Bacillus thuringiensis*. In: LG Copping (ed). *Crop Protection Agents from Nature: Natural Products and Analogues.* The

Duran N and Menck CF (2001) *Chromobacterium violaceum*: a review of pharmacological and industrial perspectives. *Critical Reviews in Microbiology*, **27**: 201-222.

Dutky SR (1940) Two new spore forming bacteria causing milky diseases of Japanese beetle larvae. *Journal of Agricultural Research* **61**: 57-68.

Engelhard EK, Kam-Morgan LN, Washburn JO and Volkman LE (1994) The insect tracheal system: a conduit for the systemic spread of *Autographa californica* M nuclear polyhedrosis virus. *Proceedings of Naiotnal Academy of Sciences*, USA **91**: 3224-3227.

Entwistle PF, Cory JS, Bailey MJ and Higgs S (1993) *Bacillus thuringiensis, An Environmental Biopesticide: Theory and Practice.* John Wiley & Sons, Chichester, UK.

Favert EM and Yousten AA (1985) Insecticidal activity of *Bacillus laterosporus. Journal of Invertebrate Pathology* **45**: 195-203.

Ferguson CM, Barton DM, Harper LA, Swaminathan J, Van Koten C and Hurst MRH (2012) Survival of *Yersinia entomophaga* MH96 in a pasture ecosystem and effects on pest and non-target invertebrate populations. *New Zealand Plant Protection* **65**: 166-173.

Glare T, Caradus J, Gelernter W, Jackson T and Keyhani N *et al.* (2012) Have biopesticides come of age? *Trends in Biotechnology* **30**: 250-257.

Gonzalez JMJ, Brown BJ and Carlton BC (1982) Transfer of *Bacillus thuringiensis* plasmids coding for delta-endotoxin among strains of *Bacillus thuringiensis* and *Bacillus cereus. Proceedings of National Academy of Sciences*, USA **79**: 6951-6955.

Green BR (2011) Chloroplast genomes of photosynthetic eukaryotes. *Plant Journal* **66**: 34-44.

Gurtler H and Peterson A (1991) Mutants or variants of *Bacillus thuringiensis* producing high yields of delta endotoxin. Patent WO 91/07481, World Intellectual Property Organization.

Hajek AE (2007) Introduction of a fungus into North America for control of Gypsy moth. In: C Vincent, MS Goettel and G Lazarovits (eds), *Biological control: A Global Perspective.* CABI, Wallingford, pp. 53-62.

Harrison R and Hoover K (2012) Baculoviruses and other occluded insect viruses. In: FE Vega and HK Kaya (eds), *Insect Pathology*. Second Edition. Academic Press, London.

Hawtin RE, Zarkowska T, Arnold K, Thomas CJ, Gooday GW, King LA, Kuzio JA and Possee RD (1997) Liquefaction of *Autographa californica* nucleopolyhedrovirus-infected insects is dependent on the integrity of virus-encoded chitinase and cathepsin genes. *Virology* **238**: 243-253.

Hernandez-Martinez P, Naseri B, Navarro-Cerrillo G, Escriche B, Ferre J and Herrero S (2010) Increase in midgut microbiota load induces an apparent immune priming and increases tolerance to *Bacillus thuringiensis. Environmental Microbiology* **12**: 2730-2737.

Hofmann O (1891) *Die Schlaffsucht (Flacherie) der Nonne (Liparis monacha) nebst einem Analog. Insektenotende Pilze mit besonderer Berucksichtigung der Nonne.* P. Weber, Frankfurt.

Hofte H and Whitely HR (1989) Insecticidal crystal proteins of *Bacillus thuringiensis. Microbiology Reviews* **53**: 242-255.

Horton HM and Burand JP (1993) Saturable attachment sites for polyhedron-derived baculovirus on insect cells and evidence for entry via direct membrane fusion. *Journal of Virology* **67**: 1860-1868.

Hoshino T (2011) Violacein and related tryptophan metabolites produced by

Chromobacterium violaceum: biosynthetic mechanism and pathway for construction of violacein core. *Applied Microbiology and Biotechnology* **91**: 1463-1475.

Huber J (1986) Use of baculoviruses in pest management programs. In: RR Granados and BA Federici (eds), *The Biology of Baculoviruses. Vol II. Practical Application for Insect Control*. CRC Press, Boca Raton, pp. 181-202.

Hughes GL and Rasgon JL (2012) *Wolbachia* infections in arthropod hosts. In: FE Vega and HK Kaya (eds), *Insect Pathology* Second Edition. Academic Press, London, UK, pp. 351-366.

Humber RA (1997) Fungi: identification. In: LA Lacey (ed.), *Manual of Techniques in Insect Pathology*. Academic Press, San Diego, pp. 153-185.

Hunter-Fujita FR, Entwistle PF, Evans HF and Crook NE (eds) (1998) *Insect Viruses and Pest Management*. John Wiley & Sons, Chichester.

Hurst MRH, Becher SA, Young SD, Nelson TL and Glare TR (2011a) *Yersinia entomophaga* sp. Nov., isolated from the New Zealand grass grub *Costelytra zealandica*. *International Journal of Systematic and Evolutionary Microbiology* **61**: 844-849.

Hurst MRH, Jones SA, Binglin T, Harper LA, Jackson TA and Glare TR (2011b) The main virulence determinant of *Yersinia entomophaga* MH96 is a broad-host-range toxin complex active against insects. *Journal of Bacteriology* **193**: 1966-1980.

Inceoglu AB, Kamita SG and Hammock BD (2006) Genetically modified baculoviruses: a historical overview and future outlook. *Advances in Virus Research* **68**: 323-360.

Jackson TA (2007) A novel bacterium for control of grass grub. In: C Vincent, MS Goettel and G Lazarovits (eds), *Biological Control: A Global Perspective*. CABI, Wallingford, UK, pp. 160-168.

Jackson TA, Pearson JF, O'Callaghan M, Mahanty HK and Willcocks MJ (1992) Pathogen to product-development of *Serratia entomophila* (Enterobacteriaceae) as a commercial biological agent for the New Zealand grass grub (*Costelytra Zealandica*). In: TA Jackson and TR Glare (eds), *Use of Pathogens in Scarab Pest Management*. Intercept Press, Andover, pp. 191-198.

Jayaraj S, Amanthakrishnan TN and Veeresh GK (1994) *Biological Pest Control in India, Progress and Per- -tives.* RGICS Project No. 2, Rajiv Gandhi Institute of Contemporary Studies, New Delhi.

Jehle JA, Blissard GW, Bonning BC, Cory JS, Hernion EA, Rohrmann GF, Theilmann DA, Thiem SM and Vlak JM (2006) On the classification and nomenclature of baculoviruses, a proporal for revision. *Archives in Virology* **151:** 1257-1266.

Ji X, Sutton G, Evans G, Axford D, Owen R and Stuart DI (2010) How baculoviruses polyhedra fir square pegs into round holes to robustly package viruses. *EMBO Journal* **29**: 505-514.

Jurat-Fuentes J and Jackson TA (2012) Bacterial entomopathogens. In: FE Vega and HK Kaya (eds) *Insect Pathology*. Second Edition. Academic Press, London, pp. 265-350.

Kaur S (2000) Molecular approaches towards development of novel *Bacillus thuringiensis* biopesticides. *World Journal of Microbial Biotechnology* **16**: 781-793.

Kaya HK, Aguilera MM, Alumai A, Chaoo HY, de la Torre M *et al* (2006) Status of entomopathogenic nematodes and their symbiotic bacteria from selected countries or regions of the world. *Biological Control* **38**: 134-155.

Khachatourians GG, Valencia EP and Miranpuri GS (2002) *Beauveria bassiana* and other entomopathogenic fungi in the management of insect pests. In: O Koul and GS Dhaliwal (eds), *Microbial Biope-*

sticides. Taylor & Francis, London, pp. 239-276.

Klein MG (1981) Advances in the use of *Bacillus popilliae* for pest control. In: HD Burges (ed.), *Microbial Control of Pests and Plant Diseases 1970-1980*. Academic Press, London, pp. 183-192.

Koivunen M, Chanbusarakum L, Fernandez L, Asolkar R, Tan E, Wallner D and Marrone P (2009) Development of a new microbial insecticide based on *Chromobacterium subtsuage*. *IOBC/wprs Bulletin* **45**: 183-186.

Koppenhofer AM and Kaya HK (2002) Entomopathogenic nematodes and insect pest management. **In:** O Koul and GS Dhaliwal (eds), *Microbial Biopesticides*. Taylor & Francis, London, pp. 277-306.

Lacey LA (1990) Persistence and formulation of *Bacillus sphaericus*. In: H de Barjac and DJ Sutherland (eds), *Bacterial Control of Mosquitoes & Black Flies: Biochemistry, Genetics & Application of Bacillus thuringiensis & Bacillus sphaericus*. Rutzers University, Press, New Brunswick, pp. 285-294.

Lacey LA (2007) *Bacillus thuringiensis* serovariety israelensis and *Bacillus sphaericus* for mosquito control. *Journal of American Mosquito Control* Association **23**: 133-163.

Lampel JS, Canter GL, Dimock MB, Kelly JL, Anderson JJ, Uratani BB, Faulke JS and Turner JT (1994) Integrative cloning, expression and stability of *Cry1AC(c)* gene from *Bacillus thuringiensis* subsp. *kurstaki* in a recombinant strain of *Clavibacter xyli* subsp. *cynodontis*. *Applied and Environmental Microbiology* **60**: 501-508.

Lange CE (2005) The host and geographical range of the grasshopper pathogen *Paranosema (Nosema) Locustae* revisited. *Journal of Orthoptera Research* **14:** 137-141.

Langewald J and Kooyman C (2007) Green Muscle, a fungal biopesticide for control of grasshoppers and locusts in Africa. In: C Vincent, MS Goettel and G Lazarovits (eds), *Biological Control: A Global Perspective*. CABI, Wallingford, pp. 311-318.

Lasa R, Ruiz-Portero C, Alcazar MD, Belda JE, Williams T and Caballero P (2007) Efficacy of optical brightner formulations of *Spodoptera exigua* multiple nucleopolyhedrovirus (SeMNPV) as a biological insecticide in green houses in southern Spain. *Biological Control* **40**: 89-96.

Laubach AC (1916) Studies on aerobic, sporebearing, non-pathogenic bacteria. Spore bearing organism in water. *Journal of Bacteriology* **1**: 505-512

Lewis EE and Clarke DJ (2012) Nematode parasites and entomopathogens. In: FE Vega and HK Kaya (eds), *Insect Pathology*. Second Edition, Academic Press, London, pp. 395-424.

Liu B and Sengonca C (2003) Conjugation of delta-endotoxin from *Bacillus thuringiensis* with abamectin of *Streptomyces avermitilis* as a new type of biocide, GCSC-BtA, for the control of agricultural insect pests. *Anzeiger-fur-Schadlingskunde* **76**: 44-49.

Lomer CJ, Bateman RP, Johnson DL, Langewald J and Thomas M (2001) Biological control of locusts and grasshoppers. *Annual Review of Entomology* **46**: 667-702.

Mahafee WE, Moor WJ and Kloepper JW (1994) Bacterial endophytes genetically engineered to express the CryIIA delta-endotoxin from *Bacillus thuringiensis* subsp. *kurstaki*. In: MH Ryder, PM Stevens and GD Brown (eds), *Improving Plant Productivity with Rhizosphere Bacteria*. CSIRO, Australia, pp. 245-246.

Marshall SDG, Hares MC, Jones SA, Harper LA, Vernon JR, Harland DP, Jackson TA and Hurst MRH (2012) Histopathological effects of the Yen-Tc toxin complex from *Yersinia entomophaga* MH96 (Enterobacteriaceae) on the *Costelytra zealandica*

(Coleoptera: Scarabaeidae) larval midgut. *Applied and Environmental Microbiology* **78**: 4835-4847.

Martin PAW, Hirose E and Aldrich JR (2007a) Toxicity of *Chromobacterium subtsuage* to Southern sting bug (Heteroptera: Pentatomidae) and corn rootworm (Coleoptera: Chrysomelidae). *Journal of Economic Entomology*, **100**: 680-684.

Martin PAW, Gundersen Rindal D, Blackburn M and Buyer J (2007b) *Chromobacterium subtsugae* sp. Nov. and use fo control of insect pests. *US Patent Application Publication*, 2007/0172463 A1.

Martin PAW, Shrophire ADS, Gundersen-Rindal DE and Blackgurn MB (2007c) *Chromobacterium subtsugae* sp. nov. and use for control of insect pests. US45 Patent Application Publication, 2007/1012463A1

McCray AH (1917) Spore-forming bacteria in the apiary. *Journal of Agricultural Research* **8**: 399-420.

Miller LK (1998) Genetically improved baculoviruses for insect pest management. In: VL Chopra, RB Singh and A Varma (eds), *Crop Productivity and Sustainability. Shaping the Future. Proc Second International Crop Science Congress*, Oxford & IBH Publ Co. Pvt. Ltd., New Delhi, pp. 269-276.

Miller LK, Lingg AJ and Bulla LA, Jr (1983) Bacterial, viral and fungal insecticides. *Science* **219**: 715-721.

Milner RJ (1981) Identification of *Bacillus poppliae*, group of insect pathogens. In: HD Burges (ed), *Microbial Control of Pests and Plant Diseases 1970-1980*. Academic Press, London, pp. 45-59.

Mohamed HAA, Abd El-Fatah MR, Sabbour MM, El-Shackawey AZ and El-Sayed RS (2010) Genetic modification of *Bacillus thuringiensis* var. *kurstaki* HD-73 to overproduce melanin, UV resistance and their insecticidal potency against potato tuber moth. *International Journal of Academic Research* **2**: 73-83.

Molnar I, Gibson DM and Krasnoff SB (2010) Secondary metabolites from entomopathogenic Hypocrealean fungi. *Natural Product Report* **27**: 1241-1275.

Moore D (2008) A plague on locusts-the Lubilosa story. *Outlook of Pest Management* **19**: 14-17.

Morales L, Moscardi F, Sosa-Gamez DR, Paro FE and Soldorio IL (1997) Enhanced activity of *Anticarsia gemmatalis* Hub. (Lepidoptera: Noctuidae) nuclear polyhedrosis virus by boric acid in the laboratory. *Annales de la Société entomologique de France* **26**: 115-120.

Moscardi F (1999) Assessment of the applications of baculoviruses for control of Lepidoptera. *Annual Review of Entomology* **44**: 257-289.

Moscardi F (2007) A nucleopolyhedrovirus of the velvetbean caterpillar in Brazilian soybeans. In: C Vincent, MS Goettel and G Lazarovits (eds), *Biological Control: A Global Perspective*. CABI, Oxfordshire, pp. 344-352.

Moscardi F, Allen GE and Greene GL (1981) Control of the velvetbean caterpillar by nuclear polyhedrosis virus and insecticides and impact of treatments on the natural incidence of the entomopathogenic fungus *Nomuraea rileyi*. *Journal of Economic Entomology* **74**: 480-485.

Nambiar PTC, Ma SW and Iyer VN (1990) Limiting an insect infestation of nitrogen fixing root nodules of the pigeonpea (*Cajanus cajan*) by engineering the expression of an entomocidal gene in its root nodules. *Applied and Environmental Microbiolgy* **56**: 2866-2869.

Narayanan K (2002) Microbial control of insect pests: Role of genetic engineering and tissue culture. In: O Koul and GS Dhaliwal (eds), *Microbial Biopesticides*. Taylor & Francis, London, pp. 117-179.

Nester EW, Thomashow LS, Metz M and Godon M (2002) 100 years of *Bacillus thuringiensis*: *A Critical Scientific Assessment*. American Academy of Microbiology, Washington 22 pp.

Oerke EC (2006) Crop losses to pests. *Journal of Agricultural Science* **144**: 31-43.

Ohkawa T, Volkman LE and Welech MD (2010) Action-based motility drives baculovirus transit to the nucleus and cell surface. *Journal of Cell Biology* **190**: 187-195.

Oliveira EJ, Rabinovitch L, Monnerat RG, Passos LK and Zahner V (2004) Molecular characterization of *Brevibacillus laterosporus* and its potential use in biological control. *Applied Environmental Microbiology* **70**: 6657-6664.

Padmanaban N (2001) Evaluation of native nuclear polyhedrosis virus isolates for the management of *Helicoverpa armigera* (Hubner). M.Sc. Thesis, Punjab Agricultural University, Ludhiana.

Padmanabhan N and Arora R (2002) Field evaluation of native nucleopolyedrovirus for the management of tomato fruit borer *Helicoverpa armigera* (Hubner). *Pesticide Research Journal* **14**: 113-119.

Pusztai M, Fast, P, Gringorten H, Kaplan H, Lessard T and Carey PR (1991) The mechanism of sunlight mediated inactivation of *Bacillus thuringiensis* crystals. *Biochemistry Journal* **273**: 43-47.

Raymond B, Johnston PR, Wright DJ, Ellis RJ, Crickmore N and Bonsall MB (2009) A mid-gut microbiota is not required for the pathogenicity of *Bacillus thuringiensis* to diamondback moth larvae. *Environmental Microbiology* **11**: 2556-2563.

Rehner SA (2009) Molecular systematics of entomopathogenic fungi. In: SP Stock, J Vandenberg, I Glazer and N Boemare (eds), *Insect Pathogens: Molecular Approaches and Techniques*. CABI, Wallingford, UK, pp. 145-165.

Rivers L, Vann CN, Zimmack HL and Dean DH (1991) Mosquitocidal activity of *Bacillus laterosporus*. *Journal of Invertebrate Pathology* **58**: 444-447.

Roberts DW and Humber RA (1981) Entomogenous fungi. In: GT Cole and B Kendrick (eds), *Biology of Conidial Fungi*. Vol 2. Academic Press, New York, pp. 201-236.

Rosell R, Quero C, Coll J and Guerrero A (2008) Biorational insecticides in pest management. *Journal of Pesticide Science* **33**: 103-121.

Roy HE, Steinkraus DC, Eilenberg J, Hajek AE and Pell JK (2006) Bizarre interactions and endgames: entomopathogenic fungi and their arthropod hosts. *Annual Review of Entomology* **51**: 331-357.

Ruiu L, Delrio G, Ellar DJ, Floris I, Paglietti B, Rubino S and Satta A (2006) Lethal and sub-lethal effects of *Brevibacillus laterosporus* on the housefly (*Musca domestica*). *Entomologia Experimentalis et Applicata* **118**: 137-144.

Ruiu L, Floris I, Satta A and Ellar DJ (2007) Toxicity of a *Brevibacillus laterosporus* strain lacking parasporal crystals against *Musca domestica* and *Aedes aegypti*. *Biological Control* **43**: 136-143.

Ruiu L, Satta A and Floris I (2008) Immature house fly (*Musca domestica*) control in breeding sites with a new *Brevibacillus laterosporus* formulation. *Environmental Entomology* **37**: 505-509.

Ruiu L, Satta A and Floris I (2011) Comparative applications of azadirachtin and *Brevibacillus laterosporus* based formulations for house fly management experiments in dairy farms. *Journal of Medical Entomology*, **48**: 345-350.

Ruiu L, Satta A and Floris I (2012) Observations on house fly larvae midgut ultrastructure after *Brevibacillus lateros-porus* ingestion. *Journal of Invertebrate Pathology*, **111**: 211-216.

Ruiu L, Satta A and Floris I (2013) Emerging entomopathogenic bacteria for insect pest managemet. *Bulletin of Insectology* 66: 181-186.

Saikia R, Gogoi DK, Mazumder S, Yadav A, Sarma RK, Bora TC and Gogoi BK (2011) *Brevibacillus laterosporus* strain BPM3, a potential biocontrol agent isolated from a natural hot water spring of Assam, India. *Microbiological Research* **166**: 216-225.

Sanahuja G, Banakar R, Twyman RM, Capell T and Christou P (2011) *Bacillus thuringiensis*: a centry of research, development and commercial applications. *Plant Biotechnology Journal* **9**: 283-300.

Sanchis V (2000) Biotechnological improvement of *Bacillus thuringiensis* for agricultural control of insect pests: benefits and ecological implications. In: JF Charles, A Delecluse, C Neilson-Leroux (eds), *Entomopathogenic Bacteria: From Laboratory to Field Applications*. Kluwer Academic Publishers, Dordrecht, The Netherlands, pp. 441-459.

Sanchis V (2011) From microbial sprays to insect-resistant transgenic plants: history of the biopesticide *Bacillus thuringiensis*. *Agronomy for Sustainable Development* **31**: 217-231.

Saxena D, Ben DE, Manasherob R, Barak Z, Boursiba S and Zaritsky A (2002) A UV tolerant mutant of *Bacillus thuringiensis* subsp. *kurstaki* producing melanin. *Current Microbiology* **44**: 25-30.

Shapiro Ilan D, Han R and Dolinksi C (2012) Entomopathogenic nematode production and application technology. *Journal of Nematology* **44**: 206-217.

Shapiro M (1992) Use of optical brightners as radiation protectants for gypsy moth (Lepidoptera: Lymantriidae) nucleopolyhedrosis virus. *Journal of Economic Entomology* **82**: 1682-1686.

Shapiro M and Argauer R (1997) Components of the stilbene brightner Tinopal LPW as enhancers for the Gypsy moth (Lepidoptera: Lymantriidae) baculovirus. *Journal of Economic Entomology* **90**: 899-904.

Shapiro M, Preisler HK and Robertson JL (1987) Enhancement of baculovirus activity on gypsy moth (Lepidoptera: Lymantriidae) by chitinase. *Journal of Economic Entomology* **85**: 1120-1124.

Sharma HC (2009) *Biotechnological Approaches for Pest Management and Ecological Sustainability*. CRC Press, Boca Raton, FL, USA.

Sheppard RF and Stairs GR (1977) Dosage-morality and time-mortality studies of a granulosis virus in a laboratory strain of the codling moth, *Laspeyresia pomonella*. *Journal of Invertebrate Pathology* **29**: 216-221.

Shi W, Guo Y, Xu C, Tan S, Miao J, Feng Y, Zhao H and Fang W (2014) Unveiling the mechanism by which microsporidian parasites prevent locust swarm behavior. *Proceedings National Academy Sciences*, USA **111**: 1343-1348.

Singer S (1981) The utility of strains of morphological group II *Bacillus*. In: LN Saul and I L Allen (eds), *Advances in Applied Microbiology*. Vol. 42, Academic Press, San Diego, pp. 219-261.

Singer S (1996) The utility of morphological group II *Bacillus*. *Advances in Applied Microbiology* **42**: 219-261.

Singh AP, Arora R and Battu GS (2001a) Effect of two *Bacillus thuringiensis* based biopesticides on larval development of diamondback moth *Plutella xylostella* (Linnaeus). *Symposium on Biocontrol Based Pest Management for Quality Crop Protection in the Current Millennium*. July 18-19, 2001. Punjab Agricultural University, Ludhiana.

Singh S (1999) Use of native nuclear polyhedrosis virus against *Helicoverpa armigera* (Hubner) on tomato. M.Sc. Thesis, Punjab Agricultural University, Ludhiana.

Singh S, Battu GS and Arora R (2001) Interaction of carbaryl, endosulfan and fenvalerate with *Nucleopolyhedrovirus* of the tomato fruit borer *Helicoverpa armigera* (Hubner). *Indian Journal of Ecology* **28**: 83-85.

Skot L, Harrison SP, Nath A, Mytton LR and Clifford BC (1990) Expression of insecticidal activity in *Rhizobium* containing delta-endotoxin gene cloned from *Bacillus thuringiensis* subsp. *tenebrionis*. *Plant Soil* **127**: 285-295.

Smirnoff WA (1961) A virus disease of *Neodiprion swanei* Middleton. *Journal of Insect Pathology* **3**: 29-46.

Smith CR, Heinz KM, Sansone CG and Flexnor JL (2000) Impact of recombinant baculoviruses on target heliothines and non target predators in cotton. *Biological Control* **19**: 201-214.

Solter LF, Becnel JJ and Di DH (2012) Microsporidion entomopathogens. In: Vega FE and Kaya HK (eds), *Insect Pathology*. Second edition, Elsevier, London, pp 221-263.

St. Leger RJ (1991) Integument as a barrier to microbial infections. In: K Binnington and A Retnakaran (eds), *Physiology of Insect Epidermis*. CSIRO, Australia, pp. 284-306.

St. Leger RJ and Wang C (2009) Entomopathogenic fungi and the genomics era. In: SP Stock, J Vandenberg, I Glazer and N Boemare (eds), *Insect Pathogens: Molecular Approaches and Techniques*. CABI, Wallingford, pp. 365-400.

Stairs GR, Fraser T and Fraser M (1981) Changes in growth and virulence of nuclear polyhedrosis virus from *Choristoneura fumiferana* after passage in *Trichoplusia ni* and *Galleria mellonella*. *Journal of Invertebrate Pathology* **38**: 230-235.

Summers MD (1971) Electron microscopic observations on granulosis virus entry, uncoating and replication processes during infection of midgut cells of *Trichoplusia ni*. *Journal of Ultrastructure Research* **35**: 606-625.

Sun X and Peng H (2007) Recent advances in biological control of pest insects by using viruses in China. *Virology Sinica* **22**: 158-162.

Sun X, Wua D, Sun X, Jin L, Mab Y, Bonning BC, Peng H and Hu D (2009) Impact of *Helicoverpa armigera* nucleopolyhedrovirus expressing a cathepsin L-like protease on target and non target insect species on cotton. *Biological Control* **49**: 77-83.

Szolajska E, Poznonski J, Ferfer ML, Michalik J, Gout E, Fender P, Bailey I, Dablet B and Chroboczek J (2004) Poneratoxin a neuro toxin from insect venom. Structure and expression in insect cells and construction of a bioinsecticide. *European Journal of Biochemistry*. **271:** 2127-2136.

Tanada Y and Kaya HK (1993) *Insect Pathology*. Academic Press, San Diego.

Tadeu M, Silva B and Moscardi F (2002) Field efficacy of the nucleopolyhedrovirus of *Anticarsia gennatalis* Hubner (Lepidoptera : Noctuidae) : effect of formulations, water pH, volume and time of application, and the type of spray nozzle. *Neotropical Entomology* **31:** 75-83.

Udayasuriyan V, Nakamura A, Masaki H and Uozumi T (1995) Transfer of an insecticidal protein gene of *Bacillus thuringiensis* into plant colonizing Azospiricllium. *World Journal of Microbial Biotechnology* **11**: 163-167.

Vail PV and Tebbets (1991) The granulosis virus of the Indian meal moth: recent developments. In: *Proc Fifth International Working Conference on Stored Product Protection.* Bordeaux, France, pp. 1247-1254.

Van Frankenhuyzen K, Liu Y and Tonon A (2010) Interactions between *Bacillus thuringiensis* subsp. *kurstaki* HD-1 and midgut bacteria in larvae of gypsy moth and spruce budworm. *Journal of Invertebrate Pathology* **103**: 124-131.

Vasconcelos SD, Cory JS, Wilson KR, Sait SM and Hails RS (1996) Modified behaviour in baculovirus-infected lepidopteran larvae and its impact on the spatial distribution of inoculum. *Biological Control* **7**: 299-306.

Vega FE and Kaya HK (eds) (2012) *Insect Pathology* Second Edition. Academic Press, London, 490 pp.

Vega FE, Meyling NV, Luangsa-ard JJ and Blackwell M (2012) Fungal Entomopathogens. In: FE Vega and HK Kaya (eds), *Insect Pathology.* Second Edition. Academic Press, London, pp. 171-220.

Wang C and St. Leger RJ (2007) The MAD1 adhesion of *Metarhizium anisopliae* links adhesion with blastospore production and virul-ence to insects, and the MAD 2 adhesion enables attachment to plants. *Eukaryotic Cell* **6**: 808-816.

Wang P and Granados RR (1997) An intestinal mucin is the target substrate for a baculovirus enhancin. *Proceedings of National Academy of Sciences,* USA **94**: 6977-6982.

Wood HA and Hughes PR (1996) Recombinant viral insecticides: delivery of environmentally safe and cost effective products. *Entomophaga* **41:** 361-373.

Wood HA, Hughes Pr., Johnson JD and Landgridge WHR (1981) Increased virulence of *Autographa californica* nuclear polyhedrosis virus by mutagenesis. *Journal of Invertebrate Pathology* **38:** 236-241.

Yap HH (1990). Field trials of B. sphaericus for mosquito control. In: H de Barjac and DJ Sutherland (eds), *Bacterial Control of Mosquitoes & Black Flies: Biochemistry, Genetics & Application of Bacillus thuringiensis & Bacillus sphaericus.* Rutzers University, Press, New Brunswick, pp. 307-320.

Young SY and Yearian WC (1974) Persistence of *Heliothis* NPV on foliage of cotton, soybean and tomato. *Environmental Entomology* **3**: 253-260.

Zare R and Gams W (2001) A revision of *Verticillium* section *Prostrata* IV. The genera *Lecanicillium* and *Simplicillium* gen. nov. *Nova Hedwigia* **73**: 1-50.

Zhu YJ and Sengonka C (2006) Efficiency of GCSC-Bt A, as a new type of biocide, on different arthropod pests and its side effects on some predators. *Mitteilungen der deutschen Gesellschaft fur allgemeine und angwandte Entomologie* **15**: 309-313.

Zimmermann G (1993) The entomopathogenic fungs *Metarhizium anisopliae* and its potential as a biocontrol agent. *Pesticide Science.* **37:** 375-379.

Zubasheva MV, Ganushkina LA, Smirnova TA and Aziz Bekyan RR (2010) Larvicidal activity of crystal-forming strains of *Brevibacillus laterosporus. Applied Biochemistry and Microbiology* **46**: 755-762.

Zubasheva MV, Ganushkina LA, Smirnova TA and Aziz Bekyan RR (2011) Enhancement of larvicidal activity of *Brevibacillus laterosporus* by bioincapsulation in Protozoa *Tetrahymena pyriformis and Entamoeba moshkovskii. Applied Biochemistry and Microbiology* **47**: 762-766.

Biological and Molecular Approaches in Pest Management
Balwinder Singh, Ramesh Arora & S S Gosal (Eds)
Scientific Publishers (India), Jodhpur, 2015, pp 153-160.

Chapter - 7

Towards Durable Gall Midge Resistance in Rice

J S Bentur

Directorate of Rice Research, Rajendranagar, Hyderabad 500 030, India,
E-mail: jbentur@yahoo.com

7.1 INTRODUCTION

Rice production suffers an average annual yield loss of about 10-15 per cent due to insect pests, in the country. Major among them are stem borers, gall midge, plant-hopper and leaf folder. Direct feeding injury due to planthoppers is compounded with transmission of viral diseases. Among various components of insect pest management, genetic approach through development and deployment of insect resistant elite rice lines is the most economically acceptable and ecologically welcome proposition. Identification and utilization of natural resistance available in the large pool of primary and secondary gene pool has already resulted in significant economic gains during the past one-half of the century. Large scale and often unplanned deployment of resistant rice cultivars has often led to unforeseen problems like development of virulent populations of the pest – biotypes – capable of overcoming the host resistance and emergence of non-target minor pests as major pests. Recent advances in molecular biology and availability of biotechnological tools have brought in a paradigm shift in our approach to explore and deploy plant resistance against insect pests towards more rational ways to achieve multiple and durable resistance. Our experience and research on rice-gall midge system over the past four decades have led to better understanding and provided an insight towards durable gall midge resistance in rice, as recounted here.

7.2. PLANT RESISTANCE AND BIOTYPES

Though report of gall midge resistance in rice is over 50 years old, systematic evaluation of rice germplasm against the pest was taken up during 70s. Greenhouse rearing and screening protocols (Kalode *et al.,* 1979) accelerated germplasm evaluation. To date over 25,000 accessions of primary pool of rice germplasm have been evaluated and over 500 accessions have been reported to be resistant to the gall midge (Bentur *et al.,* 2003; Bentur *et al.,* 2011). Even during the early phase of field screening, differences in reaction were seen against the same rice variety at different locations, suggesting presence of biotypes. Analysing the 13 year data on field evaluation from different locations, Kalode and Bentur (1989) characterized three distinct gall midge biotypes that prevailed in the country prior to 1986. The first gall midge resistant rice variety was `Kakatiya' bred and released during 1975. Since then over 100 rice varieties released for cultivation have been claimed to be gall midge resistant (Rani *et al.,* 2011).

With the report of emergence of a new biotype, gall midge biotype (GMB) 4, as a direct consequence of extensive cultivation of the resistant rice variety `Phalguna' in the north-eastern part of the Andhra Pradesh (Bentur *et al.,* 1986) second phase of resistance breeding started. During the period 1986 to 1996, three more new biotypes were reported (Nair and Devi 1994, Sinivas *et al.,* 1994, Singh 1996). More recently, seventh biotype, GMB4M was reported (Vijayalakshmi *et al.,* 2006). Concomitantly, genetics of gall midge resistance was being intensively investigated at IGAU, Raipur leading to report of ten resistance genes *Gm1* through *Gm10* (Kumar *et al.,* 2005). Thus it became apparent that most of the gall midge resistant rice varieties bred so far carried one of these genes and development of virulence in the insect population was evident in those areas where such rice varieties were extensively cultivated. More recently *Gm11* gene was identified from the breeding line CR57-MR1523 (Himabindu *et al.,* 2010). Thus, so far 11 plant resistance genes and seven distinct gall midge biotypes have been reported. Evidences strongly suggest gene-for-gene relationship between plant R genes and virulence genes in biotypes. The range and pattern of resistance displayed by rice gene differential varieties against the known biotypes are presented in Table 1. This spectrum of resistance conferred by the gall midge resistance genes helped to categorize the R genes into four groups. It has also been reported that resistance to gall midge can be associated with expression of hypersensitive reaction (HR+type) or can be independent of it (HR-type) (Bentur and Kalode, 1996). Only two of the 11 known genes, *Gm1* and *Gm8,* confer HR- type resistance while other genes provide HR+ type of resistance.

7.3. TAGGING AND MAPPING GALL MIDGE RESISTANCE GENES IN RICE

The author's group at Directorate of Rice Research, Hyderabad, India was the first to map and tag naturally occurring gall midge resistance genes in rice and design molecular markers for the marker-aided selection of several resistance genes found in the rice germplasm (Nair *et al.,* 2011). Eight of the 11 genes have been tagged and mapped so far. Four of the genes, viz., *Gm2, gm3, Gm6* and *Gm7* have been mapped on chromosome 4 located very close to each other, while two genes *Gm4* and *Gm8* have been mapped on chromosome 8 but at two distinctly separate loci. *Gm1* has been reported on Chromosome 9 and *Gm11* on chromosome 12 (Fig. 1) (Yasala *et al.,* 2012). With the help of closely linked and flanking markers, it has become now feasible to transfer any of these R genes into elite background through marker assisted selection (MAS). Recent studies have aimed to fine map the genes and develop more precise functional markers. We now have functional markers for *Gm2, gm3, Gm4* and *Gm8* genes.

7.4. GENE PYRAMIDING FOR DURABLE RESISTANCE

One of the major problems encountered in the development of rice varieties resistant to gall midge is that the resistance is governed by a single gene and over a period of time gall midge develops virulence, through the evolution of a new biotype, against the deployed resistance genes. It has been suggested that pyramiding two or more genes into a single plant and developing such gene pyramided variety will provide durable gall midge resistance (Cohen *et al.,* 2004). However, which genes in combination will provide durability needs to be determined. It is suggested that pyramided genes should be with different mechanisms of resistance. So far, most of the gall midge resistant rice varieties developed derive their resistance mainly from *Gm1, Gm2, Gm4* and *Gm11* genes. Thus, these genes are less likely candidates for pyramiding, though *Gm1* gene has shown durability over 30 years of deployment and *Gm4* carrying variety 'Abhaya' has not been cultivated widely. However, virulence against both *Gm2* and *Gm11* has been reported from different locations across the country.

Table 1. Characteristic reaction patterns of seven gall midge biotypes (GMB) against the set of 11 gene differential rice varieties

Group	Designation	Resistance Gene	Nature of resistance	Reaction pattern against gall midge biotypes						
				GM B1	GM B2	GM B3	GM B4	GM B4	GM B5	GM B6
I	W1263	*Gm1*	HR-	R	S	R	S	S	R	R
II	Phalguna	*Gm2*	HR+	R	R	S	S	S	S	S
	ARC5984	*Gm5*	HR+	R	R	S	S	S	S	S
	Dukong 6	*Gm6*	HR+	R	R	S	S	S	S	S
	RP2333156-8	*Gm7*	HR+	R	R	S	S	S	S	S
	Madhuri L9	*Gm9*	HR+	R	R	S	S	S	S	S
	BG 380-2	*Gm10*	HR+	R	R	S	S	S	S	S
III	CR57-MR1523	*Gm11*	HR+	R	R	R	R	S	S	S
IV	RP2068-18-3-5	*gm3*	HR+	R	R	R	R	R	S	S
	Abhaya	*Gm4*	HR+	R	R	R	R	R	S	S
	Aganni	*Gm8*	HR-	R	R	R	R	R	S	S
V	TN1	*None*	--	S	S	S	S	S	S	S

Source: Nair *et al.* (2011)

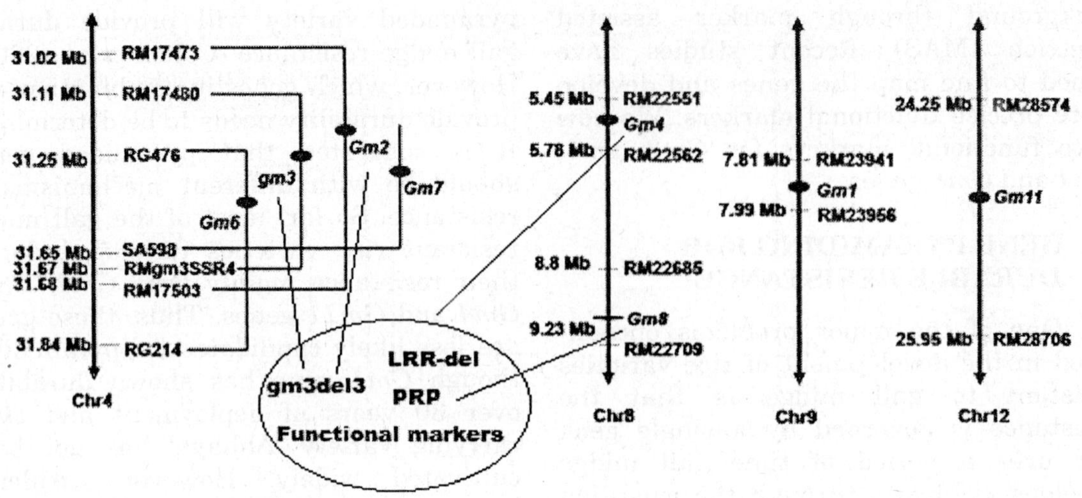

Fig. 1. Rice gall midge resistance genes have been mapped, and linked and functional markers developed for marker assisted introgression of genes and their pyramiding for durable gall midge resistance. (Modified from Yasala *et al.*, 2012)

7.5. VIRULENCE MONITORING IN GALL MIDGE POPULATIONS

A modified F_2 screen method has been developed for monitoring virulence in the gall midge populations (Bentur *et al.*, 2008, Andow and Bentur, 2010). Studies using this method have shown high level of virulence against *Gm2* mediated resistance across the country. Further, rate of virulence development in gall midge populations at Warangal against *Gm2* gene was shown to be much higher than virulence against *Gm1* resistance. Recent studies have shown high levels of virulence against *Gm11* gene and low levels against *Gm8* gene. These studies thus form basis for selecting the genes for pyramiding. Inheritance of virulence in the gall midge biotypes suggested sex linked inheritance of virulence conferred by a single recessive gene *vGm2* against *Gm2* gene conferred resistance (Bentur *et al.*, 1982). Such virulence gets fixed in the population faster than the autosomal inherited virulence gene and is responsible for less durability of the plant resistance gene.

7.6. MOLECULAR BASIS OF RESISTANCE

An understanding of the molecular basis of rice-gall midge interactions would provide basic information on nature and durability of plant resistance and insect virulence (Fig. 2). Resistance genes differ in the nature of resistance. Various gene expression profiling and validation techniques have been employed to study such interactions. It is evident from these studies that HR+ type of resistance is typically gall midge inducible plant defence as in several plant-pathogen systems (Rawat *et al.*, 2010, Rawat *et al.*, 2013). However, several of the genes like *Gm2*, *gm3* and *Gm4* known to confer this type of resistance differ in their spectrum of resistance across gall midge biotypes and in nature of inheritance, either as dominant or as recessive traits. Some of the genes like *gm3* and *Gm4* have been

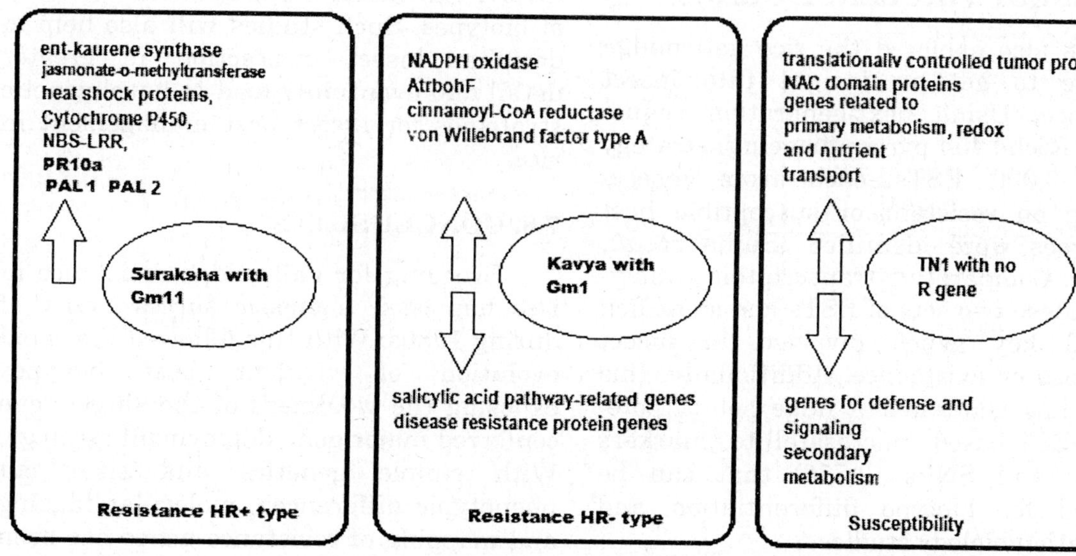

Fig. 2. Sets of genes either up or down regulated during three distinct rice-gall midge interactions

closely mapped within the narrow genomic regions and possible candidate genes have been identified. Cloning and sequencing these genes would help us understand the basis of wide spectrum resistance.

In contrast, HR- type of resistance in rice variety Kavya having *Gm1* gene displayed a novel mechanism (Rawat *et al.*, 2012a). This gene does not induce any hypersensitive response to the pest attack and it appears to be constitutively expressed. Though closely mapped on chromosome 9, this gene as yet eludes proper identification and elaboration in terms of structure and function. In view of the proven track of durability of this gene, further intensive studies would provide us the basis of durability of resistance. Another known gene, *Gm8*, also expresses HR- type resistance but has wider spectrum of resistance. It is important to know whether resistance conferred by this gene is similar to that by *Gm1* and look for more novel genes with HR- type resistance.

We probed further into the molecular basis of plant susceptibility (Rawat *et al.*, 2012b). As could be expected, all the known defense related genes were found to be down regulated during a compatible interaction while set of genes related to primary metabolism, nutrient transport, nutrient relocation, cell organization and DNA synthesis were upregulated. Development of gall is not a normal part of the plant morphogenesis and hence involves activation of genes unique to this process. Transcriptionally controlled tumor protein (TCTP) coding gene could be one candidate gene involved in the process of gall development. These studies, however, did not bring out key susceptibility genes. More intensive research is suggested to discover such susceptibility genes that are now being explored as novel class of target genes in host-plant resistance. Induction or detection of mutant forms or silencing existing susceptibility genes can confer novel mechanisms of host-plant resistance.

7.7. INSECT VIRULENCE GENES

We also explored the rice gall midge genome to get an insight into insect virulence. Using next-generation sequencing (Roche 454 pyrosequencing) strategy over 80,000 ESTs each from insects feeding on resistant or susceptible host genotypes were identified (Sinha *et al.*, 2012a). Comparative transcriptome analysis of these two sets of ESTs has identified several key genes involved in insect virulence or avirulence. Additionally, this study has generated a huge set of potential EST-based microsatellite markers (2,303) and SNPs (2,756) that can be utilized for biotype differentiation and population biology studies.

Investigating a set of serine proteases as target genes in the insect, two genes, *OoprotI* and *OoprotII* have been cloned and sequenced (Sinha *et al.*, 2011a). Quantitative real time PCR analysis revealed that both the genes were significantly upregulated in larvae feeding on resistant cultivar than in those feeding on susceptible cultivar suggesting their role in detoxification of plant resistance factors. Another gene oligosaccharyltransferase (*OoOST*), coding for one of the secretary salivary proteins, was also cloned and characterized (Sinha *et al.*, 2011b). Tissue-specific analysis of the expression of *OoOST* revealed over 7-fold increase in the transcripts of the gene in salivary glands of maggots feeding in susceptible plants than in those feeding in resistant hosts, thus implicating their role in insect virulence. Likewise, another gene coding for nucleoside diphosphate kinase (OoNDPK) was found over expressed in insects feeding on susceptible plants (Sinha *et al.*, 2012b). Intensification of research along these lines will lead to identification of more genes and pathways involved in insect virulence and evolution of biotypes. Such studies will also help to decipher insect interaction in greater detail and eventually lead to novel genetic strategies for insect pest management in rice.

7.8. CONCLUSIONS

Breeding for gall midge resistance in rice has been a classic success story of during 1980s. With this followed the rapid evolution of virulent pest biotypes exposing the weakness of the single gene conferred major gene deployment strategy. With simple genetics and clear cut phenotypic differences, molecular tagging and mapping of resistance genes has been another major success story of 1990s. Now stage is set for gene pyramiding for durable resistance. Recent studies have provided molecular basis of insect plant interactions and basis for selecting the right combination of genes that provide durable gall midge resistance. Empirically, combinations of *Gm4+Gm8* or *gm3+Gm8* or ideally *gm3+Gm4+Gm8* are the right choice. Further, these studies have opened new avenues of exploring novel genes for gall midge resistance.

REFERENCES

Andow DA and Bentur JS (2010) Pedigreed crosses to estimate recessive virulence allele frequencies in natural populations of gall midges. *Entomologia Experimentalis et Applicata* **135**: 18-36.

Bentur JS, Srinivasan TE and Kalode MB (1987) Occurance of a virulent gall midge (GM) *Orseolia oryzae* Wood-Mason biotype(?) in Andhra Pradesh, India. *International Rice Research Newsletter* **12**(3): 33-34.

Bentur JS, Cheralu C and Ram Mohan Rao P (2008) Monitoring virulence in Asian rice gall midge populations in India. *Entomologia Experimentalis et Applicata* **129**: 96-106.

Bentur JS, Padmakumari AP, Jhansi Lakshmi V, Padmavathi Ch, Kondala Rao, Y, Amudhan S and Pasalu IC (2011) *Insect Resistance in Rice.* DRR Technical Bulletin 51. Directorate of Rice Research, Hyderabad, 86 pp.

Bentur JS, Pasalu IC and Kalode MB (1992) Inheritance of virulence in rice gall midge, *Orseolia oryzae. Indian Journal of Agricultural Science* **62**: 492-493

Bentur JS, Pasalu IC, Sarma NP, Prasada Rao U and Mishra B (2003) *Gall Midge Resistance in Rice.* DRR Research Paper Series 01, Directorate of Rice Research, Hyderabad, 20 pp.

Bentur JS and Kalode MB (1996) Hypersensitive reaction and induced resistance in rice against Asian rice gall midge *Orseolia oryzae. Entomologia Experimentalis et Applicata* **78**: 77-81

Cohen MB, Bentur JS and Gould F (2004) Durable deployment of gall midge resistant varieties. In: Bennett J, Bentur: JS, Pasalu K and Krishnaiah K. (Eds), *New Approaches to Gall Midge Resistance in Rice.* ICAR and IRRI, Philippines. pp. 153-164.

Himabindu K, Suneetha K, Sama VSAK and Bentur JS (2010) A new rice gall midge resistance gene in the breeding line CR57-MR1523, mapping with flanking markers and development of NILs. *Euphytica* **174**: 179-187.

Kalode MB and Bentur JS (1989) Characterization of the Indian biotypes of rice gall midge *Orseolia oryzae* (Wood-Mason) (Diptera: Cecidomyiidae). *Insect Science and its Application* **10**: 219-224.

Kalode MB, Pophaly DJ, Kasi Viswanathan PR and Sreeramulu M (1979) Studies on resistance and mass rearing of rice gall midge, Orseolia oryzae (Wood-Mason). *Madras Agricultural Journal* **64**: 733-739.

Kumar A, Jain A, Sahu RK, Shrivastav MN, Nair S and Mohan M (2005) Genetic analysis of resistance genes for the rice gall midge in two rice genotypes. *Crop Science* **45**: 1631-1635.

Nair KVP and Devi A (1994) Gall midge biotype 5 identified in Moncompu, Kerala, India. *International Rice Research Notes* **19**(4):11.

Nair S, Bentur JS and Sama VSAK (2011) Mapping gall midge resistance genes: Towards durable resistance through gene pyramiding. In: Muralidharan K and Siddiq EA (Eds) *Genomics and Crop Improvement: Relevance and Prospects.* Institute of Biotechnology, ANGR Agricultural University, Hyderabad, India. pp. 256-264.

Rani NS, Veraprasad GS, Subba Rao LV, Sudarshan I, Hariprasad AS, Prasad ASR, Ram T, Babu VR, Padmavathi G, Bhadana V, Suneetha K and Viraktamath BC (2011) *High Yielding Varieties of India.* Technical Bulletin No. 55/2011, DRR, Hyderabad, 192 pp.

Rawat N, Sinha DK, Rajendrakumar P, Shrivastava P, Neeraja CN, Sundaram RM, Nair S and Bentur JS (2010) Role of pathogenesis-related genes in rice-gall midge interactions. *Current Science* **99**: 1361-1368.

Rawat N, Neeraja CN, Sundaram RM, Nair S and Bentur JS. 2012a. A novel mechanism of gall midge resistance in the rice variety Kavya revealed by microarray analysis. *Functional Integrated Genomics* **12:** 249-264.

Rawat N, Neeraja CN, Nair S and Bentur JS (2012b) Differential gene expression in gall midge susceptible rice genotypes revealed by suppressive subtraction hybridization (SSH) cDNA libraries and microarray analysis. *Rice* **5**: 8.

Rawat N, Himabindu K, Neeraja CN, Nair S and Bentur JS (2013) Suppressive subtraction hybridization reveals that rice gall midge attack elicits plant-pathogen like responses in rice. *Plant Physiology and Biochemistry* **63**: 122-130.

Singh MP (1996) A virulent gall midge biotype in Manipur. *International Rice Research Notes* **21**(1): 31-32.

Sinha DK, Lakshmi M, Anuradha G, Rahman S,Siddiq EA, Bentur JS and Nair S (2011a) Serine Proteases-like genes in the rice gall midge show differential expression in compatible and incompatible interactions with rice. *International Journal of Molecular Science* **12**: 2842-2852.

Sinha DK, Bentur JS and Nair S (2011b) Compatible interaction with its rice host leads to enhanced expression of gamma subunit of oligosaccharyl transferase (*OoOST*) in the Asian rice gall midge (*Orseolia oryzae*). *Insect Molecular Biology* **20**: 567-575.

Sinha DK, Nagaraju J, Tomar A, Bentur JS and Nair S (2012a) Pyrosequencing-based transcriptome analysis of the Asian rice gall midge rev als differential response during compatible and incompatiblein-teraction. *International Journal of Molecular Science* **13**: 13079-13103.

Sinha DK, Atray I, Bentur JS and Nair S (2012b) Expression of nucleoside diphosphate kinase (OoNDPK) is enhanced in rice gall midge feeding on susceptible rice hosts and its over-expression leads to salt tolerance in E. coli. *Insect Molecular Biology* **21**(6): 593–603.

Sinivas C, Reddy VN, Rao PS and Ramesh P (1994) Rice gall midge Orseolia oryzae (Wood-Mason) biotype in Karimnagar District, Andhra Pradesh, India. *International Journal of Molecular Science* **19**(2): 14-15.

Vijayalakshmi P, Amudhan S, Himabindu K, Cheralu C and Bentur JS (2006) A new biotype of the Asian rice gall midge *Orseolia oryzae* (Diptera: Cecidomyiidae) characterized from the Warangal population in Andhra Pradesh, India. *International Journal of Tropical Insect Science* **26**: 207-211

Yasala AK, Rawat N, Sama VSAK, Himabindu K, Sundaram RM and Bentur JS (2012) *In silico* analysis for gene content in rice genomic regions mapped for the gall midge resistance genes. *Plant Omics Journal* **5**: 405-413

Biological and Molecular Approaches in Pest Management
S S Gosal, Ramesh Arora & Balwinder Singh (Eds)
Scientific Publishers (India), Jodhpur, 2015, pp 161-201.

Chapter - 8

Molecular Marker Systems in Entomological Research

Prasad S Burange[1], Sujan Mamidi[2], P C Pathania[1] and Uma Kanta[1]
[1]*Department of Entomology, Punjab Agricultural University, Ludhiana 141004, India*
[2]*Genomics and Bioinformatics Program, North Dakota State University, Fargo, ND 58102, USA*
E-mail: prasadburange@yahoo.com

8.1. INTRODUCTION

Insects are the largest group of arthropods and occupy most of the terrestrial and freshwater biological niches. An estimate suggests that insects account for the highest number of species i.e. over 70 per cent of recorded animal species and 60 per cent of identified eukaryote species in the world (Grimaldi & Engel, 2005). These recorded insect species are found in every ecological adaptation (Behura, 2006). It is observed that the diversity of insects is vast among the 900,000 species of insects that have been reported. In addition insect populations vary in morphology and behavior within a species because of the interaction and adaptation to their environment. Insects affect human life directly or indirectly because insects are beneficial; acting as pollinating agents, natural enemies of damaging pests, and producing useful products such as honey, silk and wax. Conversely, insects are major pests of food crops, vectors for transmitting diseases, cause damage to our urban infrastructure, environment, forests and natural resou-rces, and even interfering in commerce, international trade and economic affairs.

With the diverse applications of molecular markers for taxonomy, population studies and host interactions, it is important to understand the molecular markers used in the study of insect populations. Initially, phenotypic markers like eye color, body spots, bands, hairs and spines were used for studying patterns of dispersal, mating behavior and inheritance of genetic traits in insects (Bartlett *et al.*, 1968; Fay & Craig 1969; Bond *et al.*, 1970; Bartlett & Butler, 1975 and Hughes, 1997). These markers were extensively used for taxonomy, but have limited usage in other applications because of minor variations within a species that are often hard to score.

With the advent of advanced biochemical methods, protein markers, known as allozymes, became more popular. Protein markers were successfully used in studying insecticide resistance (Maa & Terriere, 1983), detection of prey in insect predators (Solomon *et al.*, 1996) to name a few. The development of DNA-based marker systems became popular because they exhibited a greater level of polymorphism than protein markers. Compared to protein samples, DNA samples have shown more stability and ease of detection. These DNA based markers are defined as any gene or DNA sequence from the chromosomal or extra chromosomal genome. With the development of low cost Sanger sequencing, the use of gene fragments including regions of mitochondria, introns and coding regions became popular. In recent years, molecular marker technology has advanced with the ability to accurately sample thousands of individuals for thousands of markers in a relatively short time frame (few hours to a day). The assay costs are decreasing dramatically because of the high competition among companies and increased usage by researchers. The present trend is to use single nucleotide polymorphisms (SNP) along with insertion and deletion polymorphisms (indels) to identify the major changes in the genome.

8.2. TYPES OF MOLECULAR MARKERS

Molecular markers are classified into different categories based on the methodology used to identify the variations in DNA. Briefly they are divided into i) Restriction endonuclease based, such as restriction fragment length polymorphism (RFLP) markers; ii) PCR based, which include RAPDs, AFLPs, Microsatellites/SSR, SCAR and Inter Simple Sequence Repeats (ISSR); iii) A combination of PCR and Restriction digestion that includes Cleaved Amplified Polymorphic Sequences (CAPS)/Derived Cleaved Amplified Polymorphic Sequences (dCAPS). The characteristics of major types of molecular markers are summarized in Table 1. The advantages, strengths and weaknesses of commonly used molecular markers are presented in Table 2.

Other than identifying polymorphisms at a single base pair level or to a small fragment that extends a few nucleotides, fragment sequencing of genes helps identify distinct sequence signatures. This includes sequencing of mitochondrial genes, nuclear genes like ribosomal RNA (rRNA) and Protein-Coding Genes (PCGs). SNPs and indels form the present core of markers used to study variation within or between samples, populations or species.

8.2.1. Restriction Digestion Based Markers

In 1960s, Arber, Smith and Nathans discovered and isolated restriction endonucleases which form the basis for RFLPs (Acquaah, 2012). Botstein *et al.* (1980) for the first time used the RFLP molecular marker technique to record DNA polymorphism and to develop genetic maps. In RFLPs, restriction enzymes (RE) precisely cleave the DNA fragment at specific "restriction sites" (e.g. *Eco*RI enzyme cleaves GAATTC site). Initially, the genomic DNA was cleaved with a restriction enzyme and the banding pattern was analysed to identify the specific polymorphic signatures. But currently, RFLPs

Table 1. Characteristics of major types of molecular markers

Characteristic/ Marker	Isozyme	RFLP	RAPD	CAPS	SSR	SCAR	ISSR	SNP	AFLP	SRAP
Reference	Hunter and Merkert (1957)	Grodzicker *et al.* (1974)	Williams *et al.* (1990)	Akopyanz *et al.* (1992)	Akkaya *et al.* (1992)	Paran and Michelmore (1993)	Zietkiewicz *et al.* (1994)	Jordan and Humphries (1994)	Vos *et al.* (1995)	Li and Quiros (2001)
Marker generation	First	First	Second	Second	Second	Second	New generation	New generation	New generation	New generation
No. of loci	30-50	100s	Unlimited	10s	10s	10s	1000s	10s	Unlimited	Multiple
Locus specificity	Yes	Yes	No	Yes	No	Yes	Yes	Yes	No	Yes
Allelism	Co-dominant	Co-dominant	Dominant	Co-dominant	Co-dominant	Co-dominant	Dominant	Co-dominant	Dominant	Co-dominant
Degree of polymorphism	Low-medium	Medium high	Medium high	High	High	High	Medium-high	High	Medium-high	High
Quantity of DNA sample required	Few mg	~ 10 mg	~ 10 ng	~ 25 ng	~ 50 ng	~ 25 ng	~ 10 ng	~ 50 ng	~ 25 ng	~ 25 ng
Reproducibility	High	High	Low	High	High	High	Low-medium	High	Good	High
Assay cost	Cheap	Expensive	Average	Moderate	Average	Moderate	Moderate	Expensive	Average	Average
Major application	Population genetics	Physical mapping	Gene tagging	Allelic diversity	Genetic diversity	Gene tagging and physical mapping	Genetic finger-printing, gene tagging, phylogenetic analysis	Gene tagging, genotyping, diversity analysis	Gene tagging	Map construction, gene tagging and genetic diversity studies

Source: Modified after Maheswaran (2004), Agarwal *et al.* (2008) and http://www.lifetechnologies.com

are used after PCR (PCR-RFLP) in which alleles with unidentical sequence are detected at specific restriction site. A PCR product (amplicon) is mixed with a specific restriction enzyme which cleaves only a specific allelic form. This digested DNA fragment can be seen using electrophoresis ((Marsjan & Oldenbroek, 2007).

RFLP is an extensively utilized hybridization-based molecular marker (Semagn *et al.*, 2006). It is a versatile technique which provides information on sequence variations within nuclear and mitochondrial DNA (mtDNA) (Dowling *et al.*, 1996). To date, over 1400 restriction enzymes are known which cut DNA at specific sites within a specific recognition sequence. Most recognition sequences fall in the range of 4-6 bp (base pairs) although a few of them may be up to 12 bp long. Restriction enzymes are specific, giving rise to DNA discrete fragments

Table 2: Summary attributes of selected commonly used markers

Method	Advantages	Disadvantages	Strength(s)	Weakness(s)
(1)	(2)	(3)	(4)	(5)
Isozymes	• Useful for evolutionary studies • Isolation lot easier than that of DNA • Can be used across species • No radioactive labeling • No need for sequence information	• Laborious • Limited in polymorphism • Expensive (each system is unique) • Have to know the location of the tissue • Not easily automated	• Does not require DNA extraction or the availability of sequence information, primers or probes • Quick and easy to use Simple analytical procedures • Codominant markers that have high reproducibility • Zymograms can be readily interpreted in terms of loci and alleles	• Relatively low abundance & low level of polymorphism • Their selective neutrality may be in question • Often considered molecular markers since they represent enzyme variants • Being phenotypic markers, these may be affected by environmental conditions
RFLP	• High genomic abundance • Co-dominant markers • Highly reproducible • Can use filters many times • Good genome coverage • Can be used across species • No sequence information needed • Can be used in plants reliably (well-tested) • Needed for map based cloning	• Need large amount of good quality DNA • Laborious (compared to RAPD) • Difficult to automate • Need radioactive labeling • Cloning and characterization of probe are required	• Inexpensive • Needs only simple equipment and training	• Assays only one or few base pairs at a time • Low amount of information relative to time input
RAPD	• Cheap • Produces a large number of bands, which can then be further characterized individually (for example, converted into single locus markers)	• Low reproducibility • Mainly dominant • Difficult to analyse • Difficult to automate • Cross-study comparisons are difficult	• Inexpensive • Produces numerous markers • No need for prior genetic information	• Poor reproducibility • Needs good quality DNA • Difficult to compare across studies
Microsatellites / SSRs	• Highly informative (large number of alleles, high heterozygosity) • Low ascertainment bias • Easy to isolate	• High mutation rate • Complex mutation behaviour • Not abundant enough	• Highly variable • Occur throughout genome	• Prone to 'slippage' errors • Difficult to obtain from some insects (e.g., Lepidoptera)

contd...

(1)	(2)	(3)	(4)	(5)
		• Difficult to automate • Cross-study comparisons require special preparation		
SNPs	• Low mutation rate • High abundance • Easy to type • New analytical approaches are being developed at present • Cross-study comparisons are easy; data repositories already exist	• Substantial rate of heterogeneity among sites • Expensive to isolate • Ascertainment bias • Low information content of a single SNP	• Easy to score • Easily automated • Ubiquitous in genome	• Moderately expensive • Limited information per site • Time consuming to identify and develop numerous loci
AFLP	• High genomic abundance • High polymorphism • No need for sequence information • Can be used across species • Work with smaller RFLP fragments • Useful in preparing contig maps	• Very tricky due to changes in patterns with respect to materials used • Cannot get consistent map (not reproducible) • Need to have very good primers	• Produces numerous markers • No need for prior genetic information	• Medium reproducibility • Needs good quality DNA • Difficult to compare across studies

Source: Modified after Schlötterer (2000), Foottit *et al.* (2009), Kumar (2009) and http://www.wageningenur.nl/en/show/Allozymes.htm

after its complete digestion. Fragment size and number varies due to changes in DNA sequence caused by rearrangements (inversions, tandem duplications, inverted duplications) or through the process of addition/deletion, or substitution of specific bases during DNA synthesis (Hoy, 2013). The RFLP technique could be implemented to effectively analyze multiple scenarios such as clonal populations, heterozygosity, geographic variation, hybridization rates and phylogeny.

The major strengths of RFLP markers are high reproducibility, codominant inheritance, no requirement for prior sequence information, ease in working among different species, ease of scoring large sized fragments etc. RFLPs could also be used to detect heterozygotes and allow fine mapping of many loci (Semagn *et al.*, 2006). RFLPs could also be used to detect heterozygotes and allow fine mapping of many loci. The limitations for RFLP analysis include a requirement for high quantity of quality DNA (e.g., Potter & Jones 1991 and Roy *et al.*, 1992), or the need to develop specific probe libraries for the species. Automation is not possible since polymorphism level is low and only few loci can be detected per assay. The time required to complete a RFLP analysis could be 7-10 days. In addition this technique is laborious and expensive (Yu *et al.*, 1993), with the requirement of labeled DNA probes (radioactive or immunologic) for detecting the polymorphism (Fakrudin *et al.*, 2006).

RFLP analyses of mitochondrial DNA and ribosomal DNA have been used for population genetics, biogeographical surveys and phylogenetics (Avise, 1994). For example, these DNA sequence polymorphisms were used for genetic mapping of a temperature-sensitive mutation of human adenoviruses serotypes wherein RFLP analyses of mitochondrial DNA and ribosomal DNA have been used for population genetics, biogeographical surveys and phylogenetics of nuclear genome variability was detected by employing restriction enzymes (Grodzicker *et al.*, 1974).

8.2.2. PCR Based Markers

The invention of polymerase chain reaction (PCR) by Kary Mullis in 1983 was the main impetus for the development of DNA-based molecular markers (Saiki *et al.*, 1985) in which a specific genomic region was amplified in the absence of cloning or isolating large amounts of genomic DNA. Immature insects, puparium and adult insects could be subjected to PCR amplification through which rapid identification is possible (Mazzanti *et al.*, 2010).

8.2.2.1. Random amplified polymorphic DNA (RAPD)

The differential PCR amplification of genomic DNA is the basis of the RAPD technique. It uses short random oligonucleotide sequences (generally 10 bp long) to ascertain DNA polymorphism which is produced by processes such as deletions or rearrangements between the oligonucleotide primer binding sites in the genome (Welsh & McClelland, 1990 and Williams *et al.*, 1990). Genetic variation within the insect genome can be deduced from the banding patterns.

The benefits include being inexpensive (Penner *et al.*, 1993). This approach does not require prior knowledge of the genome, and so can be employed across species using universal primers. The drawbacks of using RAPDs are that these are dominant markers therefore heterozygotes may not be identified and lack repeatability. This condition is avoided by analysing haploid males or after determining the genotype of females through testing their male progeny. For reliability, experiments need to be repeated, reactions must be optimized and protocols must be standardized for consistent results (Edwards & Hoy, 1993).

The usefulness of RAPDs was documented in the discrimination of cryptic species and the examination of the hybridization process, species boundaries and variations in clones. RAPDs could potentially differentiate closely related species (Alvarez & Hoy, 2002). Hadrys *et al.* (1992) suggested the use of RAPD in detecting insect biotypes, assessing close kinship, and even analyzing paternity. The genetic variability in the walnut aphid parasite, *Trioxys pallidus* (Halliday) and leafminer parasite, *Diglyphus begini* (Ashmead) (Hymenoptera: Braconidae) were studied using RAPD-PCR (Edwards & Hoy, 1993). It is employed to study population structure, gene flow, and behavioral studies of biotypes (Kambhampati *et al.*, 1992 and Edwards & Hoy, 1995). It could provide an affordable technique to detect or reveal cryptic species. Two populations of the citrus leafminer parasitoid, *Ageniaspis citricola* Logvinovskaya (Hymenoptera: Encyrtidae) were evaluated with RAPD-PCR indicating two genetically distinct populations (Hoy *et al.*, 2000).

8.2.2.2. Amplified Fragment Length Polymorphism (AFLP; Selective Restriction Fragment Amplification)

AFLP is a PCR-based method for developing large numbers of markers for population analyses. The AFLP markers are based on selective PCR amplification of restriction enzyme-digested DNA fragments (Vos *et al.*, 1995). The AFLP technique combines the simplicity of the RFLP technique with the versatility of PCR-based technology by ligating adaptors (primer recognition sequences) to restricted DNA (Lynch & Walsh, 1998). Generation of AFLP markers could be done from any life form without initially investing in the development of primer/probe as well as sequence analysis. As with RAPD-PCR and microsatellite analyses, AFLP-PCR screens multiple regions of the genome. Details of the AFLP methodology have been reviewed by Ridout & Donini (1999).

Usefulness of AFLP markers has been greatly recognized in various genetic studies such as assessment of genetic differences at individual, population and species levels (Mueller & Wolfenbarger, 1999). The AFLP markers have aided in the detection of many DNA polymorphisms. These markers have been successfully employed in identification and mapping resistance to the cotton bollworm, *Helicoverpa armigera* (Hübner) (Lepidoptera: Noctuidae) resistance in chickpea, *Cicer arietinum* L. (Lawlor *et al.*, 1998); the brown plant-hopper (BPH), *Nilaparvata lugens* (Stål) (Hemiptera: Delphacidae) (Jairin *et al.*, 2005); the Asian rice gall midge, *Orseolia oryzae* (Wood-Mason) (Diptera: Cecidomyiidae) (Sardesai *et al.*, 2002) in rice, *Oryza sativa* L.

AFLP analysis can be more efficient than using RFLPs. It is a highly reproducible technique (Jones *et al.*, 1997). No prior knowledge of DNA sequence of organism is required. A single primer can be used to analyze a large number of polymorphic loci simultaneously as compared to other molecular markers such as RAPDs, RFLPs, and microsatellites (Russell *et al.*, 1997). Co-migrating AFLP amplificons are generally homologous as well as locus specific (Qi *et al.*, 1998).

AFLP also has its own limitations. The technique requires a greater number of steps to produce the result and the template DNA needs to be free of inhibitor compounds that interfere with the restriction enzyme. AFLP requires the use of polyacrylamide gels for DNA band separation in combination with $AgNO_3$ staining, radioactivity, or fluorescent methods of detection. AFLP is more costly and labor intensive than agarose gels, but that may be offset by the greater number of markers (bands) per experimental run. Like RAPDs, most AFLP loci are dominant and are incapable of distinguishing homozygotes from heterozygotes. This lessens the preciseness of these markers in marker assisted selection or marker aided selection (MAS), population genetic analysis and genetic mapping (Mishra *et al.*, 2014).

8.2.2.3. Sequence Characterized Amplified Region (SCAR)

A SCAR marker results from using a pair of specific oligonucleotide primers (15-30 bp) along with PCR amplification of a genomic DNA fragment (McDermott *et al.*, 1994). These markers are created by cloning and sequencing the two ends of a single RAPD fragment which in turn appeared to have diagnostic value. For example, a RAPD band absent in disease susceptible lines and present in resistant lines could be chosen. The advantages of SCARs over RAPD markers include capacity to detect only a single locus, amplification is less dependent on reaction conditions, and their potential to convert into codominant markers (Paran & Michelmore, 1993). Since longer PCR primers are used, SCARs can overcome the problem of low reproducibility associated with RAPDs. However, SCARs may be dominant markers when the sequence variation site gets partially overlapped by one or both primers, which can be detected later through gel electrophoresis. SCARs markers are fast and not difficult to use, have high reproducibility and locus-specificity, are PCR based, and a small quantity of template DNA is needed. They have a better reproducibility than RAPDs. The main disadvantage is that sequence data is needed for designing the PCR primers (Spooner *et al.*, 2005).

SCARs are usually dominant markers. But, some of these SCARs can be converted to codominant markers by restriction enzyme digestion. Polymorphism then can be deduced by either denaturing gel electrophoresis or Single Strand Confirmation Polymorphism (SSCP). SCARs are more informative for genetic mapping than dominant RAPDs, and can be used for map-based cloning. Additionally SCARs permit studies involving comparative mapping or homology between related species, hence making them versatile. SCARs have been used to identify and map genes for resistance to *O. oryzae* (Sardesai *et al.*, 2002) and *N. lugens* (Renganayaki *et al.*, 2002).

8.2.2.4. Microsatellites Simple Sequence Repeats (SSR)

The advantages of PCR technology are fully utilized in microsatellites or SSR (Litt & Luty, 1989). The SSRs are tandem arrays of 2-5 base repeat units which are present in eukaryotic DNA (Sharopova *et al.*, 2002). Most microsatellite loci can be amplified by a standard PCR because the region between the primers which contains the repeats is usually less than 300 bp. Microsatellites are distributed in euchromatic region of genomes. Microsatellites are polymorphic and abundant in nature which makes this marker a preferred genetic marker for paternity testing, mapping, and population genetics (Goldstein & Schlötterer, 1999).

SSR development is expensive since extensive sequence data from targeted species is required initially. The cloned DNA libraries can be used to generate sequence data with or without enrichment for a specific SSR motif (Sharma, 2008). SSR markers produce highly polymorphic and allele-specific markers from known DNA sequences (Hu & Vick, 2003). However, their use is limited for high-density map construction (Gupta & Varshney, 2000).

Microsatellites are isolated from a genomic library by screening clones for the availability of a microsatellite motif. Microsatellites above a certain repeat number are considered to be polymorphic (Schlötterer, 2000). With the techniques of next generation sequencing (NGS) (discussed in section 8.2.5), more and more sequences are being accumulated that contain SSR markers (Cook *et al.*, 2004). Semi-codominance was identified in Russian wheat aphid, *Diuraphis noxia* (Mordvilko) (Hemiptera: Aphididae) using microsatellites (Liu *et al.*, 2001). SSR markers have identified insect resistance genes in many crops (Malik *et al.*, 2003; Cook *et al.*, 2004 and Zhu *et al.*, 2004). Resistance to sorghum shoot fly, *Atherigona soccata* Rondani (Diptera: Muscidae) (Folkertsma *et al.*, 2003); resistance to greenbug, *Schizaphis graminum* (Rondani) (Hemiptera: Aphididae) in the wheat (Weng & Lazar, 2002) and soybean genomes (Narvel *et al.*, 2000) have been studied using polymorphic SSRs.

8.2.2.5. *Inter Simple Sequence Repeat (ISSR)*

In this technique, the SSR based primers are used in amplification of inter-simple sequence repeat (ISSR) DNA sequences (Srivastava & Narula, 2004). Genomic DNA is amplified by anchoring a few microsatellites at the 3′ end so as to increase their specificity. Though ISSRs are considered as dominant markers, a few of them have proven to be codominant too. A limitless number of primers can be produced for multiple combinations such as di-, tri-, tetra-, and penta-nucleotides which could be utilized for an extensive range of applications in various plant species (Zietkiewicz *et al.*, 1994).

In cases where time and processing costs prevent development of more robust markers, the ISSRs can be more useful than locus-specific SSRs. Polymorphisms and detection protocols can be elaborated by ISSRs, which is valuable addition to PCR-based markers in studies involving genomic fingerprinting as well as in screening the genetic variability in animal populations on a large scale (Abbot, 2001). Genetic variation in yellow fever mosquito, *Aedes aegypti* (Linnaeus) (Diptera: Culicidae) and two aphid genera, *Acyrthosiphon pisum* Harris and *Pemphigus obesinymphae* Moran (Hemiptera: Aphididae) (Abbot, 2001) were studied using ISSR markers.

8.2.3. Restriction and PCR Based Markers

Cleaved Amplified Polymorphic Sequences (CAPS)/derived Cleaved Amplified Polymorphic Sequences (dCAPS)

The CAPS technique is a blend of PCR and RFLP which was initially termed PCR-RFLP (Maeda *et al.*, 1990). In this technique, a target DNA is amplified through PCR and then digested with restriction enzymes (Konieczny & Ausubel, 1993 and Michaels & Amasino, 1998). Differences in the patterns of restriction enzyme digestion of PCR amplicons, which are caused by nucleotide polymorphism between samples, is the basis of CAPS marker system.

The advantages of CAPS markers include simplicity and less time-consuming technique than alternate methods. For comparative mapping studies, primers developed from Expressed Sequence Tags (ESTs) or specific genes are more applicable than other markers derived from non-functional sequences such as microsatellite markers.

CAPS are co-dominant markers (Matsumoto & Tsumura, 2004). The ability of CAPS to detect DNA polymorphism is lesser than SSRs and AFLPs since DNA polymorphism by CAPS is detected using nucleotide changes affecting restriction sites. Creation or disruption of restriction enzyme recognition site due to mutations leads to development of CAPS markers. However, an alternate marker system called derived-CAPS (dCAPS) has been developed. This technique generates mismatches in a PCR primer, subsequently a polymorphism is created based on the target mutation (Michaels & Amasino, 1998 and Neff *et al.*, 1998).

8.2.4. DNA Sequencing

DNA sequencing is the process of defining the order of nucleotides that make up the DNA molecule. The process involves identifying the order of the four nucleotides, adenine (A), guanine (G), cytosine (C), and thymine (T) in a single strand of DNA. The discovery of DNA sequencing methods has accelerated research in both biological and medical fields. The sequencing of the RNA bacteriophage MS2 is considered to be a significant milestone of nucleic acid sequencing. Fiers *et al.* (1976) sequenced the complete genome and identified complete gene set. Later, Sanger *et al.* (1977) developed a DNA sequencing method at Cambridge, UK.

With advancements in the Sanger process, sequencing of short fragments of nuclear and orgenellar DNA came into existence in late 1990s. The typical length of these early sequences varied between 200-500 bp. The sequences of many organellar genes that have distinct evolutionary signatures were used to understand phylogeny, population genetics and other related fields. More recently, the sequencing technology has advanced to a large-scale that even whole chromosomes can be sequenced. In this case, long DNA fragments are cut with restriction enzymes or sheared with mechanical forces into shorter fragments. This fragmented DNA is then cloned into a DNA vector. Purified short DNA fragments from individual clones are then sequenced individually and assembled into a large sequence (http://pustun.tk/DNA_sequencing).

"*De novo* sequencing" is a method of sequencing a DNA molecule with no previously known sequence such as the primers which are used in Sanger sequencing. When large DNA molecules like bacterial artificial chromosome (BAC) clones, yeast artificial chromosome (YAC) clones or whole chromosomes are sequenced, multiple gaps are obtained either due to lack of sequence or assembly issues. These gaps are filled by a procedure called primer walking which involves making primers from the sequence before the gap and using those primers for Sanger sequencing. This method is also called chromosome walking (Chinault and Carbon, 1979). There are multiple methods for sequencing and each has its own advantages and disadvantages in terms of speed, quality and accuracy in assembly. The methods used to sequence the whole genome and assemble it are complex and difficult.

Amplification of individual DNA molecule in most DNA sequencing techniques is achieved through addition of a cloning step (Williams *et al.*, 2006). In emulsion PCR (ePCR) procedure, PCR reaction is performed in which an individual DNA molecule is isolated with primer-coated beads within droplets of water-in-oil emulsion (Taly *et al.*, 2012). A PCR reaction then coats each bead with copies of the DNA molecule, which are later used for sequencing. Emulsion PCR is the process that is being used in many present day sequencers, including 454 Life Sciences and SOLID sequencing by Life Technologies, USA. The Illumina Hiseq™ system currently determines billions of paired 100 bp reads in a single run and is a primary platform for high-throughput sequencing (HTS) (Sadava *et al.*, 2012). This system employs an *in situ* polymerase chain reaction (PCR) on a flow-cell surface for template amplification. This is followed by sequencing-by-synthesis using fluorescently labeled nucleotides. In addition to tranditional methods of sequencing genes or fragments of DNA, HTS methods can be used for sequencing whole genome, exomes, transcriptome, and amplicons (Boyd, 2013). The large amount of HTS data contains information whose meaning is not immediately understood but optimistically it can be used in the near future to correlate with additional research (Boyd, 2013).

8.2.4.1. Mitochondrial genes

In Mitochondrial DNA (mtDNA) sequencing, portions of the mitochondrial genome are analyzed and compared. This has been a very productive avenue of research. The mitochondrion is a source of genetic material separate from the nuclear chrosomal DNA, MtDNA is found in multiple copies outside of the nucleus in the cytoplasm which makes it useful for analyzing DNA from old and degraded as well as from small insects or insect parts. This makes mtDNA a preferred choice for the DNA analysis of forensically important insects (Alessandrini *et al.*, 2008), and also for genetic diversity and evolutionary studies (Othman, 2012).

Mitochondria are cytoplasmically inherited and are transmitted through female gametes. These are self-replicating organelles found in eukaryotic cytoplasm and descended from aerobic eubacterium that became endosymbionts of eukaryotes (Kobayashi, 1998). The size of eukaryotic mitochondrial genomes (mitogenome) ranges from 6 kb to 2 Mb (Sogin, 1997). Insect mitogenome are circular, supercoiled, and double-stranded (ds) DNA molecules (Wolstenholme, 1992 and Boore, 1999) of 14,503 bp to 19,571 bp in size (Hu *et al.*, 2009). Besides these 37 genes, a A+T-rich non-coding region too exists which has a role in initiatiation of transcription and replication processes (Wolstenholme, 1992). Introns are absent from the insect mitochondrial genes and they usually have small or no intergenic regions. Mitochondria have their own genetic code which differs slightly from the standard nuclear genetic code. The first insect mitochondrial genome to be completely sequenced was from fruit fly, *Drosophila yakuba* (Burla) (Diptera: Drosophilidae). It codes for a total of 37 genes such as two ribosomal RNA (rrnL and rrnS), 22 transfer RNA (tRNA), and 13 PCGs (Wolstenholme, 1992 and Boore, 1999) which code for subunits of different enzymes involved in electron transport (ETS) or ATP synthesis (Clary & Wolstenholm, 1985 and Gray, 1989). The evolutionary rate of mitochondrial DNA is faster than single-copy nuclear DNA because of inefficient repairing errors during DNA replication or after DNA damage. As an example, the mitochondrial DNA of Hawaiian *Drosophila* evolved approximately 3 times faster compared to genes of its nuclear DNA (Moritz *et al.*, 1987).

The presence of multiple copies in each cell enables easy amplification of mtDNA in total DNA extracts through the PCR. The mtDNA is easier to purify from cells as compared to a specific segment of nuclear DNA due to its high density and copy number within cells, its position within an organelle, which makes mtDNA a useful molecular tool in systematics or population genetics studies. Havill *et al.* (2006) analyzed mitochondrial DNA from hemlock woolly adelgid (HWA), *Adelges tsugae* (Annand) (Homoptera: Adelgidae) which was introduced into eastern North America. Their study indicated presence of cryptic speciation and the possible sources of pest introduction.

Mitochondrial genes are advantageous for phylogenetic analysis due to multiple reasons. When compared to nuclear genes, the mtDNA genes are generally easy to amplify and have readily available selection of conserved mitochondrial primers (Simon *et al.*, 1994 and Simon *et al.*, 2006). Carrying out phylogenetic analysis is easier since mitochondrial genes are clonally inherited and non-recombining, making recombination, paralogy, and heterozygosity less of an issue (Sunnucks & Hales, 1996).

Some disadvantages of mtDNA are as follows. All mitochondrial genes are linked on the same chromosome which might create doubt about the independent estimate of phylogeny compared to nuclear genes (Harrison, 1989). Trautwein *et al.* (2012) noted that mtDNA can be inefficacious at detecting deep divergencies. Higher substitution rates may not resolve divergences of over 5-10 million years. Mitochondrial genes may show high levels of homoplasy (parallel evolution or convergence) that can result in misleading phylogenetic inferences (Mooers & Holmes, 2000).

8.2.4.1.1. DNA Barcoding - DNA barcoding is a molecular technique for the identification of specimens through sequencing of a short DNA fragment of the highly conserved mit-8-chondrial Cytochrome c oxidase subunit I (COI or cox1). It was championed by Hebert *et al.* (2003a, b; 2004a) and is currently used by conservation biologists, taxonomists, ecologists, regulatory agencies etc. (Herbert *et al.*, 2003a, 2003b and 2004a). The 5' end of the *cox1* gene is currently being used as a standardized DNA "barcode" to identify species in eukaryotes (Floyd *et al.*, 2009; Smith *et al.*, 2009; Jinbo *et al.*, 2011). The goal of DNA barcoding is to identify various species (DeSalle, 2007). Barcode sequence data originating from taxonomically unidentified samples are compared to an online DNA barcodes reference library that contains data of currently identified species. The unknown DNA may show itself to be a match to a known species or it may be different, but similar to a group of species (Wilson, 2010). MtDNA has been successfully employed in phylogenetic inference studies at multiple taxonomic levels (Tautz *et al.*, 2003 and Savolainen *et al.*, 2005).

DNA barcode identification process involves a number of steps. Sequences are firstly associated with species names that have been taxonomically identified (Floyd *et al.*, 2009). These sequences are incorporated into the barcode library (The Barcode of Life Database or BOLD, http://www.boldsystems.org/) (Ratnasingham & Hebert, 2007) which consists of three modules; data portal, a database of barcode clusters, and a workbench for data collection. Additional barcode sequences exist in other major DNA sequence databases. In 2004, DNA barcoding was begun at the University of Guelph, Guelph, Ontario, Canada. The International Barcode of Life (iBOL) project was initiated in 2010 (http://ibol.org). DNA barcoding helped usher in a golden age of molecular taxonomy since more species could be identified rapidly through this technique, thereby throwing more light on biodiversity at the global level. Algorithms to discover DNA barcodes from sequence databases were developed by Zahariev *et al.* (2009). Taylor and Harris (2012) thought CoxI sequences might not properly identify all animal species and proposed to adapt NGS technology in order to avoid becoming obsolete or irrelevant. Folmer *et al.* (1994) and Jinbo *et al.* (2011) observed that a large number of entomologists have adopted DNA barcoding. It appears that barcoding for the identification of arthropods exceeds its use for other animal groups (Taylor and Harris, 2012).

DNA barcoding is considered to be a standard approach to study the diversity of living organisms (Hajibabaei *et al.,* 2005) including insects (Floyd *et al.,* 2009). DNA barcodes could help resolve numerous constraints to taxonomic resolution, routine species identification, identification of cryptic species, and determination of lineages that are host-specific (Miller & Foottit, 2009). Park *et al.* (2011) assessed the effectiveness of barcodes in a study of 344 species belonging to 178 genera of Heteroptera in the Canadian National Collection of Insects. Barcodes were successfully amplified from even 35 years old specimens. DNA barcodes can also be used for identification of insects in applied conditions, such as associating immature forms with adults in pest management, for identifying eggs in phytosanitary applications, for identifying fragmentary remains to assess food quality, and conducting ecological analyses of biodiversity. Lately, Jung *et al.* (2011) yielded COI barcode sequences for East Asian Heteroptera. These authors demonstrated the effectiveness of *cox1* barcoding by identifying 139 species of true bugs collected from Korea, Japan, Northeastern China and Far East Russia. Studying the process they reported a new cryptic species of true bug.

DNA barcoding has been successfully applied to many animal groups, including moths and butterflies (Hebert *et al.,* 2004b and Hajibabaei *et al.,* 2006), aphids (Foottit *et al.,* 2008 and Lee *et al.,* 2010), birds (Hebert *et al.,* 2004a and Yoo *et al.,* 2006), fish (Ward *et al.,* 2005), and flies (Smith *et al.,* 2006). Emery (2009) employed this technique for identifying lepidopteran larvae that regularly parasitized the flower buds of beach evening primrose, *Camissoniopsis cheiranthifolia* (Onagraceae). DNA barcoding has enabled the identification of morphologically cryptic or immature insects (Miller *et al.,* 2005 and Steeves *et al.,* 2008). Ellis (2006) successfully distinguished cryptic species among two bumble bee species i.e. *Bombus ruderatus* (Fabricius) and *B. hortorum* (Linnaeus) (Hymenoptera: Apidae). Frézal and Leblois (2008) discussed a few crucial pitfalls of DNA barcoding such as the presence an under-described fraction of biodiversity thereby complicating the identification of unknown specimens, or that mitochondrial inheritance may result in over or underestimating divergence of a sample and leading to lack of conclusions on species status.

Langhoff *et al.* (2009) studied 200 specimens of 12 species of New Zealand leafroller moth genera, *Planotortrix* Dugdale and *Ctenopseustis* Meyrick (Lepidoptera: Tortricidae) to distinguish the two genera using a 468-bp DNA barcoding segment. Smith *et al.* (2005) studied ant diversity in Madagascar through DNA barcoding, which would have been a monotonous and arduous task if traditional morphological identification procedures were solely utilized. Virgilio *et al.* (2012) identified African fruit flies (Diptera: Tephritidae) with incomplete DNA Barcode libraries.

On the other hand, this technique has its own failures. For example, Meier *et al.* (2006, 2008) indicated a low rate of identification in Diptera (< 70%) through DNA barcoding. However, different sources of error could not be distinguished. Virgilio *et al.* (2010) evaluated insect DNA barcoding and "minibarcoding" through simulations performed on 15,948 DNA barcodes of six insect orders and implemented four identification criteria. Authors investigated relationships between the level of identification success and taxon coverage in insects, and suggested that the combined effects of false positives (i.e. producing an erroneous positive identification) and false negatives (i.e. erroneously discarding a query) may affect dependability of this technique in insect study. Many researchers (Kankare *et al.*, 2005; King *et al.*, 2008; Valentini *et al.*, 2009 and Ferri *et al.*, 2009) forecasted that DNA barcoding may soon emerge as a preferred tool for species identification, species discovery and delimiting species at global level in order to supplement traditional taxonomy through reaffirming and clarifying identifications.

The Barcode of Life project has been launched for members of the Lepidoptera, Trichoptera, Formicidae, and Apidae taxa. Different barcoding databases are available for important groups of insecta that includes Lepidoptera, Formicidae, Trichoptera, honey bee, mosquito etc. under interantional barcode of life. At least 79, 320 insects have been barcoded (Jinbo *et al.*, 2011). The most useful results have been to accurately identify larval stages of species and to identify small organisms that require microscopic examination for morphological traits or that have a variety of life stages.

8.2.4.2. Nuclear genes

Ribosomes, the translation machinery that translates messenger RNA (mRNA) into proteins, are composed of two subunits each comprised of rRNA and proteins. In eukaryotes, the large subunit (LSU) contains the 28S, the 5.8S and the 5S rRNAs while the small subunit (SSU) includes the 18S rRNA (SSU rRNA) (Paule & White, 2000). The genes encoding these two rRNA subunits are clustered as tandem repeats in the nucleolus-organizing regions of the nuclear chromosomes. In animals, the ribosomal copy number varies between 100-500 tandemly repeated transcription units in the nuclear genome. However, it is as low as 45 in the fungus fly, *Sciara coprophila* Lintner (Diptera: Sciaridae) to over 3000 in the migratory locust, *Locusta migratoria* (Linnaeus) (Orthoptera: Acrididae) while the copy numbers range between 39 to 1023 in 30 species of mosquitoes (Kumar & Rai, 1990).

The rRNA genes undergo concerted evolution due to unequal crossing over, gene conversion, and illegitimate recombination processes. Therefore, the sequence similarity of an rRNA family member is higher within a species than between species (Ohta, 2000). Insect phylogeny has been extensively based on both the 18S and 28S rRNA gene sequences (Misof *et al.*, 2007) and the mitochondrial gene sequences (Zhang *et al.*, 2008). Taxon sampling, sequence alignment, and analysis methods can influence the rRNA gene results (Ogden & Whiting, 2003 and Kjer, 2004). Therefore, rRNA is not very useful for investigating interordinal relationships of insects.

8.2.4.3. Nuclear Protein-Coding Genes

A number of protein-coding loci have been successfully employed in molecular systematics studies (Caterino *et al.*, 2000). Nuclear genes show a variety of evolutionary rates. Some arthropod species have entire genome sequenced, which has increased the number and type of genes that can be used in phylogenetic analyses. Nuclear DNA sequences used for phylogenetic analysis have few constraints. Nuclear genes may be heterozygous. Their low copy number could create difficulty in PCR amplification. Additionally, many genes have large intron which makes amplification difficult unless Reverse Transcriptase-PCR (RT-PCR) is employed. Pseudogenes (inactive forms of a gene) may interfere when comparisons are carried out between genes from different taxa.

Nuclear genes have some advantages over mtDNA. Nuclear genes have less biased base composition and have introns (i.e. more rapidly evolving regions) as well as exons (i.e. slowly evolving regions). Generally, these genes evolve more slowly compared to mitochondrial genes (Brower & DeSalle, 1994 and Friedlander *et al.*, 1994). The low copy number and paralogs may cause problems in phylogenetic analysis (Schubert *et al.*, 2000).

8.2.5. Next Generation Markers (NGS)

Next generation sequencing has so far been limited to just a few model species. But advancements in sequencing technology accompanied by reduction in costs, suggests that these markers will play a prominent role in the future.

8.2.5.1. Single nucleotide polymorphisms (SNP)

SNPs (pronounced as snips) are the most abundant markers in a genome. It is estimated that SNPs are the most common source of genetic variations, and they comprise about 90 per cent of all identified human genetic variations. In human genome, roughly one SNP is present for every 1200 bp of DNA. SNPs have been extensively reported in plants (Varala *et al.,* 2011), animals (Kerstens *et al.,* 2009 and Bovine HapMap Consortium, 2009), and insects (Miller *et al.*, 2012 and Kuhn *et al.*, 2013). The latest release of a high-density SNP database of *D. melanogaster* i.e. FLYSNPdb (http://flysnp. imp.ac.at/ flysnpdb.php) has ~ 27,000 SNPs, i.e one for every 50.3 kb for *Drosophila*. Spotter *et al.* (2012) recently identified 44,000 SNPs in honey bee, *Apis mellifera* Carnica (Hymenoptera: Apidae).

These SNPs can be present anywhere in the genome, in genes, intergenic regions, chromosomal elements like promoters, 3' untranslated regions, repetitive elements etc. If an SNP is present in a gene and causes a change in amino acid, it is referred to as a non-synonymous substitution or missense variant; and if a change in nucleotide does not cause change in amino acid, it is called a synonymous substitution or silent. In general the third nucleotide position in an amino acid can be synonymous or non-synonymous, but the other positions (first and second positions) are non-synonymous substitutions. The other rare forms are nonsense variants which cause premature termination of translation and frame-shift variants which cause frame-shift during translation. The SNPs that are present in other than coding regions also have an effect (Human Encode Project, www.encodeproject.org/ ENCODE/).

The current release of dbSNP (database for human SNPs) has more than 100 million variants of which 88 per cent are SNPs and 11 per cent are indels. These variants are distributed in all functional categories of the genome with 34 per cent of variants in introns, two per cent in coding regions, one per cent in UTR regions, and 62 per cent in unknown/ other regions. The SNPs in coding regions are further categorized into silent (36.1%), missense (59.1%), frameshift (3.1%) and nonsense (1.7%). It is also important to note that most SNP-related mutations are mostly transitions, i.e. A/G and C/T changes, rather than transversions (A/T, A/C, G/T, G/C polymorphisms). Two-thirds of SNPs include a C (cytosine) to T (thymine) substitution (Sung, 2010).

SNPs initially arise from mistakes during DNA replication in individual germ-line cells. They are then passed from generation to generation. Sister chromatids of homologous chromosomes exchange genetic materials and this leads to recombination during meiosis. The offspring is the product of the union of two gametes and is genetically different from the parents. These regions of the chromosome (segments of chromosomes) that are unique are called haplotypes (The International HapMap Consortium, 2005). These are shared by other individuals with a common ancestry and get passed on to following generations. SNPs are biallelic, i.e. consist of two alleles. They occur more frequently in the population than other markers like SSRs and are pass stably from generation to generation (Barreiro *et al.*, 2008). This makes SNPs unique and very useful for evolutionary research.

SNPs research can broadly be divided into a two step process: 1) Identification; 2) Genotyping.

SNP Identification: The first and foremost step is the identification of SNPs through sequencing. Fragments of DNA are sequenced in multiple samples and aligned with each other to compare the sequences. Any variation in nucleotide observed is defined as a SNP. EST sequences are the most commonly used sequences for this purpose. For most model species, they are readily available and have the advantage of being present in the coding regions.

Applying NGS of whole genomes or transcriptomes, has resulted in many SNPs being identified. The genomic DNA is fragmented with restriction enzymes, followed by sequencing the fragments using specific adapters. Millions of sequences are obtained by this approach. For example, an Illumina GAIIx which uses sequencing-by-synthesis produces ~640 million reads totaling 95 GB of data, whereas Illumina Hiseq 2500 produces 6 billion reads of 100bp length totaling 600 GB data. These reads from each sample are then mapped/aligned to a reference genome (if available) and compared to each other for nucleotide variations. If the reference genome sequence is not available, tags of varied length are compared with each other to identify SNPs (Elshire *et al.,* 2011). For this purpose many bioinformatics tools and procedures have been developed to minimize the false positives. This technology is not limited to DNA, but can also be carried out with RNA, which yields SNPs from the coding portions of the genes. The general criterion used to identify a SNP that can be used for other samples is that the SNP should have an allele frequency of at least 10 per cent.

SNP Genotyping: The identified SNP markers can be further utilized for other samples and populations. There are many methods available and the choice depends on the throughput necessary and the available resources for the species of interest. Genotyping is either electrophoresis or fluorescence based. The common methods used for genotyping individuals for SNPs include:

(a) CAPS/dCAPS/PCR-RFLP: Cleaved Amplified Polymorphic Sequence (CAPS) is a low throughput SNP detection method and is highly reliable (Agarwal *et al.*, 2008). When a SNP of interest falls within a restriction site, it can be detected by digesting the PCR fragment with the appropriate restriction endonuclease and then analysing the size profile of digested fragments on an agarose gel. The genotypes with the restriction site show an additional fragment on the gel versus a single fragment for the genotypes that do not have the restriction site. The derived cleaved amplified polymorphic sequence (dCAPS) is a variation of CAP technique where a mismatch in a PCR primer is used to create a polymorphism (Neff *et al.*, 1998). The SNP in the restriction enzyme recognition site is altered in the PCR by using a primer that contains one or more mismatches to template DNA. The modified PCR amplicon is then digested with a restriction enzyme. Based on the fragment size on gel, it is determined if the sample has the SNP (Agarwal *et al.*, 2008). This is one of the most reliable marker systems used and the only issue is that some restriction enzymes are costly. SNPs have been effectively used to develop a linkage map for common bean (*Phaseolus vulgaris* L.) (McConnell *et al.*, 2010), and for silkworm, *Bombyx mori* Linnaeus (Lepidoptera: Bombycidae) (Miao *et al.*, 2005). In addition SNPs were used to identify the marker associated with root-knot nematode resistance in sugarbeet (Weiland & Yu, 2003), and for understanding phosphine resistance in the lesser grain borer, *Rhyzopertha dominica* (Fabricius) (Coleoptera: Bostrichidae) (Kaur *et al.,* 2013). Black and Vontas (2007) did a comprehensive review on SNP assays for genotyping with special emphasis on insects.

(b) Allele-specific primers are used to amplify targeted DNA samples. Primers are designed that only amplify one or the other of the alternate alleles. Results are scored as the presence or absence of the allele. A separate PCR reaction is run for each of the alternate SNP nucleotides and products can be checked on agarose gels. Considerations for designing primers include the GC content, fragment length, annealing temperature, etc. Not all SNPs are suitable for this kind of analysis.

(c) Melting curve analysis (Papp *et al.*, 2003) is an extension of the above assay that incorporates long GC rich 5'-tails of two different lengths into the allele-specific primers. The long tailed product has a higher melting temperature than the short tailed product. PCR is carried out in the presence of SYBR Green, a fluorescent dye whose intensity is enhanced when binding to double stranded DNA. Products are gradually heated after the PCR and the decline in fluorescence is simultaneously measured on a real-time PCR machine. The shape of the resulting fluorescence curve is diagnostic for the alternate alleles and the associated genotypes. Verhaeghen *et al.* (2006) developed a melting curve assay in *Anopheles gambiae* Giles (Diptera: Culicidae) that detects both knock down resistance (*kdr*) alleles in one assay on a real-time PCR machine. Andreev *et al.* (1994) developed the PCR amplification of specific allele (PASA) technique for the serine allele of *Rdl* (codes for a subunit of a γ-aminobutyric acid (GABA) which provides resistance to dieldrin) in the red flour beetle, *Tribolium castaneum* (Herbst) (Coleoptera: Tenebrionidae).

(d) Fluorescent probes are oligonucleotide hybridization probes that become fluorescent only when bound to a specific DNA sequence. Sequence-specific oligonucleotide probe (SSOP) hybridization, reverse-hybridization, and molecular beacons belong to this category. This is costly, but with advancements in fluorescent technology many other fluorescent assays have been developed and costs may decline.

(e) Taqman is one of the most reliable and quick methods for detection of SNPs. Taqman® also uses fluorescence for SNP detection. The Taqman® probe is complementary to the SNP-containing region. The 5' end of the probe is labelled with a fluorescent reporter molecule and the 3' end contains a fluorescence quenching molecule. During PCR amplification of the SNP-containing region, the thermal stable polymerase digests the probe, cleaving the reporter dye from the probe and resulting in increased fluorescence. The polymerase digestion occurs in every cycle, resulting in an increase of fluorescence proportional to the amount of PCR product. However, probe digestion only occurs if the SNP region and the Taqman® probe are perfectly complementary.

(f) DNA microarrays could be utilized in SNP genotyping through SNP chips. The working principle of a SNP microarray is similar to microarrays. About 40bp probes are generated through restriction enzyme digestion, followed by adapter ligation and PCR. The probes are then placed on to the glass slide (referred to as an array). When DNA from a genotye is hybridized, the hybridization signal indicates which allele is carried by the individual (Gupta *et al.*, 2001 and Twyman, 2004). The Affymetrix GeneChip is a standard in present day human research and is capable of genotyping 500,000 SNPs. Similarly illumina beadchips are capable of genotyping 55,000 SNPs (De Donato *et al.*, 2013).

(g) With cheaper next generation sequencing, many methods include identification and genotyping in the same step. For example, Genotype by Sequencing (GBS) (Elshire *et al.*, 2011) is a technique in which 96 barcoded DNA samples are combined and sequenced to get a low coverage of a genome, these are further aligned with a reference genome (or assembled into contig sequences, if no reference is available) and SNPs are identified. In addition to GBS, reads from Illumina, PacBio, or 454 sequences can be directly used to map to a reference genome and identify the SNPs which are further used for studies like QTL mapping, association mapping and population genetics. For example, SNP detection has facilitated association mapping studies in the fruit fly, *Drosophila melanogaster* Meigen (Diptera: Drosophilidae) (Robin *et al.*, 2002). A number of studies have documented SNPs associated with insecticide resistance, especially with mutations conferring reduced target site sensitivity (Ffrench-Constant *et al.*, 1993; Andreev *et al.*, 1994 and Brun-Barale *et al.*, 2005).

8.2.5.2. Insertions and deletions (indels)

Indels are molecular markers that in simple terms consist of an insertion or a deletion of nucleotides (Kondrashov & Rogozin, 2004). It has different classification based on the area of study. For example, an indel in evolutionary biology refers to a mutation class that include both insertions and deletions, as well as combinations of these that have been separated by many years (Gregory, 2004 and Mank *et al.*, 2004). In general mutation studies, an indel is described as a special mutation class that results in a localized deletion or insertion with a net loss or gain in nucleotides. Microindels are defined as indels that results in either an addition or loss of 1 to 50 nucleotides (Scaringe *et al.*, 2008).

Although they are less common than SNPs, indels are still a significant source of variation that are widely distributed throughout the genome (Moghaddam *et al.*, 2014). Mechanisms such as errors occurring during the integration and excision of transposable elements, slippage in simple sequence replication, and unequal crossover can lead to the formation of indels (Britten, 2003). Indels can be utilized as user-friendly markers because they can be separated easily based on their size. These could be used in phylogenetic studies of natural populations (Väli *et al.*, 2008). Indels have been successfully utilized in several genetic studies in plants as well as in animals (Hayashi *et al.*, 2006; Väli *et al.*, 2008 and Vasemägi *et al.*, 2010). In coding regions of the genome, if the length of an indel is not a multiple of three nucleotides, the indel produces a frameshift. In general, Indels that do not have a size of three or multiples of three nucleotides are common in non coding regions because, in coding regions, they change the frame of translation regions (Zheng *et al.*, 2011). It is estimated that there are about 192-280 frameshifting indels in humans (1000 Genomes Project Consortium *et al.*, 2010). Indels can be used as genetic markers in natural populations for association mapping, phylogenetic studies, population studies that include differentiation, gene-flow, divergence and they have applications in various other research areas. Genomic regions with multiple indels could be utilized in procedures for species-identification (Pereira *et al.*, 2010). The identification of indels is easier with the next-generation sequencing. For example, in the common bean (*Phaseolus vulgaris* L.) Indel markers have a distribution of one indel marker every 200 kb on average across all market classes (pintos, kidneys, black, navy etc). (Moghaddam *et al.*, 2014). Park *et al.* (2010) reported an indel frequency of 4.83 indel/kb in field mustard, *Brassica rapa* L. (belongs to the Cruciferae (Brassicaceae) family) using 557 BAC sequences. The complete genome of humans contains 139 indels/Mb and this number is 1,050/Mb and 151/Mb for rice and *Arabidopsis thaliana* (Brassicaceae), respectively (Mills *et al.*, 2006; Shen *et al.*, 2004 and Jander *et al.*, 2002). Lü *et al.* (2013) detected 581 indels with a density of 14.1 indels/kb and 9.2 indels/contig in cotton whitefly, *Bemisia tabaci* (Gennadius) (Hemiptera: Aleyrodidae) Middle East-Asia Minor 1 (MEAM1). Zaidi *et al.* (2011) studied six species of blowflies in China wherein a 184 bp long deletion was detected in one of the sequences of *Chrysomya megacephala* F. (Diptera: Calliphoidae).

8.2.5.3. Copy number variation (CNV)

Copy-number variation/variant (CNV) is a form of structural variation (other than indels, variable numbers of repetitive sequences, and genomic structural alterations) or alteration of th DNA such that one or more sections of the DNA are in abnormal number in the genome (Sebat *et al.*, 2004). CNVs can be large regions of the genome that have been deleted or duplicated on chromosomes leading to an abnormal copy number. CNVs influence more nucleotides than SNPs. CNVs are generally absent in coding regions, because CNVs are caused by unequal recombination. Different estimates reeal that the variation roughly accounts for 12% in humans. The range of this variation is one kilobase (1,000 nucleotide bases) to several megabases in size

(Kallioniemi *et al.*, 1992; Weiss *et al.*, 1999 and Strachan & Read, 2010).

CNVs were initially detected using comparative genomic hybridization (CGH) approach. In CGH, two different flourophores or two orthogonal tags (biotin and digoxigenin) are used to label the samples. They are then hybridized onto metaphase chromosomal preparations. Each chromosome will flouresce as the mixture of two labeled hybridization flourophores. Differences due to either gain or loss of chromosome segments, in comparison with the control treatment are identified by differences in flourescence intensity of the two flourophores (Kallioniemi *et al.*, 1992).

It is difficult to produce well-spread metaphase chromosome slides and to define the subsequent karyotype. It is also difficult to standardize these preparations. The range of resolution of deleted or amplified regions lies between 5 to 10 Mb. However, by spotting DNA clones on a microarray, the resolution can be improved substantially to 30 kb. Copy DNA arrays were the first probes used for CNV detection, but they suffered from low sensitivity of the CNV (Weiss *et al.*, 1999 and Strachan & Read, 2010).

Probe normalization and background noise subtraction steps are performed to reduce false positives. Then comparison of reads of samples with the reference cells is performed to detect deviations. For instance, when a SNP signal from a sample is twice the average level, the former contains double the number of copies i.e. four. More methods are still under development for identification of CNVs.

8.3. APPLICATIONS OF MOLECULAR MARKERS IN ENTOMOLOGY

Accurate taxonomic identification is a prerequisite for developing pest management strategies. Molecular markers such as RAPDs, SSRs, and allozymes have been employed in various IPM studies. RAPD markers differentiated various populations of coccinnellid beetles from multiple geographical regions (Roehrdanz & Flanders, 1993). Similarly, Black *et al.* (1992) used RAPD markers to identify two wasps species, *Diaeretiella rapae* McIntosh and *Lysiphlebius testaceips* (Cresson) (Hymenoptera: Braconidae), which are endoparasitic on aphid genera. RAPDs showed differences in life cycle and behavior between different colonies of the citrus leafminer parasitoid, *Ageniaspis citricola* Logvin (Hymenoptera: Encyrtidae) (Hoy *et al.*, 2000). RAPDs distinguished asexually reproducing aphid species (Lushai *et al.*, 1997) and defined the population structure of parasitoids, *Aphidius ervi* (Daza-Bustamante *et al.*, 2002) and *Diaeretiella rapae* (McIntosh) Haliday (Hymenoptera: Braconidae) (Vaughn & Antolin, 1998). Vaughn and Antolin (1998) successfully studied the population structure and dynamics of the hymenopteran parasitoid, *Diaeretiella rapae* by using RAPD-PCR markers with the cabbage aphid, *Brevicoryne brassicae* (Linnaeus) and Russian wheat aphid, *Diuraphis noxia* Kurdjumov (Hemiptera: Aphididae). Insect ecological research generates important data related to speciation events, population structure, gene flow and genetic diversity. In addition, this research helps to understand the interaction of insects with environmental factors, either biotic (including other biological species) or abiotic (Arunkumar *et al.*, 2012). Hadrys *et al.* (1993) used RAPD markers to determine paternity in two odonate species of anisopteran dragonflies, the lesser emperor, *Anax parthenope* (Sélys) (Odonata: Aeshnidae) and the keeled skimmer, *Orthetrum coerulescens*

(Fabricius) (Odonata: Libellulidae). Bretman *et al.* (2004) studied polyandry in the two-spotted cricket, *Gryllus bimaculatus* de Geer (Orthoptera: Gryllidae) and suggested that high success rate in gaining paternity was observed in unrelated males compared to related males of these Gryllids. Molecular markers were also useful in investigating the interactions at genetic level between disease-causing agents and their insect vectors (Crampton *et al.*, 1997). The kissing bug, *Triatoma infestans* (Klug) (Hemiptera: Reduviidae) is a major insect vector of Chagas' disease in South America that transmits *Trypanosoma cruzi* Chagas, a protozoan parasite. Pinto *et al.* (1998) used RAPD markers to study thc vectorial ability of the insects. In another study, Severson *et al.* (2001) examined the vectorial ability of mosquitoes using molecular markers.

Studies on the migration of insect populations were carried out by utilizing RFLP and mtDNA markers (Hall, 1998 and Nielsen *et al.*, 2000). PCR-RFLP of the mtDNA *cox1* gene was employed to ascertain parasitism by the parasitic wasp, *Peristenus* spp. (Hymenoptera: Braconidae) in the tarnished plant bug, *Lygus lineolaris* (Palisot de Beauvois) (Hemiptera: Miridae) (Tilmon *et al.,* 2000). Researchers (Laroche *et al.*, 1996; Hoy *et al.*, 2000 and Hufbauer *et al.*, 2004) used various techniques (RFLP, RAPD, microsatellite, and mtDNA-based markers) to separate biotypes and sympatric species of insects. ITS2 and 16S markers were used to measure the infection rate of a species specific parasitoid, *Aphelinus hordei* Kurdjumov (Hymenoptera: Aphelinidae) that was being used in the control of Russian wheat aphid, *Diuraphis noxia* (Mord.) (Hemiptera: Aphididae) (Prinsloo *et al.*, 2002) and *Anaphes iole* Girault (Hymenoptera: Mymaridae), the egg parasitoid of *L. lineolaris* bugs in North America (Zhu & Williams, 2002). Australian and Taiwanese populations of the polyembyronic parasitoid, *Ageniaspis fuscicollis* (Dalman) (Hymenoptera: Encyrtidae) were distinguished using the ITS2 region (Alvarez & Hoy, 2002).

One of the most important applications of markers in insect studies is in efforts to understand insect-plant interactions, which involves identification of Quantitative trait loci/genes in plants that are effected and associated due to insect damage. This can be understood either by QTL mapping or association mapping techniques based on the population used. QTL mapping involves a biparental population and is used to study one or a few recombinant events, whereas, association mapping uses a population sample (gene bank collections, land races, breeding populations, varieties) and targets multiple recombination events present in the evolution of those genotypes. For example, Tétard-Jones *et al.* (2012) mapped QTLs for a rhizobacteria-aphid indirect effect in barley in addition to QTLs for an increase and decrease in aphid population size. Kloth *et al.* (2012) described in detail how the use of association mapping can enhance our understanding of plant defense mechanisms against insects.

Insecticide resistance is an important focus in entomological research because of its medical and agricultural implications. For the identification and mapping of resistance genes against insecticides, molecular markers were abundantly used in insects (Behura, 2006). For example, in malaria control programs many difficulties arose because of emerging resistance to DDT in the mosquito vectors. It was found that DDT resistance is associated with an increased metabolism in the major malaria vector, *A. gambiae* in Africa (Ranson *et al.*, 2000). The use of SSR markers in mapping experiments has identified QTLs in *A. gambiae* that determine DDT resistance (Ranson *et al.*, 2000). Knipple *et al.* (1994) employed RFLP markers to discover DDT resistance in houseflies, *Musca domestica* Linnaeus. He suggested that the resistance is due to the 'knockdown' of the kdr trait, which is associated with the reduced neuronal sensitivity to synthetic pyrethroids. Brandt *et al.* (2002) examined the expression pattern of cytochrome P450 genes near a DDT resistance gene in *Drosophila* sp. by using gene-specific PCR markers. Random amplified DNA markers were used to map genetic loci in the lesser grain borer, *Rhyzopertha dominica* (Fabricius) (Coleoptera: Bostrichidae) that determine high-level resistance to phosphine (Schlipalius *et al.*, 2002). AFLP markers were used for identification of resistance loci to pyrethroids in the Colorado potato beetle, *Leptinotarsa decemlineata* Say (Coleoptera: Chrysomelidae) (Hawthorne, 2001) and to *Bacillus thuringiensis* Berliner toxins in the two strains of diamondback moth, *Plutella xylostella* (Linnaeus) (Lepidoptera: Plutellidae) from Hawaii, USA (Heckel *et al.*, 1999). Also, biochemical markers have been used successfully for molecular diagnostic purposes for screening resistance to methyl-parathion in Nebraska Western corn rootworm, *Diabrotica virgifera virgifera* LeConte (Coleoptera: Chrysomelidae) (Zhou *et al.*, 2002) and neonicotinoid cross-resistance in cotton whitefly, *Bemisia tabaci* (Gennadius) (Hemiptera: Aleyrodidae) (Rauch & Nauen, 2003).

DNA markers have been utilzed in insects to provide raw information which ecologists use to make estimates of genetic diversity and gene flow between species, identify haplotypes and lineages, or predict migration and colonization history (Llewellyn *et al.*, 2003 and Bosio *et al.*, 2005). Molecular data can also provide the means to differentiate sympatric species from allopatric species and parapatric species (Ayres *et al.*, 2003 and Margonari *et al.*, 2004). Marker data can provide information on population structure, gene flow and genetic variations in intra- and interspecific insect studies (Mendelson & Shaw, 2005). Natural selection is a major factor driving population diversity. Molecular markers were utilized by Prasad and coworkers (2005) to understand phylogeny and biogeography of various insect populations as well as to study their evolutionary modes and pathways. Diagnostic markers that are associated to traits are used to identify individual insects (Kethidi *et al.*, 2003 and Ullmann *et al.*, 2003). Many times two parapatric populations coexist, but cross-breeding does not take place perhaps because behavioural mechanisms prevent gene flow between these populations which results in reproductive isolation. Some of the known factors such as insect mating behaviour, behavioural plasticity their interactions with hosts and biocontrol agents, plus other as yet unknown factors, play crucial roles in adding to their ecological diversity (Behura, 2006).

Insect population structures have been deciphered using various molecular techniques to estimate gene flow between different populations occupying different habitats or feeding on diverse host species. Phylogenetic studies have been conducted using both rDNA and mtDNA. The relationship between the Aphidinae and other members of the aphid family were ascertained by using the same molecular tools (Smith *et al.*, 1999). Dowton and Austin (2001) reported that molecular (16S rDNA, 28S rDNA and Cox genes) and morphological data together resolved the relationship among the apocritan (Hymenoptera) wasps. The mitochondrial *cox1* gene (Howland & Hewitt, 1995) and the nuclear small subunit rRNA gene (Caterino *et al.*, 2002) sequences were used to infer basal relationships in the order Coleoptera. However, they failed to resolve the main phylogenetic relationship.

The biogeography and population structure of the egg parasitoid, *Gonatocerus ashmeadi* Girault of the glassy winged sharp-shooter, *Homalodisca coagulata* (Say) were successfully described using ISS-PCR markers (de Leon & Jones, 2005). Five polymorphic loci were characterized from two genomic DNA libraries of the cotton bollworm, *H. armigera* in China (Ji *et al.*, 2003).

Phylogenetic studies have also been conducted using the nuclear 18S rDNA marker (Sanchis *et al.*, 2000) and 28S rDNA (Mardulyn & Whitfield, 1999) as well as mitochondrial DNA including 12S and 16S (Cameron & Williams, 2003), *cox1* (Machado *et al.*, 2001), *cox2* (Despres *et al.*, 2002), and Cytochrome B (Kerdelhue *et al.*, 1999). Similarly, Belshaw and Quicke (1997) and Thomson *et al.* (2003) reported on the usefulness of nuclear coding genes like elongation factor 1α and DNA internal transcribed spacer regions (ITS) in parasitoids, respectively. Relationships between closely related taxa such as host races of the pea aphid, *Acyrthosiphon pisum* (Harris) (Hemiptera: Aphididae) were detected using microsatellites (Simon *et al.*, 2003). Recently, Wong *et al.* (2012) used microsatellites to perform parentage analysis in the yellow fever mosquito, *Aedes aegypti* (Linnaeus) (Diptera: Culicidae) from Iquitos, Peru. Likewise, Quicke and Belshaw (1999) and Schmidt *et al.* (2001) mentioned multiple studies that utilized rRNA (18S, 28S) and mtDNA markers (16S, *cox1*, Cytochrome B, and NDI) for parasitic Hymenoptera. Using microsatellite markers, Brouat *et al.* (2003) investigated genetic structure of the predatory carabids, *Carabus nemoralis* Muller and *C. punctatoauratus* Germar (Coleoptera: Carabidae) which were separated by 13.6 km. This study revealed that populations of the specialist predator, *C. punctatoauratus* were more structured than those of the generalist predator, *C. nemoralis*.

Due to the general abundance and neutrality of microsatellites, entomologists have utilized these markers in studying the population structure and dynamics of primary parasitoid wasps (Hufbauer *et al.*, 2004), carabid beetles (Brouat *et al.*, 2003) and honeybees, *A. mellifera* (De la Rua *et al.*, 2003), and studying the phylogeography of braconid parasitoids (Althoff & Thompson, 2001 and Hufbauer *et al.*, 2004), honey bees, *A. mellifera* (De la Rua *et al.*, 2001), bumble bees, *Bombus* spp. (Estoup *et al.*, 1996 and Widmer *et al.*, 1998), and the yucca moth, *Prodoxus quinquepunctellus* (Chambers) (Lepidoptera: Prodoxidae) (Althoff and Thompson, 2001). MtDNA has been used effectively in population studies because it exhibits polymorphism at the intra-specific level. The mtDNA marker revealed spatial distribution of geographical lineages in various insects such as bumble bees (Widmer *et al.*, 1998), honey bees (De la Rua *et al.*, 2001), braconid parasitoids (Hufbauer *et al.*, 2004), yucca moth, *Prodoxus quinquepunctella* (Chambers)

(Lepidoptera: Prodoxidae) (Althoff & Thompson, 2001) and tarnished plant bug, *L. lineolaris* (Burange *et al.*, 2012). Solomon *et al.* (1996) found that PCR-based probes were more efficient and precise than SCAR markers when used in detecting *H. armigera* (Agusti *et al.*, 1999a) and whitefly, *Trialeurodes vaporariorum* (Westwood) in predator grubs (Agusti *et al.*, 1999b). Symondson (2002) discussed the identification of gut contents of predatory insects using specific probes and markers markers including enzyme electrophoresis and PCR-based methods to understand predator-prey interactions.

Another important application of markers is in establishing a phylogeny of insects and supporting the understanding of insect relationships at multiple taxonomic levels. For example, Wiegmann *et al.* (2009) used six single copy nuclear genes to resolve the phylogeny of 29 insect species that represent 11 orders of holometabola. Molecular markers such as barcoding can help to solve taxonomic problems by distinguishing cryptic species (Hebert *et al.*, 2004a; Smith *et al.*, 2006 and Roe & Sperling, 2007b). A single approach cannot answer all questions and over reliance on barcodes has been criticized (Brower *et al.*, 2006). Some researchers (Moritz & Cicero, 2004 and Roe & Sperling, 2007a) have suggested that the investigation of additional regions could improve the application of molecular data to species diagnosis. In addition, an integrative approach involving classical taxonomy, molecular-taxonomy and the DNA barcoding methodologies has been suggested to deal with taxonomic problems such as species boundaries (Roe & Sperling, 2007b). Multiple marker systems could be employed for carrying out combined analyses that would include ecological, morphological, and molecular characters (Mallet & Willmott, 2003).

Advances in insect molecular biology have enabled researchers to do comparative analyses of complete genomes of insects (Savard *et al.*, 2006; i5K Consortium) and complete mitochondrial DNA genomes (Cameron *et al.*, 2009). With the addition of new evidence, the phylogenetic placement of insect orders (Savard *et al.*, 2006) and phylogenetic arrangements of major insect subgroups (Longhorn *et al.*, 2010) has become possible.

8.4. CONCLUSIONS

Molecular markers are versatile in nature and are being used successfully in various studies including insect identification, population genetics, and ecological processes. Their usefulness ranges from helping to understand the taxonomy of natural enemies, host preference, available biotypes, migrations, and intra- and inter-specific interactions and response to novel insecticides. The gathered information becomes extremely important to the success of IPM strategies. The same data also gives insights on overall population dynamics which helps formulate effective and eco-friendly pest management strategies. The next generation molecular markers have great potential to act as powerful diagnostic tools for detecting insect pests, elucidating the interaction of newly introduced parasitoids and predators, mapping changing insect migration pattern, identifying polymorphisms, tracking behavioral responses to hosts, and furthering insect molecular ecological studies.

Even though many marker systems are available, the use of the marker system depends on the availability of genomic resources, throughput required, extant variability in the population under study, and the cost. The choice of markers is also based on the species under study. For example, model insects like drosophila have many genomic resources available. Random markers like AFLP, RFLP, and RAPD have to be used for many other insect species where even EST sequences are not available. If closely related species are involved, a few gene based markers or conserved sequences such as SSRs can be tried. Of the many marker systems, sequencing of genes, AFLP, and SSR once dominated insect molecular research. Now with low cost genotyping, SNPs and indels are forming the basis for the majority of new studies despite some difficulties encountered in analyzing the vast amount of data being generated.

ACKNOWLEDGEMENTS

Authors are thankful to Dr. Richard L. Roehrdanz and Sheila G. Sears Wichmann, Insect Genetics and Biochemistry Research Unit, USDA-ARS RRVARC, 1605 Albrecht Blvd., Fargo, ND 58105 USA and Dr. G.T. Behere, Senior Scientists, Division of Entomology, ICAR Research Complex for NEH Region Umroi Road, Umiam (Barapani), Meghalaya - 793 103, India for their useful comments and advices. The authors sincerely acknowledge the diligent efforts of Ms. Navneet Kaur Arora, Research Fellow, UGC Project and Mr. Shashi Kumar Chauhan, Research Fellow, DBT project. Department of Entomology, PAU, Ludhiana in document processing and editing assistance.

REFERENCES

Abbot P (2001) Individual and population variation in invertebrates revealed by intersimple sequence repeats (ISSRs). *Journal of Insect Science* **1:** 8.

Acquaah G (2012) *Principles of Plant Genetics and Breeding*. John Wiley & Sons, Ltd., West Sussex, UK.

Agarwal M, Shrivastava N and Padh H (2008) Advances in molecular marker techniques and their applications in plant sciences. *Plant Cell Reports* **27**: 617-631.

Agusti N, De Vicente MC and Gabarra R (1999a) Development of sequence amplified characterized region (SCAR) markers of *Helicoverpa armigera*: A new polymerase chain reaction-based technique for predator gut analysis. *Molecular Ecology* **8:** 1467-1474.

Agusti N, De Vicente MC and Gabarra R (1999b) Developing SCAR markers to study predation on *Trialeurodes vaporariorum*. *Insect Molecular Biology* **9:** 263-268.

Alessandrini F, Mazzanti M, Onofri V, Turchi C and Tagliabracci A (2008) MtDNA analysis for genetic identification of forensically important insects. *Forensic Science International: Genetics Supplement Series* **1:** 584-585.

Althoff DM and Thompson JN (2001) Geographic structure in the searching behavior of a specialist braconid: combining molecular and behavioral approaches. *Journal of Evolutionary Biology* **14:** 406-417.

Alvarez JM and Hoy MA (2002) Evaluation of the ribosomal ITS2 DNA sequences in separating the closely related populations of the parasitoid *Ageniaspis* (Hymenoptera: Encyrtidae). *Annals of the Entomological Society of America* **95:** 250-256.

Arunkumar KP, Sahu AK, Mohanty AR, Awasthi AK, Pradeep AR, Raje Urs S and Nagaraju J (2012) Genetic diversity and population structure of Indian golden silkmoth (*Antheraea assama*). *PLoS ONE* **7:** e43716.

Andreev D, Rocheleau T, Phillips TW, Beeman RW, and ffrench-Constant RH (1994) A PCR diagnostic for cyclodiene insecticide resistance in the red flour beetle, *Tribollium castaneum*. *Pesticide Science* **41:** 345-349.

Avise J C (1994) *Molecular Markers, Natural History and Evolution*, Chapman and Hall, New York, USA.

Ayres DR, Strong DR and Baye P (2003) *Spartina foliosa*: A common species on the road to rarity? *Madroño* **50:** 209-213.

Barreiro LB, Laval G, Quach H, Patin E and Quintana-Murci L (2008) Natural selection has driven population differentiation in modern humans. *Nature Genetics* **40:** 340-345.

Bartlett AC and Butler GD Jr (1975) Genetic control of the cabbage looper by a recessive lethal mutation. *Journal of Economic Entomology* **68:** 331-335.

Bartlett AC, Wilson NM and Mattix EB (1968) The fate of genetic markers in populations of boll weevils. *Journal of Economic Entomology* **61:** 808-812.

Behura SK (2006) Molecular marker systems in insects: current trends and future avenues. *Molecular Ecology* **15:** 3087-3113.

Belshaw R and Quicke DLJ (1997) A molecular phylogeny of the Aphidinae (Hymenoptera: Braconidae). *Molecular Phylogenetics and Evolution* **7:** 281-293.

Black WC, DuTeau NM, Puterka GJ, Nechols JR and Pettorini JM (1992) Use of the random amplified polymorphic DNA polymerase chain reaction (RAPD-PCR) to detect DNA polymorphisms in aphids (Homoptera: Aphididae). *Bulletin of Entomological Research* **82:** 151-159.

Black IV WC and Vontas JG (2007) Affordable assays for genotyping single nucleotide polymorphisms in insects. *Insect Molecular Biology* **16:** 377-387.

Bond HA, Craig GB Jr and Fay RW (1970) Field mating and movement of *Aedes aegypti*. *Mosquito News* **30:** 394-402.

Boore JL (1999) Animal mitochondrial genomes. *Nucleic Acids Research* **27:** 1767-1780.

Bosio CF, Harrington LC, Jones JW Sithiprasasna R, Norris DE and Scott TW (2005) Genetic structure of *Aedes aegypti* populations in Thailand using mitochondrial DNA. *American Journal of Tropical Medicine and Hygiene* **72:** 434-442.

Botstein D, White RL, Skolnick M and Davis RW (1980) Construction of a genetic linkage map in man using restriction fragment length polymorphisms. *American Journal of Human Genetics* **32:** 314-333.

Bovine HapMap Consortium (2009) Genome-wide survey of SNP variation uncovers the genetic structure of cattle breeds. *Science* **324:** 528-532.

Boyd SD (2013) Diagnostic applications of high-throughput DNA sequencing. *The Annual Review of Pathology: Mechanisms of Disease* **8:** 381-410.

Brandt A, Scharf M, Pedra JH, Holmes G, Dean A, Kreitman M and Pittendrigh BR (2002) Differential expression and induction of two Drosophila cytochrome P450 genes near the Rst (2) DDT locus. *Insect Molecular Biology* **11:** 337-341.

Bretman A, Wedell N and Tregenza T (2004) Molecular evidence of post-copulatory inbreeding avoidance in the field cricket Gryllus bimaculatus. *Proceedings of the Royal Society of London. Series B, Biological Sciences* **271:** 159-164.

Britten RJ (2003) Majority of divergence between closely related DNA sequences is due to indels. *Proceedings of the National Academy of Sciences of the United States of America* **100:** 4661-4665.

Brouat C, Sennedot F, Audiot P, Leblosis R and Rasplus JY (2003) Fine-scale genetic structure of two carabid species with contrasted levels of habitat specialization. *Molecular Ecology* **12:** 1731-1745.

Brower A and DeSalle R (1994) Practical and theoretical considerations for choice of DNA sequence region in insect molecular systematics, with a short review of published studies using nuclear gene regions. *Annual Review of Entomology* **87:** 702-716.

Brower AVZ, Freitas AVL, Lee MM, Silva-Brandão KL, Whinnett A and Willmott KR (2006) Phylogenetic relationships among the Ithomiini (Lepidoptera: Nymphalidae) inferred from one mitochondrial and two nuclear gene regions. *Systematic Entomology* **31:** 288-301.

Brun-Barale A, Bouvier JC, Paruon D, Berge JB and Benoit S (2005) Involvement of a sodium channel mutation in pyrethroid resistance in *Cydia pomonella* Linn., and development of a diagnostic test. *Pest Management Science* **61:** 549-554.

Burange PS, Roehrdanz RL and Boetel MA (2012) Geographically based diversity in mitochondrial DNA of North American *Lygus lineolaris* (Hemiptera: Miridae). *Annals of Entomological Society of America* **105:** 917-929.

Cameron SA and Williams PH (2003) Phylogeny of bumble bees in the New World subgenus *Fervidobombus* (Hymenoptera: Apidae): congruence of molecular and morphological data. *Molecular Phylogenetics and Evolution* **28:** 552-563.

Cameron SL, Sullivan J, Song H, Miller KB and Whiting MF (2009) A mitochondrial genome phylogeny of the Neuropterida (lace-wings, alderflies and snakeflies) and their relationship to the other holometabolous insect orders. *Zoologica Scripta* **38:** 575-590.

Caterino MS, Cho S and Sperling FAH (2000) The current state of insect molecular systematics: a thriving Tower of Babel. *Annual Review of Entomology* **45:** 1-54.

Caterino MS, Shull VL, Hammond PM and Vogler AP (2002) Basal relationships of Coleoptera inferred from 18S rDNA sequences. *Zoologica Scripta* **31:** 41-49.

Chinault AC and Carbon J (1979) Overlap hybridization screening: isolation and characterization of overlapping DNA fragments surrounding the *leu2* gene on yeast chromosome III. *Gene* **5:** 111-126.

Clary DO and Wolstenholm DR (1985) The mitochondrial DNA molecule of *Drosophila yakuba*: nucleotide sequence, gene organization, and genetic code. *Journal of Molecular Evolution* **22:** 252-271.

Cook JP, Wichman DM, Martin JM, Bruckner PL and Talbert LE (2004) Identification of microsatellite markers associated with a stem solidness locus in wheat. *Crop Science* **44:** 1397-1402.

Crampton JM, Beard CB and Louis C (1997) *Molecular Biology of Insect Disease Vectors: a Methods Manual,* Chapman & Hall, London, UK.

Daza-Bustamante P, Fuentes-Contreras E, Rodriguez LC, Figueroa CC and Niemeye HM (2002) Behavioural differences between *Aphidius ervi* populations from two tritrophic systems are due to phenotypic plasticity. *Entomologia Experimentalis et Applicata* **104:** 321-328.

De Donato M, Peters SO, Mitchell SE, Hussain T, Imumorin IG (2013) Genotyping-by-sequencing (GBS): a novel, efficient and cost-effective genotyping method for cattle using next-generation sequencing. *PLoS ONE* **8:** e62137: doi: 10.1371/journal.pone.0062137

De la Rúa P, Galián J, Serrano J and Moritz RF (2003) Genetic structure of Balearic honeybee populations based on microsatellite polymorphism. *Genetics Selection Evolution* **35:** 339-350.

De La Rua, P Galian J, Serrano J and Moritz RFA (2001) Molecular characterization and population structure of the honeybees from the B'alearic Islands (Spain). *Apidologie* **32:** 417-427.

de Leon JH and Jones WA (2005) Genetic differentiation among geographic populations of *Gonatocerus ashmeadi*, the predominant egg parasitoid of glassy sharp-shooter, *Homalodisca coagualata. Journal of Insect Science* **5:** 1-9.

DeSalle R (2007) Phenetic and DNA taxonomy; a comment on Waugh. *BioEssays* **29:** 1289-1290.

Despres L, Pettex E, Plaisance V and Pompanon F (2002) Speciation in the globeflower fly *Chiastocheta* spp. (Diptera: Anthomyiidae) in relation to host plant species, biogeography, and morphology. *Molecular Phylogenetics and Evolution* **22:** 258-268.

Dowling TE, Moritz C, Palmer JD and Rieseberg LH (1996) Nucleic acids III: Analysis of fragments and restriction sites, pp. 249-320. **In:** Hillis D M, Moritz C and Mable B K (eds.), *Molecular Systematics*, Sinauer Associates, Sunderland, Massachusetts, USA.

Dowton M and Austin AD (2001) Simultaneous analysis of 16S, 28S, COI and morphology in the Hymenoptera: Apocrita - evolutionary transitions among parasitic wasps. *Biological Journal of the Linnean Society* **74:** 87-111.

1000 Genomes Project Consortium, Abecasis GR, Altshuler D, Auton A, Brooks LD, Durbin R M, Gibbs RA, Hurles ME and McVean GA (2010) A map of human genome variation from population-scale sequencing. *Nature* **467:** 1061-1073.

Edwards OW and Hoy MA (1993) Polymorphism in two parasitoids detected using random amplified polymorphic DNA (RAPD) PCR. *Biological Control* **3:** 243-257.

Edwards OW and Hoy MA (1995) Random amplified polymorphic DNA markers to monitor laboratory-selected, pesticide-resistant *Trioxys pallidus* (Hymenoptera: Aphidiidae) after release into three California walnut orchards. *Environmental Entomology* **24:** 487-496.

Ellis JS, Knight ME, Carvell C and Goulson D (2006) Cryptic species identification: A simple diagnostic tool for discriminating between two problematic bumblebee species. *Molecular Ecology Notes* **6:** 540-542.

Elshire RJ, Glaubitz JC, Sun Q, Poland JA, Kawamoto K, Buckler ES and Mitchell SE (2011) A robust, simple genotyping-by-sequencing (GBS) approach for high diversity species. *PLoS ONE* **6:** e19379.

Emery VJ, Landry JF and Eckert CG (2009) Combining DNA barcoding and morphological analysis to identify specialist floral parasites (Lepidoptera: Coleophoridae: Momphinae: Mompha). *Molecular Ecology Resources* **9:** 217-223.

Estoup A, Solignac M, Cornuet JM, Goudet J and Scholl A (1996) Genetic differentiation of continental and island populations of *Bombus terrestris* (Hymenoptera: Apidae) in Europe. *Molecular Ecology* **5:** 19-31.

Fakrudin B, Vijaykumar, Kambalpally KB, Patil BV and Kuruvinashetti MS (2006) DNA-based marker systems and their utility in entomology. *Entomologica Fennica* **3:** 21-33.

Fay RW and Craig GB (1969) Genetically marked Aedes aegypti in studies of field populations. *Mosquito News* **29:** 121-127.

Ferri E, Barbuto M and Bain O (2009) Integrated taxonomy, traditional approach and DNA barcoding for the identification of filaroid worms and related parasites (Nematoda). *Frontiers in Zoology* **6:** 1-27.

Ffrench-Constant RH, Rocheleau TA, Steichen JC, Chalmers AE (1993) A point mutation in a *Drosophila* GABA receptor confers insecticide resistance. *Nature* **363:** 449-451.

Fiers W, Contreras R, Duerinck F, Haegeman G, Iserentant D, Merregaert J, Min Jou W, Molemans F, Raeymaekers A, Van den Berghe A, Volckaert G and Ysebaert M (1976) Complete nucleotide sequence of bacteriophage MS2 RNA: Primary and secondary structure of the replicase gene. *Nature* **260:** 500-507.

Floyd RM, Wilson JJ and Hebert PDN (2009) DNA barcodes and insect biodiversity, pp. 417-32. **In:** Foottit R G and Adler P H (eds.), Insect Biodiversity: Science and Society. Wiley-Blackwell, Oxford, UK.

Folkertsma RT, Sajjanar GM, Reddy BVS, Sharma HC and Hash CT (2003) Genetic mapping of QTL associated with sorghum shoot fly (*Atherigona soccata*) resistance in sorghum (*Sorghum bicolor*), pp. 42. **In:** Final Abstracts Guide, Plant and Animal Genome XI, Jan 11-15, 2003, San Diego, CA.

Folmer O, Black M, Hoeh W, Lutz R and Vrijenhoek R (1994) DNA primers for amplification of mitochondrial cytochrome c oxidase subunit I from diverse metazoan invertebrates. *Molecular Marine Biology and Biotechnolology* **3:** 294-299.

Foottit RG, Maw HEL and Von Dohlen CD (2008) Species identification of aphids (Insecta: Hemiptera: Aphididae) through DNA barcodes. *Molecular Ecology Resources* **8:** 1189-1201.

Frézal L and Leblois R (2008) Four years of DNA barcoding: Current advances and prospects. *Infection Genetics and Evolution* **8:** 727-736.

Friedlander TP, Regier JC and Mitter C (1994) Phylogenetic information content of five nuclear gene sequences in animals: initial assessment of character sets from concordance and divergence studies. *Systematic Biology* **43:** 511-525.

Goldstein D and Schlötterer C (1999) *Microsatellites: Evolution and Applications*, Oxford University Press, Oxford, UK.

Gray MW (1989) Origin and evolution of mitochondrial DNA. *Annual Review of Cell Biology* **5:** 25-50.

Gregory TR (2004) Insertion-deletion biases and the evolution of genome size. *Gene* **324:** 15-34.

Grimaldi D and Engel MS (2005) *Evolution of the Insects*. Cambridge University Press, New York, USA.

Grodzicker T, Williams J, Sharp P, Sambrook J (1974) Physical mapping of temperature sensitive mutants of adenovirus. *Cold Spring Harbor Symposia on Quantitative Biology* **39:** 439-446.

Gupta PK and Varshney RK (2000) The development and use of microsatellite markers for genetic analysis and plant breeding with emphasis on bread wheat. *Euphytica* **113:** 163-185.

Gupta PK, Roy JK and Prasad M (2001) Single nucleotide polymorphisms: A new paradigm for molecular marker technology and DNA polymorphism detection with emphasis on their use in plants. *Current Science* **80:** 524-535.

Hadrys H, Balick M and Schierwater B (1992) Applications of random amplified polymorphic DNA (RAPD) in molecular ecology. *Molecular Ecology* **1:** 55-63.

Hadrys H, Schierwater B, DeSalle R, Dellaporta SL and Buss LW (1993) Determination of paternity in dragonflies by Random Amplified Polymorphic DNA fingerprinting. *Molecular Ecology* **2:** 79-87.

Hajibabaei M, deWaard JR, Ivanova NV, Ranasingham S and Dooh RT (2005)

Critical factors for assembling a high volume of DNA barcodes. *Philosophical Transactions of the Royal Society of London B: Biological Sciences* **360:** 1959-1967.

Hajibabaei M, Janzen DH, Burns JM, Hallwachs W and Hebert PDN (2006) DNA barcodes distinguish species of tropical Lepidoptera. *Proceedings of the National Academy of Sciences of the United States of America* **103:** 968-971.

Hall HG (1998) PCR amplification of a locus with RFLP alleles specific to African honey bees. *Biochemical Genetics*. **36:** 351-361.

Harrison RG (1989) Mitochondrial DNA as a genetic marker in population and evolutionary biology. *Trends in Ecology and Evolution* **4**: 6-11.

Havill NP, Montgomery ME, Yu G, Shiyake S and Caccone A (2006) Mitochondrial DNA from hemlock wooly adelgid (Hemiptera: Adelgidae) suggests cryptic speciation and pinpoints the source of the introduction to eastern North America. *Annals of the Entomological Society of America* **99:** 195-203.

Hawthorne DJ (2001) AFLP-based genetic linkage map of the Colorado potato beetle Leptinotarsa decemlineata: sex chromosomes and a pyrethroid-resistance candidate gene. *Genetics* **158:** 695-700.

Hayashi K, Yoshida H and Ashikawa I (2006) Development of PCR based allele-specific and indel marker sets for nine rice blast resistance genes. *Theoretical and Applied Genetics* **113**: 251-260.

Hebert PDM, Cywinska A, Ball SL and deWaard JR (2003a) Biological identifications through DNA barcodes. *Proceedings of the Royal Society of London. Series B, Biological Sciences* **270:** 313-321.

Hebert PDN, Penton EH, Burns JM, Janzen DH and Hallwachs W (2004b) Ten species in one: DNA barcoding reveals cryptic species in the Neotropical skipper butterfly *Astraptes fulgerator*. *Proceedings of the National Academy of Sciences of the United States of America* **101:** 14812-14817.

Hebert PDN, Ratnasingham S and deWaard JR (2003b) Barcoding animal life: cytochrome c oxidase subunit 1 divergences among closely related species. *Proceedings of the Royal Society of London* **270:** 596-599.

Hebert PDN, Stoeckle MY and Zemlak TS (2004a) Identification of birds through DNA barcodes. *PLoS Biology* **2:** e312.

Heckel DG, Gahan LJ, Liu YB and Tabashnik BE (1999) Genetic mapping of resistance to Bacillus thuringiensis toxins in diamondback moth using biphasic linkage analysis. *Proceedings of the National Academy of Sciences of the United States of America* **96:** 8373-8377.

Howland DE and Hewitt GM (1995) Phylogeny of the Coleoptera based on mitochondrial cytochrome oxidase I sequence data. *Insect Molecular Biology* **4:** 203-215.

Hoy MA (2013) Insect Molecular Genetics: An Introduction to Principles and Applications, Elsevier Inc., San Diego, California, USA.

Hoy MA, Jeyaprakash A, Morakote R, Lo PKC and Nguyen R (2000) Genomic analyses of two populations of *Ageniaspis citricola* (Hymenoptera: Encyrtidae) suggest that a cryptic species may exist. *Biological Control* **17:** 1-10.

Hu J and Vick BA (2003) Target region amplification polymorphism: A novel marker technique for plant genotyping. *Plant Molecular Biology Reporter* **21:** 289-294.

Hu L, Jianyu G, Haiyu L and Wanzhi C (2009) Progress in the researches on insect mitochondrial genome and analysis of gene order. *Science Foundation in China* **17:** 39-45.

Hufbauer RA, Bogdanowicz SM Harrison RG (2004) The population genetics of a biological control introduction: mitochondrial DNA and microsatellie variation in native and introduced populations of *Aphidus ervi*, a parasitoid wasp. *Molecular Ecology* **13:** 337-348.

Hughes C (1997) Interacting molecular techniques with field methods in studies of social behavior: A revolution results. *Evolution* **79:** 383-399.

i5K Consortium. The i5K Initiative: Advancing Arthropod Genomics for Knowledge, Human Health, Agriculture, and the Environment; http://jhered.oxfordjournals.org/content/104/5/595.full

Jairin J, Toojindab T, Tragoonrung S, Tayapat S and Vanavichit A (2005) Multiple genes determining brown planthopper (*Nilaparvata lugens* (Stål)) resistance in backcross introgressed lines of thai jasmine rice 'KDML105'. *ScienceAsia* **31:** 129-135.

Jander G, Norris SR, Rounsley SD, Bush DF, Levin IM and Last RL (2002) Arabidopsis map-based cloning in the post-genome era. *Plant Physiology* **129:** 440-450.

Ji YJ, Zhang DX, Hewitt GM, Karg L and Li DM (2003) Polymorphic microsatellite loci for the cotton bollworm, *Helicoverpa armigera* (Lepidoptera: Noctuidae) and some remarks on their isolation. *Molecular Ecology Notes* **3:** 102-104.

Jinbo U, Kato T and Ito M (2011) Current progress in DNA barcoding and future implications for entomology. *Entomological Science* **14:** 107-124.

Jones CJ, Edwards KJ, Castaglione S, Winfield MO, Sala F, van de Wiel C, Bredemeijer G, Vosman B, Matthes M, Daly A, Brettschneider R, Bettini P, Buiatti M, Maestri E, Malcevschi A, Marmiroli M, Aert R, Volckaert C, Rueda J, Lina cero R, Vazquez A and Karp A (1997) Reproducibility testing of RAPD, AFLP, and SSR markers in plants by a network of European laboratories. *Molecular Breeding* **3:** 381-890.

Jung S, Duwal RK and Lee S (2011) COI barcoding of true bugs (Insecta, Heteroptera). *Molecular Ecology Resources* **11:** 266-270.

Kallioniemi A, Kallioniemi OP, Sudar D, Rutovitz D, Gray JW, Waldman F and Pinkel D (1992) Comparative genomic hybridization for molecular cytogenetic analysis of solid tumors. *Science* **258:** 818-821.

Kambhampati S, Black WC and Rai KS (1992) A RAPD-PCR based method for the identification and differentiation of mosquito populations: techniques and statistical analyses. *Journal of Medical Entomology* **29:** 939-945.

Kankare M, Stefanescu C, Van Houhuys S and Shaw MR (2005) Molecular phylogeny of *Cotesia* Cameron, 1891 (Insecta: Hymenoptera: Braconidae: Microgastrinae) parasitoids associated with Melitaeini butterflies (Insecta: Lepidoptera: Nymphalidae: Melitaeini). *Journal of the Linnean Society* **86:** 45-65.

Kaur R, Ebert PR, Walter GH, Swain AJ and Schlipalius DI (2013) Do Phosphine Resistance Genes Influence Movement and Dispersal Under Starvation? *Journal of Economic Entomology* **106:** 2259-2266.

Kerdelhue C, Le Clainche I and Rasplus JY (1999) Molecular phyllogeny of *Leratosolen* species pollinating *Ficus* of the subgenus *Sycomorus* senu stricto. Biogeographical history and origins of the species specificity breakdown cases. *Phyllogenetics and Evolution* **11:** 401-414.

Kerstens HHD, Kollers S, Kommadath A, Rosario M, Dibbits B, Kinders SM, Crooijmans RP and Groenen MAM (2009) Mining for Single Nucleotide

Polymorphisms in Pig genome sequence data. *BMC Genomics* **10:** 4.

Kethidi DR, Roden DB, Ladd TR, Krell PJ, Retnakaran A and Feng Q (2003) Development of SCAR markers for the DNA-based detection of the Asian long-horned beetle, *Anoplophora glabripennis* (Motschulsky). *Archives of Insect Biochemistry and Physiology* **52:** 193-204.

King RA, Tibble AL and Symondson WO (2008) Opening a can of worms: unprecedented sympatric cryptic diversity within British lumbricid earthworms. *Molecular Ecology* **17:** 4684-4698.

Kjer KM (2004) Aligned 18S and insect phylogeny. *Systematic Biology* **53:** 506-514.

Kloth KJ, Thoen MP, Bouwmeester HJ, Jongsma MA and Dicke M (2012) Association mapping of plant resistance to insects. *Trends in Plant Science* **17:** 311-319.

Knipple DC, Doyle KE, Marsella-Herrick PA *and S*oderlund DM (1994) Tight genetic linkage between the kdr insecticide resistance trait and a voltage-sensitive sodium channel gene in the house fly. *Proceedings of the National Academy of Sciences of the United States of America* **91**: 2483-2487.

Kobayashi I (1998) Selfishness and death: raison d'etre of restriction, recombination and mitochondria. *Trends in Genetics* **14:** 368-373.

Kondrashov AS and Rogozin IB (2004) Context of deletions and insertions in human coding sequences. *Human Mutation* **23:** 177-185.

Konieczny A and Ausubel FM (1993) A procedure for mapping Arabidopsis mutations using co-dominant ecotype-specific PCR-based markers. *Plant Journal* **4:** 403-410.

Kuhn KL, Duan JJ and Hopper KR (2013) Next-generation genome sequencing and assembly provides tools for phylogenetics and identification of closely related species of *Spathius*, parasitoids of *Agrilus planipennis* (emerald ash borer). *Biological Control* **66:** 77-82.

Kumar A and Rai KS (1990) Chromosomal localization and copy number of 18S + 28S ribosomal RNA genes in evolutionarily diverse mosquitoes (Diptera: Culicidae). *Hereditas* **113:** 277-289.

Langhoff P, Authier A, Buckley TR, Dugdale JS, Rodrigo A and Newcomb RD (2009) DNA barcoding of the New Zealand leafroller moth genera, *Ctenopseustis* and *Planotortrix*. *Molecular Ecology Resources* **9:** 691-698.

Laroche A, Declerck-Floate RA, Lesage L, Floate KD and Demeke T (1996) Are *Altica carduorum* and *Altica cirsicola* (Coleoptera: Chrysomelidae) different species? Implications for the release of *A. cirsicola* for the biocontrol of Canada thistle in Canada. *Biological Control* **6:** 306-314.

Lawlor HJ, Siddique KHM, Sedgley RH and Thurling N (1998) Improving cold tolerance and insect resistance in chickpea and the use of AFLPs for the identification of molecular markers for these traits. *Acta Horticulturae* **461:** 185-192.

Lee W, Kim H and Lim J (2010) Barcoding aphids (Hemiptera: Aphidadae) of the Korean Peninsula: updating the global dataset. *Molecular Ecology Resources* **11:** 32-37.

Litt M and Luty JA (1989) A hypervariable microsatellite revealed by in vitro amplification of a dinucleotide repeat within the cardiac muscle actin gene. *American Journal of Human Genetics* **44:** 397-401.

Liu XM, Smith CM, Gill BS and Tolmay V (2001) Microsatellite markers linked to six Russian wheat aphid resistance genes in wheat. *Theoretical and Applied Genetics* **102:** 504-510.

Llewellyn KS, Loxdale HD, Harrington R, Brookes CP, Clark SJ and Sunnucks P (2003) Migration and genetic structure of the grain aphid (Sitobion avenae) in Britain related to climate and clonal fluctuation as revealed using microsatellites. *Molecular Ecology* **12:** 21-34.

Longhorn SJ, Pohl HW and Vogler AP (2010) Ribosomal protein genes of holometabolan insects reject the Halteria, instead revealing a close affinity of Strepsiptera with Coleoptera. *Molecular Phylogenetics and Evolution* **55:** 846-859.

Lü ZC, Sun HB, Wan FH, Guo JY and Zhang GF (2013) High variation in single nucleotide polymorphisms (SNPs) and insertions/deletions (Indels) in the highly invasive *Bemisia tabaci* (Gennadius) (Hemiptera: Aleyrodidae) Middle East-Asia Minor 1 (MEAM1). *Neotropical Entomology* **42:** 521-526.

Lushai G, Loxdale HD, Brookes CP, Von Mende N, Harrington R and Hardie J (1997) Genotypic variation among different phenotypes within aphid clones. *Proceedings of the Royal Society, London, Biological Sciences (Series B)* **264:** 725-730.

Lynch M and Walsh B (1998) Genetics and analysis of quantitative traits. Sinauer Associates, Sunderland, Massachusetts, USA.

Maa WCJ and Terriere LC (1983) Age-dependent variation in enzymatic and electrophoretic properties of house fly (M. domestica) carboxylesterases. *Comparative Biochemistry and Physiology* **74:** 461-467.

Machado CA, Jousselin E, Kjellberg F, Compton SG and Herre EA (2001) Polygenetic relationships, historical biogeography and character evolution of fig pollinating wasps. *Proceedings of the Royal Society of London, Biological Sciences (Series B)* **268:** 685-694.

Maeda M, Uryu N, Murayama N, Ishii H, Ota M, Tsuji K and Inoko H (1990) A simple and rapid method for HLA-DP genotying by digestion of PCR-amplified DNA with allele specific restriction endonucleases. *Human Immunology* **27:**111-121.

Malik R, Smith CM, Harvey TL and Brown-Guedira GL (2003) Genetic mapping of wheat curl mite resistance genes Cmc3 and Cmc4 in common wheat. *Crop Science* **43:** 644-650.

Mallet J and Willmott K (2003) Taxonomy: Renaissance or Tower of Babel? *Trends in Ecology and Evolution* **18:** 57-59.

Mank R, Wilson MD, Rubio JM and Post RJ (2004) A molecular marker for the identification of Simulium squamosum (Diptera: Simuliidae). *Annals of Tropical Medicine and Parasitology* **98:** 197-208.

Mardulyn P and Whitfield JB (1999) Phylogenetic signal in the COI, 16S, and 28S genes for inferring relationships among genera of Microgastrinae (Hymenoptera; Braconidae): Evidence of a high diversification rate in this group of parasitoids. *Molecular Phylogenetics and Evolution* **12:** 282-294.

Margonari CS, Fortes-Dias CL and Dias ES (2004) Genetic variability in geographical populations of Lutzomyia whitmani elucidated by RAPD-PCR. *Journal of Medical Entomology* **41:** 187-192.

Marsjan PA and Oldenbroek JK (2007) Molecular markers, a tool for exploring genetic diversity, pp. 359-379. **In:** Rischkowsky B and Pilling D (eds.), The State of the World's Animal Genetic Resources for Food and Agriculture. FAO, Rome, Italy; ftp://ftp.fao.org/docrep/fao../010/a1250e/a1250e17.pdf.

Matsumoto A and Tsumura Y (2004) Evaluation of cleaved amplified polymorphic sequence markers. *Theoretical and Applied Genetics* **110:** 80-91.

Mazzanti M, Allessandrini F, Tagliabracci A, Wells JD and Campobasso CP (2010) DNA degradation and genetic analysis of empty puparia: Genetic identification limits in forensic entomology. *Forensic Science International* **195**: 99-102.

McConnell M, Mamidi S, Lee R, Chikara S, Rossi M, Papa R and McClean P (2010) Syntenic relationships among legumes revealed using a gene-based genetic linkage map of common bean (Phaseolus vulgaris L.). *Theoretical and Applied Genetics* **121:** 1103-1116.

McDermott JM, Brandle U, Dutly F, Haemmerli UA, Keller S, Muller KE and Wolf MS (1994) Genetic variation in powdery mildew of barley: Development of RAPD, SCAR and VNTR markers. *Phytopathology* **84:**1316-1321.

Meier R, Shiyang K, Vaidya G and Ng PKL (2006) DNA barcoding and taxonomy in Diptera: a tale of high intraspecific variability and low identification success. *Systematic Biology* **55:** 715-728.

Meier R, Zhang G and Ali F (2008) The use of mean instead of smallest interspecific distances exaggerates the size of the barcoding gap and leads to misidentification. *Systematic Biology* **57:** 809-813.

Mendelson TC and Shaw KL (2005) Use of AFLP markers in surveys of arthropod diversity. *Methods in Enzymology* **395:** 161-177.

Miao XX, Xub SJ, Li MH, Li MW, Huang JH, Dai FY, Marino SW, Mills DR, Zeng P, Mita K, Jia SH, Zhang Y, Liu WB, Xiang H, Guo QH, Xu AY, Kong XY, Lin HX, Shi YZ, Lu G, Zhang X, Huang W, Yasukochi Y, Sugasaki T, Shimada T, Nagaraju J, Xiang ZH, Wang SY, Goldsmith MR, Lu C, Zhao G and Huang Y (2005) Simple sequence repeat-based consensus linkage map of *Bombyx mori*. *Proceedings of the National Academy of Sciences of the United States of America* **102:** 16303-16308.

Michaels SD and Amasino RM (1998) A robust method for detecting single-nucleotide changes as polymorphic markers by PCR. *Plant Journal* **14:** 381-385.

Miller GL and Foottit RG (2009) The Taxonomy of Crop Pests: The Aphids, pp. 463-473. **In:** Foottit RG and Adler PH (eds.), *Insect Biodiversity: Science and Society*, Wiley-Blackwell, Oxford, UK.

Miller NJ, Sun J and Sappington TW (2012) High-Throughput Transcriptome Sequencing for SNP and Gene Discovery in a Moth. *Environmental Entomology* **41:** 997-1007.

Miller KB, Alarie Y, Wolfe W and Whiting M (2005) Association of insect life stages using DNA sequences: the larvae of *Philodytes umbrinus* (Motschulsky) (Coleoptera: Dytiscidae). *Systematic Entomology* **30:** 499-509.

Mills RE, Luttig CT, Larkins CE, Beauchamp A, Tsui C, Pittard WS and Devine SE (2006) An initial map of insertion and deletion (INDEL) variation in the human genome. *Genome Research* **16**:1182-1190.

Mishra KK, Fougat RS, Ballani A, Thakur V, Jha Y and Bora M (2014) Potential and application of molecular markers techniques for plant genome analysis. International Journal of Pure & Applied Bioscience. **2:** 169-188.

Misof M, Niehuis O, Bischoff I, Rickert A, Erpenbeck D and Staniczek A (2007) Towards an 18S phylogeny of hexapods: accounting for group-specific character covariance in optimized mixed nucleotide/doublet models. *Zoology* **110:** 409-429.

Moghaddam SM, Song Q, Mamidi S, Schmutz J, Lee R, Cregan P, Osorno JM and McClean PE (2014) Developing market class specific InDel markers from next generation sequence data in *Phaseolus vulgaris* L. *Frontiers in Plant Science* **5:** 1-13.

Mooers AØ and Holmes EC (2000) The evolution of base composition and phylogenetic inference. *Trends in Ecology and Evolution* **15:** 356-365.

Moritz C and Cicero C (2004) DNA Barcoding: promise and pitfalls. *PLoS Biology* **2:** 1529-1531.

Moritz C, Dowling TE and Brown WM (1987) Evolution of animal mitochondrial DNA: relevance for population biology and systematics. *Annual Review of Ecology and Systematics* **18:** 269-292.

Mueller UG and Wolfenbarger LL (1999) AFLP genotyping and fingerprinting. *Trends in Ecology and Evolution* **14:** 389-393.

Narvel JM, Chu W, Fehr W, Cregan PB and Shoemaker RC (2000) Development of multiplex sets of simple sequence repeat DNA markers covering the soybean genome. *Molecular Breeding* **6:** 175-183.

Neff MM, Neff JD, Chory J and Pepper AE (1998) dCAPS, a simple technique for the genetic analysis of single nucleotide polymorphisms: Experimental applications in *Arabidopsis thaliana* genetics. *Plant Journal* **14:** 387-392.

Nielsen DI, Ebert PR, Page RE Jr, Hunt GJ and Guzmán-Novoa E (2000) Improved polymerase chain reaction-based mitochondrial genotype assay for identification of the Africanized honey bee (Hymenoptera: Apidae). *Annals of the Entomological Society of America* **93:** 1-6.

Ogden TH and Whiting MF (2003) The problem with "the Paleoptera Problem:" sense and sensitivity. *Cladistics* **19:** 432-442.

Ohta T (2000) Evolution of gene families. *Gene* **259**: 45-52.

Othman El-Mahdy Othman (2012) Mitochondrial DNA as a marker for genetic diversity and evolution. *Advances in Genetic Engineering & Biotechnology* **1:** 1.

Papp AC, Pinsonneault JK, Cooke G, Sadée W (2003) Single nucleotide polymorphism genotyping using allele-specific PCR and fluorescence melting curves. *Biotechniques* **34:** 1068-1072.

Paran I and Michelmore RW (1993) Development of reliable PCR-based markers linked to downy mildew resistance genes in lettuce. *Theoretical and Applied Genetics* **85:** 985-993.

Park D-S, Foottit R, Maw E, Hebert PDN (2011) Barcoding Bugs: DNA-based identification of the true bugs (Insecta: Hemiptera: Heteroptera). *PLoS ONE* **6:** e18749.

Park S, Yu H-J, Mun J-H, Lee S-C (2010) Genome-wide discovery of DNA polymorphism in *Brassica rapa*. *Molecular Genetics and Genomics* **283:** 135-145.

Paule MR and White RJ (2000) Survey and summary: transcription by RNA polymerases I and III. *Nucleic Acids Research* **28:** 1283-1298.

Pereira F, Carneiro J, Matthiesen R, van Asch B, Pinto N, Gusmao L and Amorim A (2010) Identification of species by multiplex analysis of variable-length sequences. *Nucleic Acids Research* **38:** e203-e203.

Penner GA, Bush A, Wise R, Kim W, Domier L, Kasha K, Laroche A, Scoles G, Molar SJ and Fedak G (1993) Reproducibility of random amplified polymorphic DNA (RAPD) analysis among laboratories. *PCR methods and applications* **2:** 341-345.

Pinto AS, de Lana M, Bastrenta B, Barnabé C, Quesney V, Noël S and Tibayrenc M (1998) Compared vectorial transmissibility of pure and mixed clonal genotypes of Trypanosoma cruzi in Triatoma infestans. *Parasitology Research* **84:** 348-353.

Potter RH and Jones MGK (1991) Molecular analysis of genetic stability, pp. 71-91. **In:** Dodds J H (eds.), *In vitro methods for* *conservation of plant genetic resources*. Chapman and Hall, London.

Prasad MD, Muthulakshmi M, Madhu M, Archak S, Mita K and Nagaraju J (2005) Survey and analysis of microsatellites in the silkworm, *Bombyx mori*: frequency, distribution, mutations, marker potential and their conservation in heterologous species. *Genetics* **169:** 197-214.

Prinsloo G, Chen Y, Giles KL and Greenstone MH (2002) Release and recovery in South Africa of the exotic aphid parasitoid *Aphelinus hordei* verified by the polymerase chain reaction. *Biocontrol* **47:** 127-136.

Qi X, Stam P and Lindhout P (1998) Use of locus-specific AFLP markers to construct a high-density molecular map in barley. *Theoretical and Applied Genetics* **96:** 376-384.

Quicke DLJ and Belshaw R (1999) Incongruence between morphological data sets: An example from the evolution of endoparasitism among parasitic wasps (Hymenoptera: Braconidae). *Systematic Biology* **48:** 436-454.

Ranson H, Jensen B, Wang X, Prapanthadara L, Hemingway J and Collins FH (2000) Genetic mapping of two loci affecting DDT resistance in the malaria vector *Anopheles gambiae*. *Insect Molecular Biology* **9:** 499-507.

Ratnasingham S and Hebert PDN (2007) BOLD: The Barcode of Life Data System; http://www.barcodinglife.org. *Molecular Ecological Notes* **7:** 355-364.

Rauch N and Nauen R (2003) Identification of biochemical markers linked to neonicotinoid cross resistance in *Bemisia tabaci* (Hemiptera: Aleyrodidae). *Archives of Insect Biochemistry and Physiology* **54:** 165-176.

Renganayaki K, Fritz AK, Sadasivam S, Pammi S, Harrington SE, McCouch SR, Kumar SM and Reddy AS (2002) Mapping and progress toward map-based cloning of brown planthopper biotype-4 resistance gene introgressed from *Oryza officinalis* into cultivated rice, *O. sativa*. *Crop Science* **42:** 2112-2117.

Ridout C and Donini P (1999) Use of AFLP in cereals research. *Trends in Plant Science* **4:** 76-79.

Robin C, Lyman RF, Long AD, Langley CH and Mackay TFC (2002) Hairy: a quantitative trait locus for *Drosophila* sensory bristle number. *Genetics* **162:** 155-164.

Roe AD and Sperling FAH (2007a) Patterns of evolution of mitochondrial cytochrome *c* oxidase I and II DNA and implication for DNA barcoding. *Molecular Phylogenetics and Evolution* **44:** 325-345.

Roe AD and Sperling FAH (2007b) Population structure and species boundary delimitation of cryptic Dioryctria moths using an integrative approach. *Molecular Ecology* **16:** 3617-3633.

Roehrdanz RL and Flanders RV (1993) Detection of DNA polymorphisms in predatory Coccinellids using polymerase chain reaction and arbitrary primers (RAPD-PCR). *Entomophaga* **38:** 479-491.

Roy A, Frascaria N, Mackay J and Bousquet J (1992) Segregating random amplified polymorphic DNA (RAPDs) in *Betula alleghaniensis*. *Theoretical and Applied Genetics* **85:** 173-180.

Russell JR, Fuller JD, Macaulay M, Hatz BG, Jahoor A, Powell W and Waugh R (1997) Direct comparison of levels of genetic variation among barley accessions detected by RFLPs, AFLPs, SSRs and RAPDs. *Theoretical and Applied Genetics* **95:** 714-722.

Sadava DE, Hillis DM, Heller HC and Berenbaum M (2012) Life: The Science of Biology. W. H. Freeman & Company, New York, USA.

Saiki RK, Scharf S, Faloona F, Mullis KB, Horn GT, Erlich HA and Arnheim N

(1985) Enzymatic amplification of ß-globin genomic sequences and restriction site analysis for diagnosis of sickle cell anemia. *Science* **230:** 1350-1354.

Sanchis A, Latorre A, Gonzalez-Candelas F and Michelena JM (2000) An 18S rDNA-based molecular phylogeny of Aphidiinae (Hymenoptera: Braconidae). *Molecular Phylogenetics and Evolution* **14:** 180-194.

Sanger F, Nicklen S and Coulson AR (1977) DNA sequencing with chain-terminating inhibitors. *Proceedings of the National Academy of Sciences of the United States of America* **74:** 5463-5467.

Sardesai N, Kumar A, Rajyashri K R, Nair S and Mohan M (2002) Identification and mapping of an AFLP marker linked to Gm7, a gall midge resistance gene and its conversion to a SCAR marker for its utility in marker-aided selection in rice. *Theoretical and Applied Genetics* **105:** 691-698.

Savard J, Tautz D, Richards S, Weinstock GM, Gibbs RA, Werren JH, Tettelin H and Lercher MJ (2006) Phylogenomic analysis reveals bees and wasps (Hymenoptera) at the base of the radiation of Holometabolous insects. *Genome Research* **16:** 1334-1338.

Savolainen V, Cowan RS, Vogler AP, Roderick GK and Lane R (2005) Towards writing the encyclopaedia of life: an introduction to DNA barcoding. *Philosophical Transactions of the Royal Society of London. Series B, Biological Sciences* **360:** 1805-1811.

Scaringe WA, Li K, Gu D, Gonzalez KD, Chen Z, Hill KA, Sommer SS (2008) Somatic microindels in human cancer: the insertions are highly error-prone and derive from nearby but not adjacent sense and antisense templates. *Human Molecular Genetics* **17:** 2910-2918.

Schlipalius DI, Cheng Q, Reilly PE, Collins PJ and Ebert PR (2002) Genetic linkage analysis of the lesser grain borer Rhyzopertha dominica identifies two loci that confer high-level resistance to the fumigant phosphine. *Genetics* **161:** 773-782.

Schlötterer C (2000) Evolutionary dynamics of microsatellite DNA. *Chromosoma* **109:** 365-371.

Schmidt S, Naumann ID and De Barro PJ (2001) *Encarsia* species (Hymenoptera: Aphelinidae) of Australia and the Pacific Islands attacking *Bemisia tabaci* and *Trialeurodes vaporariorum* (Hemiptera: Aleyrodidae): A pictorial key and descriptions of four new species. *Bulletin of Entomological Research* **91:** 369-387.

Schubert M, Holland LZ, Holland ND and Jacobs DK (2000) A phylogenetic tree of the *Wnt* genes based on all available full-length sequences, including five from the cephalochordate *Amphioxus. Molecular Biology and Evolution* **17:** 1896-1903.

Sebat J, Lakshmi B, Troge J, Alexander J, Young J, Lundin P, Månér S, Massa H, Walker M, Chi M, Navin N, Lucito R, Healy J, Hicks J, Ye K, Reiner A, Gilliam TC, Trask B, Patterson N, Zetterberg A and Wigler M (2004) Large-scale copy number polymorphism in the human genome. *Science* **305:** 525-528.

Semagn K, Bjørnstad Å and Ndjiondjop MN (2006) An overview of molecular marker methods for plants. *African Journal of Biotechnology* **5:** 2540-2568.

Severson DW, Brown SE and Knudson DL (2001) Genetic and physical mapping in mosquitoes: Molecular approaches. *Annual Review of Entomology* **46:** 183-219.

Sharopova N, McMullen MD, Schultz L, Schroeder S, Sanchez H, Gardiner J, Bergstrom D, Houchins K, Polacco M, Edwards KJ, Ruf T, Register JC, Brower C, Thompson R, Chin E, Lee M, Liscum III, Cone E, Davis G and Coe EH Jr (2002) Development and mapping of SSR markers for maize. *Plant Molecular Biology* **48:** 463-481.

Sharma HC (2008) Biotechnological Approaches for Pest Management and Ecological Sustainability, CRC Press, Florida, USA.

Shen YJ, Jiang H, Jin JP, Zhang ZB, Xi B, He YY, Wang G, Wang C, Qian L, Li X, Yu QB, Liu HJ, Chen DH, Gao JH, Huang H, Shi TL and Yang ZN (2004) Development of genome-wide DNA polymorphism database for map-based cloning of rice genes. *Plant Physiology* **135:** 1198-1205.

Simon C, Frati F, Beckenbach A, Crespi B and Liu H (1994) Evolution, weighting, and phylogenetic utility of mitochondrial gene sequences and a compilation of conserved polymerase chain reaction primers. *Annals of the Entomological Society of America* **87:** 651-701.

Simon JC, Carre S, Boutin M, Prunier-Leterme N, Sabatermunoz B, Latorre A and Bournoville R (2003) Host-based divergence in populations of the pea aphid: Insights from nuclear markers and the prevalence of facultative symbionts. *Proceedings of the Royal Society of London (Series B), Biological Sciences* **270:** 1703-1712.

Simon C, Buckley TR, Frati F, Stewart JB and Beckenbach AT (2006) Incorporating molecular evolution into phylogenetic analysis, and a new compilation of conserved polymerase chain reaction primers for animal mitochondrial DNA. *Annual Review of Ecology, Evolution, and Systematics* **37:** 545-579.

Smith PT, Kambhampati S, Volkl W and Mackauer M (1999) A phylogeny of aphid parasitoids (Hymenoptera: Braconidae: Aphidiinae) inferred from mitochondrial NADH 1 dehydrogenase gene sequence. *Molecular Phylogenetics and Evolution* **11:** 236-245.

Smith MA, Fisher BL and Hebert PDN (2005) DNA barcoding for effective biodiversity assessment of a hyperdiverse arthropod group: the ants of Madagascar. *Philosophical Transactions of the Royal Society B: Biological Science* **360:** 1825-1834.

Smith MA, Woodley NE, Janzen DH, Hallwachs W and Hebert PDN (2006) DNA barcodes reveal cryptic host-specificity within the presumed polyphagous members of a genus of parasitoid flies (Diptera: Tachinidae). *Proceedings of the National Academy of Sciences of the United States of America* **103:** 3657-3662.

Smith MA, Fernandez-Triana J, Roughley R and Hevert PDN (2009) DNA barcode accumulation curves for understudied taxa and areas. *Molecular and Ecological Research* **9:** 208-216.

Sogin ML (1997) History assignment: when was the mitochondrion founded? *Current Opinion in Genetics & Development* **7:** 792-799.

Solomon MG, Fitzgerald JD and Murray RA (1996) Electrophoretic approaches to predatorprey interactions, pp. 457-68. **In:** Symondson W O C and Liddell J D (eds.), *The Ecology of Agricultural Pests: Biochemical Approaches, Systematics Association. Special Volume No. 53*, Chapman and Hall, London, UK.

Spötter A, Gupta P, Nürnberg G, Reinsch N and Bienefeld K (2012) Development of a 44K SNP assay focussing on the analysis of a *Varroa*-specific defence behaviour in honey bees (*Apis mellifera carnica*). *Molecular Ecology Resources* **12:** 323-332.

Spooner D, van Treuren R, de Vicente MC (2005) Molecular markers for genebank management. IPGRI Technical Bulletin No. 10. International Plant Genetic Resources Institute, Rome, Italy; http://www.bioversityinternational.org/uploads/tx_news/Molecular_markers_for_genebank_management_1082.pdf

Srivastava S and Narula A (2004) *Plant Biotechnology and Molecular Markers*, Anamaya Publishers, New Delhi, India.

Steeves R, Nazari V, Landry JF and Lacroix CR (2008) Molecular identification of a larval pre-dispersal seed predator of the Gulf of Saint Lawrence Aster. *The Canadian Entomologist* **140:** 297-305.

Strachan T and Read AP (2010) *Human Molecular Genetics*, Garland Science, New York, USA.

Sunnucks P and Hales HF (1996) Numerous transposed sequences of mitochondrial cytochrome oxidase I-II in aphids of the genus *Sitobion* (Hemiptera: Aphididae). *Molecular Biology and Evolution* **13:** 510-524.

Sung W-K (2010) Algorithms in Bioinformatics: A Practical Introduction. Chapman & Hall/CRC, Florida, USA.

Symondson WOC (2002) Molecular identification of prey in predator diets. *Molecular Ecology* **11:** 627-641.

Taly V, Pekin D, El Abed A and Laurent-Puig P (2012) Detecting biomarkers with microdroplet technology. *Trends in Molecular Medicine* **18:** 405-416.

Tautz D, Arctander P, Minelli A, Thomas RH and Vogler AP (2003) A plea for DNA taxonomy. *Trends in Ecology and Evolution* **18:** 70-74.

Taylor HR and Harris WE (2012) An emergent science on the brink of irrelevance: a review of the past 8 years of DNA barcoding. *Molecular and Ecology Resource* **12:** 377-388.

Tétard-Jones C, Kertesz MA and Preziosi RF (2012) Identification of plant quantitative trait loci modulating a rhizobacteria-aphid indirect effect. *PLoS ONE* **7:** e41524.

The International HapMap Consortium (2005) A haplotype map of the human genome. *Nature* **437:** 1299-1320.

Thomson LJ, Rundle BJ, Carew ME and Hoffmann AA (2003) Identification and characterization of *Trichogramma* species from southeastern Australia using the internal transcribed spacer 2 (ITS-2) region of the ribosomal gene complex. *Entomologia Experimentalis et Applicata* **106:** 235-240.

Tilmon KJ, Danforth BN, Day WH and Hoffmann MP (2000) Determining parasitoid species composition in a host population: A molecular approach. *Annals of the Entomological Society of America* **93:** 640-657.

Trautwein MD, Wiegmann BM, Beutel R, Kjer KM and Yeates DM (2012) Advances in insect phylogeny at the dawn of the postgenomic era. *Annual Review of Entomology* **57:** 449-468.

Twyman RM (2004) SNP discovery and typing technologies for pharmacogenomics. *Current Topics in Medicinal Chemistry* **4:** 1421-1429.

Ullmann AJ, Piesman J, Dolan MC *and Black WC* (2003) A preliminary linkage map of the hard tick, Ixodes scapularis. *Insect Molecular Biology* **12:** 201-210.

Valentini A, Pompanon F and Taberlet P (2009) DNA barcoding for ecologists. *Tree* **24:** 110-117.

Väli Ü, Brandström M, Johansson M and Ellegren H (2008) Insertion-deletion polymorphisms (indels) as genetic markers in natural populations. *BMC Genetics* **9:** 8.

Varala K, Swaminathan K, Li Y and Hudson ME (2011) Rapid genotyping of soybean cultivars using high throughput sequencing. *PLoS ONE* **6:** e24811.

Vasemägi A, Gross R, Palm D, Paaver T, Primmer CR (2010) Discovery and application of insertion-deletion (INDEL) polymorphisms for QTL mapping of early life-history traits in Atlantic salmon. *BMC Genomics* **11:** 156.

Vaughn TT and Antolin MF (1998) Population genetics of an opportunistic parasitoid in an agricultural landscape. *Heredity* **80:** 152-162.

Verhaeghen K, Van Bortel W, Roelants P, Backeljau T and Coosemans M (2006) Detection of the East and West African kdr mutation in *Anopheles gambiae* and *Anopheles arabiensis* from Uganda using a new assay based on FRET/Melt Curve analysis. *Malaria Journal* **5:** 16.

Virgilio M, Backeljau T, Nevado B and De Meyer M (2010) Comparative performances of DNA barcoding across insect orders. *BMC Bioinformatics* **11:** 206.

Virgilio M, Jordaens K, Breman FC, Backeljau T and De Meyer M (2012) Identifying insects with incomplete DNA barcode libraries, African fruit flies (Diptera: Tephritidae) as a test case. *PLoS ONE* **7:** e31581: doi:10.1371/journal.pone.0031581.

Vos P, Hogers R, Bleeker M, Reijans M, Lee TVD, Homes M, Frijters A, Pot J, Kuiper M and Zabeau M (1995) AFLP: A new technique for DNA fingerprinting. *Nucleic Acids Research* **23:** 4407-4414.

Ward RD, Zemlak TS and Innes BH (2005) Barcoding Australia's fish species. The *Philosophical Transactions of the Royal Society of London B, Biological Sciences* **360:** 1847-1857.

Weiland JJ and Yu MH (2003) A cleaved amplified polymorphic sequence (CAPS) marker associated with root-knot nematode resistance in sugarbeet. *Crop Science* **43:** 814-881.

Weng Y and Lazar MD (2002) Amplified fragment length polymorphism- and simple sequence repeat-based molecular tagging and mapping of greenbug resistance gene *Gb3* in wheat. *Plant Breeding* **121**: 218-223.

Widmer A, Schmid-Hempel P, Estoup A and Scholl A (1998) Population genetic structure and colonization history of *Bombus terrestris* s.l. (Hymenoptera: Apidae) from the Canary Islands and Madeira. *Heredity* **81:** 563-572.

Wiegmann BM, Trautwein MD, Kim J-W, Cassel BK, Bertone MA, Winterton SL and Yeates DK (2009) Single-copy nuclear genes resolve the phylogeny of the holometabolous insects. *BMC Biology* **7:** 34.

Weiss M, Hermsen M, Meijer G, Van Grieken N, Baak J, Kuipers E and Van Diest P (1999) Comparative genomic hybridization. *Molecular Pathology* **52:** 243-251.

Welsh J and McClelland M (1990) Fingerprinting genomes using PCR with arbitrary primers. *Nucleic Acids Research.* **18:** 7213-7218.

Williams JGK, Kubelik AR, Livak KJ, Rafalski JA and Tingey SV (1990) DNA polymorphisms amplified by arbitrary primers are useful as genetic markers. *Nucleic Acids Research* **18:** 6531-6535.

Williams R, Peisajovich SG, Miller OJ, Magdassi S, Tawfik DS and Griffiths AD (2006) Amplification of complex gene libraries by emulsion PCR. *Nature methods* **3:** 545-50.

Wilson JJ (2010) Assessing the value of DNA barcodes and other priority gene regions for molecular phylogenetics of Lepidoptera. *PLoS ONE* **5:** e10525.

Wolstenholme DR (1992) Genetic novelties in mitochondrial genomes of multicellular animals. *Current Opinion in Genetics & Development* **2:** 918-925.

Wong J, Chu YY, Stoddard ST, Lee Y, Morrison AC and Scott TW (2012) Microsatellite-based parentage analysis of *Aedes aegypti* (Diptera: Culicidae) using nonlethal DNA sampling. *Journal of Medical Entomology* **49:** 85-93.

Yoo HS, Eah JY, Kim JS, Kim YJ and Min MS (2006) DNA barcoding Korean birds. *Molecules and Cells* **22:** 323-327.

Yu TP, Rothschild M and Fand Tuggle CK (1993) A MspI restriction fragment length polymorphism at the swine PIT- 1 locus. *Journal of Animal Science* **71:** 2275.

Zahariev M, Dahl V, Chen W and Levesque CA (2009) Efficient algorithms for the discovery of DNA oligonucleotide barcodes from sequence databases. *Molecular and Ecology Research* **9:** 58-64.

Zaidi F, Wei SJ, Shi M and Chen XX (2011) Utility of multi-gene loci for forensic species diagnosis of blowflies. *Journal of Insect Science* **11:** 1-12: doi: http://dx.doi.org/10.1673/031.011.5901.

Zhang J, Zhou C, Gai Y, Song D and Zhou K (2008) The complete mitochondrial genome of *Parafronurus youi* (Insecta: Ephemeroptera) and phylogenetic position of the Ephemeroptera. *Gene* **424:** 18-24.

Zheng LY, Guo XS, He B, Sun LJ, Peng Y, Dong SS, Liu TF, Jiang S, Ramachandran S, Liu C M and Jing HC (2011) Genome-wide patterns of genetic variation in sweet and grain sorghum (*Sorghum bicolor*). *Genome Biology* **12:** R114.

Zhou X, Scharf ME, Parimi S, Meinke LJ, Wright RJ, Chandler LD and Siegfried BD (2002) Biochemical diagnostic assays on western corn rootworm (Coleoptera: Chrysomelidae) based on esterase-mediated resistance mechanisms. *Journal of Economic Entomology* **95:** 1261-1266.

Zhu LC, Smith CM, Fritz A, Boyko E and Flinn MB (2004) Genetic analysis and molecular mapping of a wheat gene conferring tolerance to the greenbug (*Schizaphis graminum* Rondani). *Theoretical and Applied Genetics* **109:** 289-293.

Zhu YC and Williams L (2002) Detecting the egg parasitoid *Anaphes iole* (Hymenoptera: Mymaridae) in tarnished plant bug (Heteroptera: Miridae) eggs by using a molecular approach. *Annals of the Entomological Society of America* **95:** 359-365.

Zietkiewicz E, Rafalski A and Labuda D (1994) Genome fingerprinting by simple sequence repeat (SSR)-anchored polymerase chain reaction amplification. *Genomics* **20:** 176-183.

Biological and Molecular Approaches in Pest Management
Balwinder Singh, Ramesh Arora & S S Gosal (Eds)
Scientific Publishers (India), Jodhpur, 2015, pp 202-227

Chapter - 9

Transgenic Cotton in India: Ten Years and Beyond

Vijay Kumar, A K Dhawan and P S Shera
Department of Entomology, Punjab Agricultural University, Ludhiana-141 004, India
*Email: vijay_ento@pau.edu

9.1. INTRODUCTION

India is the only country in the world where all the four cultivated species of cotton, viz. *Gossypium arboreum* L., *G. hirsutum* L., *G. herbaceaum* L. and *G. barbadense* L. along with intra-and inter-specific hybrids are cultivated along the diverse agroclimatic conditions, varying from 8-32° N latitude and 70-80° longitude on commercial scale spanning around 10.80 million hectares. Among the cotton growing countries of the world, India has a place of prominence with respect to area under this crop contributing about 34 per cent of the world cotton area, but with regard to production, it is ranked next to China, the top producer. The production increased from a meagre 2.8 million bales (170 kg lint/bale) in 1947-48 to a high of 17.6 million bales in 1996-97 and an all time record of 34.5 million bales during 2011-12. Among cotton growing states, Gujarat leads in production with 8.5 million bales followed by Maharashtra (7.4 million bales) and Andhra Pradesh (7.2 million bales) in 2012-13. However, the productivity of cotton in India (496 kg ha^{-1}) is still far less than other major cotton growing countries of the world, viz. Australia (2055 kg ha^{-1}), Brazil and Turkey (1415 kg ha^{-1}), China (1403 kg ha^{-1}), USA (971 kg ha^{-1}), Pakistan (697 kg ha^{-1}) and Uzbekistan (729 kg ha^{-1}) (AICCIP, 2013).

Besides other causes, major bottleneck in cotton cultivation is biotic stresses due to attack of insect pests, diseases and competition from weeds, which plays a significant role in achieving optimum yield potential. Cotton crop harboured 1326 species of insects from sowing to maturity in different cotton growing areas of the world (Hargreaves, 1948) and 162 species have been reported on the crop in India (Sundramurthy and Chitra, 1992). More than 130 species of arthropods including 54 pestiferous species were found to be associated with cotton crop in Punjab. Among these, the bollworm complex comprising of American bollworm, *Helicoverpa armigera* (Hübner), spotted bollworm, *Earias vittella* (Fabricius); spiny bollworm, *E. insulana* (Boisduval) and pink bollworm, *Pectinophora gossypiella* (Saunders) remained as key pests in the northern regions before the introduction of Bt cotton. The monetary value of yield losses due to insect pests has been estimated to be Rs. 2, 87, 600/- million annually (Dhaliwal *et al.,* 2004). The insect pest complex of cotton crop broadly divided into three categories, viz. sucking insect pests, foliage feeders and bollworms causing damage to various plant parts at different growth stages through out the cropping season.

Due to non-availability of resistant cultivars and effective bioagents along with constraints in adoption of cultural practices, insecticides remained the mainstay of integrated pest management (IPM) in cotton. But insecticide resistance has rendered insecticides ineffective, thus increasing need for repeated applications, wastage of resources and consequent environmental pollution (Kranthi *et al.,* 2009). The total consumption of pesticides in Punjab has increased from 3200 MT technical grade in 1980-81 to 7600 MT in 1995-96 and decrease to 3800 MT in 2001-02, cotton being the maximum shareholder. However, there has been reduction in pesticide load ever since transgenic cotton has been commercialized in Punjab. The total consumption of pesticides was 4500 MT during 2005-06 in Punjab and got reduced to 2600 MT in 2010 (Anonymous, 2011). Besides increasing cost of production and environmental problems, the excessive and indiscriminate use of insecticides for the control of these pests has resulted in development of insecticidal resistance particularly in *H. amigera* (Kapoor *et al.,* 2002; Gill and Dhawan, 2006), decline in natural enemies' population and resurgence of the pests like whitefly, *Bemisia tabaci* (Gennadiaus) (Dhawan and Simwat, 1997; Butter and Kular 1999; Virk *et al.,* 2004) and jassid, *Amrasca biguttula biguttula* (Ishida) (Dhawan and Simwat,1997). Therefore, for a country that consumed 50 per cent of insecticide on cotton crop alone despite its covering only 5 per cent of cropped area, any technology that reduced dependence upon chemicals was an immediate priority. Bt cotton has now been playing a major role in effectively protecting the crop from bollworms, especially the *H. armigera*, thus preventing yield losses. Indian cotton farmers, who were displeased with the intractable problem of bollworms in cotton, responded positively to Bt cotton and strongly supported the technologies, to the extent that in the history of agriculture, there has never been ever before, an example of such a rapid technology adoption.

9.2. TRANSGENIC BT COTTON

The first generation Bt cotton was developed by Monsanto, USA who released their best event MON 531 as Bollgard event. *Bt* cotton was developed by incorporating *cry1Ac* gene from soil bacterium, *Bacillus thuringiensis* (Bt) Berliner expressing delta endotoxin into the cultivar Coker 312 through genetic engineering using *Agribacterium tumefasciens.* The first Bt transgenic cotton varieties called *Bt* cotton, expressing the gene construct of *cry lAc,* were commercially released in Australia as INGARD and in the United States as BOLLGARD in 1996. Later it was released in Mexico, Columbia, Indonesia, Argentina, South Africa and India. Bt cotton was approved for commercial cultivation during 2002 in India and the country has 10 years (2002 to 2012) of impressive benefits from Bt cotton. Since 2002, Genetic Engineering Approval Committee (GEAC) has approved the cultivation of total 780 Bt cotton hybrids (including a *Bt* variety) belonging to six transgenic events in India (Choudhary and Gaur, 2010).

9.3. INTRODUCTION OF BT COTTON IN INDIA

The sequence of events that led to the development and release of the *Bt* cotton hybrids in India is given below (Rathore *et al.*, 2008).

- Myhco applied to DBT for permission to import a small stock of Bollgard (Bt cotton) seeds from Monsanto Company, USA in 1995 and DBT gave the permission.
- After getting permission from DBT in 1996, Mahyco imported the first Bt-cotton variety (US Coker-312) from Monsanto, USA and backcrossed it with "elite" Indian varieties to produce 40 locally adapted Bt cotton varieties with the *cry 1Ac* gene (MON531).
- As per guidelines of regulatory authorities established by Government of India, all biosafety tests including all aspects of pollen flow, aggressiveness, gene stability, allergi-nicity, toxicity to small and large animals, fish studies, protein expression, toxin in byproducts, influence on beneficial microorganisms and baseline susceptibility studies were carried out between 1997-2001 (Kranthi *et al.,* 2009).
- In 1998-99, multi-location field trials were conducted at 40 locations in nine states to assess the agronomic performance of these Bt cotton varieties and bio-safety aspect.
- Field trials were repeated again in 1999-2000 at 10 locations in six states.
- Genetic Engineering Approval Committee gave approval in July 2000 for conducting large-scale field trials on Bt cotton in about 85 ha and also to undertake seed production on 150 ha. Consequently, large-scale field trials were conducted in 2001 in 10 ha area. In addition, field trials were also conducted under All India Coordinated Cotton Improvement Project of ICAR.
- In 2002, Mahyco was the first to receive approval for three Bt cotton hybrids, i.e. MECH 12, MECH 162 and MECH 184 from GEAC for commercial cultivation in Central and South zones.
- In North Zone (Punjab), for the first time, two Bt cotton hybrids, MECH 162 Bt and MECH 915 Bt were evaluated in research trials during 2002. However, these were poor yielder and highly susceptible to cotton leaf curl disease, hence were withdrawn from further testing in the state.
- During 2003-2004, different Bt hybrids were tested in ICAR-AICCIP research trials, RSGM trials and large-scale trials in the state.
- The GEAC gave first official approval for the commercial cultivation of Bt cotton hybrids in Northern region in 2005.

9.4. APPROVED EVENTS AND BT COTTON HYBRIDS IN INDIA

Since commercialization during 2002, the number of events, Bt cotton hybrids and companies marketing approved hybrids have all increased significantly in India. The events, MON 531 Bollgard I (BG I) having single gene *(cry 1 Ac)* and MON 15985 Bollgard II (BG II) having dual genes *(cry 1Ac + cry 2 Ab)* have been developed by Mahyco/ Monsanto and were approved in 2002 and 2006, respectively. These two events have been licensed to other seed companies. The third event, Event 1 (modified *cry 1Ac*) has been developed by IIT Kharagpur and being utilized by JK Agri. Genetic Limited. The fourth event, GFM event (fusion *cry 1Ac / cry 1Ab*) has been sourced from China and being utilized by Nath Seeds Company. In the landmark achievement, the fifth event, BNLA-601 (*cry 1Ac* gene) developed by CICR Nagpur (ICAR) and UAS Dharwad, Karnatka was approved for commercial cultivation in an indigenous public sector variety *G. hirsutum "Bikaneri Narma* Bt" in contrast to the above four events, which were all incorporated in cotton hybrids. In 2009, a public sector developed Bt cotton hybrid NHH 44 based on same event was also approved by GEAC. The sixth event, MLS-9124 (synthetic *cry 1C* gene) has been developed indigenously by Metahelix Life Sciences and was approved for commercial cultivation in Central and South in 2009 (Chaudhary and Gaur, 2011). The commercial deployment of the first five events in hybrids and sixth event in both variety and hybrids in India is summarized in Table 1.

In 2010, four new cotton events expressing insect resistance and herbicide tolerance in the cotton plant were field-tested in different cotton growing locations in the country. Two out of the four events were developed by stacking two biotech traits including insect resistance (IR) with single/multiple genes and herbicide tolerance (HT) to effectively control both the bollworm complex and control of weeds.

Table 1. Commercial release of different *Bt* cotton events by GEAC in India

Event	Genes	Year of approval	Developer
MON531	*cry1Ac*	2002	Mahyco/ Monsanto
MON15985	*cry1Ac, cry2Ab*	2006	Mahyco/ Monsanto
Event-1	Truncated *cry1Ac*	2006	JK Agri-Genetics/ IIT Kharagpur
GFM Event	*cry1Ab* and *cry1Ac*	2006	Nath seeds/CAS China
BNLA-601	Truncated *cry1Ac*	2008	CICR, Nagpur and UAS Dharwad
MLS-9124	*cry1c*	2009	Metahelix Life Sciences

Source: ISAAA (2010)

The Indian Council of Agricultural Research in 2009 reported that weeds account for a major loss in cotton crop as much as 47.5 per cent reduction in cotton yield due to weeds. Traditionally, cotton farmers have used various methods including mechanical, physical, cultural and chemical methods to control weeds. An age-old physical method of weed control by hand that employs farm laborers, especially women laborers, is now considered a farm drudgery by many farm laborers and is in gradual decline. Growing use of herbicides is being witnessed as an important new component of integrated weed management in today's agricultural systems. Both selective and non-selective herbicides including diuron, paraquet, fluchloralin and pendimethalin are being used to control weeds so, caution has to be exercised in their application because they will also damage the cotton crop if not applied properly. Herbicide tolerant (HT) cotton plants including Bollgard II Roundup Ready Flex (BG-II RRF) cotton, which is at an advanced stage of regulatory review and subject to approval, be commercially released in coming years, could provide effective control of weeds without damaging the cotton crop. These new events are undergoing field-testing and pending approval has the potential to be commercialized sometime between 2014 and 2015 (Table 2).

Table 2. Bt and Bt/Ht events field-tested and pending approval for commercialization in India

Event	Developer	Status
MON 15985 x MON 88913	Mahyco/ Monsanto	Field Tested
Widestrike Event 3006 – 210-23 and Event 281-24-236	Dow Agro Sciences	Field Tested
Event 1 and Event 24	JK Agri-Genetics	Field Tested
2mEPSPS gene	Bayer Biosciences	Field Tested

Source: ISAAA (2010)

9.4.1. Approved Events of Bt Cotton

First Event: MON531, Bollgard I (BG I), featuring the *cry1Ac* gene was developed by Maharashtra Hybrid Seeds Company Limited (Mahyco), sourced from Monsanto and approved for sale in 2002 in central and South India and in 2005 in northern India. During 2010, a total of 210 hybrids were released for use in the North, Central and South zones in 2010. The number of hybrid for sale was 80 BG I hybrids in 2009, 141 BG I hybrids in 2008, 96 BG I hybrids in 2007 and 48 BG I hybrids in 2006.

Second Event: MON15985, Bollgard II (BG II), featured the two genes *cry1Ac* and *cry2Ab* was also developed by Mahyco and sourced from Monsanto and was approved for sale for the first time in 2006 in a total of seven hybrids for use in the Central and South zones. This event was approved for commercial cultivation for the first time in the Northern zone in 2007 and the number of hybrids for sale increased from 7 in 2006, 21 in 2007, 94 in 2008, 248 in 2009 and further increased significantly to 438 BG II cotton hybrids in 2010 in the North, Central and South zones.

Third Event (Event-1): It was developed by JK Seeds featuring the *cry1Ac* gene, sourced from IIT Kharagpur, India. The event was approved for sale for the first time in 2006 in a total of four hybrids for use in the North, Central and South zones. Whereas this event was approved in only four hybrids in 2006, in 2008 it increased to 15 hybrids, 27 in 2009 and 41 Bt cotton hybrids in 2010.

Fourth Event (GFM event): It was developed by Nath Seeds, sourced from China, and features the fused genes *cry1Ab* and *cry1Ac*. It was approved for sale for the first time in a total of three hybrids in 2006, one in each of the three regions of India, 4 in 2010, the number of hybrids for sale increased three-fold from 24 in 2008 to 63 in 2009 and 87 Bt cotton hybrids in 2010 in three zones.

Fifth Event: In contrast to the above four events, the fifth event known as BNLA-601 was approved for commercial sale in an indigenous publicly-bred cotton variety named Bikaneri Nerma (BN) expressing the *cry1Ac* gene. It was first indigenous Bt cotton event developed by the Central Institute of Cotton Research (CICR), Nagpur along with University of Agricultural Sciences, Dharwad, Karnataka. It was approved for commercial release in the North, Central and South cotton growing zones in India during *kharif*, 2008.

Sixth event (MLS-9124): It was developed indigenously by Metahelix Life Sciences and features a synthetic *cry1C gene*. In 2009, two Bt cotton hybrids namely MH-5125 and MH-5174 expressing the synthetic *cry1C gene* (MLS-9124) were approved for commercial sale for Central and Southern zones (ISAAA, 2010).

9.4.2. New Events yet to be Commercialised

Seventh event: This event is known as Bollgard II (BG II) Roundup Ready Flex (BGIIRRF) and is being developed by Mahyco and sourced from Monsanto, first time features stacking of two events in India including insect resistance and herbicide tolerance in cotton. BGIIRRF expresses three genes; *cry1Ac* and *cry2Ab* to confer insect resistance and CP4EPSPS genes to impart herbicide tolerance. In 2010, four BG II RRF cotton hybrids including two hybrids for North zone and two for Central and South zones were approved for seed production in an area of 25 acres per hybrid. This event is yet to be approved for its commercial release.

Eighth event: It is known as "Widestrike" is being developed by Dow AgroSciences expressing multiple genes including *cry1F gene* and *cry1Ac* (Event 3006-210-23 and Event 281- 24-236) and has two genes for insect protection.

Ninth event: It is a combination of Event-1 x Event-24 is being developed by JK Seeds featuring two genes *cry1Ac* and *cry1Ec*, sourced from IIT Kharagpur and NBRI Lucknow, India. Hybrids expressing Event-1 x Event-24 event have been field tested in the North, Central and South zones in 2010.

Tenth event: It is known as Glytol cotton, expresses herbicide tolerance in cotton and is undergoing elite event selection in field trials in 2010. This event was developed by Bayer Biosciences and contains the *2mEPSPS gene* conferring tolerance to cotton hybrids sprayed with the herbicide Glyphosate (Table 2).

The number of Bt cotton hybrids as well as the number of companies offering Bt cotton hybrids in India has increased dramatically over the last ten years since the first commercialization in 2002. In 2010, the number of Bt cotton hybrids increased substantially to 780 introductions (779 hybrids and one variety) from 522 in 2009, 274 hybrids in 2008, 131 hybrids in 2007, 62 hybrids in 2006, 20 hybrids in 2005, 4 hybrids in 2004 and 3 hybrids in 2003 and 2002, respectively (Table 3).

Table 3. Zone wise approved Bt cotton events and hybrid/variety in India (2010)

Event	NZ	CZ	SZ	NZ/CZ	NZ/SZ	CZ/SZ	NZ/CZ/SZ	Total
MON 531	42	45	42	14	1	53	13	210
MON 15985 (BGII)	104	55	74	6	6	140	53	438
Event-I	9	8	6	0	0	17	1	41
GFM Event	20	28	14	4	0	20	1	87
BNLA-601	0	0	0	0	0	1	1*	2
MLS-9124	0	0	0	0	0	2	0	2
Total	175	136	136	24	7	233	69	**780**

Source: ISAAA (2010); * Bt cotton variety; NZ-North zone; CZ – Central Zone; SZ- South Zone

Importantly, this increase in number of hybrids has provided much more choice year after year to farmers in the North, Central and Southern regions, where specific hybrids have been approved for cultivation in specific regions. Of the 780 Bt cotton hybrids approved for commercial cultivation, 271 hybrids featuring five events were sold by 31 companies in the Northern zone, 459 hybrids featuring six events were sold by 35 companies in the Central zone and 444 hybrids featuring six events were sold by 35 companies in the Southern zone (Table 4). There has been a substantial increase in the area and number of hybrids with two genes for pest resistance, the BG II event, in 2010. The BG II cotton hybrids more than doubled to 438 in 2010 from 248 in 2009, 94 in 2008 and only 21 hybrids in 2007 (Table 5 & 6). This trend is due to the multiple benefits that double genes offer in terms of more effective control of *H. armigera* and *Spodoptera litura* and offer an effective tool in insect resistance management to Indian cotton growers. For this reason, the BG II hybrids are preferred by farmers across all three different cotton-growing zones.

Table 4. Name of companies marketing Bt cotton seeds in India

SN	Name of company	Place and State
(1)	(2)	(3)
1	Ajeeth Seeds Limited	Aurangabad, Maharashtra
2	Amar Biotech Limited	Hyderabad, Andhra Pradesh
3	Ankur Seeds Private Limited	Nagpur, Maharashtra
4	Bayer Biosciences Limited	Hyderabad, Andhra Pradesh
5	Bioseed Research India Limited	Hyderabad, Andhra Pradesh
6	Ganga Kaveri Seeds Private Limited	Hyderabad, Andhra Pradesh
7	Green Gold Private Limited	Aurangabad, Maharashtra
8	J.K. Agri Genetics Limited	Hyderabad, Andhra Pradesh
9	Kaveri Seeds Limited	Hyderabad, Andhra Pradesh
10	Kohinoor Seed Fields India Limited	New Delhi
11	Krishidhan Seeds Limited	Pune, Maharashtra
12	Mahyco (Maharashtra Hybrid Seeds Company Limited)	Jalna, Maharashtra
13	Mahyco Monsanto Biotech Limited (MMB)	Mumbai, Maharashtra
14	Metahelix Life Sciences Private Limited	Bengaluru, Karnataka
15	Monsanto Genetics Private Limited	Mumbai, Maharashtra
16	Namdhari Seeds Private Limited	Bengaluru, Karnataka
17	Nandi Seeds Private Limited	Ahmedabad, Gujarat
18	Nath Seeds Limited	Aurangabad, Maharashtra
19	Navkar Hybrid Seeds Private Limited	Ahmedabad, Gujarat
20	Nuzineedu Seeds Limited	Hyderabad, Andhra Pradesh
21	Palamoor Seeds Private Limited	Hyderabad, Andhra Pradesh
22	Prabhat Agri Biotech Limited	Hyderabad, Andhra Pradesh
23	Pravardhan Seeds Limited	Hyderabad, Andhra Pradesh
24	Rassi Seeds Private Limited	Attur, Tamilnadu

(1)	(2)	(3)
25	R J Biotech Private Limited	Aurangabad, Maharashtra
26	Safal Seeds and Biotech Limited	Aurangabad, Maharashtra
27	Seed Works India Private Limited	Hyderabad, Andhra Pradesh
28	Solar Agrotech Private Limited	Rajkot, Gujarat
29	Super Seeds Private Limited	Hisar, Haryana
30	Tulasi Seeds Private Limited	Hyderabad, Andhra Pradesh
31	Uniphos Enterprises Limited	Hyderabad, Andhra Pradesh
32	Vibha Agrotech Limited	Hyderabad, Andhra Pradesh
33	Vikki Agrotech Limited	Hyderabad, Andhra Pradesh
34	Vikram Seeds Private Limited	Ahmedabad, Gujarat
35	Xylem Seeds Private Limited	Hyderabad, Andhra Pradesh
36	Yashoda Hybrid Seeds Private Limited	Wardha, Maharashtra
37	Zuari Seeds Limited	Hyderabad, Andhra Pradesh
Public Institutes		
1	Central Institute of Cotton Research (CICR)	Nagpur, Maharashtra
2	University of Agricultural Sciences (Dharwad)	Dharwad, Karnataka

9.5. ADOPTION AND IMPACT ANALYSIS OF BT COTTON

The commercial exploitation of transgenic Bt cotton hybrids has given a new impetus to cotton cultivation in the country by substantially increasing area, adoption rate, seed cotton yield along with lowering the pesticide load on the crop, thereby lowering the risks pertaining to environmental hazards, human health and production costs.

Table 5. Zone wise number of approved *Bt* cotton events and hybrid/variety by companies/ institutions in India (2002 to 2010)

Year	North Zone			Central Zone			South Zone		
	Events	Hybrids	Companies	Events	Hybrids	Companies	Events	Hybrids	Companies
2002	-	-	-	1	3	1	1	3	1
2003	-	-	-	1	3	1	1	3	1
2004	-	-	-	1	4	1	1	4	1
2005	1	6	3	1	12	4	1	9	3
2006	3	14	6	4	36	15	4	31	13
2007	4	32	14	4	84	23	4	70	22
2008	4	62	15	4	148	27	4	149	27
2009	5	164	26	6	296	35	6	294	35
2010	5	271	31	6	459	35	6	444	35

Source: ISAAA (2010)

Table 6. Year wise number of approved Bt cotton events and hybrid/variety by companies/ institutions in India (2002 to 2010)

Event	2002	2003	2004	2005	2006	2007	2008	2009	2010
Hybrids	03	03	04	20	62	131	274	522*	780
Events	01	01	01	01	4	4	4	6	6
Companies	01	01	01	03	15	24	30	35	35

Source: ISAAA (2010)

9.5.1 Increase in Area

Remarkably, for the tenth consecutive year, the hectarage, adoption rate and the number of farmers using *Bt* cotton in India, all continued to soar to record highs. The increase in area from 50,000 hectares in 2002-03, (when *Bt* cotton was first commercialized) to 11.00 million ha in 2012-13 or 93 per cent of the total area and represents an unprecedented 188-fold increase in eleven years (James, 2013). Bt cotton has transformed cotton production in India by increasing yield, decreasing insecticide applications and through welfare benefits contributed to the alleviation of poverty for over 6 million small resource poor farmers in 2010 alone (James, 2010). There were 54,000 farmers, which grew 50, 000 hectares of officially approved *Bt* cotton hybrids for the first time in 2002 which doubled to approximately 100,000 hectares in 2003. The *Bt* cotton area increased again four fold in 2004 to reach half a million hectares. in 2005, the area planted to Bt cotton in India continued to climb reaching 1.3 million hectares, an increase of 160 per cent over 2004. In 2006, the adoption record increases, which continued with almost a tripling of the area of *Bt* cotton to 3.8 million hectares. In 2006, India's *Bt* cotton area (3.8 million hectares) exceeds for the first time that of China's 3.5 million hectares. In 2007, the Indian cotton sector continued to grow with a record increase of 63 per cent in Bt cotton area from 3.8 to 6.2 million hectares, to become the largest hectarage of Bt cotton in any country in the world. In 2008, the area again increased to a record 7.6 million hectares which further increased to 8.4 million hectares in 2009 (Fig 1). Numbers of small farmers growing cotton increased from a few thousand in 2002-03 to 7 million small farmers in 2011-12. Hybrid cotton area doubled from 45 per cent in 2001-02 to 8 per cent in 2011-12 led by widespread use of *Bt* cotton hybrids.

The state-wise cotton crop data released by the ministry indicated that *Bt* cotton has emerged as the largest cotton crop in all the major cotton producing states like Maharashtra (3.71 million hectares) representing 40 per cent of all Bt cotton in India in 2010, followed by Gujarat (1.78 million hectares or 19%), Andhra Pradesh (1.65 million hectares or 18%), Northern Zone comprising Punjab, Haryana and Rajasthan (1.16 million hectares or 12 per cent) Madhya Pradesh (610000 hectares or 7%) and balance in Karnatka, Tamil Nady and other states (Table 6). In Punjab, more than 95 per cent area is under *Bt* cotton hybrids (approved and unapproved). In 2006, a second generation of genetically modified cotton, i.e. Bollgard II developed by Mahyco/Monsanto has been approved by Genetic Engineering Approval Committee.

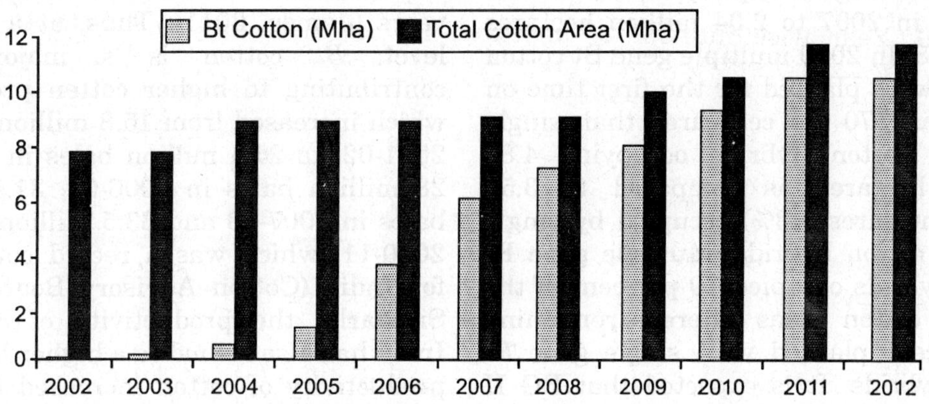

Figure 1. Area of *Bt* cotton in India

Table 6. Statewise adoption of *Bt* cotton in India (thousand hectares)

State	2002	2003	2004	2005	2006	2007	2008	2009	2010
Maharashtra	25	30	200	607	1840	2800	3130	3396	3710
Gujarat	10	36	122	150	470	908	1360	1682	1650
Northern region	-	-	-	60	215	682	840	1243	1780
Andhra Pradesh	8	10	75	280	830	1090	1320	1049	610
Madhya Pradesh	2	13	80	146	310	500	620	621	1162
Karnataka	3	4	18	30	85	145	240	273	370
Tamil Nadu	2	7	5	27	45	70	90	109	110
Others	-	-	-	-	5	5	5	8	8
Total	50	100	500	1300	3800	6200	7605	8381	9400

Source: ISAAA (2010)

Since 2002, *Bt* cotton scenario was dominated by Bollgard I hybrids but now, there is increasing tendency among farmers to adopt dual gene hybrids (cry 1 Ac + cry 2Ab) which confer in-built resistance to tobacco caterpillar as well as bollworms over single gene hybrids (cry 1 Ac) which provide resistance to bollworms only. The area under BG II hybrids was 0.15 million hectares (4%) in 2006, the year of its release as compared to 3.65 million hectares under BG I which has now increased to 4.82 million hectares (57%) as against 3.58 million hectares (43%) under BG I in 2009. In the first year 2006-07, the multiple gene Bt cotton hybrids were planted on 0.15 million hectares whilst single gene Bt cotton hybrids occupied 3.65 million hectares equivalent to 96 per cent of all the *Bt* cotton planted. The area under Single gene *Bt* cotton hybrids increased to 5.74 million hectares in 2007 and then registered a decline of 5.56 million hectares in 2008 and 3058 million hectares in 2009 and 2.8 million hectares in 2010. During this period, multiple gene

Bt cotton area grew rapidly to 0.46 million hectares in 2007 to 2.04 million hectares to in 2008. In 2009 multiple gene Bt cotton hybrids were planted for the first time on more than 570 per cent area than single gene *Bt* cotton hybrids occupying 4.82 million hectares as compared to 3.58 million hectares (43%) occupied by single gene *Bt* cotton hybrids. Multiple gene Bt cotton hybrids occupied 70 per cent of the total *Bt* cotton areas whereas remaining 30 per cent planted with single gene *Bt* cotton hybrids. It is expected that BG II hybrids will completely dominate *Bt* cotton scenario is India in coming years (Table 7).

Table 7. Area and percent adoption of BG-I and BG II cotton hybrids in India

Hybrids	2005	2006	2007	2008	2009	2010
BG - I	1.3 (100)	3.65 (96)	5.74 (92)	5.56 (73)	3.58 (43)	6.66 (70)
BG II	-	0.15 (4)	0.46 (8)	2.04 (27)	4.82 (57)	2.80 (30)
Total	1.3 (100)	3.80 (100)	6.20 (100)	7.60 (100)	8.40 (100)	9.40 (100)

Source: ISAAA (2010): Area in million hectares and figures in parenthesis are per cent adoption

9.5.2. Increase in Production and Productivity

Bt cotton has literally revolutionized cotton production in India. In 2011, India celebrated a decade of successful cultivation of *Bt* cotton. Remarkably, for the tenth consecutive year, the hectarage, adoption rate and the number of farmers using *Bt* cotton in India, all continued to soar to record highs. The increase in area from 50,000 hectares in 2002, (when *Bt* cotton was first commercialized) to 10.6 million hectares in 2011 represents an unprecedented 212-fold increase in ten years (James, 2011). Thus, at a national level, *Bt* cotton is a major factor contributing to higher cotton production, which increased from 15.8 million bales in 2001-02, to 24.4 million bales in 2005-06, 28 million bales in 2006-07, 31.5 million bales in 2007-08 and 33.5 million bales in 2010-11, which was a record cotton crop for India (Cotton Advisory Board, 2010). Similarly, the productivity of cotton in India has also rallied to a higher level. The productivity of cotton increased from 308 kg per hectare in 2001-2002 to 526 kg per hectare in 2008-09, major share being attributed to Bt cotton and continue to sustain above 500 kg/ha in 2010-11 season (Choudhary and Gaur, 2010). The deployment of *Bt* cotton over the last ten years has resulted in India becoming the number one exporter of cotton globally as well as the second largest cotton producer in the world. The global study of benefits generated by biotech crops conducted by Brookes and Barfoot (2012), estimates that India enhanced farm income from Bt cotton by US$ 9.4 billion in the period 2002 to 2010 and US$ 2.5 billion in 2010 alone. Besides, it has halved insecticide requirements, contributed to the doubling of yield and transformed India from a cotton importer to a major exporter. Large-scale cultivation of *Bt* cotton has resulted in significant reduction of insecticide use to the tune of 40 to 60 per cent less than the intensity on the corresponding non-transgenic varieties.

9.5.3. Pest Scenario on *Bt* Cotton

The adoption of *Bt* cotton has not only changed the cultivation profile, but also the pest scenario. While there is a decline in the pest status of bollworms; the sap feeders, viz. aphids, jassid, whitefly, mirids and mealy bugs are emerging as serious pests (Vennila, 2008). The pest status of bollworm complex has declined owing to the introduction of genetically engineered Bollgard cotton having gene from *Bacillus thuringenesis* Berliner expressing delta endotoxin conferring in-built resistance to bollworms. Besides this, weather conditions, IPM as well as IRM strategies and insecticides with new chemistry, i.e. spinosad and indoxacarb have also played a major role. Though Bt cotton (Bollgard with *cry1Ac* gene) provides effective management of bollworms requiring less number of sprays, it is not effective against sucking pests and tobacco caterpillar, which continue to possess a threat to its cultivation. Since 2003, tobacco caterpillar [*Spodoptera litura* (Fabricius)], a polyphagous pest, has also started causing serious damage to cotton crop. Moreover, the intensity of *S. litura* is likely to further increase under the potential climate change, as it has been found to consume more than 30 per cent cotton leaves at elevated CO^2 levels (Kranthi *et al.,* 2009). To combat the problem of *S. litura*, a second generation of genetically modified cotton, i.e. Bollgard II (cry 1 *Ac* + cry 2 *Ab*) containing an additional gene conferring in-built resistance to both bollworms as well as tobacco caterpillar has been approved by Genetic Engineering Approval Committee (GEAC).

Table 8. Changing status of cotton insect pests in cotton-wheat cropping belt of Northern region

Year	Major pest	Minor pest	New pest	Pest outbreak
Up to 1970	Jassid, PBW	Whitefly, Aphid, Thrips	SBW	ABW in some pockets (1978)
1971-1980	Jassid, PBW, SBW	Whitefly, Aphid, Thrips	ABW	ABW (1983)
1981-1985	Jassid, PBW, SBW	Whitefly, Aphid, Thrips	-	ABW (1990)
1986-1990	Jassid, PBW, SBW	Whitefly, Aphid, Thrips	Tobacco Caterpillar	ABW (1995)
1991-1995	Jassid, PBW, SBW, ABW	Whitefly, Aphid, Thrips	Leaf Miner, CLCuV	Whitefly (1995)
1996-2000	Jassid, Whitefly, SBW, ABW, PBW,	Aphid, Thrips	-	ABW (1997, 1998)
2001-2004	Jassid, Whitefly, SBW, ABW	Aphid, Thrips	-	ABW (2001)
2004-2006	Whitefly, Jassid, Tobacco Caterpillar	Aphid, Thrips	Mealy bug	Tobacco Caterpillar in some pockets
2006-2012	Whitefly, Jassid, Tobacco caterpillar,	Aphid, Thrips, Mealy bug	Lint Stainers (Dusky & Red Cotton bug)	Tobacco Caterpillar (2005) Mealy bug in some pockets (2006, 2007), whitefly in 2010 & 2012-2013 (in few pockets)

Source: Modified after Dhawan *et al.,* (2008); ABW–American bollworm; SBW-Spotted bollworm; PBW-Pink bollworm

Besides economically important pests like jassid and whitefly, other sucking pests like thrips, *Thrips tabaci* (Lindemann) hitherto occurring during May-June and aphids [*Aphis gossypii* (Glover)] that occur at the end of the crop season are also gaining importance. During 2006, a new sucking pest, mealybug, *Phenacoccus solenopsis* (Tinsley) has appeared on cotton crop. *P. solenopsis*, which was hitherto not reported to occur in India is an exotic species, which originated from the USA and now appears to be widespread on cotton in almost all cotton-growing states of the country (Nagrare *et al.* 2009). First report of mealy bugs was in 1997 on desi and hybrid varieties of cotton in Wagad area of Kutchh district. But with the cultivation of Bt cotton it has became havoc in cotton since 2006. During 2007-08, the pest caused serious damage in Punjab with an approximate yield loss upto 0.2 million bales worth Rs 1.59 billion. Similar situation was noticed in Haryana also. Other affected states with this pest include Rajasthan, Sourashtra (Gujarat), Maharashtra, Andhra Pradesh etc. Unfortunately, farmers are forced to rely on heavy use of insecticides once again in cotton to control mealy bugs. *Phenacoccus solenopsis* (Tinsely) (Pseudococcidae: Homoptera) is largely appearing species throughout India which has been confirmed both by conventional taxonomy and molecular analyses. *Phenacoccus solani* and *Maconellicoccus hirsutus* (Pseudococcidae: Homoptera) are also found to appear in isolated patches.

In Punjab, it appeared in few pockets of Bathinda, Ferozepur and Muktsar districts during 2006, but in 2007, it also spread to other cotton growing areas of Punjab and emerged in a serious proportion causing 30 to 40 per cent losses in the yield of cotton (Dhawan *et al.,* 2007). The integrated strategy has been recommended involving prevention of carry-over of pest, eradication of weeds, regular monitoring and need based application of insecticides for its management. Similarly, mirid bug has also emerged as an emerging threat on *Bt* cotton resulting in avoidable loss upto 2.90 q/ha (11.7%) compared to no protection in Dharwad (Karnataka) (Udikeri *et al.*, 2009). A green mirid, *Creontiades biseratense* (Distant) (Miridae: Hemiptera) is appearing since 2005 in Karnataka and causing considerable damage to *Bt* cotton. This was also recorded in Tamilnadu, Andhra Pradesh and Maharastra. Apart from this *Campylomma livida* (Reuter) has also been reported from Maharastra. The mirid pests cause heavy shedding squares and small sized bolls. Taxonomic identity of the three mirids, *Campylomma livida* Reuter, *Creontiades biseratense* Distant and *Hyalopeplus lineifer* Walker] occurring on cotton was established. While *C. livida* and *C. biseratense* are specific to central and south cotton production systems, respectively, *H. lineifer* is common to both regions (AICCIP, 2010). The damage potential of mirids has now been noticed in Sirsa area of North zone also (NCIPM Newsletter, Jan-June 2010). In addition, some of the minor pests like shoot weevil and stem weevil are becoming serious on Bt cotton (Sarode *et al.*, 2009).

9.5.3.1. Reduction in bollworm incidence

Results from the extensive field trials in India conducted under the aegis of ICAR clearly indicated that *Bt* cotton hybrids were highly effective in controlling bollworms thereby resulting in higher retention of first formed bolls due to low fruiting point and boll damage. Bambawale *et al.* (2004) reported 50 per cent overall reduction in *H. armigera* larval population in Bt cotton compared to non-*Bt* cotton. Further, the locule damage caused by pink bollworm was found to be 58 per cent lesser in *Bt* cotton. Udikeri *et al.* (2003) also showed that Bt hybrids were able to reduce larval population of *H. armigera* up to 40 per cent, spotted bollworm up to 30-40 per cent and pink bollworm up to 60-80 per cent in South India. Field studies conducted by Vennila *et al.* (2004) using five *Bt* cotton hybrids viz. RCH 2 Bt, RCH 20 *Bt*, RCH 134 Bt, RCH 138 Bt and RCH 144 Bt with cry 1Ac gene along with their non-Bt showed lesser square damage, green boll, open boll and locule damage than their non-Bt counterparts and conventional hybrids. Sharma and Pampapathy (2006) observed that transgenic hybrids suffered low shoot damage (0-9.6%) by *E. vittella* as compared to non-transgenic counterparts (8.7-46.7%). Differences in shoot damage across protected regimes were not very large as very little insecticide was applied during vegetative phase. No significant differences in oviposition of *H. armigera* between transgenic and non-transgenic cultivars were observed under protected and unprotected conditions. The larval numbers were significantly lower on transgenic hybrids under high infestation but the differences in larval density between transgenic and non-transgenic hybrids under low levels of infestation were quite small. Bollworm damage in squares and bolls was significantly lower in transgenic hybrids than in non-transgenic. Differences in seed cotton yield between transgenic and non-transgenic hybrids were not significant under unpro-tected conditions at moderate levels of infestation. However, significant differen-ces in seed cotton yield were observed under heavy bollworm infestation.

Udikeri *et al.* (2007) observed that Bt genotypes with one gene (cry 1Ac) recorded *E. vittella* larval population in fraction and those genotypes with two genes (cry 1Ac + cry 2Ab) did not record larval population at 50 and 65 days after sowing. BG-II was proved to be superior over BG-I by recording lower *H. armigera* larval population. However, the average incidence was high on conventional cotton. All the Bt hybrids were effective in containing pink bollworm incidence. Bal and Dhawan (2008) observed that population of spotted bollworm larvae within the unsprayed plots was significantly low in *Bt* cotton (0.10/plant) as compared to non-*Bt* cotton (0.33/plant), while the sprayed plots of Bt and non-*Bt* were at par with each other and within *Bt* and non-*Bt* plots, sprayed and unsprayed plots were at par with each other. The incidence in intact fruiting bodies was significantly lower in Bt sprayed (0.94%). There was significantly higher incidence in non-Bt cotton (7.73%) as compared to Bt cotton (3.27%) with in sprayed treatments. The sprayed and unsprayed treatments of non-*Bt* were at par with each other. Mann *et al.* (2010) demonstrated that transgenic cultivars showed bollworm damage as compared to conventional cultivars. Similarly, overall mean boll infestation was observed to be comparatively low in *Bt* cotton hybrids ranging from 0.0 to 1.51 and 0.0 to 0.92 per cent on boll and loculi basis, respectively as against 12.37 and 6.70 per cent in non-*Bt* check LHH 144 (Shera *et al.*, 2010).

9.5.3.2. Variability in toxin expression

The temporal and spatial variation in expression of cry protein and insecticidal activity of *Bt* cotton on *H. armigera* has been widely studied and well understood in many countries including India. The studies have showed that the quantitative levels of cry 1Ac toxin and the seasonal decline in expression differed significantly among different commercial *Bt* hybrids and also highly variable among different plant parts.

9.5.3.2.1. Variability with age of the plant

For the Bt cotton to be sustainable, it is important that the toxin protein be expressed in adequate quantitative in appropriate plant parts at requisite time of the season to afford protection against major target insect pests. However, number of studies conducted in Australia, USA, China and India have indicated that the level of Bt protein in cotton fluctuate during the whole growing season and may logically cause variation in the efficacy of Bt cotton against lepidopteran pests. Kranthi *et al.* (2005) reported that cry 1Ac expression decreased consistently as the plant aged and decline was more rapid in some Bt cotton hybrids as compared to others. Further, despite the variability in toxin expression, the pest control properties are unlikely to be affected significantly at least until the crop becomes 110-115 days old, after which, the toxin levels may decline below 1.8 μg/g in some plant parts. Kaur (2008) also reported that the expression of two insecticidal proteins (Cry 1Ac and Cry 2Ab) in Bollgard II varies among hybrids, plant parts and both declined with age of the plant from 60 to 180 days. The Cry 1Ac declined from 4.70 to 2.54 μg/g in leaves, 4.26 to 2.86 μg/g in squares and 3.74 to 2.99 μg/g in bolls and Cry 2Ab expression declined from 19.01 to 17.04 μg/g in leaves, 20.22 to 19.31 μg/g in squares and 19.69 to 19.24 μg/g in bolls.

9.5.3.2.2. Variability in plants parts

Intra-plant variability in cry 1 Ac expression in Indian Bt cotton hybrids has been reported. The expression of the cry1Ac throughout season in Bt cotton plants was high in terminal leaves, moderate in squares and slightly less in bolls. However, the expression in bolls was adequate for bollworm control. The cry 1Ac expression ranged from 0.01 to 19 μg/g in various plant parts. Though 2-3 fold differences in cry 1Ac levels were common between hybrids during the early phase of the crop growth, variability up to seven fold was also observed at times. The increasing levels of *H. armigera* were correlated with the toxin levels decreasing below 1.8 μg / g in the plant parts, below which *H. armigera* would be able to survive. The toxin expression in the boll rind, square bud and ovary of flowers was clearly inadequate to confer full protection to the fruiting parts. Thus, some larval survival was observed on fruiting structures and leaves of older Bt cotton plants in laboratory bioassay and on the field crop (Kranthi *et al.*, 2005). Kaur (2008) reported that the maximum expression of cry 1Ac and cry 2Ab was found in leaves and squares, respectively. Moreover, the expression of cry 2Ab is 3 to 6 times more than that of cry 1Ac expression in Bollgard II genotypes.

9.5.3.3. Economic Losses Due to Sucking Pests in Bt Cotton

Prior to the introduction of transgenic cotton, bollworms and sucking pests were the two groups against which most of the pesticide use was directed on cotton. The commercial exploitation of transgenic *Bt* cotton hybrids has substantially lowered the pesticide load on the crop against bollworms. However, it is not effective against sucking pests, which still pose a great threat to cultivation of *Bt* cotton and necessitate insecticidal applications to avoid yield losses. Many studies have reported that no differences occur in transgenic and non-transgenic hybrids in their density and relative susceptibility to sucking insect pests (Sharma and Pampapathy, 2006; Mann *et al.,* 2010; Shera *et al.* 2010). However, their tolerance or resistance is mainly dependent on morphological or genetic base. Sharma and Pampapathy (2006) and Surulivelu *et al.* (2009) reported lower jassid number on MECH 184 because of its hairiness as compared to other hybrids. Yield losses due to jassid was observed to be less in MECH 184 (4.22%) followed by MRC 6918 Bt (9.01%), RCH 2 Bt (9.14%), Bunny Bt (10.34%), Mallika Bt (10.92%), RCH 20 Bt (11.31%) and RCHB708 Bt (12.18%). The extent of avoidable losses due to sucking pests has been reported to be 2.52 q/ha (11.9%) in MRC 6301 Bt as against 2.63 q/ha (11.2%) in RCH 134 Bt and 2.90 q/ha (16.2%) in NCS 145 BG II (Shera *et al.,* 2010). Based on multilocation trials, the avoidable losses were observed to vary from 1.1 to 6.4 per cent on different *Bt* hybrids due to sucking pests on *Bt* cotton (Dhawan *et al.*, 2008).

In last few years, a new sucking pest, mealybug has appeared on cotton crop. *P. solenopsis*, which was hitherto not reported to occur in India, is an exotic species, which originated from the USA and now appears to be widespread on cotton in almost all cotton-growing states of the country (Nagrare *et al.* 2009). In Punjab, it appeared in few pockets of Bathinda, Ferozepur and Muktsar districts during 2006, but in 2007, it also spread to other cotton growing areas of Punjab and emerged in a serious proportion causing 30 to 40 per cent losses in the yield of cotton (Dhawan *et al.,* 2007). The integrated strategy has been recommended involving prevention of carry-over of pest, eradication of weeds, regular monitoring and need based application of insecticides for its management. This non-chemical strategy suggested by Punjab Agricultural University, Ludhiana has been adopted at the national level for the control of this pest. Similarly, mirid bug has also emerged as an emerging threat on *Bt* cotton resulting in avoidable loss upto 2.90 q/ha (11.7%) compared to no protection in Dharwad (Karnataka) (Udikeri *et al.*, 2009).

9.5.4. Reduction in Insecticide Usage

Apart from yield advantages, reduction in insecticidal sprays was found to contribute significantly in paving the way for a healthier environment and conserving the natural biodiversity. Traditionally, cotton accounted for around 50 per cent of pesticide consumption in the country despite being grown on an area of 5 per cent of total cropped area. The pesticide consumption in India has shown consistent downward trend from 48,350 metric tons in 2002 (first year of commercialization) to 43,860 in 2008-09 (www.indiaagristat.com) estimated at a minimal reduction of approximately 4500 metric tons technical grade active ingredient coinciding with increase in area under *Bt* cotton. The pesticide consumption in different cotton growing states of the country is presented in Table 9.

A comparison of cotton market from 1996 to 2006 revealed that of the total pesticide market in India in 1998 valued at US$ 770, 30 per cent was for cotton pesticides only, which were equal to 42 per cent of total insecticide, market for all crops in India. Subsequent to the

Table 9. Statewise pesticide consumption in India (Metric tons active ingredient)

Year	2001-02	2002-03	2003-04	2004-05	2005-06	2006-07	2007-08	2008-09	2009-10
Punjab	7200	7200	6780	6900	5610	5975	6080	5760	5810
Haryana	5020	5012	4730	4520	4560	4600	4390	4288	4070
Rajasthan	4628	3200	2303	1628	1008	3567	3804	3333	3527
Maharashtra	3135	3725	3385	3030	3198	3193	3050	2400	4639
Gujarat	4100	4500	4000	2900	2700	2670	2660	2650	2750
Andhra Pradesh	3850	3706	2034	2135	1997	1394	1541	1381	1015
Karnatka	2500	2700	1692	2200	1638	1362	1588	1675	1647
Tamil Nadu	1576	3346	1434	2466	2211	2048	3940	2317	2335

Source: www.indiaagristat.com

Table10. Impact of Bt cotton in India (2002 to 2010)

Publication	**Bennett *et al* (2004; 2006)**	**Vaidya (2005)**	**Gandhi and Namboodiri (2006)**	**ICAR, (2006)**	**Ramgopal (2006)**	**Dev & Rao (2007)**	**AIC-CIP (2008)**	**Subramaniaum & Qaim (2009)**	**Sadashivappa and Qaim (2009)**	**Qaim *et al* (2009)**	**Subramanian & Qaim 2010**
(1)	**(2)**	**(3)**	**(4)**	**(5)**	**(6)**	**(7)**	**(8)**	**(9)**	**(10)**	**(11)**	**(12)**
Area of study	Maharashtra, MP, Karnatka, Gujarat	Maharashtra	Gujarat, Maharashtra, AP, Tamil Nadu	Multilocation in India	Andhra Pradesh	Andhra Pradesh	Punjab	-	-	Maharashtra AP, Karnataka TN	
Period studied	2002 & 2003	2003	2004	2005	2006	2004 & 2005	2007	2004 & 2005	2006 & 2007	1998 to 2006	2006-07
Yield increase (%)	45-63	-	31	30.9	46.5	32	39	30-40	43	37	43
Reduction in no. of sprays (%)	67	-	39	-	55	25	65	50	21	41	21
Average increase in profit/ hectare	-	Rs. 14090 (US$ 320.2)	Rs. 9821-13821 (US$ 224-314)	-	Rs. 17347 (US$ 394.3)	(US $ 225)	Rs. 12708 (US$ 288)	(US$ 156 or more -	(US$ 148 or more -	(US$ 131 or more -	(US$ 161 or more -
Increased profit (%)	50% or more gross margins	79.2	88	-	66	83	57	-	70	89	134

Source: Karihaloo and Kumar, (2009); Chaoudhary and Gaur, (2010); AICCIP report, ISSAYA, 2011

introduction of Bt cotton, cotton consumed only 18 per cent of the total pesticide market, in 2006 valued at US$ 900 million as compared to a much higher 30 per cent in 1998. Similarly, the market share for cotton insecticides declined from 42 per cent in 1998 to 28 per cent in 2006, the year in which Bt cotton occupied 3.8 million hectares area (42%) (Choudhary and Gaur, 2010). The reduction in number of sprays on Bt cotton has been reported to vary from 21 to 75 per cent by many workers in studies conducted on impact of Bt cotton (Table 10). Reports showed that yields increased significantly by adopting *Bt* cotton and farmers in India were able to reduce at least 2-3 insecticide applications due to low bollworm infestation. However, the benefits of *Bt* cotton were more in other countries where bollworm infestation was high. Insecticide applications on Bt cotton varieties were reduced up to 14 applications in China, 7 in South Africa and 5-6 in Indonesia and Australia (Kranthi *et al.,* 2009).

In Punjab, before the introduction of *Bt* cotton (2002), in IRM villages of six cotton growing districts i.e. Mansa, Muktsar, Ferozepur, Bathinda, Faridkot and Barnala, mean number of sprays given for the control of insect pests was 9.4 in IRM and 15.0 in Non-IRM villages. However, after the introduction of Bt cotton in Punjab, the average number of insecticide sprays for IRM and Non-IRM villages was 4.18 and 6.18, respectively in 2008 and 3.61 and 5.53 in 2009, respectively. On the average, approxi-mately 55-60 per reduction in insecticidal sprays were observed with cultivation of Bt cotton. Bt cotton halved insecticide usage from 46 per cent of total insecticides used in 2001-02 to 21% of total insecticide use in 2010-11.

9.5.5. Economic Benefits

The deployment of Bt cotton over the last ten years has resulted in India becoming the number one exporter of cotton globally as well as the second largest cotton producer in the world. Bt cotton contributed to India's farm economy a whooping US$9.4 billion in the period 2002 to 2010-11 (first nine year) and US$2.5 billion in 2010-11 alone halved insecticide requirements, contributed to the doubling of yield and transformed India from a cotton importer to a major exporter (James, 2010). The global study of benefits generated by biotech crops conducted by Brookes and Barfoot (2010), estimates that India enhanced farm income from *Bt* cotton by US$ 5.1 billion in the period 2002 to 2008 and US$ 1.8 billion in 2008 alone. The analysis of the different economic studies on the impact of *Bt* cotton (Naik, 2001; ICAR, 2002; Bennett *et al.,* 2004; Bambawale *et al.,* 2004; Qaim, 2006; Bennett *et al.,* 2006; Gandhi and Namboodiri, 2006; Ramgopal, 2006; ICAR, 2006; Dev and Rao, 2007; Sadashivappa and Qaim, 2009; Subramaniaum & Qaim (2009), Subramanian & Qaim 2010, Qaim *et al.,* 2009) conducted by public sector institutes in different parts of the India over the period 1998 to 2010 have consistently confirmed yield advantages, reduction in insecticidal sprays and high economic returns as against corresponding non transgenic cotton. These studies have been initially carried out by seed companies before commercialization of Bt cotton as a part of regulatory approval procedures and later by research centers as well as civil society organizations (Karihaloo and Kumar, 2009).

Economic studies on impact of *Bt* cotton before commercialization (1998 to 2002) analyzed by different workers (Naik, 2001; ICAR, 2002; Qaim and Zilberman,

Table 11. Impact of IPM on Bt cotton in Punjab

Parameter	2002	2003	2004	2005**	2006	2007	2008	2009	2010	2011	2012
No. of villages	32	30	45	121	213	230	225	225	53	80	45
No. of IPM farmers	300	567	900	10044	14804	27196	12919	12120	2282	3122	1145
Area under IPM (ha)	1152	28282	19429	[illegible]	[illegible]4848	45262	31735	35182	14048	11235	9872
Seed cotton yield (q/ha)	15.85	19.69	24.41	24.52	24.35	22.60	24.33	25.86	22.05	18.84	21.78
Number of sprays*	10.6	9.5	5.4	3.4	3.1	4.7	4.2	3.6	3.8	5.57	6.08
Spray cost (Rs/ha)	5460	7937	4021	1793	2163	2714	1761	1955	2076	3779	5366
Addition Profit (Rs./ha)	7670	7844	13111	5509	5371	9864	12456	15699	13732	8106	11199

The average number of sprays was 15-20 in 2001;
** Bt cotton was adopted as component of IPM from 2005 onwards

2003; Qaim 2006) have showed 34 to 92 percent increase in yield, 50 to 75 per cent reduction in insecticidal sprays and 68 to 77 per cent increase in profit from Bt cotton, equivalent to US$ 76 to US$ 236 per hectare. It is noteworthy that the benefits recorded in pre-commercialization field trials are consistent with the actual experience of farmers in post commercialization of *Bt* cotton during 2002 to 2009 (Chaoudhary and Gaur, 2010). Multi-location studies conducted on large numbers of *Bt* cotton farmers after commercialization of *Bt* cotton (2002 to 2010) in different parts of the country confirmed yield gain analyzed to be 31 to 63 per cent resulting in 50 to 110 per cent increase in profit which accrued average profit ranging from US$ 131 to US$ 250 per hectare. Besides increasing yield, the reduction in number of sprays occurred from 21 to 55 per cent as compared to non-*Bt* cotton. These profits have accrued to small and resource-poor cotton farmers in the various cotton-growing states of India (Table 11).

In Maharashtra, Bennett *et al.*, (2004 and 2006) demonstrated the performance of *Bt* cotton under on-farm study from 9000 farmers plot and observed 45 and 63 per cent increase in yield for *Bt* over non-Bt versions in 2002 and 2003, respectively. The number of sprays was 1.44 on *Bt* and 3.84 on non-*Bt* during 2002 and 0.71 on Bt and 3.11 on non-*Bt* during 2003. However, three-year studies conducted in four cotton-growing districts of Andhra Pradesh concluded that *Bt* cotton yielded 30 per cent less than non-*Bt* and earning with non-*Bt* cultivation was 60 per cent more than with *Bt* cultivation (Qayum and Sakkhari, 2005). Vaidya, (2005) reported significantly higher profits of US $ 725/ha in Bt as against US $ 404/ha in non-Bt

cotton cultivation in a study conducted in Yavatmal and Budhana districts of Maharashtra. Multilocation front line demonstrations covering 1200 demonstration and farmer's plot conducted in India demonstrated 31 per cent increase in seed cotton yield over non-*Bt* hybrids (ICAR, 2006). A study conducted in two districts of Andhra Pradesh (Ramgopal, 2006) demonstrated 46 and 47 per cent yield advantages and 55 and 16 per cent less sprays in Guntur and Warangal districts, respectively. The net income yielded by *Bt* farmers was US $ 600/ha as compared to US $ 206/ha by non-*Bt* farmers. A survey conducted involving 694 cotton growing farmers (Gandhi and Naamboodiri, 2006) in Gujarat, Maharashtra, Andhra Pradesh and Tamil Nadu showed US $ 347 to US $729/ha as compared to US $ 123 to US $ 415 /ha. The yield gain was 31 per cent with 39 per cent reduction in number of sprays. Dev and Rao (2007) in a study conducted by Centre for Economic and Social Studies (CESS) in Warangal, Nalgonda, Guntur and Kurnool districts of Andhra Pradesh during 2004-05 and 2005-06 reported 83 per cent increase in net income, 18 per cent reduction in insecticidal cost and 32 per cent yield gains over non-*Bt* hybrids. The trials conducted at Ludhiana and Fardikot districts of Punjab resulted in 39 per cent increase in yield leading to net profit of Rs. 22286/ha in *Bt* cotton as against Rs. 9578/ha in non- *Bt* cotton besides saving in 5 sprays (AICCIP, 2008). A case study by Subramanian and Qaim, (2009) showed that *Bt* cotton increased the yield between 30-40 per cent besides 50 per cent reduction in insecticide usage resulting in an additional income of US $ 156 per ha or more. In the same way, Sadashivappa and Qaim (2009) demonstrated 40 per cent reduction in pesticides, 30-40 per cent yield increase leading to net profit of US $ 148 hectare or more. The pre and post commercialization farm surveys conducted in Maharashtra, Andhra Pradesh, Karnataka and Tamil Nadu demonstrated 89 per cent or US $ 135 per ha increase in net profit by growing Bt cotton (Qaim *et al.*, 2009).

9.5.6. Impact of IPM Strategies in Bt Cotton

For sustaining the production and productivity of cotton crop, an imperative necessity is to adopt management practices according to their modification in relative pest status. Bt cotton has literally revived cotton cultivation in India. The commercial exploitation of *Bt* cotton in India has given a new impetus to IPM programme with significant reduction in pesticide load on cotton crop. In transgenic cotton, the need to spray against bollworms especially *Helicoverpa* spp. is dramatically reduced thereby allowing other insects, normally controlled inadvertently by sprays targeting *Helicoverpa* spp., to emerge as pests. This includes aphids, jassids, whiteflies, mealybug, mirids and other foliage feeders and lint stainers. Use of insecticides against these surviving and emerging pests may disrupt natural enemies, creating outbreaks of mites, whitefly, thrips or aphids. Hence, in both conventional (non-transgenic) and transgenic cotton there remains a strong incentive to use IPM to help reduce reliance on insecticides. Furthermore, an IPM approach helps to avoid creating situations that increase the risk of losses due to these pests. The integration of *Bt* cotton as a component of IPM module augmented with suitable cultural, mechanical control strategies, sowing of refugia around the main field and need-based application of pesticides has not

only reduced pesticide load and make cotton cultivation economical but will also help in management of resistance in insects against Bt cotton. Considerable effort has been directed toward such alternatives, and new technology has been implemented. In this context, studies have been conducted in different parts of the country to validate and prom IPM technology in Bt cotton through farmer's participatory approach to sustain productivity of cotton.

In a comprehensive study, conducted in all cotton growing districts of Punjab, IPM module developed by Punjab Agricultural University, Ludhiana for Bt-cotton pest management was implemented and evaluated in different cotton growing villages of Punjab during 2005 to 2012 (Table 12). In six cotton-growing districts i.e. Mansa, Muktsar, Ferozepur, Bathinda, Faridkot and Barnala, 53-225 villages covering an area of 14048 to 44848 ha were adopted during different years. Non-IPM villages were also selected for ascertaining the impact of adopted strategies.

In Punjab, the impact of adoption of IPM strategies resulted in the reduction in insecticidal sprays and increase in seed cotton yield in IPM villages over non-IPM villages. The average net profit per ha was more in IPM villages (Rs. 46491 & 55026 kg/ha) as compared to non-IPM villages (Rs. 35660 & 37236 kg/ha). The additional net profit of IPM villages over non-IPM villages varied from Rs. 5371 to 15699 per ha (Dhawan *et al.,* 2011). The adoption of insect pests resistance management strategies on Bt cotton in 112 and 120 villages of Ferozepur, Bathinda and Mansa districts covering an area of 2591 and 5720 hectares during 2005 and 2006, respectively resulted in additional profit of Rs. 4725 and 4267 per hectare to the farmers of IPRM villages as against non IPRM villages. The impact of adoption of IRM strategies on Bt cotton in 225 cotton growing villages of Punjab during 2008-09 and 2009-10 resulted in the reduction of insecticidal sprays (32.36 & 33.62%) in IRM villages over non-IRM villages during 2008 and 2009, respectively. Seed cotton yield was higher where IRM strategies were adopted (2423 & 2586 kg/ha) over non-IRM villages (2096 & 2043 kg/ha) during 2008 and 2009, whereas yield obtained during 2002-03 was (1587.5 & 1412 kg/ha) in IRM and non-IRM villages, respectively. The average net profit per ha was more in IRM villages (Rs. 46491 & 55026 kg/ha) as compared to non-IRM villages (Rs. 35660 & 37236 kg/ha). The additional profit of IPM villages over non-IPM villages was Rs. 10831 & 17790 per ha (IRM report, 2009).

Bambawale *et al.* (2004) reported performance of MECH-162 Bt along with non-Bt MECH-162 and a conventional variety/hybrid under integrated pest management in farmer's participatory field trials conducted in Maharashtra. Seed cotton yield in MECH-162 Bt (12.4 q/ha) was much higher than that of non-Bt MECH-162 (9.8 q/ha) and conventional cotton (7.1 q/ha). Net returns after taking into account cost of production and protection were Rs. 16,231 (US$ 368.9)/ha in MECH-162, Rs.12,433 (US$ 282.6)/ha in non-Bt MECH- 162 and Rs. 10,507 (US$ 238.8) /ha in conventional cotton. Tayade *et al* (2009) reported that adoption of IRM strategies through farmers participatory mode in six tehsils of Nagpur during 2006-07 resulted in 12 per cent yield gain, reduction of insecticidal sprays from 3.42 to 1.57 with net profit of Rs. 11209/ha in IRM villages as against Rs. 6148/ha in non-IRM villages. In Coimbatore (Tamil

Nadu), the adoption of IRM strategies in Bt cotton resulted in reduction in number of insecticidal sprays, quantity of insecticide used and reduction in plant protection cost from 3.26 to 1.44, 903.66 to 388.99 g a.i./ha and Rs. 2638 to Rs. 1387/ha, respectively. An average increase in yield was 27 per cent over non-IRM villages with net profit of Rs. 41,318/ha as against 26248/ha (Dharajothi *et al.*, 2009). Udikeri *et al.* (2009) reported that average number of sprays given by IPRM farmers were 3.4 as against 5.9 sprays by non-IPRM farmers leading to saving of insecticides to the tune of 20 to 53.3 per cent in different districts besides yield advantage of 1.29 q/ha with intervention of IPRM strategies on Bt cotton in Karnataka.

9.6. CONCLUSIONS

The commercial exploitation of Bt cotton in India has significantly established its positive impact in all cotton-growing areas of the country. The pesticide load on the cotton crop has substantially declined owing to effective control of bollworm complex, which in turn has led to the reduction in cost of cultivation and a decline environmental pollution. Besides this, social and economical status of farmers has been uplifted due to yield advantage, net profit and higher economic returns. Farmers' earnings and profitability from Bt cultivation have been significantly higher than those from cultivation of non-Bt cotton. That the impacts have been widespread is evident from the national level cotton statistics. It has been observed that the Bt cottons are early in maturity as compared to their non-Bt counterparts. The earliness ranges from 20-30 days in different cotton hybrids tested in India. The deployment of Bt cotton over the last ten years has resulted in India becoming the number one exporter of cotton globally as well as the second largest cotton producer in the world. However, the potential concern among the researchers now is sustainability of Bt technology keeping in view widespread and prolonged exposure of Bt toxin to target species. The resistance to transgenic cotton can not be ruled out therefore continuous efforts are required from research point of view w.r.t monitoring and managing resistance through development and implementation of proper insect resistance management strategies for Bt cotton. Therefore, proper monitoring for forewarning of pest adaptation is of utmost necessity on transgenics in order to conserve this effective technology for longer period. The concept of the multi gene cotton is to enhance the expression of toxins and delay resistance development and also to increase the range of pests targeted by cry toxins. It is emphasized that pest problems should be tackled with IPM approach based on sound ecological principles involving integration of Bt cotton and all the available control strategies with minimum use of insecticides so that pest population can be managed below damaging level for achieving sustained production in the cotton ecosystem. In addition, insect resistance management strategies as a part of IPM module will help in delaying the resistance in insects to transgenic cotton. Genetically modified cotton having genetic resistance provides innate control of bollworms and tobacco caterpillar but sucking pests are still posing a great threat. Therefore, special attention is required for the validation and promotion of IPM technologies in transgenic cotton keeping in view the up-liftment of new pests, i.e. mealybugs and secondary pest problems like thrips, mirid bugs etc. Further, the study on ecological impact of transgenic cotton and impact analysis of IPM approach on socio-economic status of farmers' is of utmost importance.

REFERENCES

AICCIP (2008) All India Coordinated Cotton Improvement Project. Annual Report 2007-08, Central Institute for Cotton Research, Regional Station, Coimbatore. pp 3-5.

AICCIP (2010) *All India Coordinated Cotton Improvement Project*. Annual Report 2009-10, Central Institute for Cotton Research, Regional Station, Coimbatore. pp 4-6.

AICCIP (2011) All India Coordinated Cotton Improvement Project. Annual Report 2010-11, Central Institute for Cotton Research, Regional Station, Coimbatore. pp 3-5.

AICCIP (2013) *All India Coordinated Cotton Improvement Project*. Annual Report 2012-13, Central Institute for Cotton Research, Regional Station, Coimbatore. pp 3-5.

Anonymous (2011) *Agriculture at a glance 2010-11.* Department of Agriculture, Punjab, Chandigarh.

Bal H and Dhawan AK (2008) Effect of Intercropping on population of pests and natural enemies in Bt and non-Bt cotton under sprayed and unsprayed conditions. *Indian Journal of Ecology* **35**: 59-63.

Bambawale OM, Singh A, Sharma OP, Bhosle BB, Lavekar RC, Dhandapani A, Kanwar V, Tanwar RK, Rathod KS, Patange NR and Pawar VM (2004) Performance of Bt cotton (MECH-162) under integrated pest management in farmers's participatory field trial in Nanded district, Central India. *Current Science* **86**: 1628-33.

Bennett R, Kambhampati U, Morse S and Ismail Y (2006) Farm level economic performance of genetically modified cotton in Maharastra India. *Review of Agricultural Economics* **28**: 59-71.

Bennett R, Ismael Y, Kambhampatti U and Morse S (2004) Economic impact of genetically modified cotton in India. *AgBio Forum* **7**: 96-100.

Brookes G and Barfoot P (2010) GM Crops: Global Socio-economic and Environmental Impacts 1996-2008. P.G. Economics Ltd, Dorchester, UK. (http:// www. pgeconomics.co.uk/ pdf / 2010-global-gm-crop-impact-study-final-April-2010.pdf)

Butter NS and Kular JS (1999) Resurgence of whitefly in cotton and its management *Indian Journal of Entomology* **61:** 85-90.

Choudhary B and Gaur K (2010) *Bt Cotton in India: A Country Profile*. ISAAA Series of Biotech Crop Profiles. ISAAA: Ithaca, NY.

Choudhary B and Gaur K (2011) *Bt Cotton Events & Hybrids in India, 2002 to 2010.* ISAAA Biotech Information Centre, International Service for the Acquisition of Agri-Biotech Applications (ISAAA), New Delhi, India.

Cotton Advisory Board (2010) Finalization of area and production of cotton 2008-09 cotton season, further revision of area under cotton cultivation and production of cotton for cotton season 2009-10 and to protect state wise area and cotton production for the forth coming cotton season 2010-11. Cotton Advisory Board, Office of Textile Commissioner, Ministry of Textile, September 2010.

Cotton Advisory Board (2008) Minutes of the meeting for the cotton year 2007-08 and cotton projection for 2008-09, Office of the Textile Commissioner, Ministry of Textile, Government of India. URL: http://www. txcindia.com/html/production,%20yield%2 02007089 % 20cabmeeting% 2020102008.pdf.

Dev SM and Rao NC (2007) *Socio-Economic Impact of Bt Cotton*, CESS Monographs,

Centre for Economic and Social Studies (CESS), Hyderabad, A.P.

Dhaliwal GS, Arora R and Dhawan AK (2004) Crop losses due to insect pests in Indian agriculture. *Indian Journal of Ecology* **31**: 1-7.

Dharajothi B, Gopalakrishnan N, Surulivelu T and Manjula TR (2009) Impact analysis of IRM strategies on Bt cotton in the irrigated tracts of Tamil Nadu. (Abstract). National Symposium on "Bt cotton: Opportunities and Prospects" November 17-19, 2009. Central Institute for Cotton Research , Nagpur. p 99.

Dhawan AK and Simwat GS (1997) Insecticide induce resurgence of sucking pests in upland cotton in Punjab, India. In: *Third International Congress of Entomological Sciences*, March 18 20, 1997 (Abstract). National Agricultural Centre, Islamabad.

Dhawan AK, Kumar V, Singh K and Saini S (2011) *Pest Management Strategies in Cotton*. Society for Sustainable Cotton Production, Department of Entomology, Punjab Agricultural University, Ludhiana

Dhawan AK, Shera PS, Jindal V and Aggarwal N (2008) Changing scenario of cotton insect pests and their management strategies in the Punjab. *Cotton Research in Punjab*. Department of Plant Breeding, Genetics & Biotechnology, Punjab Agricultural University, Ludhiana. pp. 81-99.

Dhawan AK, Singh K, Saini S, Mohindru B, Kaur A, Singh G and Singh S (2007) Incidence and damage potential of mealybug *Phenacoccus solenopsis* Tinsley on cotton in Punjab. *Indian Journal of Ecology* **34**: 166-72.

Gandhi V and Namboodiri NV (2006) *The adoption and economics of Bt cotton in India: Preliminary results from a study*. IIM Ahmedabad working paper no. 2006-09-04,. pp 1-27.

Gill HK and Dhawan AK (2006) Global status of insecticide resistance in *Helicoverpa armigera* on cotton. *Journal of Cotton Research and Development* **20**: 226-31.

Hargreaves H (1948) *List of Recorded Cotton Insects of the World*. pp 50. Commonwealth Institute of Entomology, London.

ICAR (2002) Report on 2001 IPM trial cost benefit analysis. ICAR, New Delhi, India.

ICAR (2006) Front line demonstrations on cotton 2005-06. Mini Mission II, Technology Mission on Cotton. Indian Council of Agricultural Research (ICAR), New Delhi, India.

James C (2009) *Global Status of Commercialized Biotech/GM Crops:* 2009. ISAAA Brief No.41. ISAAA: Ithaca, NY.

James C (2010) *Global Status of Commercialized Biotech/GM Crops:* 2010. ISAAA Brief No.42. ISAAA: Ithaca, NY.

James C (2011) *Global Status of Commercialized Biotech/GM Crops:* 2009. ISAAA Brief No.43. ISAAA: Ithaca, NY.

James C (2013) *Global status of commercialized biotech/GM crops:* ISAA Briefs No. 46 ISAAA: International Service for Acquisition in Agribiotech Applications. New York, USA.

Kapoor SK, Singh J, Russel D, Singh B and Kalra RL (2002) Susceptibility change of *Helicoverpa armigera* (Hubner) to different insecticides in Punjab. *Pesticide Research Journal* **14**: 177-80.

Karihaloo JL and Kumar PA (2009) *Bt Cotton in India-A Status Report* (Second Edition). Asia-Pacific Consortium on Agricultural Biotechnology (APCoAB), New Delhi, India.

Kaur M (2008). Impact of Bt cotton expressing two insecticidal proteins of *Bacillus thuringiensis* Berliner on the management of noctuid larvae. Ph.D. dissertation, Punjab Agricultural University, Ludhiana.

Kranthi KR, Kranthi S, Gopalakrishnan N, Asokan R and Mayee CD (2009) *Bt* resistance -its management and prospects in the current context of climate change In: Ramamurthy, V V; Gupta, G P and Puri, S N (ed.) *Proceedings. National Symposium on IPM strategies to Combat Emerging pests in the Current Scenario of Climate Change.* pp 8-13, College of Horticulture and Forestry, Central Agricultural University, Pasighat, Arunachal Pradesh.

Kranthi KR, Naidu S, Dhawad CS, Tatwawadi A, Mate K, Patil E, Bharose AA, Behere G T and Kranthi S (2005) Temporal and intra-plant variability of CrylAc expression in Bt cotton arid its influence on the survival of the cotton bollworm, Helicoverpa armigera (Hubner) (Noctuidae: Lepidoptera). *Current Science* **89**: 291-98.

Mann RS, Gill RS, Dhawan AK and Shera PS (2010) Relative abundance and damage by target and non-target insects on Bollgard and Bollgard II cotton cultivars. *Crop Protection* **29:** 793-801.

Nagrare VS, Kranthi S, Biradar VK, Zade NN, Sangode V, Kakde G, Shukla RM, Shivare D, Khadi BM and Kranthi KR (2009) Widespread infestation of the exotic mealybug species, *Phenacoccus solenopsis* (Tinsley) (Hemiptera: Pseudococcidae), on cotton in India. *Bulletin of Entomological Research* **99**: 537-41.

Naik G (2001) An analysis of socio-economic impact of Bt technology on Indian cotton farmers. Centre for Management in Agriculture, IIMA, India.

Qaim M (2006) Adoption of Bt cotton and impact variability: insights from India, *Review of Agricultural Economics* **28**: 48-58.

Qaim M and Zilberman D (2003) Yield effects of genetically modified crops in developing countries. *Science* **299**: 900-02.

Qaim M, Subramanian A and Sadashivappa P (2009) Commercialized GM crops and yield, Correspondence, *Nature Biotechnology* **27**: 803-04

Qayum A and Sakkhari K (2005) *Bt cotton in Andhra Pradesh- A three-year assessment.* Deccan Development Society, A.P.

Ramgopal N (2006) *Economics of Bt cotton vis-à-vis Traditional cotton varieties.* Agro-Economic Research Center, Andhra University, A.P.

Rathore P, Gill J S, Shera PS, Jindal V and Aggarwal N (2008) Bt cotton in Punjab: Problems and Prospects. In: *Cotton Research in Punjab.* Department of Plant Breeding, Genetics & Biotechnology, Punjab Agricultural University, Ludhiana. pp. 36-43

Sadashivappa P. and Qaim M (2009) Bt Cotton in India: Development of Benefits and the Role of Government Seed Price Interventions. *AgBio Forum* **12**: 1-12.

Sarode SV, Kohle AV and Sable VR (2009) IPM strategies for cotton in relation to climate change. In: Ramamurthy, VV; Gupta, GP and Puri, SN (ed.) *Proceedings. National Symposium on IPM strategies to Combat Emerging pests in the Current Scenario of Climate Change.* pp 8-13, College of Horticulture and Forestry, Central Agricultural University, Pasighat, Arunachal Pradesh.

Sharma HC and Pampapathy G (2006) Influence of transgenic cotton on the relative abundance and damage by target and non-target insect pests under different protection regimes in India. *Current Science* **25**: 800-13

Shera PS, Arora R and Dhawan AK (2010) Extent of avoidable losses due to sucking pests in transgenic cotton (Abstract). In: Verma, A.K.; Gupta, P.R.; Bhardwaj, S.P. and Sharma, R (eds) "National Symposium on Perspectives and Challenges of Integrated Pest Management for Sustainable Agriculture", November 19-21, 2010 held at Dr. Y S Parmar University of

Horticulture & Forestry, Nauni, Solan, Himachal Pradesh 173 230, India. p 5-6.

Shera PS, Dhawan AK, Sekhon PS, Gill MS, Sohu RS and Gill BS (2010) Evaluation of Bt transgenic cotton genotypes for pest infestation and yield performance. Presented at 3rd Conference Biopesticides in Food and Environment Security (BET 2010), October 20-22, 2010. Hisar, India.

Subramanian A and Qaim M (2009) Village-wide Effects of Agricultural Biotechnology: The Case of Bt Cotton in India. *World Development* **37**: 256–267.

Subramanian A, and Qaim, M (2010) The impact of Bt cotton on poor households in rural India. *Journal of Development Studies* **46**(2): 295-311.

Sundramurthy and Chitra (1992) Integrated Pest management in cotton. *Indian Journal of Plant Protection* **20**: 1-17.

Surulivelu T, Dhaeajhothi B and Manjula TR (2009) Relative occurrence of leafhopper and assessment of yield loss due to sucking pests in the popular hybrids of Bt cotton grown in South India. National Symposium on "Bt cotton: Opportunities and Prospects" November 17-19, 2009. Central Institute for Cotton Research, Nagpur. pp 99-100.

Tayade AS, Gupta RR, Wasnik SM, Patil SS, Singh G, Galkate UV, Chauhan SN, Deulkar PB, Kubde RM and Kranthi KR (2009) Impact of IRM strategies on productivity, economic viability, equitability and sustainability of Bt cotton. National Symposium on "Bt cotton: Opportunities and Prospects" November 17-19, 2009. Central Institute for Cotton Research, Nagpur. p 101.

Udikeri SS, Patil BV, Khadi BM, Vamadevaiah HM, Basavangoud K and Kulkarni K A (2007) Performance of new generation Bt cotton hybrids in Indian rainfed ecosystem. Available at http://wcrc.confex.com /wcrc/2007 /tecprogram/ P1697.htm

Udikeri SS, Patil SB, Nadaf AM and Khadi BM (2003) Performance of Bt cotton genotypes under unprotected conditions. In: Swanepoel, A. (ed.) *Proceedings World Cotton Research Conference–3.* Agricultural Research Council, IIC, Cape Town, South Africa. pp 1282-1283,

Udikeri SS, Ventateshalu, Hanumanthaswami BC, Ydahalli KB, Chougala, Venkatesh H, Matti PV, Shaila HM, Patil SB, Nagaraj MV, Shahsikumar S and Khadi B M (2009) Adoption of insect pest resistance management strategies in *Bt* cotton in Karnataka. *National Symposium on "Bt cotton: Opportunities and Prospects"* November 17-19, 2009. Central Institute for Cotton Research, Nagpur. p 124.

Vaidya A (2005) Important pointers to future of Bt cotton. *The Times of India,* Pune. 6 June 2005.

Vennila S (2008) Pest management for cotton ecosystem or ecosystem management for cotton production. *Current Science* **94**: 1351-1352.

Vennila S, Biradar VK, Gadpayle JG, Panchbhai PR, Ramteke MS, Deole SA and Karanjkar PP (2004) Field evaluation of Bt transgenic cotton hybrids against sucking pests and bollworms. *Indian Journal of Plant Protection* **32**: 1-10.

Virk JS, Brar KS and Sohi AS (2004) Impact of insecticides on the resurgence of jassid and whitefly in cotton. *Indian Journal of Entomology* **66**: 319-22.

Biological and Molecular Approaches in Pest Management
S S Gosal, Ramesh Arora & Balwinder Singh (Eds)
Scientific Publishers (India), Jodhpur, 2015, pp 228-248.

Chapter - 10

A Status Update on the use of Biotechnological Techniques for Combating Insect Pests of Fruit Crops

Krishan Kumar[1], P K Arora[1] and Kuldeep Singh[2]
[1] *PAU Regional Research Station, Abohar-152116*
[2] *School of Agricultural Biotechnology, Punjab Agricultural University, Ludhiana-141004, India*
E-mail: kkshorti@pau.edu

10.1. INTRODUCTION

Fruits are known as protective foods because of their richness in vitamins, mineral and antioxidants and their daily consumption protects mankind from various kinds of diseases. At global level, the current fruit production is 646.75 million metric tons from an area of 57.26 million ha. China, India and Brazil are the three leading fruit growing countries in terms of area and production (Anonymous 2013). The global population is projected to rise to 9.0 billion by 2050 from the existing count of 7.1 billion (2012) and consequently, a parallel increase in the global fruit production will also be required to feed this population. The FAO report projects that the fruit production will have to be increased by 33 per cent from the production level of 2005-07 for meeting the demand of world population by 2050 (Linehan *et al.* 2012). This could be achieved by bringing additional area into cultivation under fruit crops and managing various kinds of losses associated with biotic and abiotic stresses. There are various biotic and abiotic stresses which cause reduction in the yield of fruit crops. In the category of biotic stresses, apart from diseases, insect pests cause heavy yield losses. As per 1996 estimates, insects cause 6 per cent fruit crop losses despite the use of insecticides and in the absence of insecticide protection, these losses reach 23 per cent (Krattiger 1997). The insects besides causing direct reduction in the yield of fruit crops (by causing a loss to different parts of the fruit trees namely foliage, twigs, flowers and fruits), also work as carriers for different disease caus-ing pathogens. A huge amount of money is invested on insecticides and the excessive use of pesticides becomes hazardous not only for the consumer but also for the environment.

Breeding insect pest resistant fruit varieties is the only environment friendly and permanent solution to overcome the losses caused by insect pests. However, long juvenile phase, pre and post pollination barriers and polyembryony are some of the hurdles that limit the success of conventional breeding in developing insect pest resistant genotypes in fruit crops. The situation becomes even more complicated for the fruit breeders, when the genes for conferring resistance to insect pests are not available in the primary gene pool and from the secondary gene pool, these are difficult to transfer alone without the supplementation of other undesirable traits.

The biotechnological approaches like marker assisted breeding, genetic transformation and RNA interference offer solutions for these limitations of conventional breeding. Marker assisted breeding can expedite the pace of varietal development, genetic transformation can bring improvement in the single trait of the existing elite variety and RNA interference can be used in insect pest management by impairing the function of vital genes of insects metabolism. These techniques with their experimental or commercial application in fruit crops are described hereunder.

10.2. MARKER ASSISTED BREEDING

It is an indirect selection approach, in which individuals with desirable trait(s) are selected on the basis of the linkage of gene of interest with the molecular marker. Marker aided breeding holds great promise for developing insect pest resistant varieties. The availability of tightly linked markers to the gene/ quantitative trait locus (QTL) conferring resistance to a target insect pest can expedite the pace of varietal development by helping in early and efficient screening of the individuals. The susceptible genotypes can be discarded at seedling stage itself and only the resistant types can be carried forward and thereby, improves cost, space and time efficiency. Moreover, in the artificial infestation based screening methods, certain susceptible individuals could even escape the pest attack due to uneven build up of the insect population and could be recorded as resistant types. Such individuals can readily be identified with the aid of DNA markers and can be rogued off at initial stages. The pre-requisites for the application of marker assisted selection are linkage map, mapping of the gene/QTLs of interest on it and high resolution mapping of the gene containing region.

10.2.1. Linkage Map

Linkage map is the assembly of different linkage groups. Molecular markers on these linkage groups are arranged according to their relative genetic distances. The production of linkage map is based on the principle that genes (markers/loci) segregate via chromosome recombination during meiosis. During this process, the markers located on the different chromosomes (unlinked) assort independently while the markers present on the same chromosome tend to remain linked and gets arranged on the same linkage group. As during meiosis, two non sister chromatids are involved in strand exchange, recombination does take place between the DNA segments. The chance of a cross over producing recombination between genes/ markers is related to the genetic distance between them. Closer they are situated, lesser is the chance of cross over between them (Semagn *et al.* 2006). The steps involved in the development of linkage map are discussed below.

a. Development of mapping population: For developing mapping popula-tions, the contrasting genotypes *i.e.* pest resistant and susceptible are chosen. The fruit crops are highly heterozygous in nature, have long juvenile phase. Due to these factors, it is very difficult to develop ideal mapping populations viz., recomb-inant inbred lines (RILs), doubled haploids (DH), near isogenic lines (NILs) in these crops. In these crops, F_1 populations, where marker data can be analysed in two way pseudo test cross manner; (Rosa *et al.* 2003; Mehlenbacher *et al.* 2006; Gisbert *et al.* 2009; Gulsen *et al.* 2010), half or full sib derived populations (Brennan *et al.* 2008) and in some cases F_2 (Dillon *et al.* 2006; Sargent *et al.* 2006; Blas *et al.* 2009) populations have been used for mapping purpose (Table 1). The size of the population may vary from 50 to 250 for initial map development, but for high resolution mapping, larger sized populations are required.

b. Identification of polymorphic markers between the parents and genotyping of the population with polymorphic markers: Available DNA markers are screened against the genomes of the parents to find out the polymorphic ones (that can differentiate the two parents). Generally, different authors have used different markers viz., RFLPD, RAPD, AFLP, SSRs, SNPs etc. for the development of linkage map in fruit crops. The markers should have higher information content, co-dominant inheritance, reprodu-cibility, easy transferability and ease of automation. SSRs represent all these advantages (Gupta *et al.* 1999). The polymorphic markers are screened against the individuals of the population, and data is recorded according to its similarity to the banding pattern of the parents.

Table 1. Fruit crops with available molecular maps

Fruit crop	Parents/ populations used	Marker type and Number	Total Map length (cM) & linkage groups (LG)	Reference
(1)	(2)	(3)	(4)	(5)
Straw-berry	*Fragaria vesca* x *F. nubicola* derived F_2 population	175 SSRs, 6 gene specific and 1 SCAR marker	424.3cM, 7 LG	Sargent *et al.* (2006)
Sweet cherry	56 Microspore callus culture individuals of Emperor Francis	2 allozymes, 90 RAPDs	503.3cM, 10 LG	Stockinjer *et al.* (1996)
Apricot	81 F_1 individuals of cross 'Goldrich' x 'Valenciano'	AFLP, RAPD, RFLP and SSR markers	Goldrich-(511 cM) Valenciano-(467.2cM)	Hurtado *et al.* (2002)
	F_1 progeny of 'Stark Early Orange' x 'Tyrinthos'	180 AFLPs, 29 SSRs	602 cM	Vilanova *et al.* (2003)

Contd...

(1)	(2)	(3)	(4)	(5)
Guava	Three mapping populations	AFLPs and SSRs	1379 cM, 11 LG	Rodriguez *et al.* (2007)
European Hazelnut	'OSU252.146' x 'OSU 414.062' F_1 progeny	249 RAPDs and 20 SSRs	OSU252.146 - 661cM OSU 414.062 - 812 cM and 11 LG	Mehlenbacher *et al.* (2006)
Papaya	F_2 progeny of interspecific cross 'AU9' x 'Sun Up'	712 SSRs, 21 AFLP and 1 morphological marker	Major LG : 9, Minor LG : 5	Blas *et al.* (2009)
Grapes	'Syrah' x 'Pinot Noir'	483 SNPs, 132 SSRs and 379 AFLP markers	1245 cM, 19 LG	Troggio *et al.* (2007)
	'Syrah' x 'Pinot Noir', 'Syrah' x 'Grenache', 'Cabernet Sauvignon' x 'Riesling'	1134 markers (350 AFLP, 501 SNPs, 283 SSRs)	1443 cM, 19 LG	Vezzulli *et al.* (2008)
Apple	'Fiesta' x 'Discovery' progeny	475AFLPs, 235 RAPDs, 129 SSRs and 1 SCAR	1371 cM, 17 LG	Liebhard *et al.* (2003)
Prunus	Almond cv. 'Texas' x Peach cv. 'Earlygold'	235 RFLPs, 11 isozymes and 96 SSRs	522 cM, 8 LG	Joobeur *et al.* (1998); Aranzana *et al.* (2002)
	Prunus persicae x *P. ferganensis*) x *P. persica* BC_1 progeny	78 RFLPs, 63 AFLPs, 57 SSRs, 16 RAPDs and two morphological markers	665 cM LG : 8	Verde *et al.* (2005)
Olive	F_1 progeny of 'Leccino' x 'Dolce Agogia'	61 RAPDs, 21 AFLPs, 8SSRs and 4 RFLPs	Lecino: 2765 cM (22 major and 17 minor LG) Dolce Agogia:2445 cM (27 major & 3 minor LG)	Rosa *et al.* (2003)
Mango	Progeny of 'Keitt' x 'Tommy-Atkins'	34 AFLP markers	161.5 cM, 13 LG	Kashkush *et al.* (2001)
Litchi	'Maguili' x 'Jiaohesanyuehong' F_1 population	312 (169 AFLPs and 143 RAPDs) markers	1040.43 cM, 16 LG	Liu *et al.* (2008)
Loquat	'Algerie' x 'Zaozhong-6' Progeny	SSRs and AFLPs	Algerie -900cM, 17 LG, Zaozhong-6 - 870cM	Gisbert *et al.* (2009)

Contd...

(1)	(2)	(3)	(4)	(5)
Citrus	F_1 progeny of 'Clementine mandarin' x 'Orlando tangelo'	609 [(385 SRAPb, 97 RAPD, 95 SSR, 18 ISSR peroxidase gene polymorphism (POGP) and 2 resistance gene analog (RGA)] markers	Clementine- 760cM, 9 LG Orlando tangelo- 740 cM, 9 LG	Gulsen *et al.* (2010)
Black Currant	Full sibling progeny between lines (SCRIS36/1/100) x (EMRS B1834)	AFLP, SSR (both genomic and EST based) and SNP markers	---	Brennan *et al.* (2008)

c. Linkage analysis and map construction: Linkage analysis between the markers is generally calculated by computer programs such as MAPMAKER (Lander *et al.* 1987), JoinMap (Stam 1993), GMENDEL (Echt *et al.,* 1992) etc. The genetic distance between them is calculated on the basis of mapping functions such as Haldane and Kosambi.

The linkage maps based on molecular markers are available in most of the fruit crops except sapota, pomegranate and ber. Highly saturated maps are available in fruit crops such as apple, grapes, *Citrus*, *Prunus* etc. but, the linkage map in commercial crops such as mango are still in preliminary phase (Kashkush *et al.* 2001).

10.2.2. Gene/QTL Mapping

After the development of linkage map, the next step is to locate the gene / quantitative trait loci (QTLs) controlling traits of importance on the either linkage group(s) of the map. For this, the correlation between the genotypic (marker data) and phenotypic data is established using software packages like Map Manager QTX (for single marker analysis, Manly *et al.* 2001), Map Maker/QTL [for Simple Interval Mapping (SIM), Lincoln *et al.* 1993] or QTL Cartographer [for Composite Interval Mapping (CIM), Basten *et al.* 2001]. Mapping gives information regarding the position of the gene controlling resistance to particular insect pest/ disease or other trait. If the trait is controlled by a major gene, it could be present at any of the linkage group and if, it is quantitative in nature, then various QTLs governing it would be present either on different linkage groups or at different positions of the same linkage group of the linkage map. Genes/ QTLs conferring resistance to certain insect pests have been mapped on the linkage map of few fruit crops (Table 2). Among these, the genes for aphid and rust mite resistance in apple and gall mite resistance in black currant are the major genes while the resistance for leaf miner in citrus is under the control of polygenes.

10.2.3. High Resolution Mapping

Once the linkage between a gene/QTL with the marker is established, the next step is to fine map the gene containing region with additional markers as the genetic distance (cM) on the linkage map does not reflect the true physical distance in the genome. But, it is quite difficult to delimit the gene containing region with the small population size.

Table 2. Insect pests resistant genes/QTLs mapped with molecular markers in fruit crops

Fruit crop	Pest	Population used	Mapped with or between marker(s) on linkage group	Reference
Apple	Aphid (*Dysaphis devecata*)	'Prima' (aphid susceptible) x 'Fiesta' (aphid resistant) F_1 progeny	*Sd1* gene between a SCAR marker 2B12a (0.4 cM) and SSR marker SdSSRa (0.9 cM) on LG7	Cevik and King (2002)
Apple	Rust mite (*Aculus schlechtendali*)	'Fiesta' x 'Discovery' F_1 progeny	The AFLP marker E35M42-0146 (20.2 cM) and the RAPD marker AE10-400 (45.8 cM) on LG7 of 'Fiesta'	Stoeckli *et al.* (2009)
Black currant	Gall mite (*Cecidophyopsis ribis*)	Full sibling progeny between gall mite susceptible (SCRIS36/1/100) X gall mite resistant (EMRS B1834) lines	*gmr* gene at a distance of 4.0 cM from AFLP marker E41M88-280	Brennan *et al.* (2008a)
Citrus	Leaf miner (*Phyllocnistis citrella*)	Maps: *Poncirus trifoliata* (Pa) based on 63 markers, *Citrus aurantium* based on 157 markers	- One antibiosis QTL with marker CR7 on LG 7 of Pa map - another antibiosis QTL with marker S2-AS4. 800 on - Sour orange linkage map Six antixenosis QTLs also mapped	Bernet *et al.* (2005)

To obtain the new linked markers within the previously mapped region, the size of the population is either increased or other additional progenies sharing the same resistant parent in their pedigree are screened to find out any possible recombinant between the gene and previously linked markers. This strategy has been followed for fine mapping the *Sd1* gene conferring resistance to biotype 1 and 2 of aphid in apple (Cevik and King 2002).

For increasing the density of the markers in the vicinity of the gene containing region, most of the researchers have used AFLP markers, as these are easy to construct and no prior sequence information is required for their development. Once the tightly linked AFLPs are found, they are converted into PCR amenable SCAR (Sequence characterized amplified region) markers (Brennan *et al.* 2008a).

The marker(s) showing linkage with the gene of trait are suitable for marker assisted selection, if they either co-segregate or show close linkage (at less than 1cM ideally) with it. Ideally, they should be PCR amenable and their results should be reproducible across the laboratories.

Presently, marker assisted selection is being followed for imparting resistance against dagger nematode (*Xiphinema index*), powdery mildew and pierce diseases in grapes (Xu *et al.*, 2008) and for pyramiding scab resistance genes (*Vf, Vr, Vb, Va, Vm* and *Vbj*) in cultivated apples (Gygax *et al.*, 2004). The markers can be used in selection of the genotypes resistant to insect pests efficiently (Brennan *et al.* 2008), nonetheless, so far, it has not substituted the phenotype based screening in fruit crops.

10.3. TRANSGENICS

The technology of introduction of gene(s) in the host genome is called genetic transformation. The process of introduction, integration and expression of trans (foreign) gene(s) in the host is called trans-genesis and so raised fruit crops are called transgenics. The first transgenic was produced in tobacco (1983), closely followed by a fruit crop (papaya), which was later commercialized in 1998. The genetic transformation requires gene construct, a method of gene introduction and selection, regeneration and testing of transgenic plants.

10.3.1. Gene Construct

Gene construct is made up of gene of interest (trans gene) and the selectable marker gene. Both the two mentioned genes have their own promoters, coding regions and terminator sequences as in normal genes.

a. Gene of interest

The transformation of various fruit crops with genes, imparting resistance to insect pest has been described in Table 3. The fruit crops have been transformed with numerous genes encoding insecticidal proteins of bacterium (McGranahan *et al.* 1988; Dandekar *et al.* 1992, 1994; James *et al.*, 1993); plants origin (Graham *et al.*, 1997; Yang *et al.*, 2000) and synthetic reconstructs (Tao *et al.*, 1997) to effectively control the target insects.

As it is evident from the table 3, the genes of the bacterium *Bacillus thuringiensis* has commonly been employed for generation of insect pest resistant transgenic fruit crops. The insecticidal property of bacterium *Bacillus thuringiensis* lies in its Cry and Cyt series of toxins. These two types of toxic proteins have selective preference for insects of different orders. Cry proteins are effective against the insects of Lepidoptera, Coleoptera, Hymenoptera and Diptera orders, while Cyt toxins are useful against dipteran insects. Both these two series of toxins are pore forming proteins and their primary action is to lyse midgut epithelial cells by inserting into the target membranes. The crystal inclusions of Cry toxins after its ingestion by susceptible larvae gets dissolved in the alkaline environment of the gut, and the so solubilized but inactive protoxins upon cleavage by midgut proteases yields 60–70 kDa protease resistant active proteins (Bravo *et al.* 2007). The activated toxin then binds to specific receptors on the brush border membrane of the midgut epithelium columnar cells (de Maagd *et al.* 2001; Bravo *et al.* 2007) and finally gets inserted into the membrane. Toxin insertion leads to the formation of lytic pores in microvilli of apical membranes (Aronson and Shai 2001; Bravo *et al.* 2007). This causes cell lysis and the released cell contents provides spores a germinating medium leading to a severe septicemia and insect death (de Maagd *et al.* 2001; Bravo *et al.* 2007).

Table 3. Fruit crops transformed with insect pest resistance genes

Fruit crop	Insect	Transgene	Effect of the expression of the trans gene	Reference
Apple	Coddling moth (*Cydia pomonella*)	*Cry1Ac* and *ICP*	Low level of expre-ssion of the target gene	Dandekar *et al.* (1992); James *et al.* (1993)
Craneberry	Blackheaded fireworm (*Rhopobota naevana*)	*Btk-ICP* from *B. thuringenesis* var kurstaki	Non effective control of the insect during bioassays	Serres *et al.* (1992)
Grapefruit	Aphid	*Gna*	-	Yang *et al.* (2000)
Juneberry	-	*Bacillus thuringiensis* var. *kurstaki* gene encoding for toxin I-ID73	-	Hajela *et al.* (1993)
Persimon	Oriental moth (*Monema flavescens*)	Synthetic *cry1Ac*	Significant mortality of insect in the bioassays	Tao *et al.* (1997)
Strawberry	Vine weevil (*Otiorhynchus sulcatus*)	*CpTI*	-	Graham *et al.* (1997)
Walnut	Coddling moth (*Laspeyresia pomonella*)	*Cry1Ac*	Increased level of larval mortality	McGranahan *et al.* (1988); Dandekar *et al.* (1994)

The Cry proteins comprise 50 subgroups with more than 200 members. Boncheva *et al.* (2006) evaluated toxicity potential of 9 different Cry δ endotoxins (Cry1Aa, Cry1Ab, Cry1Ac, Cry1Ba, Cry1Ca, Cry1Cb, Cry1Da, Cry1Fa, and Cry1Ia), and a hybrid protein SN19 against the larvae of coddling moth (*Cydia pomonella*), a lepidopteran insect. The Cry1Da was found to be significantly more toxic than the other tested prototoxins and was followed by Cry1Ab, Cry1Ba, and Cry1Ac, which in turn were significantly more active then Cry1Aa, Cry1Fa, Cry1Ia, and SN19.

The Cyt toxins also belong to the category of pore forming proteins but they are slightly different from the Cry toxins structurally. Unlike Cry toxins, which first bind to the receptors, the Cyt toxins directly get inserted in the membrane.

Among the genes of plant origin, *gna* (galanthus nivalis agglutinin) from snowdrop lectin (*Galanthus nivalis*) and cowpea trypsin inhibitor protein from cowpea, have been used in fruit crops. The *gna* gene is effective against the homopteran insects like aphids, against whom, the *cry* genes does not provide full control. The *gna* gene controls aphids by producing *lectin* protein, which in turn binds in the gut and causes surface lesions (Eisemann *et al.* 1994) and by virtue of this, induces mortality in the aphids. On the other hand, the gene *CpTI* inhibits the function of enzyme Trypsin (a protease) in the insects, which is required for food digestion in the insects. Consequently, due to breakage in the supply of essential amino acids, insect death occurs (Ismail *et al.* 2010).

b. Selectable marker genes

The selectable marker gene provides a competitive advantage to the transformed cells and selectively promotes their growth over the non transformed cells in the regeneration medium containing the selective agent. Various commonly used selectable agents in fruit crops for genetic transformation are mentioned in Table 4.

Table 4. Commonly used selectable marker genes along with their selective agent

Selectable marker gene	Substrate used for selection
Neomycin phospho-transferase (*nptII*)	Kanamycin, neomycin
Hygromycin phosph-otransferase (*hptII*)	Hygromycin B
Gentamycin acetyl transferase	Gentamycin
Streptomycin phospho-transferase	Streptomycin
Phosphinothricin acetyl transferase (*bar*)	L-phosphinothricin (PPT)
Phosphomannose isomerase (*manA*)	Mannose

Of the above mentioned selectable marker genes, *nptII* gene coding for resistance to antibiotic Kanamycin has been used mostly for the initial selection of the putative transgenics. But, the long presence of the antibiotic genes in the transgenics has raised concerns in the commercialization of these crops due to the potential risks associated with the transfer of these genes to environment or medically related bacteria or from transgenic plant product as food to intestinal micro-organisms (Darbani *et al.* 2007). Due to these speculated risks with the antibiotic resistance genes, the use of alternative selectable markers or marker free transgenics programme has been initiated by the researchers (Upadhyaya *et al.* 2010).

The alternative marker genes which do not have toxic effects on the environment and human health like *manA* gene from *E. coli and daoI* gene from yeast *Rhodotorula gracilis* have been utilized in fruit crops. The *manA* gene codes for phospho mannose isomerase (PMI) enzyme that helps in utilization of mannose in transformed cells by converting mannose-6-phosphate into fructose-6-phosphate, while the non transformed cells which are unable to utilize mannose-6-phosphate, starve off in the absence of suitable carbohydrate. The *manA* as a selectable marker gene has been used in papaya (Zhu *et al.* 2005), apple (Degenhardt *et al.* 2006) and citrus (Ballester *et al.* 2008) with good transformation efficiency.

The gene *dao1* encodes the enzyme D-amino acid oxidase which catalyzes the oxidative deamination of toxic D- amino acids viz., D-serine and D-alanine into non toxic compounds. It has been utilized as the selectable marker gene in apple cultivars (Hattasch *et al.* 2009). But, due to low transformation efficiency, it did not seem to be a suitable selectable marker.

Besides this, there are genes, which code for highly sensitive enzymes, which upon reaction with the substrates produce a visible colour. The *gfp* (from jelly fish) and *gus* (from E-coli) genes are the two most commonly used visual markers, which gives green and blue colour, respectively. The *gfp* compared to *gus* is more advantageous as it does not need the substrate for its expression. With the availability of *egfp* (mutant *gfp* with enhanced expression), it has been used as visual marker in several fruit crops e.g. citrus, papaya, apple etc.

In some of the cases, it has been found that the new selectable markers have not been as much effective as was kanamycin (Hattasch *et al.* 2009). Moreover, the use of new marker genes requires changes/ alterations in the already standardized transformation protocols. After differentiating the transformed cells from the non transformed one, the selectable marker is of no further use to the plant cell; instead it is a potential risk. So, now a days, stress is laid on the production of marker free transgenics, which involves selection of the putative trasformants with the help of selectable markers initially followed by their removal from the plant system. Marker free transgenics have been obtained in both highly efficient e.g. strawberry and relatively recalcitrant system e.g. apple (Schaart *et al.* 2004)

10.3.2. Methods of Gene Introduction

Out of the various methods *Agrobacterium* and particle bombardment are the methods of choice for the gene introduction or transfer. *Agrobacteium,* if can stably transfer the genes in single copy in the euchromatin region, the particle bombardment has the advantage of transferring more than one gene at a single time. Although the commercialized transgenic fruit tree (papaya) was produced with the aid of particle bombardment but most of the researchers have relied upon *Agrobacterium* for gene transfer in fruit crops. In Citrus, supplementation of the existing Agrobacterium method with sonication and vaccum infiltration has increased the transformation efficiency (Oliveira *et al.* 2009).

10.3.3. Selection, Regeneration and Testing of Transgenics

The pre-requisite for the transformation of fruit crops is the reliable tissue culture regeneration protocol. The plants regenerated in medium containing marker genes (as described in section 3.1b) are further verified for their transgenic status by PCR and western blotting techniques. Once the transgene has been confirmed in the plants, these have to pass through glasshouse screening, confined and open field trials with the permission from the regulating agency of the particular country before attaining the commercial status.

Among fruit crops, papaya varieties namely 'Rainbow' and 'Sun Up' resistant to papaya ring spot virus, has only been commercialized so far. Another transgenic 'Honey Sweet' plum, which has attained the deregulated status from Animal and Plant Health Inspection Service of US department of Agriculture in 2007 (Anonymous 2007) is also a virus resistant variety of plum. Any of the transgenic fruit variety showing resistance to insect pests has not been commercialized till date.

In a survey in USA, it has been revealed that if the native genes (with original promoters, coding sequences, introns and terminator in the same sense) of sexually crossable species are used for transformation (*cisgenesis*), their acceptance could be increased. Attempts are being made for the production of cisgenic scab resistant apple at Wageningen University, Netherlands (Jacobsen and Schouten 2007). In case of insect pests, transgenic trap crop strategy is suggested

(Dandekar *et al.* 2002). In this, insect pest resistant transgenic plants of variety or crop are grown together with commercial variety of the same or different crop. The insect pest after feeding on the transgenic plants of trap variety / crop gets killed and the main crop is saved from the insect damage. Here, the trap crop is primarily meant for protection of crop (variety) of interest from insect pest and not for the commercial income and thereby, it addresses the concerns of the consumers as well. This strategy has been followed for walnut where menace of coddling moth is controlled by using transgenic plants of apple as the trap crop (Dandekar *et al.* 2002). The coddling moth is a serious pest of apples, pears, walnuts and many other temperate fruits.

10.4. RNA INTERFERENCE

It is an homology dependent gene silencing process, where the short ds RNA (siRNA/ miRNA) mediates in the reduction or complete suppression of the expression of target gene. The phenomenon was first observed by Napoli *et al.* (1990) in their genetic transformation experiments of petunia (an annual flowering herb). However, the molecular basis of this mechanism was revealed eight years later by the study of Fire *et al.* (1998) on the nematode *Caenorhabditis elegans* and they termed this phenomenon as RNA interference (RNAi).

10.4.1. Mechanism of RNAi

The dsRNA is the trigger biomolecule for the initiation of RNAi response. The trigger molecule could be either the perfectly complementary hairpin like dsRNA structure or the primary - miRNA transcript (pri-miRNA).

These biomolecules are either introduced in the host system externally or produced endogenously by the host enzymes. These endogenous or introduced precursor molecules in the cell are recognized by the processor enzyme called Dicer (Ribonuclease type III enzyme), which cleaves them into small RNA molecules of 21-25 long nucleotide fragments with 2-3 base pair hangs at 3' end (Zamore *et al.* 2000).

After the genesis of the small RNA molecules (siRNAs / miRNAs), the process of RNAi is carried forward by the RNA induced silencing complex (RISC), which is a complex ribonucleoprotien. It has different subunits of which helicase (Stevenson 2004), argonaute, a multidomain protein having RNase H like activity (Elbashir *et al.* 2001) have so far been documented.

Now depending upon the small RNA molecule (siRNA /miRNA) and the type of organism, the gene silencing could be achieved in different ways. The short interfering RNA molecules (siRNA) cause silencing of the target gene by either degrading the mRNA transcript (Molesini *et al.* 2012) or inhibiting the transcription through methylation of the promoter region of the gene. The miRNA, on the other hand induce gene silencing by blocking translation.

10.4.2. Status and Factors Affecting the Success of RNAi

The RNAi technique has been used in wide range of areas including manipulation of the plant architecture, modulation of flower colour, development of male sterile plants, induction of seedlessness, enriching the nutritional qualities of food, removal of allergens and toxic compounds from food, engineering the plants for a specific secondary metabolite by interfering at a particular stage of biosynthesis, enhancing shelf life of the perishables such as vegetables and inducing tolerance in plants against various kinds of biotic and abiotic stresses (For details see: Jagtap *et al.* 2011; Price and Gatehouse 2008; Mansoor *et al.* 2006; Angaji *et al.* 2010). It effectively down regulates the expression of the target gene(s). In this section, the experimental progress of RNAi against the insect pests of fruit crop has been described (Table 5). The technique has been tested for a range of genes by employing different inducer molecules (dsRNA/siRNA) with different modes of introduction. So far, it has been tried against light brown apple moth, fruit fly, citrus psylla, and glassy winged sharpshooter insects of fruit crops.

Table 5: Examples of RNAi use against insects feeding on fruit crops

Insect pests	Host Fruit crop	Type of inducer molecule	Target gene (s)	Method of dsRNA introduction	RNAi response	References
Light brown apple moth (*Epiphyas postvittana*)	Apple	dsRNA	Carboxyesterase	Larval feeding	Effect persistent up to adult stage	Turner *at al.* (2006)
Oriental fruit fly (*Bactrocera dorsalis*)	Guava, Mango, Ber, Citrus etc.	dsRNA	Female specific double sex (*Bddsx*) gene	Abdominal micro-injection	-Reduced expression of *Bddsx* and Bdyp1genes -Delayed ovary development and reduced number of mature eggs - 27% of female progeny had deformed ovipositor	Chen *et al.* (2007)
			*rpl*19, *V-ATPase* 13 *noa* and *rab11*	Direct feeding and by feeding bacteria expressing these genes	- Direct feeding was more responsive than that of bacterial feeding for RNAi -Silencing of rab11 killed 20% of adult flies. Egg production was affected by dsRNA of *noa* and *rab11* genes.	Li *et al.* (2011)
Citrus psylla (*Diaphorina citri*)	Citrus	dsRNA	-	Through irrigation water	-	Hunter *et al.* (2010)
Glassy winged sharpshooter (*Homalodisaca vitripennis*)	Grapes, almond and citrus	dsRNA	*actin*	Tansfection	10 fold decrease in the mRNA of *actin* genes and altered phenotypes	Rosa *et al.* (2010)

Various factors which determine the success of this potential technique in insect pests of fruit crops are discussed below.

10.4.2.1. Target gene

The researchers have observed RNAi effects in insect pests by targeting single gene (Turner *et al.* 2006; Chen *et al.* 2007) or more than one gene (Borgio 2010; Rosa *et al.* 2010; Li *et al.* 2011).

Turner *et al.* (2006) fed the light brown moth of apple with the dsRNA of carboxyesterase gene for observing RNAi. Fruit fly (*Bactrocera* spp.) is a versatile insect pest and causes substantial damage to the fruits of many crops like guava, mango, citrus and ber. The introduction of dsRNA for female specific *Bddsx* in it caused down regulation of *Bddsx* transcript, inhibited the selective expression of yolk protein gene (*bdyp*), delayed the ovary development, reduced the egg laying capacity of the insect and the female progeny with RNAi had deformed ovipositors (Chen *et al.* 2007). Similarly, Li *et al.* (2011) tested the RNAi potential of dsRNA of four different group of genes (genes encoding for ribosomal protein Rpl19, V type ATPase D subunit, the fatty acid elongase Noa and a small GTPase Rab11) against adults of fruit fly by feeding on dsRNA solutions or on bacteria expressing dsRNA. Although a relative reduction in the expression of these genes was observed but silencing of rab11 dsRNA caused mortality of 20% of adult flies. Egg production was also affected with the ingestion of dsRNA of Noa and Rab11genes.

Borgio (2010) tested RNAi potential of *in vitro* synthesized dsRNA of five unannotated genes (GENE1, GENE2, GENE3, GENE4 and GENE5) against adults of whitefly (Homoptera). Out of these five genes, GENE2 and GENE3 caused more than 85% mortality. Rosa *et al.* (2010) transfected glassy winged sharpshooter (*Homalodisaca vitripennis*), (a serious pest of grapes, almond and citrus etc.), with dsRNA of its arginine kinase, actin, sar1 or Lian-Aa1 actin genes to study RNAi response. Best response was observed with dsRNA of actin gene.

The findings of Borgio (2010), Rosa *et al.* (2010) and Li *et al.* (2011) suggest that for observing best RNAi response, the target insect should initially be tested against different genes. This could help in finding out the key gene of insect metabolism and silencing or down regulation of which could cause mortality in the insect.

10.4.2.2. Type of inducer RNA molecule

There are mainly three inducer molecules viz., dsRNA, siRNA, miRNA and lately tasiRNA a new potential inducer has been added in this list, are being explored for RNAi.

Of the above described four trigger molecules, only dsRNA and siRNA have been applied against the insect pests affecting fruit crops. The dsRNA has been used against light brown apple moth (Turner *et al.* 2006); fruit fly (Chen *et al.* 2007; Li *et al.* 2011) and glassy winged sharpshooter (Rosa *et al.* 2010) (Table 5). Borgio (2010) used siRNA as an RNAi precursor molecule against sucking pests. There is one report of Upadhyay *et al.* (2011) against whitefly (*Bemisia tabaci*), where efficiency of both types of inducer molecules viz., siRNA and dsRNA for RNAi has been compared. Both of the molecules were found equally effective in down regulation of the studied genes (Upadhyay *et al.* 2011).

The siRNA and dsRNA induce gene silencing in a homology dependent manner. The third inducer molecule (miRNA) relies on its stem loop structure rather than its sequence for causing the silencing of the target gene. The artificial trans acting small RNAs (tasiRNAs) target the genes in *Trans* for silencing the genes. In this inducer molecule, multiple small RNAs can be pyramided in a single construct, for simultaneous down regulation of the expression of many genes. The artificial tasiRNAs hold great promise in the field of insect pest management, as knockdown of the multiple essential genes at a single time could cause mortality in the insects.

10.4.2.3. Methods of introduction

To induce RNAi in insect pests of economic importance of fruit crops, the introduction of inducer molecule has been achieved by different means, which are discussed hereunder.

a. Artificial diet

In this method of bioassays, the insects are fed with the artificial diet containing inducer molecules (dsRNA/ siRNA) of their key genes for observing the RNAi response. The dsRNA against the light brown apple moth (Turner *et al.* 2006) and against oriental fruit fly (Li *et al.* 2011) was fed through the diet to observe RNAi for the target gene. Li *et al.* (2011) tested the efficiency of direct dsRNA feeding and bacteria expressed dsRNA against fruit fly. The direct feeding proved to be more effective in knocking down the expression of the target gene. The diet based assay can become a feasible method for controlling the insect pests, as once RNAi for a particular gene is found effective; the inducer molecule could be made available as insecticide/ pesticide formulation.

b. Microinjection

The inducer molecules are injected in body cavity for facilitation of its circulation in the haemolymph which allows its quick effects to be observed in most of the receiving cells (Price and Gatehouse 2008). It was considered to be used method for inhibiting the target gene expression in insect pests until there was a belief that insects lack RNA systemic RNAi. However, new reports state that the systemic RNAi could be present in the insects, but its mechanism could be different from *C. elegans*. Still, it is a very cumbersome technique and is not feasible for the insects of very small size such as whitefly (Upadhyay *et al.* 2011). In addition, if while injecting, the insects get wounded, the injury could not be differentiated from the RNAi effects. This mode of introduction has been utilized against fruit fly in fruit crops.

c. Genetically modified organisms

In this, transgenic plants or insects for the target gene are produced. In the first approach, the plants are transformed with the target DNA sequences of the insect, which upon transcription yields dsRNA. The insects when feed on such plants, the functioning of their target endogenous gene is impaired. Since in the transgenic plants, the dsRNA is continuously expressed, the insect pests could be controlled effectively. However, there is no as such example of this approach in fruit affecting insect pests. In the second approach, transgenic insect cell lines are produced. Rosa *et al.* (2010) prepared a transgenic cell line of the insect glassy winged sharpshooter.

d. Other methods

Hunter *et al.* (2010) at USDA, ARS have demonstrated that dsRNA can also be taken up by the plants through irrigation water. They irrigated the 2.5 m tall key lime plants with 18.9 litre of water containing 250 ml (10 mg/ml) of dsRNA specific to glossy winged sharpshooter and asian citrus psyllid. The dsRNA was transported in the whole vascular system after four days. This method of introduction of inducer molecules can prove effective for both xylem as well as phloem feeder insects. The glassy winged sharpshooter and citrus psyllid were the xylem and phloem feeder, respectively in their study.

10.4.2.4. Cell autonomous versus systemic RNAi

The RNAi mediated gene silencing varies depending upon the target organism. The RNAi in the literature has been classified broadly as: Cell autonomous RNAi and Systemic RNAi. In the cell autonomous RNAi, the silencing effect of the gene is limited to the cell receiving the dsRNA molecule, whereas in systemic RNAi, the expression of the target gene is not only knocked out in the recepient cell but it is also spread systemically to the neighbouring cells and therefore affects the whole insect.

For the control of insect pests through RNAi, systemic spread of knockdown effect of the target gene in the system of the insect is essential. The initial RNAi studies on *Drosophilla melanogaster*, suggested the absence of systemic RNAi in insects. This was probably due to the lack of gene RdRP (coding for enzyme RNA dependent RNA polymerase) in insects, which was responsible for spreading the siRNA signal in *C. elegans*. It indicated that the RNAi could be used for functional genomics in insects but would have limited merit in its deployment for the control of insect pests. This was the belief until the two independent studies (Tomoyasu and Denell 2004; Bucher *et al.* 2002) on coleopteran insect *Tribolium castaneum*, gave proof of systemic RNAi in insects. However, a recent genome comparision of *C. elegans* and *Tribolium* have revealed that *Tribolium* lacks *C. elegans* like RdRP gene. Therefore, systemic RNAi in *Tribolium* could be due to either the other gene having RdRp like activity or altogether a different mechanism (Tomoyasu *et al.* 2008). Studies of Turner *et al.* (2006) on light brown apple moth, Chen *et al.* (2007) and Li *et al.* (2011) against fruit flies have given evidences for long term effects of RNAi in insects. The effects vary from reduced expression of the target gene to reduction in the egg laying potential of the insect, deformed ovipositor. Moreover, the effects are not limited to the midgut region but are expressed in other parts of the body as well. These results indicate that in insects the RNAi is not limited to a particular cell. Now, it is the question whether the RNAi is systemic in all the insects or it is limited to few insects. If it is systemic in all the insects which could be revealed by future investigations, it will be the effective tool in the management of insect pests and if the systemic RNAi is limited to only few insects, its use in the integrated insect pest management could be limited only for those insects.

10.5. CONCLUSIONS

Biotechnological approaches namely marker aided breeding, transgenics and RNAi hold great potential in reducing the insect pest associated losses in fruit crops. The marker aided selection can expedite the pace of breeding while the other two technologies can directly control the insect pests.

In marker aided breeding, the location of the loci conferring resistance against a particular insect pest and closely linked markers with it, need to be ascertained with initial basic studies. Once the location of the loci and markers tightly linked to it are found, the information can be used in expediting the breeding for resistance against that particular insect pest. Among fruit crops, the loci conferring resistance to aphid and rust mite in apple, gall mite in black currant and leaf miner in citrus has been mapped.

The transgenic technology can prove vital for improvement in a single trait of an otherwise elite variety. The transgenic papayas (Rainbow and SunUp) with resistance to ring spot virus, being cultivated commercially in Hawaii, are the direct evidence for the success of this technology in fruit crops. However, none of the insect resistant fruit crops has come up commercially. The use of alternate marker genes such *manA*, *daoI* and *egfp,* the development of marker free insect resistant transgenics, adoption of *cisgenesis* (use of insect pest resistant genes from the secondary gene pool) in the elite fruit varieties and the growing of transgenics as trap crop as demonstrated in walnut for codling moth, could enhance the acceptability of this technology for control of insect pests.

The RNA interference is being carried out by targeting various key genes in insects. So far, it has been tested for insects namely fruit fly, citrus psylla, light brown apple moth and glassy-winged sharpshooter. The outcome of RNAi is being evaluated experimentally. However, once the RNAi for key gene(s) is established in insect pests, the small RNAs for that particular gene(s) can be made available in the spray formulations or transgenic plants expressing dsRNA specific to target insect pest can be produced to effectively control the insect. In this manner, RNAi can be put into the integrated pest management schedules.

REFERENCES

Angaji SA, Hedayati SS, Poor RH, Poor SS, Shiravi S and Madani S (2010) Application of RNA interference in plants. *Plant Omics Journal* **3(3):** 77-84.

Anonymous (2007) http://www.aphis.usda.gov/ brs/ aphisdocs/ 04-26401p.pdf.

Anonymous (2013) Major fruit producing countries in the world. *Indian Horticulture Database 2013*, p 242.

Aranzana MJ, Pineda A, Cosson P, Dirlewanger E, Ascasibar J, Cipriani G, Ryder CD, Testolin R, Abbott A, King GJ, Iezzoni AF, Arffls P (2002) A set of simple-sequence repeat (SSR) markers covering the *Prunus* genome. *Theoretical and Applied Genetics* **106:** 819-825.

Aronson AI and Shai Y (2001) Why *Bacillus thuringiensis* insecticidal toxins are so effective: unique features of their mode of action. *FEMS Microbiology Letters* **195:**1-8.

Ballester A, Cervera M and Pena L (2008) Evaluation of selection strategies alternative to *nptII* in genetic transformation of citrus. *Plant Cell Reports* **27:** 1005-1015.

Basten C, Weir B, Zeng ZB (2001) QTL Cartographer. Department of Statistics, North Carolina State University, Raleigh, NC.

Bernet GP, Margaix C, Jacas J, Carbonell EA and Asins MJ (2005) Genetic analysis of citrus leafminer susceptibility. *Theoretical and Applied Genetics* **110:** 1393-1400.

Blas AL, Yu Q, Chen C, Veatch O, Moore PH, Paull RE and Ming R (2009) Enrichment of a papaya high-density genetic map with AFLP markers. *Genome* **52:** 716-725.

Boncheva R, Dukiandjiev S, Minkov I, de Maagd RA and Naimov S (2006) Activity of *Bacillus thuringiensis* - endotoxins against codling moth (*Cydia pomonella* L.) larvae. *Journal of Invertebrate Pathology* **92:** 84-87.

Borgio JF (2010) RNAi mediated gene knockdown in sucking and chewing insect pests. *Journal of Biopesticides* **3**(1 Special Issue): 386-393.

Bravo A, Gill SS and Soberon M (2007) Mode of action of *Bacillus thuringiensis* Cry and Cyt toxins and their potential for insect control. *Toxicon* **49(4):** 423-435.

Brennan RM, Jorgensen L, Hackett C, Woodhead M, Gordon SL and Russell J (2008) The development of a genetic linkage map of blackcurrant (*Ribes nigrum* L.) and the identification of regions associated with key fruit quality and agronomic traits. *Euphytica* **161:** 19-34.

Brennan R, Jorgensen L, Gordon S, Loades K, Hackett C and Russell J (2008a) The development of a PCR-based marker linked to resistance to the blackcurrant gall mite (*Cecidophyopsis ribis* Acari: Eriophyidae) *Theoretical and Applied Genetics* DOI 10.1007/s00122-008-0889-x.

Bucher, G, Scholten J and Klingler M (2002) Parental RNAi in *Tribolium* (Coleoptera). *Current Biology* **12:** R85–R86.

Cevik V and King GJ (2002) High-resolution genetic analysis of the Sd-1 aphid resistance locus in *Malus* spp. *Theoretical and Applied Genetics* **105:** 346-354.

Chen SL, Dai SM, Lu KH and Chang C (2007) Female-specific doublesex dsRNA interrupts yolk protein gene expression and reproductive ability in oriental fruit fly, *Bactrocera dorsalis* (Hendel). *Insect Biochemistry and Molecular Biology* **38:** 155-165.

Dandekar AM, McGranahan GH and James DJ (1992) Transgenic woody plants. In: Shain-dow and Wu R (eds) *Transgenic Plants*, Vol. 2. Academic, New York p. 129-151.

Dandekar AM, McGranahan GH, Vial PV, Uratsu SL, Leslie CA, Tebbets JS and Hoffman DJ (1994) Low levels of expression of cryIA(c) sequences of *Bacillus thuringiensis* in transgenic walnut. *Plant Science* **96:** 151-162.

Dandekar AM, Fisk HJ, McGranahan GH *et al.* (2002) Different genes for different folks in tree crops: What works and what does not. *HortScience* **37:** 281-286.

Darbani B, Eimanifar A, Stewart CN Jr and Camargo WN (2007) Methods to produce marker-free transgenic plants. *Biotechnology Journal* **2:** 83-90.

de Maagd RA, Bravo A and Crickmore N (2001) How *Bacillus thuringiensis* has evolved specific toxins to colonizethe insect world. *Trends in Genetics* **17:**193-199.

Degenhardt J, Poppe A, Montag J, Szankowski I (2006) The use of the phosphomannose isomerase /mannose selection system to recover transgenic apple plants. *Plant Cell Reports* **25:**1149–1156.

Dillon S, Ramage C, Ashmore S and Drew RA (2006) Development of a codominant CAPS marker linked to PRSV-P resistance in highland papaya. *Theoretical and Applied Genetics* **113:** 1159-1169.

Echt C, Knapp S and Liu BH (1992) Genome mapping with non inbred crosses using Gmendel 2.0. *Maize Genetics Cooperation Newsletter* **66:** 27-29.

Eisemann CH, Donaldson RA, Pearson RD, Cadagon LC, Vuocolo T and Pellam RL (1994) Larvicidal action of lectins on *Lucilia cuprina*; mechanism of action. *Entomologia Experimentalis et Applicata* **72:** 1-11.

Elbashir SM, Martinez J, Patkaniowska A, Lendeckel W and Tuschl T (2001) Functional anatomy of siRNAs for mediating efficient RNAi in *Drosophila melanogaster* embryo lysate. *EMBO Journal* **20:** 6877-6888.

Fire A, Xu S, Montgomery MK, Kostas SA, Driver SE, Mello CC (1998) Potent and specific genetic inteference by double-stranded RNA in *Caenorhabditis elegans. Nature* **391:** 806-811.

Gisbert AD, Martınez-Calvo J, Llacer G, Badenes ML and Romero C (2009) Development of two loquat [*Eriobotrya japonica* (Thunb.) Lindl.] linkage maps based on AFLPs and SSR markers from different Rosaceae species. *Molecular Breeding* **23:** 523-538.

Graham J, Gordon SC, Mcnicol RJ (1997) The effect of the *CpTI* gene in strawberry against attack by vine weevil (*Otiorhynchus sulcatus* F. Coleoptera: Curculionidae). *Annals of Applied Biology* **131:**133-139.

Gulsen O, Uzun A, Canan I, Seday U and Canihos E (2010) A new citrus linkage map based on SRAP, SSR, ISSR, POGP, RGA and RAPD markers. *Euphytica* **173:** 265-277.

Gupta P, Varshney R, Sharma P and Ramesh B (1999) Molecular markers and their applications in wheat breeding. *Plant Breeding* **118:** 369-390.

Gygax M, Gianfranceschi L, Liebhard R, Kellerhals M, Gessler C and Patocchi A (2004) Molecular markers linked to the apple scab resistance gene *Vbj* derived from *Malus baccata jackii. Theoretical and Applied Genetics* **109:**1702-1709.

Hajela RK, Hajela N, Bolyard MG, Barnes WM and Sticklen MB (1993) A simple transformation system using adventitious shoot multiplication of Juneberry. *HortScience* **28:** 330-332.

Hattasch C, Flachowsky H and Hanke MV (2009) Evaluation of an alternative D-amino acid/DAAO selection system for transformation in apple (*Malus x domestica* Borkh.). *Journal of Horticultural Sciences and Biotechnology* **ISAFRUIT Special Issue:** 188-194.

Hunter W, Glick E, Bextine B and Paldi N (2010) RNA-interference for reducing insect pests: plant infusion. USDA ARS, June 2010.

Hurtado MA, Romero C, Vilanova S, Abbott AG, Lacer G and Badenes ML (2002) Genetic linkage maps of two apricot cultivars (*Prunus armeniaca* L.) and mapping of PPV (sharka) resistance. *Theoretical and Applied Genetics* **105:** 182-191.

Ismail I, Lee FS, Abdullah R, Fei CK, Zainal Z, Sidik NM and Zain CRCM (2010) Molecular and expression analysis of cowpea trypsin inhibitor (*CpTI*) gene in transgenic *Elaeis guineensis* Jacq Leaves. *Australian Journal of Crop Science* **4(1):** 37-48.

Jacobsen, E and Schouten HJ (2007) Cisgenesis strongly improves introgression breeding and induced translocation breeding of plants. *Trends in Biotechnology* doi:10.1016/j. tibtech. 2007.03.008.

Jagtap UB, Gurav RG and Bapat VA (2011) Role of RNA interference in plant improvement. *Naturwissenschaften* **98:** 473-492.

James DJ, Passey AJ, Webster AD, Barbara DJ, Visa P, Dandekar AM and Uratsu S (1993) Transgenic apples and strawberries: advance in transformation, introduction of genes for insect resistance and field studies of tissue cultured plants. *Acta Horticulturae* **336:**179-194.

Joobeur T, Viruel MA, de Vicente MC, Jauregui B, Ballester J, Dettori MT, Verde I, Truco MJ, Messeguer R, Batlle I, Quarta R, Dirlewanger E and Arus P (1998) Construction of a saturated linkage map of *Prunus* using an almond x peach F2 progeny. *Theoretical and Applied Genetics* **97:**1034–1041.

Kashkush K, Jinggui F, Tomer E, Hillel J and Lavi U (2001) Cultivar identification and genetic map of mango (*Mangifera indica). Euphytica* **122:** 129-136.

Krattiger AF (1997) Insect resistance in Crops: A case study of *Bacillus thuringiensis* (*Bt*) and its transfer to developing countries. *ISAAA Briefs* No. 2. ISAAA: Ithaca, NY. pp. 42.

Lander ES, Green P, Abrahamson J, Barlow A, Daly MJ, Lincoln SE and Newburg L (1987) Mapmaker an interactive computer package for constructing primary genetic linkage maps of experimental and natural populations. *Genomics* **1:** 174-181.

Li X, Zhang M, Zhang H (2011) RNA interference of four genes in adult *Bactrocera dorsalis* by feeding their dsRNAs. *PLoS ONE* **6(3):** e17788. doi:10.1371/ journal. pone.0017788.

Liebhard R, Koller B, Gianfranceschi L and Gessler C (2003) Creating a saturated reference map for the apple (*Malus* x *domestica* Borkh.) genome. *Theoretical and Applied Genetics* **106:** 1497-1508.

Lincoln S, Daly M and Lander E (1993) Constructing genetic linkage maps with MAPMAKER/EXP. Version 3.0. Whitehead Institute for Biomedical Research Technical Report, 3rd Edn.

Linehan V, Thorpe S, Andrews N, Kim Y and Beaini F (2012) Food demand to 2050: Opportunities for Australian agriculture, ABARES conference Canberra, March.

Liu CM, Guo YS, Liu R, Zhang B, Xie JH, Hu YL, Ouyang R, Zhao YH and Fu JX (2008) Construction of a high density molecular linkage map for lychee based on AFLP and RAPD markers. *Acta Horticulturae* **63:** 87-94.

Manly KF, Cudmore Jr RH and Meer JM (2001) Map Manager QTX, cross platform software for genetic mapping. Mamm. *Genome* **12:** 930-932.

Mansoor S, Amin I, Hussain M, Zafar Y and Briddon RW (2006) Engineering novel traits in plants through RNA interference. *Trends in Plant Science* **11(11):** 559-565.

McGranahan GH, Leslie CA, Uratsu SL, Martin LA and Dandekar AM (1988) Agrobacterium mediated transformation of walnut somatic embryos and regeneration of transgenic plants. *Biotechnology* **6:** 800-804.

Mehlenbacher SA, Brown RN, Nouhra ER, Gokirmak T, Bassil NV and Kubisiak TL (2006) A genetic linkage map for hazelnut (*Corylus avellana* L.) based on RAPD and SSR markers. *Genome* **49:** 122-133.

Molesini B, Pii Y and Pandolfini T (2012) Fruit improvement using intragenesis and artificial microRNA. *Trends in Biotechnology* **30(2):** 80-88.

Napoli C, Lemieux C and Jorgensen R (1990) Introduction of chimeric chalcone synthase gene into *Petunia* results in reversible cosuppression of homologous genes in trans. *Plant Cell* **2:** 279-289.

Oliveira MLP de, Febres VJ, Costa MGC Moore GA and Otoni WC (2009) High-efficiency *Agrobacterium*-mediated transformation of citrus via sonication and vacuum infiltration. *Plant Cell Reports* **28:** 387-395.

Price DRG and Gatehouse JA (2008) RNAi-mediated crop protection against insects. *Trends in Biotechnology* **26(7):** 393-400.

Rodriguez N, Valdes-Infante J, Becker, D, Velazquez B, Gonzalez G, Sourd D, Rodriguez J, Billotte N, Risterucci AM, Ritter E and Rohde W (2007) Characterization of guava accessions by SSR markers, extension of the molecular linkage map, and mapping of QTLs for vegetative and reproductive characters. *Acta Horticulturae* **735:** 201-216.

Rosa RL, Angiolillo A, Guerrero C, Pellegrini M, Rallo L, Besnard G, Bervill A, Martin A, and Baldoni L (2003) A first linkage map of olive (*Olea europaea* L.) cultivars using RAPD, AFLP, RFLP and SSR markers. *Theoretical and Applied Genetics* **106:** 1273-1282.

Rosa C, Kamita SG, Dequine H, Wuriyanghan H, Lindbo JA and Falk BW (2010) RNAi effects on actin mRNAs in *Homalodisca vitripennis* cells. *Journal of RNAi and Gene Silencing* **6** (1): 361-366.

Sargent DJ, Clarke J, Simpson DW, Tobutt KR, Arus P, Monfort A, Vilanova S, Denoyes-Rothan B,. Rousseau M, Folta KM, Bassil NV, Battey NH (2006) An enhanced microsatellite map of diploid Fragaria. *Theoretical and Applied Genetics* **112:** 1349-1359.

Schaart JG, Krens FA, Pelgrom KTB, Mendes O and Rouwendal GJA (2004) Effective production of marker-free transgenic strawberry plants using inducible site-specific recombination and a bifunctional selectable marker gene. *Plant Biotechnology Journal* **2:** 233-240.

Serres R, Stang E, McCabe D, Russel D, Mahr D and McCown B (1992) Gene transfer using electric discharge particle bombardment and recovery of transformed craneberry plants. *Journal of American Society for Horticultural Science* **11:** 174-180.

Semagn K, Bjornstad A and Ndjiondjop MN (2006) Principles, requirements and prospects of genetic mapping in plants. *African Journal of Biotechnology* **5:** 2569-2587.

Stam P (1993) Construction of integrated genetic linkage maps by means of a new computer package: JoinMap. *Plant Journal* **3:** 739-744.

Stevenson M (2004) Therapeutic Potential of RNA Interference. *The New England Journal of Medicine* **351:** 1772-1777.

Stockinjer EJ, Mulinix CA, Long CM, Brettin TS and Iezzoni AF (1996) A linkage map of sweet cherry based on RAPD analysis of a microspore-derived callus culture population. *Journal of Heredity* **87:** 214-218.

Stoeckli S, Mody K, Patocchi A, Kellerhals M and Dorn S (2009) Rust mite resistance in apple assessed by quantitative trait loci analysis. *Tree Genetics and Genomes* **5:** 257-267.

Tao R, Dandekar AM, Uratsu PL, Vail PV and Tebbets JS (1997) Engineering genetic resistance against insects in Japanese persimmon using *cry*1A(c) gene of *Bacillus thuringiensis. Journal of American Society for Horticultural Science* **122:** 764-771.

Tomoyasu Y and Denell RE (2004) Larval RNAi in *Tribolium* (Coleoptera) for analyzing adult development. *Development Genes and Evolution* **214:** 575-578.

Tomoyasu Y, Miller SC, Tomita S, Schoppmeier M, Grossmann D and Bucher G (2008) Exploring systemic RNA interference in insects: a genome-wide survey for RNAi genes in *Tribolium*. *Genome Biology* **9:** R10.

Troggio, M , Malacarne G, Coppola G, Segala C, Cartwright DA, Pindo M, Stefanini M, Mank R, Moroldo M, Morgante M, Grando MS and Velasco R (2007) A dense single-nucleotide polymorphism-based genetic linkage map of grapevine (*Vitis vinifera* l.) anchoring Pinot Noir bacterial artificial chromosome contigs. *Genetics* **176:** 2637-2650.

Turner CT, Davy MW, MacDiarmid RM, Plummer KM, Birch NP and Newcomb RD (2006) RNA interference in the light brown apple moth, *Epiphyas postvittana* (Walker) induced by double-stranded RNA feeding. *Insect Molecular Biology* **15:** 383-391.

Upadhyaya CP, Nookaraju A, Gururani MA, Upadhyaya DC, Kim DH, Chun SC and Park SW (2010) An update on the progress towards the development of marker-free transgenic plants. *Botanical Studies* **51:** 277-292.

Upadhyay SK, Chandrashekar K, Thakur N, Verma PC, Borgio JF, Singh PK and Tuli R (2011) RNA interference for the control of whiteflies (*Bemisia tabaci*) by oral route. *Journal of Biosciences* **36:** 153-161.

Verde I, Lauria M, Dettori MT, Vendramin E, Balconi C, Micali S, Wang Y, Marrazzo MT, Cipriani G, Hartings H, Testolin R, Abbott AG Motto M and Quarta R (2005) Microsatellite and AFLP markers in the *Prunus persica* [L. (Batsch)] x *P. ferganensis* BC_1 linkage map: saturation and coverage improvement. *Theoretical and Applied Genetics* **111:** 1013-1021.

Vezzulli S, Troggio M, Coppola G, Jermakow A, Cartwright D, Zharkikh A, Stefanini M, Grando MS, Viola R, Adam-Blondon AF, Thomas M, This P and VelascoR (2008) A reference integrated map for cultivated grapevine (*Vitis vinifera* L.) from three crosses, based on 283 SSR and 501 SNP-based markers. *Theoretical and Applied Genetics* **117:** 499-511.

Vilanova S, Romero C, Abbott AG, Llcer G and Badenes ML (2003) An apricot (*Prunus armeniaca* L.) F2 progeny linkage map based on SSR and AFLP markers, mapping plum pox virus resistance and self-incompatibility traits. *Theoretical and Applied Genetics* **107:** 239-247.

Xu K, Riaz S, Roncoroni NC, Jin Y, Hu R, Zhou R and Walker MA (2008) Genetic and QTL analysis of resistance to *Xiphinema index* in a grapevine cross. *Theoretical and Applied Genetics* **116:** 305-311.

Yang ZN, Ingelbrecht IL, Louzada E, Svaria M and Mirkov TE (2000) *Agrobacterium* mediated transformation of the commercial important grapefruit cultivar Rio Red (*Citrus paradisi* Macf.). *Plant Cell Reports* **19:** 1203-1211.

Zamore PD, Tuschl T, Sharp PA, Bartel DP (2000) RNAi: doublestranded RNA directs the ATP-dependent cleavage of mRNA at 21 to 23 nucleotide intervals. *Cell* **101:** 25-33

Zhu YJ, Agbayani R, McCafferty H, Albert HH and Moore PH (2005) Effective selection of transgenic papaya plants with the PMI/*Man* selection system. *Plant Cell Reports* **24:** 426-432.

Biological and Molecular Approaches in Pest Management
Balwinder Singh, Ramesh Arora & S S Gosal (Eds)
Scientific Publishers (India), Jodhpur, 2015, pp. 249-311.

Chapter - 11

Pest- Insects Resistance to Microbial Control Agents: Current Status and Management Strategies

Arshdeep K Gill and Ramesh Arora

Department of Entomology, Punjab Agricultural University, Ludhiana- 14004, India.
E-mail: arsh.pau@gmail.com

11.1. INTRODUCTION

The insects are among the most versatile and ubiquitous of all animals and have developed more successful survival mechanisms than other organisms that share the planet with them. The biological success of insects is quite evident. As compared to more than one million identified species of insects, there are only about 62,000 species of all vertebrates including mammals, birds, reptiles, amphibians and fishes. One of the reasons for dominance of insects in the world fauna is their great adaptability which has enabled them to colonize every conceivable habitat where life can exist and to survive under unfavourable environmental conditions, whether natural or human induced. The application of highly toxic insecticides to kill unwanted insects termed 'pest insects' or 'insect pests' has exerted great selection pressure on insects. But insects have proved equal to the challenge posed by these chemicals and have developed the adaptability to survive these toxicants. The phenomenon is known as insecticide resistance in insects. The insecticide resistance action committee (IRAC), an international group of more than 150 members of crop protection industry, defines resistance as a "heritable change in the sensitivity of a pest population that is reflected in the repeated failure of a product to achieve the expected level of control when used according to the label recommendation for that pest species" (http://www.irac-online.org/about/resistance/). Tabashnik *et al.* (2014) have recently reviewed the terminology used in pesticide resistance literature and defined resistance as a genetically based decrease in susceptibility to a pesticide. Thousands of cases of insecticide resistance in several hundred species of medical, veterinary and agricultural insect pests are listed at the IRAC website. Insect resistance to insecticides has emerged as a major obstacle to the use of these chemicals for limiting pest damage. Unfortunately but predictably insects have developed the ability to develop resistance not only to chemical insecticides but also to all other pest management tactics including natural enemies (parasitoids, predators and entomopathogens), host plant resistance, insect growth regulators, sterile insect techniques and even cultural control practices (Pedigo, 2002). Of these, the microbial control agent (MCAs) based on entomopathogens have emerged as an important component of environment friendly pest management (Glare *et al.*, 2012) and are considered the desirable alternative to synthetic pesticides in biointensive IPM (Arora and Shera, 2014). But several species of insects have already been reported to have developed resistance to these microbial control agents (McGaughey, 1985; Tabashnik, 1994b; Fritsch *et al.*, 2005). Other species exposed to the selection pressure of *Bt* -toxin genes inserted into crop plants have been equally successful in their responses (Petzold-Maxwell *et al.*, 2012; Tabashnik *et al.*, 2013). It is therefore, imperative to understand the causes of insect resistance to microbial control agents and to utilize them judiciously in future pest management programmes.

11.2. STATUS OF INSECT RESISTANCE TO MICROBIAL CONTROL AGENTS

Even though microbials currently account for only a tiny fraction of the pesticides used for the management of insect pests (Glare *et al.*, 2012), several cases of resistance to these bioinsecticides have been reported from around the world.

11. 2.1 *Bacillus thuringiensis*

B. thuringiensis (*Bt*) is the most widely used microbial biopesticide and has found application in forestry and in high value horticultural crops. In addition, *Bt* genes have been inserted in cotton, corn, soybean and other crops to produce insect resistant transgenic crops. Resistance has been reported to externally applied *B. thuringiensis* sprays has well as to *Bt* transgenic crops producing Cry toxins (Tabashink *et al.*, 2014).

11.2.1.1. Resistance to exogenous *Bt* applications to crops

B. thuringiensis insecticidal crystal proteins (ICPs) have a long history of safe application as suspensions of crystal proteins and spores and continue to dominate the biological insecticide market. The highly selective activity of each crystal protein is particularly valuable for the implementation in integrated pest management programs. Only a limited number of insects are targeted whereas soil organisms (Saxena and Stotzky, 2001; Al-Deeb *et al.*, 2003) and beneficial arthropods such as predators (Romeis *et al.*, 2004) or parasitoids (Haseeb *et al.*, 2004) remain mostly unaffected. Numerous studies have shown that *B. thuringiensis* based products are safe to vertebrates (Siegel, 2001; Teshima *et al.*, 2002). Furthermore, crystal proteins are generally rapidly digested under simulated human gastric conditions (Betz *et al.*, 2000), reducing the possibility of potential allergenic reactions upon ingestion. Because of the selective activity and rapid degradation of crystal proteins by UV irradiation, *B. thuringiensis* sprays have a less disturbing effect on the environment compared to some classical chemical broad spectrum insecticides (Vettori *et al.*, 2003).

A. Field Resistance

Tabashnik *et al.* (2014) defined field-evolved resistance as a "genetically based decrease in susceptibility to a pesticide in a population caused by exposure to the pesticide in the field". The impact of field-evolved resistance on pest control can vary from none to severe

The diamondback moth, *P. xylostella* is the only insect species that has evolved high levels of resistance to *Bt* in the field. The first case of field resistance was reported from Hawaii. Populations from areas heavily treated with Dipel proved less susceptible than populations that had been treated at lower levels, with the highest level of resistance at 30-fold (Tabashnik *et al.*, 1990). A diamondback moth colony (BL) derived from a field population in the Philippines regularly exposed to Dipel (*Btk* var. HD1) showed more than 200-fold resistance to Cry1Ab (Ferre *et al.*, 1991). In another colony sampled later from the same location in the Philippines, resistance was limited to Cry1Ab and did not extend to Cry1Aa, Cry1Ac, or Cry1Ba (Ballester *et al.*, 1994). Following additional selection with Cry1Ab and later with a hybrid Cry1A protoxin, this colony (PHI) was observed to be partially resistant to Cry1Aa, Cry1Ab, and Cry1Ac, and still susceptible to Cry1Ca, Cry1Fa, and Cry1Ja (Tabashnik *et al.*, 1997a).

A *P. xylostella* colony (Loxa A) from Florida was found to be more than 1500-fold resistant to Javelin (a commercial formulation of *Btk* NRD12) in the second generation after the colony was collected from the field (Tang *et al.*, 1996). Resistance rapidly fell to about 300-fold in the absence of selection but remained stable at this level in subsequent generations. Insects with this level of stabilized resistance were more than 200-fold resistant to Cry1Aa, Cry1Ab, and Cry1Ac, but they were still fully susceptible toward Cry1Ba, Cry1Ca, Cry1Da (Tang *et al.*, 1996), and Cry9Ca (Lambert *et al.*, 1996). Whereas survival of susceptible larvae at 72 h on leaf disks from Cry1Ac-expressing broccoli was much reduced compared with survival on leaf discs from normal broccoli, survival of resistant larvae did not differ between the two types of broccoli (Tang *et al.*, 1999). In fact, resistant larvae could complete their development from egg to adult and could cycle for multiple generations equally well on both plant types (Tang *et al.*, 2001). Insect resistance in *P. xylostella* populations to *Btk* products has resulted in extensive use of *Bt* var. *aizawai* - based insecticides in certain locations. These Cry1Ca-resistant colonies were only two- to four fold less susceptible to *Bta*, somewhat less susceptible to Cry1Ab, and 50- to 130-fold less susceptible to *Btk* formulations when compared with a susceptible colony.

A *P. xylostella* colony (SERD3) with considerable levels of resistance to both *Btk* and *Bta* has been reported from Malaysia (Wright *et al.*, 1997). After rearing in the laboratory for seven generations without selection, resistance to *Btk* and *Bta* was 330-fold and 160-fold, respec-tively. Another *P. xylostella* colony (UNSE-LMEL) that developed field resistance to *Btk* products in Malaysia showed resistance to Cry1Ac, Cry1Ab, *Btk*, and *Bta* (Sayyed *et al.*, 2000). Selection with Cry1Ac (1AcSEL-MEL) during 5 generations resulted in a more than 95-fold increase in resistance to this toxin, whereas selection with Cry1Ab (1AbSEL-MEL), *Btk* (*Btk*SEL-MEL), or *Bta* (*Bta*SEL-MEL) increased resistance to these toxins or *Bt* products only ten fold or less.

Wang *et al.* (2007) monitored resistance of field populations of the diamondback moth *P. xylostella* in different provinces of China to the insecticidal protein Cry1Ac, Cry1Ba and commercial formulation of *B. thuringiensis* var. *kurstaki* (Btk). Bioassay results showed that the populations of the diamondback moth from different locations exhibited different levels of resistance, compared with a susceptible laboratory population. The Guangdong field population was 56.15- and 21.90-fold resistant to Cry1Ac and *Btk*, respectively. Shanghai, Hunan, Shandong and Zhejiang populations were 37.85-, 17.24-, 10.24- and 9.41-fold resistant to Cry1Ac, respectively, but were not resistant to *Btk*. The Hubei population did not show resistance to Cry1Ac and *Btk*. Almost all tested populations were susceptible to Cry1Ba, but the Guangdong population showed some tolerance to Cry1Ba with a LC_{50} of 0.69 μg/ml which was 6.17-fold higher than that of the susceptible population.

Another insect pest in which spatial susceptibility of different populations to *Bt* products has been studied, is the cotton bollworm, *Helicoverpa armigera*. Gujar *et al.* (2000) determined the baseline susceptibility of *H. armigera* to *Btk*. HD-1 and HD-73 strains for the populations collected from different places in India. The populations from Delhi, Raichur and Bangalore were least susceptible to HD-1 strain. On the contrary, the Hyderabad and Madurai populations were most susceptible. HD-1 with neonate mortality of 37.4 per cent at 10 ppm and 68.6 per cent at 100 ppm after 96 h of treatment. Nagpur populations were least susceptible to *Btk* HD-73 whereas Guntur, Bapatla, Hyderabad, Madurai and Vijayawada populations were most susceptible. HD-73 caused mortality of 62.3 per cent at 100 ppm and 91.7 per cent at 500 ppm after 96 h of treatment. Temporal variations of insect mortality showed that LC_{50} (96 h) of HD-1 increased from 9.47 ppm in October 1998 to 51.04 ppm in December 1998 for the insect population on pigeonpea in Delhi.

Kranthi *et al.* (2001) studied the baseline toxicity of CrylAc-endotoxins on field populations of the cotton bollworm, *H. armigera* and determined through log-dose probit analysis that all the three Cry1Ac proteins were found to be toxic to the bollworm larvae. CrylAc was found to be the most toxic followed by Cry1Aa and Cry1Ab. The LC_{50} values ranged from 0.07 to 0.99 μg/ml (14-fold) for CrylAa, 0.69 to 9.94 μg/ml (14-fold) for CrylAb and 0.01 to 0.67 μg/ml of diet (67-fold) for CrylAc. The LC_{50} values deduced from the cumulative log dose probit response of the data pooled from all assays, were 0.62 μg/ml for CrylAa, 4.43μg/ml for CrylAb and 0.100 μg/ml of diet for CrylAc. The LC_{99} values derived from the cumulative data were 515 μg/ml for CrylAa, 13385 μg/ml for CrylAb and 75 μg/ml of diet for CrylAc. Fakrudin *et al.* (2003) reported the level of resistance of *H. armigera* population to Cry1Ac in selected locations of South India. The level of resistance was 8.4-fold in Nanded population followed by 8.03, 7.70, 7.13, and 6.80 fold, respectively, for Guntur, Nalgonda, Madhurai and Raichur populations. It was as low as 1.131-fold in Kovilpatti population located in the extreme south. Authors explained that even where *Bt* sprays were used to some extent as a component of IPM programs carried out in Andhra Pradesh and Tamil

Nadu by the state department agencies, resistance factor (RF) values were not indicative of a definite trend. But there was some relationship in slope and RF value indicative of heterogeneity and levels of resistance, respectively. Heterogeneity within a geographical location is expected due to migratory nature of the *H. armigera* and lack of selection history for CrylAc toxin in these geographic populations. Jalali *et al.* (2004) collected populations of *H. armigera* from cotton fields of nine major cotton growing states in India. All populations were susceptible to CrylAc. The mean lethal concentrations, LC_{50} ranged from 0.14 to 0.71 and from 0.11 to 0.61; LC_{90} ranged from 1.17 to 6.94 and from 1.02 to 6.70 µg of CrylAc / ml of diet in 1999 and 2001 populations, respectively. The effective concentrations (EC_{50}) values ranged from 0.003 to 0.008 and from 0.0003 to 0.004; EC_{90} ranged from 0.029 to 0.076 and from 0.009 to 0.054ug of CrylAc in 1999 and 2001 populations, respectively.

Wan *et al.* (2012) reported field-evolved resistance to Cry1Ac by pink bollworm, *Pectinophora gossypiella* in the Yangtze River Valley of China. Laboratory bioassay data from 51 field-derived strains showed that the susceptibility to Cry1Ac was significantly lower during 2008 to 2010 than 2005 to 2007. The percentage of field populations yielding one or more survivors at a diagnostic concentration of Cry1Ac increased from zero per cent in 2005–2007 to 56 per cent in 2008–2010.

B. Laboratory Selection for Resistance

Keeping in view the importance of *B. thuringiensis* as a biopesticide, scientists have selected insects for resistance to *Bt* products in the laboratory. These cases are briefly discussed here (Table 1).

Table 1. Laboratory selected insect species and strains resistant to *Bt* proteins

Species	Selecting agent		Details on resistance			Resistance mechanism*	Reference
	Type	Form	ES	Formulation/ Toxin	RR		
(1)	(2)	(3)	(4)	(5)	(6)	(7)	(8)
Plodia	Dipel	FSC	13	Dipel	100		McGauhey 1985
Interpunctella				*Btk* HD-337	>100		McGaughey and
				Bte HD-198	1.1		Johnson 1987
				Bta HD-133	1.4		
			36	Dipel	>250	≡processing	McGaughey and
							Beeman 1988
				Cry 1Ab	(877)	↓binding	
				Cry 1Ca	0.3		
P. interpunctella	Bte HD-198	SC	19	HD-198	21		McGaughey and
							Johnson 1992
				HD-198	32		McGaughey and
				Cry 1Aa	10		Johnson 1994
				Cry 1Ab	27		

Contd...

(1)	(2)	(3)	(4)	(5)	(6)	(7)	(8)
				Cry 1AC	150		
				Cry 1Ba	9		
				Cry 1Ca	5		
				Cry 2A	20		
				Cry 1Ac-PT	128	↓activation	Oppert *et al.*, 1994
				Cry 1Ab-PT	264		Herrero *et al.*, 2001
				Cry 1Ab	25	Δbinding	
				Cry 1Ac	n.d	≡ binding	
P. interpunctella	*Bta* HD-133	SC	23	HD-133	61.5		McGauhey and Johnson 1992
				HD-133	94		McGaughey and Johnson 1994
				HD-1	45		
				Cry 1Aa	17		
				Cry 1Ab	226		
				Cry 1Ac	789		
				Cry 1Ba	44		
				Cry 1Ca	19		
				Cry2A	24		
P. interpunctella	Dipel	FSC	24	Dipel	70		McGaughey and Johnson 1992
				Cry 1 Ac	2816		McGaughey and Johnson 1994
				Cry 1Ab-PT	1049		Herrero *et al.*, 2001
				Cry 1Ab	290	↓binding	
Plutella xylostella	*Btk*	FSC	Field	Dipel	1		Ferre *et al.*, 1991
				Cry 1Ab	>200	↓binding	
				Cry 1Ba	2		
				Cry 1Ca	0.5		
P. xylostella	*Btk*	FSC		Dipel	0.4		Ballester *et al.*, 1994
				Cry 1Aa	1.3		Herrero *et al.*, 2001
				Cry 1Ab	236		
				Cry 1Ac	1	≡ binding	
				Cry 1Ba	1		
	Cry 1 Ab/	T/CC		Cry 1Aa	>1	≡binding	Ballester *et al.*, 1999, Tabashnik *et al.*, 1997b
	Cry 1 Ac-1 Ab			Cry 1Ab	>1	↓binding	Tabashnik *et al.*, 1997b

Contd...

(1)	(2)	(3)	(4)	(5)	(6)	(7)	(8)
				Crys 1Ac	>1	≡binding	Ballester *et al.*, 1999. Tabashnik *et al.*, 1997b
				Cry 1 Ca	(1)		Tabashnik *et al.* 1997b
				Cry 1Fa	(1)		
				Cry 1Ja	(1)		
P. xylostella	*Btk*	FSC	Field	Dipel	26.5		Tabashnik *et al.*, 1990
	Dipel	FSC	9	Dipel	820		Tabashnik *et al.*, 1991
			15	Dipel	Low		Tabashnik *et al.*, 1994b
	Dipel	FSC		Dipel	3300	≡processing	Liu and Tabashnik, 1997; Tabashnik *et al.*, 1993
				Cry 1Ac	>59	↓binding	Tabashnik *et al.*, 1994b
				Cry 1Ac	>100	Δbinding	Ballester *et al.*, 1999 Tabashnik *et al.*, 1996
				Cry 1Ab	>100	↓binding	Tabashnik *et al.*, 1996
				Cry 1Ac	>100	↓binding	
				Cry 1 Ba	3		Tabashnik *et al.*, 1996
				Cry 1Ca	2		
				Cry 1Da	3		
				Cry 1Fa			
				Cry 1 Fa	>100		
				Cry 1Ia	3		
				Cry 1Ja	>140		
				Cry 1Ac-Canola	98/0		Suresh *et al.*, 1998
	Cry 1 Ac	NO – QAGE		Cry 1Aa	>20000		Tabashnik *et al.*, 2000
				Cry 1Ab	>10000		
				Cry 1 Ac	>40000		
				Cry 1Bb	2		
				Cry 1Ca	1		
				Cry 1Da	1		
				Cry 1Fa	>10000		
				Cry 2Aa	5		
				Cry 9Ca	2		
P.xylostella	*Btk/Bta*	FSC	Field	Cry 1Ca	22		Liu and Tabashnik, 1997

Contd...

(1)	(2)	(3)	(4)	(5)	(6)	(7)	(8)
				Dipel	134		Liu *et al.*, 1996
				Xentari	3		
	Cry 1 Ca	CC	6	Cry 1Ca	62		Liu and Tabashnik, 1997
	Cry 1Ca	T	5	Cry 1Ca	19	≡binding	Liu *et al.*, 2000
			5	Cry 1Ca Crystal	48		
Spodoptera exigua	Btk HD-1	SC	20	Btk HD-1	1		Moar *et al.*, 1995
	Cry1Ca	IB/T	25	Cry1Ca	850	Δbinding	
			22	Cry1Ab	93	↑NS binding	
			34	Cry2Aa	73		
			34	Cry9Ca	12		
Spodoptera littoralis	Cry1Ca	SC	14	Cry1Ca spcry	>500		Muller-Cohn *et al.*, 1996
			14	Cry1Da spcry	7		
			14	Cry1Ea spcry	34		
			14	Cry1Fa spcry	1		
			14	Cry1Ab protoxin	3		
			14	*Bia* 7.29 spcry	7		
Pectinophora gossypiella	Cry1Ac	CC	AZP-R	Cry1Ac Cell Cap	300		Tabashnik *et al.*, 2000
				Cry1Ac-cotton	40/1.6		
				Cry1Aa protoxin	Yes		
				Cry1Aa protoxin	Yes		
				Cry1Aa protoxin	Yes		
				Cry1Bb protoxin	Weak		
				CrylCa protoxin	No		
				Cry1Da protoxin	No		
				Cry1Fa protoxin	No		
				Cry1Ja protoxin	No		
				Cry2Aa protoxin	No		
Trichoplusia ni	Cry1Ab	T	7	Cry1Ab	31		Estada and Ferre, 1994
				Cry1Aa	1		
				Cry1Ac	0.5		
Chrysomela scripta	Cry3Aa	CC	35	Cry3Aa CellCap	>3000		Bauer, 1995

Contd...

(1)	(2)	(3)	(4)	(5)	(6)	(7)	(8)
				Cry2Aa crystal	>5000		Federici and Bauer, 1998
				Cry1Ba crystal	400		
				Cry1Ba protoxin	100		
				Cryt1Aa	1.2		
Leptinotarsa decemlineata	Cry2Aa	CC	12	Cry3Aa CellCap	59		Whalon *et al.*, 1993

Source: Ferre and van Rie (2002)

***The mechanism of resistance to the particular toxin:** reduced binding (↓ binding), slightly altered binding (Δ binding), unaltered binding (≡ binding), apparently unaltered binding (≅ binding), increased non-specific binding (↑ NS binding), reduced protoxin activation (↓ activation), increased toxin degradation (↑ degradation), unaltered proteolytic processing (≡ processing), increased cell repair or cell replacement (↑ cell repair).

***Plodia interpunctella* (Hübner):** The first case of laboratory development of resistance to *Bt* was the 100-fold decrease in susceptibility (ratio between the final and initial LC_{50} values) to Dipel (*Btk* HD1) in an Indianmeal moth, *Plodia interpunctella* population from grain bins following 15 generations of laboratory selection with Dipel (McGaughey, 1985). Resistance reached 250-fold (ratio between the LC_{50} of the selected colony and the LC_{50} of a control colony) after 36 generations of selection with Dipel. Relative to the unselected control colony, 343- R was more than 800-fold resistant to Cry1Ab, but it was nearly fourfold more susceptible to Cry1Ca (Van Rie *et al.*, 1990). Selection of another colony (Dplr) with Dipel increased resistance to Cry1Ab over 200-fold without increasing susceptibility to Cry1Ca (Herrero *et al.*, 2001). *P. interpunctella* colonies have also been selected for resistance to various other *Bt* strains, including *Bt* var. *entomocidus* (*Bte*) HD-198 and *Bt* var. *aizawai* (*Bta*) HD-133 (colonies 198r and 133r, respectively) (McGaughey and Johnson,1992, 1994). Although both *Bt* strains contain Cry1C and although Cry1Ca is highly toxic to *P. interpunctella*, resistance levels against Cry1Ca were low in colonies 198r and 133r. Perhaps Cry1C-resistance alleles are less frequent in this species than are Cry1A-resistance alleles. Both resistant colonies showed relatively higher resistance levels to Cry1Ac than to Cry1Ab, although Cry1Ab, but not Cry1Ac, is present in both *Bt* strains used for selection.

***Heliothis virescens* (Fabricius):** High levels of resistance to Cry1Ac (over 10,000-fold) were obtained in a *Heliothis virescens* colony by selection with Cry1Ac protoxin (Gould *et al.*, 1995). In a *H. virescens* colony (SEL) selected on Cry1Ab, resistance to Cry1Ab was 20-fold after 14 generations (Stone *et al.*, 1989) and further increased to 71-fold by 4 additional generations of selection with Dipel (Sims and Stone, 1991). Selection of another colony of *H. virescens* with Cry1Ac resulted in a 50-fold resistance to Cry1Ac, 13-fold to Cry1Ab, and 53-fold to Cry2Aa (Gould *et al.*, 1992). Luttrell *et al.* (1999) reported that after seven generations of selection, *H. zea* developed up to 100-fold resistance to *Bt* toxin.

***Spodoptera exigua* (Hübner):** Selection of a *Spodoptera exigua* colony with

Cry1Ca resulted in significant levels of resistance to Cry1Ca (850-fold) and cross-resistance to Cry1Ab, Cry2Aa, and Cry9Ca (Moar *et al.*, 1995). Abdullah *et al.* (2000) reported *Spodoptera exigua* selected strain exhibited 5.6-fold increase in the LC_{50} of *Btk* after 12 generations.

***Spodoptera littoralis* (Boisduval):** Selection experiments on *Spodoptera littoralis* larvae using spore-crystal mixtures of a recombinant *Bt* strain expressing only Cry1Ca generated a resistant colony with greater than 500-fold resistance to Cry1Ca (Muller-Cohn *et al.*, 1996).

***Ostrinia nubilalis* (Hübner):** Decreased susceptibility to Dipel was observed in all the five colonies of *Ostrinia nubilalis* selected in the laboratory with this *Bt* formulation (Huang *et al.*, 1997). Upon 3 generations of selection with Dipel, one colony (KS-SC-R) displayed a 36-fold decrease in susceptibility; further selection for 4 generations increased the resistance ratio to 73-fold. Cry1Ac selection on other *O. nubilalis* colonies led to increased resistance to this toxin with resistance levels peaking at about 160-fold after 8 generations of selection. Continued selection resulted in a decrease rather than further increase of Cry1Ac resistance (Bolin *et al.*, 1999).

***Pectinophora gossypiella* (Saunders):** Selection of pink bollworm larvae with Cry1Ac protoxin resulted in a 300-fold-resistance level. High levels of cross-resistance were observed for Cry1Aa and Cry1Ab protoxin and only lowlevels for Cry1Bb protoxin. The colony showed no cross-resistance to Cry1Ca, Cry1Da, Cry1Fa, Cry1Ja, and Cry2Aa protoxins or to Cry9Ca toxin. Larvae from this colony (AZP-R) showed 40 per cent adjusted survival on Cry1A c-expressing cotton compared with 1.6 per cent for susceptible larvae (Tabashnik *et al.*, 2000). Following three rounds of Cry1Ac selection of APHIS-BTX, another *P. gossypiella* colony previously exposed to artificial diet containing leaf powder from Cry1Ac-cotton, resistance to Cry1Ac increased to more than 100-fold (colony APHIS-98R) in comparison with a susceptible colony (APHIS-S) (Liu *et al.*, 2001). APHIS-98R showed a narrow spectrum of cross-resistance, similar to colony AZP-R, with limited cross-resistance to Cry1Ja protoxin (Tabashnik *et al.*, 2000).

***Leptinotarsa decemlineata* (Say):** A colony of colorado potato beetle. *Leptinotarsa decemlineata*, established from adults collected from fields sprayed with *Bt* formulations containing Cry3Aa, was subjected to laboratory selection with Cry3Aa formulations for 29 generations, resulting in 293-fold resistance (Rahardja and Whalon, 1995; Whalon *et al.*, 1993). No reduction in feeding, compared with susceptible larvae on nontreated plants, was observed on potato foliage treated with the Cry3Aa formulation. However, foliage from transgenic potatoes expressing Cry3Aa virtually eliminated feeding by resistant larvae within 4 h (Altre *et al.*, 1996). Selection with Cry3Aa elicited high levels of resistance in another coleopteran species, *Chrysomela scripta* (Bauer, 1995). Cross-resistance to Cry1Ba (400-fold), but not to Cyt1Aa, was observed in this colony (Federici and Bauer, 1998).

***Helicoverpa armigera* (Hübner):** It was demonstrated through laboratory experiments in Australia (Akhurst *et al.*, 2003) and China (Liang *et al.*, 2000) that *H. armigera* was capable to develop resistance to Cry1Ac protein. In the first 4-5 regimes of selection, there was no apparent change in the susceptibility of *H. armigera* to CrylAc. However, initial indications of resistance were clear after the 6th selection regime and by the end of 10th generation resistance increased by 76- folds as reflected by the LC_{50} values. Similarly, resistance factors with respect to the EC_{50} increased to 34 fold by end of 10th generation. The slope, which was relatively steep at 1.8 in the first generation, declined to 0.68 by the end of the 11th generation, indicating an increase in the number of resistant heterozygous individuals in the final population (Kranthi *et al.,* 2000). Chandrashekar and Gujar (2004) reported that *H. armigera*, evolved 31 fold resistance to selection pressure of *B. thuringiensis* endotoxin CrylAc within six generations. The CrylAc selected larvae of *H. armigera* showed cross-resistance to CrylAa and CrylAb both in terms of mortality and growth reduction.

***Plutella xylostella* (Linnaeus):** Laboratory selection of *P. xylostella* using Dipel increased resistance rapidly to over 1000-fold (Tabashnik *et al.*, 1993, 1994a). The resulting colony (NOQA) displayed high levels of resistance to Cry1Aa, Cry1Ab, Cry1Ac, Cry1Fa, and Cry1Ja, whereas resistance to Cry1Ba, Cry1Bb, Cry1Ca, Cry1Da, Cry1Ia, and Cry2Aa was not considered significant (Tabashnik *et al.*, 1993, 1994a, 1996). Leaves of Cry1Ac expressing canola killed all susceptible larvae exposed to them, whereas 90 per cent of the larvae survived on leaves of normal canola. In contrast, NO-QA larvae developed equally well on both types of leaves (Ramachandran *et al.*, 1998).

***Chrysomela scripta* (Fabricius):** Bauer (1995) reported that median lethal concentration (LC_{50}) for second instars after a 96-h exposure to *B. thuringiensis* var. *san diego*, was 21,000 Colorado potato beetle international units per milliliter water. The LCs estimates for the third instars and adults were 20 and 40 times higher than for second instars, respectively. Adult mortality continued for approximately 14 d after initial exposure to *B. thuringiensis* var. *san diego.* The adult LC_{50} calculated at 14 d was half the 4-d estimate. Federici and Bauer (1998) showed that Cyt1Aa is highly toxic to *C. scripta* and Cyt1Aa suppresses high levels of resistance in *C. scripta* selected for resistance to Cry3Aa.

11.2.1.2 Resistance to transgenic Bt crops expressing Cry toxins

B. thuringiensis possesses many of the attributes required in an ideal biopesticide. But its utility in pest control is limited by the photodegrability of Bt deposit on the crops (Arora *et al.*, 2000). To overcome this problem, the scientists have genetically engineered crop plants to make insecticidal proteins encoded by genes from *B. thuringiensis* (*Bt*). Benefits of *Bt* crops include reduced insecticide use, pest suppression, conservation of beneficial natural enemies, increased yield and higher farmer profits. The area planted with *Bt* crops worldwide increased from 1.1 million hectares in 1996 to 175 million hectares in 2013. *Bt* corn accounted for 67 percent of corn planted in the United States during 2012 (http:// www.ers.usda.gov/Data/BiotechCrops) and *Bt* cotton accounted for 79–95 percent of cotton planted in Australia, China, India and the United States during 2010 to 2012. Bt genes that are approved include the following, singly and stacked: Cry1A.105 (MON89034), Cry1Ab (MON810), Cry1F (1507), Cry2Ab (MON89034), Cry3Bb1 (MON863 and MON88017), Cry34Ab1 (59122), Cry35Ab1 (59122), mCry3A (MIR604), and Vip3A (MIR162), in both corn and cotton (Castagnola and Jurat-

Fuentes, 2012). Corn genetically modified to produce a vegetative insecticidal protein (VIP) was first approved in the US in 2010 (Hodgson and Gassmann, 2010).

The first results concerning the transfer of *cry* genes into tobacco and tomato were published in 1987 (Vaeck *et al.*, 1987). Since then, genes encoding different *Bt* toxins with constitutive or tissue-specific promoters have been introduced into a range of crop plants, including tobacco, maize, rice, potato, apple, cotton, tomato, rapeseed, etc. Well-documented and very important fact is that *Bt* toxins have no or very few adverse effects on mammals (including humans) and birds (Goldberg and Tjaden, 1990). In human gut, *Bt* proteins behave as any other dietary proteins and pose no significant risk to human health (Mendelson *et al.*, 2003). There have been cases of pest resistance, pest replacement and secondary pest outbreaks in transgenic crops (Table 2).

Table 2. Some recent publications dealing with pest management problems in *Bt* crops.

Crop/ Region	Pest	Effect	Reference
Corn / USA (Minnesota)	*Striacosta albicosta*	Pest replacement	O'Rourke and Hutchison (2000)
Corn / USA (Illinois, Missouri)	*Striacosta albicosta*	Pest replacement	Dorhaut and Rice (2004)
Corn / USA (South Dakota)	*Striacosta albicosta*	Pest replacement	Catangui and Berg (2006)
Cotton/ China	*Helicoverpa armigera*	Higher tolerance (Cry1Ac)	Li *et al.* (2007)
Cotton / China	*Creontiades* sp.	Secondary pests	Wang *et al.* (2008)
Corn / USA (Michigan, Ohio)	*Striacosta albicosta*	Pest replacement	Di Fonzo and Hammond (2008)
Corn /Puerto Rico	*Spodoptera frugiperda*	Resistance (Cry1F)	Tabashnik *et al.* (2009)
Corn/ South Africa	*Busseola fusca*	Resistance (Cry1Ab)	Tabashnik *et al.* (2009)
Cotton/ USA	*Helicoverpa armigera*	Resistance (Cry1Ac, Cry2Ab)	Tabashnik *et al.* (2009)
Cotton/ China	Aphids, spider mites, lygus bugs	Secondary pests	Zhao *et al.* (2010)
Cotton/ China	*Creontiades* sp.	Secondary pests	Lu *et al.* (2010)
Cotton/ India	*Pectinophora gossypiella*	Resistance (Cry1Ac)	Monsanto (2010)
Corn / USA	*Diabrotica virgifera virgifera*	Resistance (Cry3Bb)	Petzold-Maxwell *et al.* (2012)
Cotton/ USA	*Heliothis zea*	Resistance (Cry1Ac)	Tabashnik *et al.* (2013)

Source: Modified after Christoph (2010)

Bt Maize

In 1996, the first genetically modified maize producing a *Bt* cry protein was approved, which killed the European corn borer and related species. Subsequent *Bt* genes were introduced that killed corn rootworm larvae. Genetically modified maize was grown for the first time in the US and Canada in 1997. Since then, the field area used for GM varieties has increased to 57.4 million hectares in 2013 primarily in the USA, Argentina, Canada, South Africa, Uruguay, Egypt, the Philippines and South America. Nearly a third of the current maize production worldwide is based on GM maize.

The first official report of resistance of a maize pest to Cry 1Ab maize was made in South Africa during 2007 (Van Rensburg, 2007). This report of field resistance of *Busseola fusca* to *Bt* maize showed that the larvae at certain locations were able to survive in the presence of the *Bt*-toxin but not without some detrimental effect on larval growth rate. Within one year of the first report of resistance of *B. fusca*, another reportedly resistant population was observed by farmers at the Vaalharts irrigation scheme, approximately 50 km from the initial site. Follow-up surveys conducted between 2010 and 2011 showed that resistant pest populations occurred in many areas within the maize production region of the country (Kruger *et al.*, 2012). Based on the incidence of farmers that sprayed insecticides for borer control on *Bt* maize, it was concluded that *B. fusca* resistance to *Bt* maize (MON810) was wide spread in the country.

Field-evolved resistance of *S. frugiperda* to *Bt* corn producing Cry1F occurred in 4 years in the United States territory of Puerto Rico (Matten *et al.*, 2008), making this the fastest documented case of field-evolved resistance to a *Bt* crop. This is also the first case of resistance leading to withdrawal of a *Bt* crop from the marketplace.

Carriere *et al.* (2010) indicated that the Western corn rootworm is evolving resistance to Cry3Bb1 maize in some populations in Iowa, USA. This is the first case of the Western corn rootworm, or any species of beetle, evolving resistance to a *Bt* toxin in the field. Gassman *et al.* (2011) reported that Western corn rootworm populations sampled from problem fields in Iowa, USA showed statistically significant, but not complete, resistance to the *Bt* maize. Resistance was found only for Cry3Bb1 maize, the type of *Bt* maize that had been grown historically in those fields, and was not present for maize that produced *Bt* toxin Cry34/35Ab1.

Since the year 2000 it has been observed that genetically engineered corn expressing the *Bt* toxin Cry1Ab is being infested by Western bean cutworm (*Striacosta albicosta*) (Rice, 2000; O'Rourke and Hutchison, 2000). Interaction between the Western bean cutworm and the corn earworm (cannibalistic to other pest insects such as the Western bean cutworm) was confirmed in 2010 (Dorhout and Rice, 2010), showing that the spread of the Western bean cutworm was in fact fostered by the cultivation of *Bt* corn expressing Cry1Ab. Damages caused by the Western bean cutworm can even exceed those caused by the European corn borer in conventional plants (Catangui and Berg, 2006).

Bt cotton

Bt cotton is the second most important transgenic insect resistant crop. Several insect resistant cotton events, which confer resistance to lepidopteran pests of cotton, have been approved in 15 countries so far (James, 2013). The era of transgenic cotton begin in 1990 with introduction of Cry 1Ab and Cry 1Ac genes into cotton plants and transformed plants showed high level of resistance to *Helicoverpa*. In USA, the first product 'bollgard' was launched in 1996 by the Monsanto company. Transgenic cottons resistant to lepidopteran pests are commercially cultivated in Australia, Mexico, China, India, USA and Argentina.

Dhrua and Gujar (2011) reported that resistance to Cry1Ac had evolved by 2008 in a population of *P. gossypiella* sampled from non-*Bt* cotton in the Amreli district of

Gujarat in India. The median lethal concentration of Cry1Ac for five-day-old larvae (LC_{50}) was significantly higher for insects derived in 2008 from Amreli than for any of the other field populations tested from four locations in India. For Cry1Ac, the mean LC_{50} for the strain derived from Amreli in 2008 was 44 times higher than for the most susceptible population. However, for seed powder of Bollgard II containing primarily Cry2Ab2, the 2008 Amreli population was only slightly less susceptible than the most susceptible population. Generally, the toxin needs to be expressed at concentrations of more than 0.2 per cent of total soluble protein in the appropriate tissue in a transgenic plant to be effective (Gatehouse, 2008). Downes *et al.* (2010) showed a significant exponential increase in the frequency of alleles conferring Cry2Ab resistance in Australian field populations of *H. punctigera* since the adoption of a second generation, two-toxin *Bt* cotton expressing this insecticidal protein. Furthermore, the frequency of Cry2Ab resistance alleles in populations from cropping areas was 8-fold higher than that found for populations from non-cropping regions. Significant decrease in susceptibility to Cry1Ac was detected in some field populations from northern China but control failures of *Bt* cotton have not been reported (Zhang *et al.*, 2011, 2012). Caccia *et al.* (2010) reported that in Australia, a transgenic variety of cotton expressing Cry1Ac and Cry2Ab (Bollgard II) comprises at least 80 per cent of the total cotton area. Prior to the widespread adoption of Bollgard II, the frequency of alleles conferring resistance to Cry2Ab in field populations of *H. armigera* and *H. punctigera*, was significantly higher than anticipated. Colonies established from survivors of F_2 screens against Cry2Ab were highly resistant to this toxin, but susceptible to Cry1Ac.

Wan *et al.* (2012) reported field-evolved resistance to Cry1Ac by pink bollworm in the Yangtze River Valley of China. Laboratory bioassay data from 51 field-derived strains show that the susceptibility to Cry1Ac was significantly lower during 2008 to 2010 than 2005 to 2007. The percentage of field populations yielding one or more survivors at a diagnostic concentration of Cry1Ac increased from 0 per cent in 2005–2007 to 56 per cent in 2008–2010. However, the median survival at the diagnostic concentration was only 1.6 per cent from 2008 to 2010 and failure of *Bt* cotton to control pink bollworm has not been reported in China.

Tabashnik *et al.* (2013) showed that *H. zea* survival on *Bt* cotton in the field was higher for resistant populations than for susceptible populations. In bioassays with *Bt* cotton leaves, a field-derived strain of *H. zea* from Mississippi (USA) with 44- fold resistance to Cry1Ac had 52 per cent survival versus zero per cent for a susceptible strain.

Field-evolved resistance to Cry1Ac in cotton bollworm and pink bollworm was detected with laboratory diet bioassays, but there were no reports of control failures of *Bt* cotton in China. the highest percentage of resistant individuals detected with bioassays in 2010 was less than 5 per cent for both pests. In 2010, the highest survival at the respective diagnostic concentration for each species was 2.6 per cent for cotton bollworm (Anyang population) (Zhang *et al.*, 2011) and 4.1 per cent for pink bollworm (Jianli population) (Wan *et al.*, 2012).

Lu *et al.* (2010) reported that field trials conducted over ten years in northern

China show that mirid bugs (Heteroptera: Miridae) have progressively increased population sizes and acquired pest status in cotton and multiple other crops, in association with a regional increase in *Bt* cotton adoption.

Zhao *et al.* (2011) surveyed 1,000 randomly selected farm households in five provinces in China and found that the reduction in pesticide use in *Bt* cotton cultivars is significantly lower than that reported in research elsewhere. This is consistent with the hypothesis suggested by recent studies that more pesticide sprayings are needed over time to control emerging secondary pests, such as aphids, spider mites, and lygus bugs.

11.2.2. Resistance to Dipteran-specific *Bacillus* Toxins

Products based on *Bacillus thuringiensis* subsp. *israelensis (Bti)* and *Lysinibacillus (=Bacillus) sphaericus (Bs)* are marketed for mosquito larval control and are widely used in the USA and Europe. *Bs*, because of its better residual activity in polluted waters, has been broadly used against *Culex* species in the US, Central America, Brazil, India, Thailand and China.

11.2.2.1. *Bacillus thuringiensis* subsp. *israelensis*

Bti forms a crystalline parasporal body composed of protein protoxins (δ-endotoxins) that are also used as a commercial biopesticide against larvae of noxious arthropod species of the suborder *Nematocera*, including mosquitoes, black flies and chironomid midges. *Bti* is much more effective against many species of mosquito and black fly larvae than any previously known bio-control agent.

Resistance to *Bti* has been extensively searched in natural populations of mosquitoes (Vasquez *et al.*, 2009, Loke *et al.*, 2010, Kamgang *et al.*, 2011) and it was reported that there was lack of *Bti* resistance in field populations even after decades of *Bti* treatments, although a few studies described a decreased susceptibility in some populations (Zhang *et al.*, 2004, Paul *et al.*, 2005, Boyer *et al.*, 2007, 2012). Recent works on *Bti*-resistance showed that while levels of resistance to *Bti* can be low in a laboratory-selected strain (3.5-fold), resistance to each of the Cry toxins tested separately in bioassays can be much higher (up to 60-fold) (Paris *et al.*, 2011, Tetreau *et al.*, 2012).

The earliest report involved the selection of four populations of *C. quinquefasciatus* against *Bti* from interior (collections 1 and 2) and coastal areas (collections 3 and 4) of southern California (Vasquez-Gomez, 1983). Regardless of the origin of the collection or the level of selection pressure, only modest levels of resistance were attained. Collection 1 reached the highest level of resistance of 16.5-fold at the LC_{95} in generation 46, after which resistance declined, staying around 12-fold. Resistance was unstable in the absence of selection pressure, declining by 50 per cent in 3 generations. Collections 2, 3, and 4 attained 4.4, 4.1, and 5.9-fold resistance at the LC_{95}. *C. pipiens* from Egypt were selected for 20 generations at the LC_{90} and a maximum of 2.8-fold resistance was observed at the LC_{50} (Saleh *et al.*, 2003). Resistance ratio (RR) in *Cx. quinquefasciatus* colony was found 13-fold in 22 generations against *B. thuringiensis* sub-species *jegathesan*. However, RR dropped to 2.3-fold and remained low in 26-40 generations (Wirth *et al.*, 2004).

C. quinquefasciatus from India showed very similar resistance levels after 20 generations of laboratory selection with 2-3 fold resistance (Mittal, 2005). Available data suggested that increasing the number of Cry toxins delayed the evolution of resistance somewhat, but the presence of the Cyt1A toxin, combined with the 3 Cry toxins was essential to delay the evolution of resistance. Paul *et al.* (2005) investigated the levels of resistance in *C. pipiens* populations collected from 2 metropolitan areas (Albany and Syracuse, NY) to 4 larvicides (methoprene, phenothrin, *Bs* and *Bti*) and 1 adulticide (phenothrin) registered for mosquito control in New York State. High levels of resistance were found only to *Bti* and only at 1 site (Syracuse). Resistance levels to the other insecticides were less than 10-fold. Vasquez *et al.* (2009) reported that no resistance was detected against *Bti* in *C. pipiens pipiens* populations on Cyprus sampled over a 6-yr period from 2002 to 2008.

Three colonies of *Aedes aegypti,* consisting of a laboratory strain from Georgia, USA and collections from Sri Lanka and Brazil, was selected with *Bti* for 15 generations at the LC_{50} (Goldman *et al.*, 1986). The resistance ratio for the laboratory colony, and the colony from Sri Lanka, reached 1.1-fold and was not significant. A slight, but statistically significant shift of 2-fold was reported for the Brazil collection. One hypothesis that developed from the early studies with *Bti* was that resistance evolved more slowly and to lower levels because *Bti* produces a complex parasporal crystal with at least 3 diverse Cry toxins and a cytolytic toxin. These diverse toxins might act at different receptors, making the evolution of resistance to all 4 materials very difficult. Jahan and Shahid (2012) reported that the field collected larvae were 10 times more resistant than susceptible population with respect to dose, while RR LT_{50} - RR LT_{90} ranged 1.97-2.22 against *Bti* (WDG) in *Aedes aegypti* larvae.

Tetreau *et al.* (2013) developed a bioassay procedure using each *Bti* Cry toxin separately to detect cryptic *Bti*-resistance evolving in field mosquito populations. Although no resistance to *Bti* was detected in any of the three mosquito species tested (*Aedes rusticus, Aedes sticticus* and *Aedes vexans*), an increased tolerance to Cry4Aa (3.5-fold) and Cry11Aa toxins (8-fold) was found in one *Ae. sticticus* population compared to other populations of the same species, suggesting that resistance to *Bti* may be evolving in this population. This study confirmed previous works showing a lack of *Bti* resistance in field mosquito populations treated for decades with this bioinsecticide. Loke *et al.* (2010) studied susceptibility status of field-collected *Ae. aegypti* from a dengue endemic area to *Bti* and temephos in 2 sites, A and B, in Malaysia. Larvae were bioassayed against *Bti* or temephos in accordance with WHO standard methods. The LC_{50} of *Ae. aegypti* larvae to *Bti* ranged between 0.08890 - 0.1814 mg/L throughout the year in both sites, showing uniform susceptibility of field larvae to *Bti*, in spite of Site A receiving 18 *Bti* treatments over a period of 7 months. *Ae. aegypti* and *Ae. albopictus* were sampled in six urban localities of Central Africa. Larval bioassays, carried out to determine the lethal concentrations (LC_{50} and LC_{95}) and resistance ratios (RR 50 and RR 95) suggested that both vector species were susceptible to *Bti* and temephos (Kamgang *et al.*, 2011).

11.2.2.2. *Lysinibacillus (=Bacillus) sphaericus*. During field trials on *Bs* water-dispersible granules (WDG) against natural populations of *Culex quinquefasciatus* in a low-income community in Thailand, control failure occurred within 4 months after 5 treatments using VectoLex WDG at the dosages of 50 - 200 mg/m^2. The resistance ratios (RR) at LC_{50}, depending on reference colonies, were 21, 100 – 28 and 100 fold against *Bs* WDG and *Bs* technical-grade material. These *Bs*-resistant mosquitoes, however, were completely susceptible to *Bti* preparations and LC_{50} ranging from 0.017 ppm for technical material with 7,000 ITU/ mg to 0.052 ppm for water-dispersible granules with 3,000 ITU/mg; but addition of *Bti* to *Bs* substantially enhanced the mosquitocidal activity (synergism) against these highly *Bs*-resistant *Cx. quinquefasciatus* (Su and Mulla, 2004).

Resistance to *B. sphaericus* has been reported in *B. sphaericus*-treated field populations of the *C. pipiens* complex in Brazil (10 fold resistance; Silva-Filha *et al.*, 1995), India (150 fold; Rao *et al.*, 1995) and France on *C. pipiens* (10,000 fold, Charles and Nielsen-LeRoux, 2000). Reports from China (25,000) and Tunisia (2,000 fold) confirmed that resistance to *Bs* may develop in the field when this bacteria is used intensively. Before records of field resistance to *Bs*, active laboratory selections for resistance had been done in two different laboratories in California (>100,000 fold, Georghiou *et al.*, 1992). Studies were done to investigate the evolution of resistance to *B. sphaericus* strains, C3-41, 2362 and IAB59, in field-collected populations of *C. quinquefasciatus* from China and Brazil under laboratory conditions. Two independent laboratory selections with California mosquitoes (*C. pipiens quinquefasciatus*) have also led to resistance. Levels of stable laboratory-selected resistance of between 35-fold and more than 100,000-fold have been reported, suggesting that there may be different resistance mechanisms (Poopathi and Abida, 2010).

11.2.3. Entomopathogenic Viruses

There are a number of reports of differences in susceptibility among geographically distinct insect host populations to baculoviruses (Briese, 1986, Battu *et al.*, 2002). In general, the levels of decreased host susceptibility to baculoviruses following selection have been low. Field observations of resistance are very rare (Cory and Myers, 2003). Abot *et al.* (1996) reported the selection of resistance to *Anticarsia gemmatalis* multiple nucleopolyhedrovirus (AgMNPV) in strains of the velvetbean caterpillar, *A. gemmatalis,* originating from Brazil and the US. The Brazilian strain reached 1000-fold resistance after 13-15 generations, whereas the strain from the US did not exceed a resistance ratio of five.

Cydia pomonella granulovirus (CpGV) resistance is the first case of field resistance in codling moth, *Cydia pomonella* against a baculovirus. A spread of CpGV resistance is a severe threat to the efficient control of the codling moth, particularly in organic farming. Beginning the 2000s some organic apple growers in south-west Germany first noticed that CpGV treatments failed to control the codling moth in their orchards. In 2003 diapausing codling moth larvae were collected in these orchards and reared to adulthood. Susceptibility of the offsprings was tested in the following year. Surprisingly, two of the three populations were more than 1000-fold less susceptible than a susceptible laboratory strain used as a reference (Fritsch *et al.*, 2005). A second collection of diapausing larvae in autumn 2004 with adjacent susceptibility tests in 2005 confirmed these results and discovered two more resistant populations at different locations (Fritsch *et al.*, 2006). In France collections of overwintering larvae were performed in autumn 2004 and tested in 2005. This study revealed that there were CpGV resistant populations existing also in France (Sauphanor *et al.*, 2006). More than 35 orchards in Germany, France, Italy, Switzerland, and the Netherlands were identified, where codling moth showed very low CpGV susceptibility. Hence, CpGV resistance is geographically widely distributed. In 2005 and 2006, susceptibility tests were performed on 21 field populations from orchards in southern Germany, where farmers reported high infestation rates despite intense CpGV treatment. Fifteen populations showed a resistance ratio of about 1000 and two populations even exceeded a value of 10,000 (Asser-Kaiser *et al.*, 2007).

Schulze-Bopp and Jehle (2013) developed a test based on a 14-day bioassay at a discriminative CpGV concentration of 2× 105 occlusion bodies/ml diet and can be applied to L1–L4 larvae. Information on virus resistance can be obtained within four weeks. In a survey, codling moth larvae were isolated and tested from 6698 samples from 10 different orchards in Germany, Austria, Switzerland, Italy and the Netherlands and they identified seven codling moth populations resistant or partly resistant to CpGV isolates (Bopp and Jehle, 2013).

11.3. CROSS-RESISTANCE AMONG TOXINS

Cross-resistance is said to occur when resistance to one insecticide confers resistance to another insecticide even where the pest has not been exposed to the latter product. Cross-resistance occurs because two or more compounds are acting on the same target site and/or are affected by the same resistance mechanism. Cross-resistance develops most commonly with compounds having the same mode of action and that are usually, but not always chemically related from the same chemical groups (FAO, 2012).

11.3.1. Cross-Resistance Between Cry Proteins

The risk of cross resistance occurring between Cry proteins is substantiated because their mode of action is similar. Though, different Cry proteins can bind to different receptors, there are certain overlapping binding sites which can bind to more than one Cry protein. In one case, one gene was demonstrated to confer resistance to four different Cry proteins (Tabashnik *et al.*, 1997b). Among the four Cry toxins, three (Cry1Aa, Cry1Ab and Cry1Ac) were closely related while the fourth (Cry1F) displayed low sequence similarity with others. Cross-resistance patterns varied considerably among different insects and different Cry toxins.

Selection for Cry1Ac resistance in *H. virescens* (YHD2) (Gould *et al.*, 1995), and *P. gossypiella* (AZP-R and APHIS-98R) (Tabashnik *et al.*, 2000) resulted in high levels of cross resistance to Cry1Aa and Cry1Ab and no or little cross-resistance to Cry1Ca or Cry2Aa. Different strains of the same insect species have different cross-resistance patterns though they were selected using the same Cry protein. Such examples can be observed in Cry1Ac resistant *H. virescens* strains; CP73-3 (>50-fold resistant strain), showed 53-fold cross-resistance to Cry2Aa (Gould *et al.*, 1992), whereas YHD2 strain with over

10,000-fold resistance to Cry1Ac did not have any cross-resistance to Cry2Aa (Gould *et al.*, 1995). In *P. gossypiella,* AZP-R with 3100-fold resistance to Cry1Ac had no cross-resistance to Cry1Ja (Tabashnik *et al.*, 2000). On the other hand, APHIS-98R with just over 100- fold resistance to Cry1Ac had little cross-resistance to Cry1Ja (Tabashnik *et al.*, 2000).

P. xylostella strain NO-QA, which was selected for resistance to a mixed formulation of Cry1A toxins, also exhibited no cross-resistance to Cry1Ca or Cry2Aa (Tabashnik *et al.*, 1996). This condition of high level (>500-fold) resistance to Cry1A toxins that does not lead to cross-resistance to Cry1Ca is defined as 'mode 1' resistance (Tabashnik *et al.*, 1998). A similar mode was observed in *P. interpunctella* strain 198r, which was selected for resistance to a *Bt* strain expressing Cry1A, Cry1C and Cry1D proteins (McGaughey and Johnson, 1994). These observations suggest two basic types of resistance, one exhibiting high-level, narrow-spectrum resistance, and the other featuring moderate-level, broad-spectrum resistance.

Tabashnik *et al.* (2009) reported that laboratory selection of pink bollworm with Cry2Ab caused up to 420-fold cross-resistance to Cry1Ac as well as 240-fold resistance to Cry2Ab. Gulzar *et al.* (2012) investigated cross-resistance mechanism in a *P. xylostella* population possessing single-gene, recessive mode of inheritance but lacking toxin-binding mechanism. Bioassays on unselected (Unsel-Karak) and Cry1Ac-selected (Sel-Karak) populations of *P. xylostella* revealed that deltamethrin, chlorpyrifos and spinosad were significantly more toxic than Cry1Ac. The resistance ratio against Cry1Ac in Sel-Karak population was more than 660-fold compared with the susceptible population (Lab-UK). However, compared to Unsel-Karak the resistance ratio against Cry1Ac in Sel-Karak population was less than10-fold. Velez (2013) indicated that Cry1F showed no significant cross-resistance to Cry1Aa, Cry1Ba and Cry2Aa. Vip3Aa was effective against *Spodoptera frugiperda* resistant and susceptible strains indicating no cross-resistance with Cry1F. In contrast, significant cross-resistance (< 20-fold) was observed for Cry1Ab and Cry1Ac.

11.3.2. Cross-Resistance to *B. thuringiensis* subsp. *israelensis*

A Cry11A resistant colony was examined for cross-resistance to a number of wild-type *Bt* strains that each produced a mixture of both Cry and Cyt toxins. Cross-resistance was detected toward *B. t. jegathesan* (6.7-fold), *B.t. kyu-shuensis* (8.2-fold), and *B.t. fukuokaensis* (8.1-fold). (Cheong *et al.,* 1997). The cross-resistance patterns to different Cry toxins and combinations of Cry toxins, including *Bti,* were reported for the 4 *Bti* selected lines (Wirth and Georghiou, 1997).

The Cq11A line showed cross-resistance at the LC_{95} to Cry4A and Cry4B (41.6-fold), to Cry4A, Cry4B and Cry11A (13.5-fold), but no resistance to *Bti* (1.1-fold). The Cq4AB line showed cross-resistance to Cry11A (350-fold), to Cry4A, Cry4B and Cry11A (16.2-fold), and a low but significant resistance to *Bti* (3.2-fold). The Cq4AB11A line showed high resistance to Cry11A (185-fold), and to Cry4A and Cry4B (12.9-fold) but no significant resistance to *Bti* (1.2-fold). The Cq4AB-11ACytA line, selected with *Bti*, showed resistance to Cry11A (30.1-fold), Cry4A and Cry4B (10.2-fold), Cry4A, Cry4B and Cry11A (8.1-fold). Resistance to *Bti* was

3.2-fold. When recombinant *Escherichia coli* expressing the insecticidal toxins and genes from *Bti* were assayed against these resistant lines, similar patterns of resistance were reported, although the levels of resistance differed (Wirth *et al.*, 2007b).

Cross-resistance toward *Bt jegathesan* and 2 recombinant strains expressing 2 of its Cry toxins was examined in the 4 *Bti* selected lines. Cross-resistance to *Bt jegathesan* was low in all lines. Resistance ratios at the LC_{95} were 6.7/2.8, 2.3, 3.5, and 5.1 for Cq11A, Cq4AB, Cq4AB11A, and Cq4AB11ACytA, respectively (Cheong *et al.,* 1997). Cross-resistance toward Cry11B was 6.8/53.1, 80.7, 347, and 11.8 for the same respective lines. Interestingly, no cross-resistance was detected toward Cry19A from *Bt jegathesan* when the resistant lines were assayed with that material. Resistance ratios were 2.4, 2.6, 3.7, and 1.0 (Wirth *et al.,* 2001). The strong cross-resistance observed among most mosquitocidal Cry toxins, and the presence of high levels of Cry toxin resistance in the *Bti*-selected line, Cq4A11A-CytA, indicate that Cry resistance does evolve in mosquitoes selected with *Bti,* but that resistance is suppressed to some degree by the interactions between the Cry toxins, and is strongly suppressed by the presence of Cyt1A. The direct impact of Cyt1A was demonstrated when *C. quinquefasciatus* were selected with Cry11A, Cyt1A or a mixture of Cry11A and Cyt1A (Wirth *et al.*, 2005). High levels of resistance evolved in the 2 lines selected with a single toxin, but only 8.1-fold resistance was observed with the mixture.

11.3.3. Cross-Resistance to *Lysinibacillus sphaericus*

For the laboratory selected low-level *Bs* resistant colony from California, cross-resistance was found in strain 2297 (Rodcharoen and Mulla, 1996). This was also the case for the field selected population from India (Poncet *et al.*, 1997). There was no cross-resistance to *Bti* within the populations resistant to *Bs* and there was even evidence for an increased susceptibility to *Bti* (Rao *et al.*, 1995, Silva- Filha *et al.*, 1995). In these cases, cross resistance seems to be very limited and extends only to ICPs binding to the same binding site. In contrast, in those cases where resistance is due to another as yet unknown modification, inheritance was found to follow an additive pattern. Levels of resistance were moderate and at least in one case, a more general cross-resistance was observed (Ferre *et al.*, 1995). *B. sphaericus* IAB872 has high toxicity against susceptible *Culex* spp. and medium larvicidal activity against binary toxin-resistant *Culex* spp. Sequence analysis revealed that the sequence of the binary toxin gene from IAB872 was totally identical to that of the reference strain 2362.

11.4. BASIS OF INSECT RESISTANCE TO MCAs

A number of morphological, biochemical and molecular differences have been observed to be related to changes in susceptibility to microbial control agents.

11.4.1. Morphological and Behavioral Basis of Resistance

Ingestion of a sublethal dose of Cry1Ac by fourth instar CP73-3 *H. virescens* larvae resulted in similar histopathological changes in columnar gut cells compared with cell damage in susceptible larvae (Martinez *et al.*, 1999). Likewise, larvae from both a susceptible colony and another resistant *H. virescens* colony (KCB) showed comparable midgut epithelium damage following Cry1Ac ingestion (Forcada *et al.*, 1999). Therefore, it is possible that resistance in CP73-3 and KCB is due to a more efficient repair (or replacement) of damaged midgut cells. Hernández Martinez *et al.* (2010) revealed that midgut proliferation was neither increased in resistant insects nor induced by exposure of susceptible larvae to Xentari™, suggesting that mechanisms other than midgut proliferation were involved in the response to *B. thuringiensis* by *S. exigua* larvae.

Ribeiro *et al.* (2013) reported that susceptible and resistant populations of *P. xylostella* exhibit different histopathology and histochemistry in the midgut upon exposure to *Bt*-based formulations. Groups of larvae from one susceptible population (SP) and one resistant population (RP) of the moth were exposed to Dipel at 1.3 mg/L (SP) and 64.9 mg/L (RP) or XenTari at 5.2 mg/L (SP) and 236.2 mg/L (PR). Midguts from fourth-instar larvae were collected at 0, 1, 6 and 12 hours after exposure and processed for analysis under light microscope. The insecticide Dipel was more aggressive in the midgut regardless of population. Resistant populations exhibited dense granules (spherites) covering the surface of the columnar epithelial cells and mucus that may be involved in the preservation of larvae midgut resistance.

Velez (2013) conducted behavior experiments in Puerto rico which suggested that there was no strong difference between resistant and susceptible phenotypes in *S. frugiperda* and *O. nubilalis* on *Bt* corn. However, behavioral differences were observed between species. *O. nubilalis* exhibited increased movement between leaf discs, with susceptible neonates avoiding Cry1F tissue. In contrast, *S. frugiperda* selected plant tissue regardless of the presence of Cry1F, suggesting that refuge in a bag might be a suitable strategy for this pest.

Zago *et al.* (2014) reported that in Brazil most populations of *P. xylostella* on *Bt* cabbage were resistant to *Bt* products, particularly to Xentari WDG (2–54-fold). Differences in walking characteristics of larvae were variable for most populations, for both Dipel WP and Xentari WDG, but not associated with resistance. Most females preferred to lay eggs on untreated surfaces and showed a reduced proportion of oviposition on treated surfaces that only correlated with resistance to Dipel®WP.

Levy *et al.* (2007) observed reduced susceptibility of *A. gemmatalis* larvae to AgMNPV correlated with heavily chitinous peritrophic membrane in resistant larvae that would prevent the virus passage from the lumen to midgut epithelium to complete the viral infective cycle. Although the peritrophic membrane may be an important barrier to infection by the AgNPV, allowing a much lesser amount of nucleocapsids reaching the midgut cells in comparison to susceptible larvae, other mechanisms of resistance may be involved at the midgut cellular level to avoid further progression of the viral infection.

11.4.2. Biochemical and Physiological Basis of Resistance

Altered proteolytic activation has been reported as mechanism of resistance in many insects including *Spodoptera littoralis, Pieris brassicae, Heliothis virescens, Plodia interpunctella, Choristoneura occidentalis, Melolontha melolontha, Ostrinia nubialis* and *Leptinotarsa decemlineata* (Loseva *et al.*, 2002). Some insects resistant to *Bt* were found to have higher proteolytic activity or a relative higher concentration of proteases in the gut. Toxin sensitivity may be affected by types and/or by the activity levels of gut proteases. Research was conducted on *P. brassicae, Mamestra brassicae* and *S. littoralis*. It was found that there was a direct correlation between the toxicity of *Bt* subsp. *thuringiensis*, gut protein concentration and protease activity (Oppert, 1999). In *Bt* subsp. *entomocidus*-resistant population of *Plodia interpunctella*, significantly lower soluble gut proteinase activities were found. When phenotypic expression of gut proteinases was compared between susceptible and resistant population to *Bt* subsp *entomocidus* the absence of a major serine proteinase activity was observed in the resistant population. This proteinase is involved in *Bt* protoxin activation. Loss of proteinase could lead to toxin resistance. *Bt* resistance and loss of proteinase is genetically linked (Oppert *et al.*, 1997). In a resistant colony of *H. virescens* (CP73-3) slower activation of Cry1Ab protoxin and faster degradation of Cry1Ab toxin was observed when compared with susceptible population (Ferre and Van Rie, 2002).

In *H. armigera* toxin degradation was suggested as the mechanism of resistance. The excessive degradation was caused by chymotrypsin-like proteases (Shao *et al.*, 1998). Excessive toxin degradation was also the suggested cause for resistance to *Bt* in *Choristoneura fumiferana* (Pang and Gringorten, 1998). Another possible mechanism of resistance is sequestration of toxin by gut proteases and this mechanism has been reported in *C.fumiferana* (Milne *et al.*, 1995). The toxic effect of 14 different *Bt* strains were studied on *P.brassicae, H.virescens* and *S.littoralis*. There was a large variation in relative toxicities. This variation in relative toxicities depended on what the insects were fed i.e. they were fed crystals or solubilized crystals or *in vitro*-activated toxins (Jacquet *et al.*, 1987).

Several resistant populations were reported to have more susceptibility towards activated toxin than protoxin. Resistant population of *P. xylostella* showed more susceptibility towards activated Cry1Ac and Cry1Ca toxin than Cry1Ac and Cry1Ca protoxin. Similarly resistant population of *P. interpunctella* and *O. nubialis* showed more susceptibility towards activated Cry1Ab toxin than Cry1Ab protoxin. The results from these populations suggest that only protoxin was affected by the resistance mechanism but not the activated toxin. The reason for this might be that resistant insects are able to reduce the rate of toxin activation or cause over-proteolysis of protoxin which lead to toxin degradation (Herrero *et al.*, 2001). Gonzalez *et al.* (2013) reported that in resistant strain (MR) of *M. unipuncta*, which is capable of completing its life cycle on *Bt* maize, resistance is mediated by alteration of toxin activation rather than to an increase in the proteolytic degradation of the protein.

Tiewsiri and Wang (2011) showed that the Cry1Ac-resistant *T. ni* strain differed from its near-isogenic susceptible strain by two proteins in the midgut brush border membrane, APN1 and APN6. In the Cry1Ac-resistant strain, APN1 expression was significantly reduced to a ratio of 0.11 at the protein level and a ratio of 0.026 at the transcript level to the susceptible strain. In contrast, APN6 was significantly increased, to a ratio of 6.0 at the protein level and a ratio of 39 at the transcript level. Furthermore, the 110-kDa APN1 protein was lacking in the resistant larvae. The differential changes of APN1 and APN6 proteins in the resistant strain were not conferred by mutations of the APN1 and APN6 genes, but instead were regulated at the transcription level by an asyet unidentified trans-regulatory mechanism. The down-regulation of APN1 expression was linked to the resistance to Cry1Ac in *T. ni*.

Tetreau *et al.* (2012) reported that four soluble metalloproteinases were found under-transcribed together with a drastic decrease of metalloproteinases activity in the resistant strain, suggesting a role in resistance by decreasing the amount of activated Cry toxins in the larval midgut. Fuentes *et al.* (2011) reported reduced levels of midgut membrane-bound alkaline phosphatase (mALP) as a common feature in strains of Cry-resistant *Heliothis virescens*, *Helicoverpa armigera* and *Spodoptera frugiperda* when compared to susceptible larvae.

Boyer *et al.* (2012) investigated whether detoxifying enzyme activities such as cytochromes P450, esterases and glutathione S-transferases can play a role in *Bti* resistance mechanisms in these populations. To examine the effect of *Bti* treatment on larval resistance and detoxifying enzyme activities, *Bti* mortality and enzymatic activities were measured in treated and untreated sites coming from four different locations in the Rhone-Alpes region, France and results revealed a significant global decrease in *Bti* mortality in treated populations, but showed low resistance ratio up to 7.9-fold between treated and untreated sites. A positive correlation has been observed between larval resistance to *Bti* and glutathione S-transferase activities, raising the question of its potential role in *Bti* resistance. Tetreau *et al.* (2012) reported that *Bti* Cry toxin receptors including alkaline phosphatases and N-aminopeptidases and toxin-binding V-ATPases exhibited altered expression levels in the resistant strain. The under-expression of putative *Bti*-receptors is consistent with *Bt*-resistance mechanisms previously described in Lepidoptera.

Haas-Stapleton *et al.* (2005) demonstrated that the oral infectivity of a virus could be influenced by the ability of the occlusion-derived visions (ODV) to bind to the midgut of insects. They observed a reduced affinity of AcMNPV ODV to midgut cells of *S. frugiperda* compared to ODVSf MNPV suggesting that SfMNPV might bind to a different receptor than AcMNPV.

11.4.3. Genetic and Molecular Basis of Resistance

Understanding the genetic basis of insect resistance to *Bt* toxins is useful for managing pest resistance to *Bt* crops (Tabashnik *et al.*, 2008). Knowledge of the level of dominance of resistance is especially important because dominance affects the success of the refuge strategy, which is the most widely adopted approach for delaying pest resistance to *Bt* crops worldwide (Tabashnik *et al.*, 2008). This strategy is based on the idea that refuges of non-*Bt* host plants near *Bt* crops provide susceptible insects to mate with resistant insects. Refuges are expected to delay resistance most effectively if resistance is inherited as a recessive trait, because the matings between homozygous-resistant and homozygous-susceptible adults produce heterozygous progeny that are killed by the *Bt* crop. Conversely, if resistance is not recessive and some of the heterozygous progeny survive on the *Bt* crop, refuges are expected to be less effective for delaying resistance. In many resistant populations of *P. xylostella* reduced binding of Cry toxins was found. NOQA population of *P. xylostella* showed high levels of resistance to Cry1Aa, Cry1Ab, Cry1Ac, Cry1Fa and Cry1Ja but not to Cry1Ba, Cry1Bb, Cry1Ca. Experiments conducted with brush border membrane vesicles (BBMV) of NOQA population and Cry1Ac showed a dramatic reduction in the binding of Cry1Ac. Binding of Cry1Ab was virtually absent but binding of Cry1Aa to BBMV was unaltered. These results are in agreement with the proposed model for binding of Cry toxins to the brush border membrane of mid-gut cells of susceptible *P. xylostella* larvae. Similar results were obtained with a *P. xylostella* colony from Pennsylvania. The population showed reduced binding to Cry1Ac and Cry1Ab but binding to Cry1Aa was unaltered. *P. xylostella* Loxa colony also showed similar results. These results showed that there is relationship between reduced binding and resistance. In NOQA and the other *Plutella* strains, although the genetic approach has so far failed to reveal the identity of the resistance gene, it has conclusively eliminated 13 known genes as candidates, including the cadherin. The genetic approach has also revealed the existence of a *cry1Ac* resistance gene distinct from both the 12-domain cadherin and the *Plutella* gene. Linkage mapping has identified a gene (*BtR-5*) on LG10 of *H. virescens* that is the Cry1Ac major resistance factor in the CxC strain (Gahan *et al.*, 2005). None of the seven aminopeptidases mapped in *H. virescens* so far map to LG10, eliminating them as candidates for *BtR-5*. Although midgut aminopeptidases have been widely studied for their ability to bind the Cry1A toxins, the evidence linking them to resistance comes from studies with the Cry1C toxins. Species in the genus *Spodoptera* are generally less sensitive to Cry1A toxins and more sensitive to Cry1C than are the heliothines. In Cry1C-susceptible cotton leafworm *Spodoptera litura*, experimental depression of mRNA levels for a Cry1C binding aminopeptidase by injection of double-stranded RNA confers some protection from Cry1C toxicity (Rajagopal *et al.*, 2002). In the beet armyworm *Spodoptera exigua*, however, there is a Cry1C-resistant strain (Moar *et al.*, 1995) in which mRNA for APN1, one of four aminopeptidases studied, could not be detected by Northern blotting (Herrero *et al.*, 2005). This suggests that low APN1 expression would co-segregate with Cry1C resistance in a linkage analysis, although such a study was not reported.

Xie *et al.* (2005) have shown that single amino acid substitutions can reduce toxin binding, although whether this could confer resistance *in vivo* is unknown. Bravo *et al.* (2004) have hypothesized that cadherin binding by the toxin results in an additional processing step whereby two alpha-helices from Domain 1 are clipped off, promoting toxin oligomerization and subsequent binding to aminopeptidases eventually resulting in pore formation. This model would be consistent with the finding of Adang *et al.* (2005) that peptide fragments from the membrane-proximal region of the cadherin protein unexpectedly increase the toxicity of Cry1A, if these fragments could accelerate the hypothesized additional toxin processing step. Zhang *et al.* (2006) argued that the cell-killing mechanism in cultured cells heterologously expressing the cadherin was not pore formation but rather a signaling pathway initiated by toxin binding to the cadherin and mediated by adenylyl cyclase and Protein Kinase A, although data on this mechanism in midgut cells was lacking.

Caccia *et al.* (2010) showed that binding site alteration, as a mechanism of resistance, is not restricted to the Cry1A class of proteins, but can also extend to the Cry2 class of insecticidal proteins. They examined multiple Cry2A-resistant strains carrying resistance alleles present in field populations and concluded that binding site alteration is the most likely means that field populations evolve resistance to Cry2 proteins in *Helicoverpa* spp. Based on the present and previous studies, it is likely that changes in midgut binding sites may prove to be the most common means that insects evolve field resistance to *Bt* insecticidal proteins.

A *P. xylostella* colony (PHI) from the Philippines showed resistance to Cry1Aa, Cry1Ab and Cry1Ac but was susceptible to Cry1Ca, Cry1Fa and Cry1Ja. This population showed reduced binding of Cry1Ab but binding to Cry1Aa and Cry1Ac was unaltered. Similar results were obtained from *P. xylostella* SERD-3 population from Malaysia. This population showed reduced binding to Cry1Ab but not to Cry1Aa, Cry1Ac and Cry1Ca. Xu and Wu, (2008) showed that BBMV of Cry1Ac resistant *H. armigera* strain GYBT had lower binding affinity to Cry1Ac as compared to the BBMV of Cry1Ac susceptible *H. armigera* strain GY. Wang *et al.* (2007) showed that BBMVs from Cry1Ac resistant strain of *T.ni* lacked specific affinity for binding to Cry1Ab and Cry1Ac. Reduced binding was also observed in resistant strains of *H. virescens* and *S.exigua* (Lee *et al.*, 1995). In pink bollworm, *P. gossypiella*, and two other major lepidopteran pests of cotton, mutations in genes encoding cadherin proteins that bind *Bt* toxins are associated with resistance to *Bt* toxins (Xu *et al.*, 2005). In pink bollworm, three mutant alleles (r1, r2, and r3) linked with resistance to *Bt* toxin Cry1Ac encode incomplete versions of a cadherin protein that binds Cry1Ac in susceptible larvae (Fabrick and Tabashnik, 2007). Khajuria *et al.* (2011) showed that there were changes in two amino acid residues of the aminopeptidase-P like gene (*OnAPP*) involving Glu305 to Lys305 and Arg307 to Leu307 in the two *O. nubilalis* Cry1Ab-resistant strains as compared with three Cry1Ab-susceptible strains. The mature *OnAPP* contains 682 amino acid residues and has a putative signal peptide at the N-terminus, a predicted glycosylphosphatidyl-inositol (GPI)-anchor signal at the C-terminal, three predicted *N*-glycosylation sites at residues N178, N278 and N417, and an *O*-glycosylation site at residue T653. Fabrick and Tabashnik (2012) reported that inheritance of resistance to a diagnostic concentration of Cry1Ac was recessive in Bt4R. In inters-train complementation tests for allelism, F1 progeny from crosses between AZP-R and *Bt4R* were resistant to Cry1Ac, indicating a shared resistance locus in the two strains. Molecular analysis of the *Bt4R* cadherin gene identified a novel 15-bp deletion (r4) predicted to cause the loss of five amino acids upstream of the Cry1Ac-binding region of the cadherin protein. Four recessive mutations in *PgCad1* are now implicated in resistance in five different strains, showing that mutations in cadherin are the primary mechanism of resistance to Cry1Ac in laboratory-selected strains of pink bollworm from Arizona. Zhang *et al.* (2012) identified that laboratory- selected resistance to *Bt* toxin Cry1Ac based on recessive mutations in a gene encoding a toxin-binding cadherin protein in three major cotton pests. Most common resistance alleles in field populations of cotton bollworm, *H. armigera*, selected with *Bt* cotton in northern China, had recessive cadherin mutations, including the deletion mutation identified via laboratory selection. In field-selected populations, recessive cadherin alleles accounted for 75–84 per cent of resistance alleles detected. However, most resistance alleles occurred in heterozygotes and 59–94 per cent of resistant individuals carried at least one nonrecessive resistance allele.

Kranthi *et al.* (2006) elucidated the mode of inheritance of resistance to Cry1Ac toxin of *B. thuringiensis* in *H. armigera* through bioassay analysis of the response of resistant, susceptible, F1 hybrid and backcross *H. armigera* progeny to Cry1Ac using semi-synthetic diet and transgenic *Bt* cotton plants. The dominance value estimates ranged between 0.42 and 0.57. Resistance was found to be monogenic, autosomal and inherited as a semi-dominant trait. Genetic studies of response of *H. armigera* to transgenic *Bt* cotton showed that the effective resistance in transgenic *Bt* cotton plants was also inherited as a semidominant trait. Previous results with laboratory-selected strains indicated that mutations disrupting a cadherin protein that binds Cry1Ac in the larval midgut are tightly linked with recessive resistance to Cry1Ac in three major cotton pests: *H. virescens*, *P. gossypiella*, and *H. armigera* (Xu *et al.*, 2005).

Mahon *et al.* (2012) reported that frequency of Vip3A resistance alleles in *H. armigera* and *H. punctigera* was uncomfortably high albeit essentially recessive in both species. *H. armigera* heterozygotes created by a cross between male susceptible insects and a homozygous Vip3A resistant female were subtly more tolerant of toxin than susceptible insects. Until now, field resistance has been reported to be inherited as an incompletely to highly recessive trait in only one pest, *S. frugiperda*. Field-evolved resistance alleles of *H. armigera* in India and *H. zea* in the United States have been reported to be semi-dominant and incompletely dominant, respectively.

Campagne *et al.* (2013) reported that resistance of *B. fusca* to *Bt* corn was dominant, which refutes the hypothesis of recessive inheritance. Survival on *Bt* corn was not lower than on non-*Bt* corn for both resistant larvae and the F1 progeny from resistant × susceptible parents. Nair *et al.* (2013) reported that variations in the *cadherin* gene sequence between Cry1Ac-resistant BH-R strain showing 227.9-fold resistance and the wild susceptible VA-S strain of *H. armigera* were identified. Amplification of exon 26 from *H. armigera cadherin* gene identified two mutant alleles, *r9* (553 bp) and *r10* (717 bp) against the wild *S allele* (588 bp). Sequence analysis of *r9* showed a deletion of 45 bp in the intron region and mutation in the 5′-splice site of exon 26 resulted in an additional amino acid. While in the r10 allele a premature stop codon resulted in the putative translation into two different proteins of 1343 and 335 amino acids relative to a single protein of 1756 amino acids of the susceptible strain. The segregation of these mutant alleles was examined in F2 progenies derived from matings of Cry1Ac-resistant and susceptible individuals and found to be associated with Mendelian principles. Cadherin genotyping showed that resistance to Cry1Ac, survival on Cry1Ac-treated diet and frequency of mutant r allele of cadherin gene were high in the selected strain than the unselected strain and were completely absent from susceptible VA-S strain. DNA-based screening of *H. armigera* collected in India failed to detect a single *r9* or *r10* allele in the populations collected from vegetable-growing areas of Delhi. However, very low frequency of mutant alleles was detected in *Bt* cotton-growing areas of Anand, indicating that these mutations were likely to be rare in the field.

Velez (2013) reported the inheritance of Cry1F resistance in a *S. frugiperda* resistant strain from Puerto Rico which displayed >387-fold resistance to Cry1F. Concentration-response bioassays of reciprocal crosses of resistant and susceptible parental populations indicated that resistance is recessive and autosomal. Bioassays of the backcross of the F1 generation crossed with the resistant parental strain suggested that a single locus was for resistance.

Bonin *et al.* (2009) have shown that the resistance of a natural *Ae. aegypti* population to *B. thuringiensis* serovar *israelensis* toxins was associated with a polymorphism in the *cadherin* gene. In addition to this polymorphism, the under-expression of this gene was observed. Paris *et al.* (2012) investigated gene transcription level variations associated with *Bti* toxin resistance in the mosquito *Ae. aegypti* using a next-generation sequencing approach. More than 6 million short cDNA tags were sequenced from larvae of two strains sharing the same genetic background: a *Bti* toxins-resistant strain and a susceptible strain. These cDNA tags were mapped with a high coverage (308 reads per position in average) to more than 6000 genes of *Ae. Aegypti* genome and used to quantify and compare the transcription level of these genes between the two mosquito strains. Among them, 86 genes were significantly differentially transcribed more than 4-fold in the *Bti* toxins resistant strain comparatively to the susceptible strain. These included gene families previously associated with *Bti* toxins resistance such as serine proteases, alkaline phosphatase and alphaamylase. Lee (2013) demonstrated that N-cadherin and alkaline phosphotase were associated with Cry11Aa resistance in *Aedes agegypti.* Alkaline phosphatase has been proposed as a secondary receptor mediating Cry11Aa toxicity with *Aedes cadherin*, and hence attenuation of its expression can lead to Cry11A resistance.

All of the *B. sphaericus* resistant *C. pipiens* populations selected on strain 2362, 1593 or C3-41 belonged to the same serotype and had identical genes encoding the binary toxin. However, there were small differences in the amino acid sequences of the *B. sphaericus* Bin toxins, which may be important in the structure and function of the toxin-receptor complex and therefore, for larvicidal activity (Nielsen-LeRoux *et al.*, 2001). Guo *et al.* (2013) studied the molecular basis of the resistance developed by *Culex quinquefasciatus* to *B. sphaericus* C3-41. The *cqm1* genes were cloned from susceptible (CqSL) and resistant (CqRL/C3-41) colonies, respectively. The sequence of the cDNA and genomic DNA derived from CqRL/C3-41 colony differed from that of CqSL one by a one-nucleotide deletion which resulted in a premature stop codon, leading to production of a truncated protein. Recombinant Cqm1S from the CqSL colony expressed in *Escherichia coli* specifically bound to the Bin toxin and had α-glucosidase activity, whereas the Cqm1R from the CqRL/C3-41 colony, with a deletion of three quarters of the receptor's C-terminal lost its α-glucosidase activity and could not bind to the binary toxin. Immunoblotting experiments showed that Cqm1 was undetectable in CqRL/C3-41 larvae, although the gene was correctly transcribed. Thus, the *cqm1R* represents a new allele in *C. quinquefasciatus* that confers resistance to *B. sphaericus*.

Mass-crossing experiments between a susceptible laboratory codling moth strain (CpS) and a field collected codling moth strain, termed CpR, which showed a 100-fold reduced susceptibility against CpGV suggested an autosomal, incomplete dominant inheritance (Eberle and Jehle, 2006). Conversely, single-pair crosses between the susceptible laboratory strain and a genetically homogeneous strain originating from CpR, termed CpRR1, showed that resistance was based on a single, dominant gene that was located on the Z-chromosome (Asser-Kaiser *et al.*, 2007). Asser-Kaiser *et al.* (2010) reported that in case of CpGV resistance monogenic dominant sex-linked mode of inheritance was previously demonstrated in single-pair crosses between a homogeneous resistant (CpRR1) and a susceptible (CpS) laboratory strain. However, resistant field populations (CpR) are more heterogeneous in their levels of resistance, and the possibility that they could harbor different resistance genes to CpRR1 has not been directly addressed. In single pair crossing experiments using a resistant codling moth strain collected from an apple orchard in the southwest of Germany, single-pair crosses within the field strain revealed a genetic basis to the heterogeneity of CpR concerning CpGV resistance. Hybrid crosses to a susceptible laboratory strain and backcrosses of the F1 generation to the resistant CpR strain confirmed that the homogeneous CpRR1 and the heterogeneous field strain CpR share the same mode of inheritance. Thus the variable levels of CpGV resistance in field populations is likely due to frequency differences of the same resistance-conferring gene, rather than different genes.

11.5. STABILITY OF BIOINSECTICIDE RESISTANCE

In most studies on resistance to *Bt* products and toxins, resistance reverted once selection ceased, most likely owing to fitness costs associated with resistance genes or with other loci closely linked to these. However, there are a few examples in which resistance levels remained stable after selection was discontinued. In the 343-R colony of *P. interpunctella*, resistance to Dipel did not decline even after 29 generations on untreated diet (McGaughey and Beeman, 1988). No fitness cost seemed to be associated to resistance in this colony either on untreated diet or on treated diet (Oppert *et al.*, 2000). *P. xylostella* provides three additional examples where resistance remained high and stable after discontinuing selection (Perez and Shelton, 1997; Tabashnik *et al.*, 1995). To measure the rate at which resistance declines, Tabashnik *et al.* (1994a) proposed to use the genetic parameter *R*, which is usually used to measure response to selection (Tabashnik *et al.*, 1994a). *R*, or the average rate of decline in resistance per generation, can be calculated as log (final LC_{50}) - log (initial LC_{50}) divided by the number of generations without exposure to the insecticide, or by an equivalent equation in which the above numerator is substituted by log (final resistance ratio/initial resistance ratio) (Tabashnik *et al.*, 1994a). The inverse of the absolute value of *R* is the number of generations required for a tenfold change in LC_{50}. In *P. xylostella* from Hawaii, a rapid decline in resistance to Dipel (*R* ranged from 0.26 to 0.30) was found in three resistant laboratory-selected colonies derived from the NO field–resistant population (Tabashnik *et al.*, 1994a). The LC_{50} of the most resistant colony (NO-Q, 2800-fold resistant) declined to a similar value to that of a susceptible colony in 13 generations without selection. The decline was much slower in an unselected colony, with 22-fold resistance, for which *R* D= 0.06. In the Loxa A colony of *P. xylostella*, derived from field-resistant insects, resistance to Javelin was unstable from the second to the third generation (*R* D= 0.50), but thereafter it reached a plateau (150-fold to 300-fold resistance) and remained practically stable for at least 7 generations (Tang *et al.*, 1997). Fast reversal of resistance was found in a number of other populations of *P. xylostella* (Hama *et al.*, 1992; Zhao *et al.*, 2000).

Similar results have been reported in other insect species. In *H. virescens* (SEL colony), resistance to Cry1Ab protoxin declined from 69-fold to 13-fold in 5 generations of nonselection (*R* D= 0.14) and seemed to remain stable thereafter (Sims and Stone, 1991). No differences in fitness were detected in a colony derived from these insects (Gould and Anderson, 1991). When selection with Cry1C was discontinued in a colony of *S. littoralis* (resistance >500-fold), resistance declined in one generation, remained relatively stable in the subsequent five generations, and then declined again in the next two generations (*R* < 0.21 for the eight generations) (Muller *et al.*, 1996). There were no significant differences between the resistant colony and a susceptible colony in sex ratio and pupal weights; however, development time was increased in the resistant insects.

In three different experiments with a resistant colony of *L. decemlineata* (80-fold to 223- fold resistant), resistance to Cry3A declined to a stable 30- to 80-fold level in four to eight generations after selection was stopped (*R* D= 0.08 to 0.11) (Rahardja and Whalon,1995). Resistance to Cry1Ac was also found to be unstable in a laboratory-selected colony of *O. nubilalis* (*R* D= 0.31) (Bolin *et al.*, 1999). Undorf-Spahn *et al.* (2012) reported that resistance against the biocontrol agent *C. pomonella* granulovirus (CpGV-M) was previously observed in field populations of codling moth in South-West Germany. Incidental observations in a laboratory reared field colony (CpR) indicated that this resistance was rather stable, even in genetically heterogeneous CM colonies consisting of both susceptible and resistant individuals.

11.6. FITNESS COSTS OF BIOINSECTICIDE RESISTANCE

The development of resistance in any insect is often associated with decreased fitness. The fitness of a resistant strain may be affected in terms of incubation period, larval duration, pupal duration, adult longevity, fecundity, fertility hatchability, diapauses survivability etc.

Reduced survival (51.5 per cent) of Cry1Ac-resistant pink bollworm (PBW) was observed on non *Bt* (NBt) cotton compared to susceptible strains; however, there was no difference in their developmental time. Crosses between resistant and susceptible strains indicated that survival costs could be dominant (Carriere and Tabashnik, 2001). Additionally, the emergence of PBW moths from over wintering pupae was greatly (71 per cent reduction) affected in the Cry1Ac-resistant strain compared to the susceptible strain (Carriere and Tabashnik, 2001). Resistant larvae feeding on *Bt* cotton required an average of 5.7 days longer to develop than susceptible larvae on N*Bt* cotton. This developmental asynchrony, therefore, favors assortative mating among resistant moths from *Bt* plants. In the field, the extent of developmental asynchrony and assortative mating would be affected by variation in toxin expression in plant tissues and crop phenology, weather and the overlapping of generations (Liu *et al.*, 1999). Reduced developmental time and pupal weight was observed in Cry1Ac-resistant *P. gossypiella* when they were tested on increasing concentrations of Cry1Ac in artificial diet. In addition, tests conducted on NBt cotton using resistant and susceptible strains indicated lower survivability, slower development, lower pupal weight and fecundity in the resistant strain compared to the susceptible strain (Liu *et al.*, 2001). Glasshouse experiments using *Bt* cotton showed that 50 and 62 per cent of BX (Cry1Ac resistant) strain of *H. armigera* could complete larval development on initial (<15 weeks) and later (>15 weeks) crop stages, respectively. However, no susceptible larvae survived to pupation on *Bt* cotton plants. However, their developmental rate was faster on NBt cotton compared to the BX strain, which took 7 days more. Though studies on NBt cotton indicated no developmental delay in F1 progeny of reciprocal crosses indicating recessive inheritance of resistance, results from studies on *Bt* cotton showed that F1 progeny can survive to later instars indicating partially dominant inheritance (Bird and Akhurst, 2004). Fitness costs can vary in resistance strains on different host plants. Dipel-resistant *T. ni* performed differently on cucumber, tomato and pepper, among which pepper is least preferred host. None of the (Dipel resistant strain) larvae survived on pepper, contrasting to more than 50 per cent survivors on two other crops (Jaanmat and Myers, 2005).

Raymond *et al.* (2007) investigated whether reduced plant quality would increase the fitness costs associated with resistance to *B. thuringiensis* in two populations of the diamondback moth. Larval development time was extended for both resistant Karak and Keluang populations on the better defended host. However, fitness costs acting on survival depended on interractions between population and enviroment, as found previously (Raymond *et al.,* 2005). Fitness costs seen in larval survival did not always increase as plant quality declined. Both the fitness and the stability experiment indicated that fitness costs were higher on the most suitable plant for one population.

Gulzar (2010) reported that treatment of *H. virescens* (WFO6, unselected) and *P. xylostella* (Karak, unselected) with a sub-lethal concentration of Vip3A was found to increase larval and pupal development time and pupal weight, reduced survival (neonate larvae to adult emergence) and mating pair success, increased fecundity and reduced egg viability and the intrinsic rate of population increase (rm). Various fitness costs were found to be associated with resistance in Vip3A-Sel *H. virescens* and some fitness costs were shown to be temperature dependent: survival (egg to adult), mating pair success, production of viable progeny, egg viability and fecundity, and rm of Vip3A resistant insects being lower at sub-optimal culture temperatures compared with unselected insects. Vip3A resistant larvae were also found to be more susceptible to entomopathogenic nematodes, and had a lower flight muscle ratio, and a greater level of fluctuating asymmetry of morphological traits compared with unselected insects.

Williams *et al.* (2011) reported that concentration of gossypol, a cotton defensive chemical, was higher in pink bollworm larvae with cadherin resistance alleles than in larvae lacking such alleles. Adding gossypol to the larval diet decreased larval weight and survival, and increased the fitness cost affecting larval growth, but not survival. Across cadherin genotypes, the cost affecting larval growth increased as the gossypol concentration of larvae increased. These results suggest that increased accumulation of plant defensive chemicals may contribute to fitness costs associated with resistance to *Bt* toxins.

Shikano and Cory (2014) demonstrated a sex difference in a fitness cost and found that *Bt*-resistant *T. ni* composed a higher ratio of protein to carbohydrate than susceptible *T. ni*. *Bt*-resistant males exhibited no fitness cost, while the fitness cost (reduced pupal weight) was present in resistant females. The absence of the fitness cost in resistant males was associated with increased carbohydrate consumption compared to females.

Paris *et al.* (2011) showed that resistance costs of *B. thuringiensis* subsp. *israelensis* (*Bti*) resistance are expressed at all the life-stages of the dengue vector *Ae. aegypti*, including egg survival, larval development time, and female fecundity. Without selection pressure resistance was quickly lost in the AgMNPV American strain because it was associated with reduced fertility (Fuxa and Richter, 1998). Milks *et al.* (1998) also found a negative correlation between TnSNPV resistance and fertility in *T. ni*.

11.7. MANAGEMENT OF INSECT RESISTANCE TO MCAs

The methods for management of resistance to microbial control agents are similar to those employed in case of chemical insecticides (Dhaliwal and Arora, 2006; FAO, 2012). Georghiou (1994) classified resistance management strategies into 3 broad categories; management by moderation, management by saturation, and management by multiple attacks.

Management by moderation recognizes the value of susceptible genes in insect populations and seeks to preserve that susceptibility by limiting the selection pressure that is applied. Management by saturation indicates saturation of insect defenses by highly targeted insecticides that can kill heterozygous resistant insects while they are relatively rare in populations. One example of saturation was the use of synergism to suppress detoxication enzymes. Management by multiple attacks involves control through the use of several independently acting materials that can be incorporated into rotations or mixture strategies.

11.7.1. Management of Resistance to *B. thuringiensis* and its Toxins

Since *B. thuringiensis* is the most widely used microbial biopesticide and *Bt* transgenic crops also have great commercial importance, substantial literature has accumulated on management of resistance to *B. thuringiensis* and its toxins (Neppl, 2000; Alphey, 2009; FAO, 2012).

11.7.1.1. Keeping a susceptible population

Resistance is a heritable property depending on the frequency of resistance alleles in an insect population. Therefore maintaining a susceptible population serves to delay the development of resistance. In case of exogenous application of *Bt* to crops some of the area may be left untreated and this serves to maintain the susceptible alleles in the population. In case of *Bt* crops, a susceptible population may be maintained by any of the following strategies.

11.7.1.1.1. Release of susceptible insects

Alternative sources of susceptible insects could in theory perform the same function as refuges, but without sacrificing valuable acreage to insect-susceptible crops. In principle, alleles conferring sensitivity to *Bt* toxins could be provided by mass-rearing susceptible insects and releasing them into the environment. Deliberate release of large numbers of the pest insects is unlikely to be viewed as a good idea. However, with modification, the strategy may have merit. The release of mass-reared insects is central to the Sterile Insect Technique (SIT) but release only occurs after the insects have been sterilized by irradiation (Dyck *et al.*, 2005; Knipling, 1955; Krafsur, 1998). Because the released insects are sterile, no viable offspring result from matings between the released insects and the wild population, so there is no introgression of sensitivity alleles into the wild population. The toxin-sensitive alleles provided by traditional SIT are not inherited and so can play no part in the evolution of resistance allele frequency in the pest population. However, in a modification to the SIT the released insects are not irradiated, but instead are homozygous for one or more dominant lethal genes (Release of Insects carrying a Dominant Lethal, RIDL) (Alphey, 2002; Alphey and Andreasen, 2002; Gong *et al.*, 2005). The lethal genes are repressible, using a dietary additive, so that mass-rearing can occur. One version of this system posits female specific dominant lethal constructs and the use of male insects homozygous for one or more of those ("RIDL males"). When RIDL males mate with wild females all of their F1 progeny would inherit a dominant female-specific lethal, so the female offspring would die but the males would live. Like the SIT, the lack of viable female offspring would reduce the reproductive potential of the wild population. The male offspring would inherit genes from their fathers, including the lethal gene and the allele conferring susceptibility to *Bt* toxins. The nature of repressible female-specific lethal genes means that a male-only population for release can be achieved by withdrawing the repressor for the generation intended for release (male-only release is generally desirable for SIT) (Marec *et al.*, 2005; Rendon *et al.*, 2004). Release of such RIDL males would therefore, provide a potential source of sensitivity allele, combined with a pest control mechanism, that could be appropriate for managing pest resistance to *Bt* crops.

11.7.1.1.2. *Refugia*

Based on the simple strategy described above, several resistance programs involve the nearby placement of a susceptible population which, it is hoped, will diffuse into the treated population to mate. This is the basis of the refuge strategy. Refugia can vary in size and placement, and serve as a reservoir of susceptible insects. Ideally, many susceptible individuals will mate with few resistant individuals, creating an overall very low rate of resistance in following generations. The success of refugia depends upon four conditions: that the resistance trait is recessive, that there is random mating, that adults will travel sufficiently between toxic plants, and that there is complete lack of insecticidal action in the refugia. If refugia are subjected to any sort of insecticide, the available susceptible population for mating with *Bt*-

exposed individuals will decrease (Tabashnik, 1997). Wan *et al.* (2012) suggested that main approach for delaying pest adaptation to *Bt* crops uses non-*Bt* host plants as "refuges" to increase survival of susceptible pests. To delay evolution of pest resistance to transgenic cotton producing *Bt* toxin Cry1Ac, the United States and some other countries have required refuges of non-*Bt* cotton, while farmers in China have relied on "natural" refuges of non-*Bt* host plants other than cotton. The "natural" refuge strategy focuses on cotton bollworm (*Helicoverpa armigera*), the primary target of *Bt* cotton in China that attacks many crops, but it does not apply to another major pest, the pink bollworm (*Pectinophora gossypiella*), which feeds almost entirely on cotton in China.

Refugia placed alongside treated areas and external to the field are more successful than rows of refuge plants planted in with rows of *Bt* plants. The larger the refugium, the longer resistance is delayed (Frutos *et al.*, 1999). However, resistance will eventually develop when migration of resistant individuals into the susceptible population finally brings the resistant population of the untreated area to too high a fraction to sustain the equilibrium refuges are designed to create (Comins, 1977). A refuge area covering five to ten per cent of the total crop area was recommended in a 1992 study (Mallet and Porter, 1992). Computer modeling using information about a lepidopteran life cycle indicated that a crop area with 10 per cent refuge content delays resistance for five to 1120 generations (Tabashnik, 1994a). A program with ten per cent refugia was also shown to help *P. xylostella* maintain susceptibility to *Bt* subspecies *aizawai*. Other species, however, did not necessarily respond well to refugia of this size (Liu and Tabashnik, 1997).

While the refuge strategy is successful in a general sense, the correlation between what occurs in laboratory studies and what happens in the ever-variable field is not easy to predict. In 1999, a study indicated that random mating might not necessarily be a safe assumption for insects in the field. *Bt*-resistant *P. gossypiella* populations took an average of 5.7 more days to develop fully than susceptible populations. Because more than 80 per cent of *P. gossypiella* mate within three days of hatching and die soon thereafter; this favors assortative, not random, mating. Susceptible individuals will mate with each other before the resistant individuals even hatch (Liu *et al.*, 1999). Of course, the degree to which this affects *P. gossypiella* refuge strategy in the field depends on generation overlap and perhaps on other factors as well. Another factor that is rather difficult to predict is the migration of adults when finding a mate. This information is helpful in deciding where to place refugia.

The refugia strategy is imperfect, but its successes are well backed by data. It is important to remember that the susceptible population is a resource that can be depleted (Wood, 1981). The refuge method is very often used in combination with other strategies to enhance the effectiveness of both. An unfortunate problem is the potential unwillingness of farmers to set aside area for refugia, which is then not usable cropland. This is a difficult decision to make when competing with other farmers who are not using refugia (Mallet and Porter, 1992). Yenagi *et al.* (2011) suggested adopting okra-chickpea or marigold-chickpea or chilli refuge cropping systems with *Bt* cotton for higher production and greater economic benefits.

In addition, a crucial condition for the effectiveness of refugia is the pest movement throughout a particular crop. In cases of pests with limited movement, such as Western corn rootworm larvae, modeling studies suggest that as grower compliance to refugia decreases, resistance emergence sharply increases (Onstad *et al.*, 2011).

11.7.1.1.3. Seed-mix refuges

Seed mixtures, like refugia, act to delay resistance by maintaining a susceptible population for mating. A field planted using this strategy will result in a random mix of *Bt* and toxin-free plants. Two major studies investigated the efficacy of seed mixtures in comparison with refugia. Mallet and Porter (1992) used computer modeling to show that seed mixtures actually hastened the development of resistance in comparison with fields of solely toxic plants (Mallet and Porter, 1992). Two years later, a contradictory study found that seed mixtures are preferable to pure *Bt* fields (Tabashnik, 1994a). Both studies agree that refugia are more successful than seed mixtures and, in many cases, even more successful when combined with seed mixtures in a single program (Mallet and Porter, 1992; Tabashnik, 1994a).

The reason for the poor performance of seed mixtures as a resistance management strategy is closely related to theoretical advantage of seed mixtures. While individuals from susceptible plants can more easily travel to and mate with *Bt*-exposed and resistant individuals, the ease of travelling for individuals also puts susceptible insects at risk for too much *Bt* exposure from neighboring plants. Working in favor of seed mixtures is the potential behavioral pattern of some insect species to preferentially choose toxin-free over *Bt* plants, thus cutting down on feeding exposure of susceptible insects to *Bt*. This behavior has been observed in laboratory populations of *H. virescens*. However, neither the extent of this behavior in the field nor the propensity of other species toward this behavior is known (Mallet and Porter, 1992). A primary concern relative to seed mixes is that larval movement between *Bt* and non-*Bt* plants might accelerate evolution of pest resistance in seed mixes. Larvae receiving a sub-lethal dose and moving from *Bt* plants to non-*Bt* plants, as well as larger less susceptible larvae moving from non-*Bt* to *Bt* plants, create scenarios where heterozygote fitness and, therefore, selection for resistance are both increased. In addition, movement of susceptible larvae off non-*Bt* plants on to *Bt* plants reduces refuge efficiency by lowering the number of susceptible insects produced by non-*Bt* plants. The impact of reduced refuge efficiency likely will be greatest for *Bt* plants with single proteins exposed to pests with highly mobile larvae, i.e., lepidopteran pests.

11.7.1.1.4. *Mosaics*

Mosaics are like refugia in that they involve a patchwork of *Bt* areas and non-*Bt* areas. However, mosaics are different in that the non-*Bt* areas, instead of being untreated, are treated with some other insecticide – a different strain of *Bt* or something completely unrelated. For this strategy to be effective, it is important that there is no cross-resistance between the two insecticides (meaning resistance to one of them cannot be positively correlated with or helpful in developing resistance to the other) (Tabashnik, 1994b). In fact, negative cross-resistance is preferable (Hoy, 1998). The model, like the refuge strategy, relies on the assumption of sufficient migration of insects between the patches of the mosaic. The mosaic strategy has seen little experimental evaluation, but models have suggested that it is not useful because one ends up simultaneously selecting for resistance to both toxins rather than delaying resistance to either (Tabashnik, 1994b; Frutos *et al.*, 1999). Only one early study provides any support for the mosaic model. This study showed that if a mosquito population is evenly distributed across a mosaic of two equally effective insecticides, and if immigration and emigration rates are equal for all plots, then mosaics can delay the development of insecticide resistance (Curtis, 1981).

11.7.1.1.5. Bioinsecticide rotations

An alternative to the mosaic strategy is the rotation method. Here, a pattern of two or more bioinsecticides or microbials is arranged temporally rather than spatially. Using two or more insecticides or bioinsecticides sequentially rather than simultaneously has been shown to delay resistance relatively (Wood, 1981). Like mosaics, rotations cannot be successful if cross-resistance is present between the insecticides. The premise of rotations is that, because there is a fitness cost associated with resistance, that whenever the population is not exposed to a certain insecticide, the frequency of resistance to that insecticide will drop (Hoy, 1998). Unfortunately, it has been shown that after removal of selection pressure for resistance to *Bt* (during periods where *Bt* is in use), the frequency of resistance in the population remains stable or only decreases very slowly (Tabashnik, 1994b). Additionally, fitness costs associated with resistance have generally been negligible (Hoy, 1998). Thus, there would probably not be considerable pressure selecting for *Bt*-susceptible individuals during the non-*Bt* periods of the rotation.

11.7.1.1.6. Tissue-specific and time-specific toxin expression

Tissue-specific and time-specific toxin expression are management strategies which seek to minimize *Bt* overexposure by engineering transgenic plants to express toxin genes only at times when needed or only in parts of the plant which are most economically important or most vulnerable (Frutos *et al.*, 1999). This is similar to refugia and seed mixtures in that the plant itself provides its own refuges (Mallet and Porter, 1992). Plants could be developed to only produce toxins after a certain threshold of damage. Or, in tissue-specific expression, studies on the interactions between a particular pest and its target plant could suggest strategies to develop. For instance, cotton plants attacked by bollworms could produce toxin only in young boll tissues, as this is the most important part of the plant. In addition to specifically protecting the critical plant tissue, this would affect only one generation of bollworms (that which lives at the time the boll is young), avoiding the constant selection pressure that hastens evolution of resistance (Gould, 1998). Tissue-specific expression would not be a viable option for pests which target nearly the entire plant, such as the European corn borer (Monsanto, 2000). For most species, not enough is known about field behavior, and therefore effective methods of carrying out tissue-specific or time-specific expression of *Bt* are not yet known. Because specific toxin expression, provides refuge in a way similar to seed mixtures, the same negative results could result with use of specific expression as have resulted with seed mixtures.

11.7.1.2. Combining insect control methods

11.7.1.2.1. Use of multiple insecticides/bioinsecticides

The first of these strategies is a simple combination of two or more insecticides on the same field, such as a *Bt* and non-*Bt* toxin. Necessary to this approach is a lack of cross-resistance between the two. However, it is difficult to assume that no degree of cross-resistance will be present between any two insecticides, as it has been shown to occur even between some insecticides with completely different modes of action (Hoy, 1998). No experimental evidence indicates that use of insecticide combinations in lethal doses delays resistance to *Bt* (Tabashnik, 1994b). However, some work with synergists looks somewhat promising. Synergists are chemicals that are not necessarily lethal or toxic alone but when used with an insecticide increase the toxicity of that insecticide. Serine protease inhibitors increase toxicity of *Bt* against some species, but there is no evidence that this greater toxicity brings with it lessened resistance development (Tabashnik, 1994b).

11.7.1.2.2. Pyramided *Bt* crops

Gene stacking is like the combination method explained above, but involves only combinations of two or more different *Bt* Cry toxins, or a *Bt* Cry and a *Bt* Cyt toxin. The toxins are delivered simultaneously but each recognizes a different binding site in the insect midgut (Frutos *et al.*, 1999). This strategy is based on the same assumptions and criteria as the combination method. Some studies show Cry and Cyt toxins working together, with Cyt toxins as a synergist. In a 1998 study, use of *Bt* CytIAa helped overcome 5000-fold resistance to *Bt* CryIIIAa toxins in *Chrysomela scripta* (Frutos *et al.*, 1999). Similarly, sublethal doses of CytA toxins, when combined with CryIV toxins, appear to suppress and/or reduce up to 1000-fold resistance in CryIV-resistant populations of *C. quinquefasciatus* (Wirth *et al.*, 1997). However, considerable evidence works against any strategy using combinations of solely Cry toxins. *H. virescens* shows cross-resistance to many strains of *Bt* Cry toxins; *P. xylostella* and *P. interpunctella* in the field and the laboratory readily evolve resistance to up to five or six *Bt* Cry toxins simultaneously (Tabashnik, 1994b). Downes *et al.* (2010) recorded five year data which showed a significant exponential increase in the frequency of alleles conferring *Cry2Ab* resistance in Australian field populations of *H. punctigera* since the adoption of a second generation, two-toxin *Bt* cotton expressing this insecticidal protein. Furthermore, the frequency of *cry2Ab* resistance alleles in populations from cropping areas was 8-fold higher than that found for populations from non-cropping regions. This report of field evolved resistance to a protein in a dual-toxin Bt-crop has precisely fulfilled the intended function of monitoring for resistance; namely, to provide an early warning of increases in frequencies that may lead to potential failures of the transgenic technology. Furthermore, it demonstrated that pyramids are not 'bullet proof' and that rapid evolution of resistance to *Bt* toxins in the Cry2 class is possible. Huang *et al.* (2012) reported that since 2010, transgenic maize expressing pyramided *Bt* genes has been planted in the US mid-Southern region and by 2011, pyramided *Bt* maize has replaced Cry1Ab maize as the dominant *Bt* maize for managing lepidopteran pests including *Diatraea saccharalis*. The timely switching from single-gene Cry1Ab maize to the pyramided *Bt* maize prevented further increases in Cry1Ab resistance allele frequency and thus ensured the continued success of *Bt* maize for managing *D. saccharalis* in the region.

Pyramiding of the *cry2Ab2* gene with a chimeric toxin gene (*cry1A.105*) composed of portions of the *cry1Ab*, *cry1Ac*, and *cry1Fa* genes, into maize plants (event MON89034, Yieldgard line of products) expanded the range of control to armyworms (*Spodoptera* spp.) and the black cutworm(*Agrotis ipsilon*). Similarly, Agrisure Viptera maize (events MIR162 and Bt11) producing Vip3Aa20 and Cry1Ab toxins can control corn earworm (*Helicoverpa zea*) and the fall armyworm (*Spodoptera frugiperda*), which display low susceptibility to Cry1A toxins. Control of these pest larvae is also accomplished in cotton varieties producing Cry1Ac and Cry1Fa (events DAS 21023-5 and DAS 24236-5, respectively, WideStrike line of products) or Cry1Ac and Cry2Ab (event 15985, Bollgard II) toxin pyramids in cotton. In addition to expanding range of activity to pest species within the same taxonomic order, second generation *Bt* maize varieties have also been developed to control both lepidopteran and coleopteran pests. For example, the Genuity Vt triple line of maize products produces Cry1Ab and Cry2Ab toxins targeting lepidopteran larvae (event MON89034-3), and Cry3Bb toxin to control root-feeding Western corn root-worm (*D. virgifera*) larvae (event MON88017-3). Reductions in mandated refuge size from 20 per cent to 5 per cent have been proposed for pyramided *Bt* crops based on the increased toxicity against targeted pests and expression of multiple *Bt* toxins with diverse mode of action.

Lira *et al.* (2013) reported the efficacy of *B. thuringiensis* Cry protein, Cry1Bh1, against several lepidopteran pests, including laboratory-selected *B. thuringiensis*-resistant strains of *O. nubilalis* and *H. virescens* and progeny of field-evolved *B. thuringiensis*-resistant strains of *P. xylostella* and *S. frugiperda*. Cry1Bh1 is active against susceptible and *B. thuringiensis*-resistant colonies of *O. nubilalis, P. xylostella*, and *H. virescens* in laboratory diet-based assays, implying a lack of cross-resistance in these insects. However, Cry1Bh1 is not active against susceptible or Cry1F-resistant *S. frugiperda*. Further, Cry1Bh1 does not compete with Cry1Fa or Cry1Ab for *O. nubilalis* midgut brush border membrane binding sites. Cry1Bh1-expressing corn, while not completely resistant to insect damage, provided significantly better leaf protection against Cry1Fa-resistant *O. nubilalis* than did Cry1Fa-expressing hybrid corn. The lack of cross-resistance with Cry1Ab and Cry1Fa along with independent membrane binding sites in *O. nubilalis* makes Cry1Bh1 a candidate to further optimize for in-plant resistance to this pest.

11.7.1.2.3. Combining *Bt* with natural enemies

Another potential strategy uses natural enemies, instead of other toxins, in combination with low doses of *Bt*. This strategy cannot be used with most insecticides, as they have broad-range effects and will kill natural enemies as well as pests (Chilcutt and Tabashnik, 1997). Natural enemies used can be predators or parasitoids. The assumption underlying this strategy is one earlier mentioned as part of the rotation strategy. Theoretically, the fitness costs of resistance will make resistant pests more susceptible to attack by natural enemies (Frutos *et al.*, 1999). Any natural enemy used must, of course, be unaffected by exposure to *Bt*. Already, as part of many integrated pest management programs, low doses of *Bt* are used in conjunction with natural enemies when *Bt* is not strong enough in the preferred dosage to control the pest population. A 1997 study investigated the toxic effects of a combination of *Bt* and the parasitoid *Cotesia plutellae* on *P. xylostella*. The interaction of the two did not significantly affect mortality of *P. xylostella* populations resistant to *Bt,* though *Bt*-susceptible populations were affected by the combination more than by either the parasitoid or *Bt* used alone (Chilcutt and Tabashnik, 1997). However, no studies in the field or the laboratory have actually shown a slowing of resistance development by using a *Bt* and natural enemy combination. As mentioned previously, *Bt* resistance has not been shown to have a detectable fitness cost. Additionally, there is no guarantee that a natural enemy will prefer to attack *Bt*-resistant pests, even if they do show detectable fitness costs. If natural enemies attack susceptible and resistant pests, or prefer susceptible pests, they may even hasten resistance development when used with *Bt* (Frutos *et al.*, 1999).

11.7.1.2.4. High dose strategy

The high dose strategy hopes to delay resistance by using a high enough dosage of toxin to kill heterozygous insects – those with alleles for both resistance and susceptibility. Insects with a resistance allele have varying degrees of resistance depending on whether they are carrying one or two alleles for resistance. Theoretically, an increasingly higher dose of toxin will kill insects with an increasing degree of resistance, with a high enough dose killing even those individuals who are resistant homozygotes (Tabashnik, 1994b). A dose high enough to be 100 per cent lethal is, however, not feasible. Therefore, the high dose strategy aims to kill just heterozygotes (and will also kill individuals homozygous for the susceptibility allele).

The success of the high dose strategy depends on rare and recessive or partially recessive resistance alleles (Huang *et al.*, 1999). Controlling the dose carefully is also key to using this strategy. If the dose is allowed to degrade, heterozygotes will survive and pass on the resistance gene. Additionally, heterozygotes must have no advantage over insects homozygous for the susceptibility allele, or the development of resistance can be hastened (Curtis, 1981). In use, the high dose strategy must be combined with refugia to provide susceptible individuals for mating with the resistant individuals that remain after high dose application. In addition, using a combination of toxins helps delay resistance further. Consideration of problems with the refugia and toxin combination strategies must be taken even when these are used in combination with the high dose strategy (Frutos *et al.*, 1999).

Resistance has generally been found to be a recessive or partially recessive trait, though some evidence indicates that some resistance alleles may be dominant or codominant (Tabashnik *et al.*, 2000). In one study, resistance to Dipel ES, a spray formulation of *Bt* using toxins of subspecies *Berliner,* appears to be an incompletely dominant trait in *O. nubilalis* (Huang *et al.*, 1999). This evidence undermines the usefulness of the high dose strategy. However, no similar results have been found in studies of resistance to transgenic plants, which use different toxins (Tabashnik *et al.*, 2000). Another potential problem with the high dose strategy involves natural enemies. If pest populations are eliminated temporarily, natural enemies in the ecosystem which fed upon these pests may leave or die. This leaves room later for a secondary pest outbreak when pests return and the population is not controlled naturally by enemies (Hoy, 1998).

To achieve high doses, synthetic development or enhancement of *Bt* toxins is possible. Another method recently studied involves expression of toxins in different parts of the cell. With tobacco plants engineered to express Cry2Aa2 toxins in the chloroplasts, resistance was dramatically reduced in populations of three different species already showing significant resistance to nuclear expression of Cry2Aa2 (Kota *et al.*, 1999).

11.7.1.3. Planting Trap plants

The trap plant strategy is somewhat similar to refugia, but differs in theory and is therefore in a category all its own. Andow and Alstad developed this theory in 1995, using computer simulations of the behavior of *O. nubilalis*. This strategy uses transgenic *Bt* crops, not spray formulations. A patch of *Bt* crop is planted as a trap, and a non-*Bt* productive crop is grown nearby. The *Bt* patch is planted earlier and matures earlier than the crop, luring in insects, which are then killed by the toxin. Later, the crop matures, but insects have been controlled already by the *Bt* trap (Alstad and Andow, 1995). This strategy is optimized for *O. nubilalis* because the moths emerge in June, an early part of the growing season, after overwintering as mature larvae. At this time they prefer the most mature plants available for laying eggs, which are the trap plants. Unfortunately, computer simulations with species other than *O. nubilalis* showed failure of this strategy. The strategy also would not be successful with species such as *H. virescens* which, as mentioned earlier, have shown some avoidance of *Bt* plants when given a choice between toxin containing and toxin-free plants (Frutos, *et al.*, 1999).

11.7.1.4. Deploying additional genes

Serine protease inhibitors. Serine protease inhibitors have been shown to increase the efficacy of insecticidal activity of Cry toxin in three different insect orders (Coleoptera, Lepidoptera and Diptera). Serine protease inhibitors are present in legume seeds. At higher concentrations serine protease inhibitors kill insects. Serine protease inhibitors when used at a concentration 105 times below their insecticidal level with Cry toxins enhance the insecticidal activity of the Cry toxins. Genetically modified tobacco plant expressing *Cucurbita maxima* trypsin protease inhibitor with Cry toxin showed six times increase in insecticidal activity of Cry toxin when compared with genetic modified tobacco plant expressing only Cry toxin against *H. virescens*. It is suggested that protease inhibitor enhances Cry toxin efficacy by inhibiting gut proteases or by preventing the degradation of membrane bound receptors (Pardo-Lopez *et al.*, 2009).

Chitinase. It has been shown that chitinase when used with Cry toxin has a synergistic effect. Chitinase increases the potency of Cry toxin by ten times. It is suggested that chitinase increases the larvicidal effect of Cry toxin by forming holes in the peritrophic membrane there by making it easy for the Cry toxin to bind to the receptors (Pardo-Lopez *et al.*, 2009).

23.3 kDa CR12-MPED (membrane – proximal extracellular domain). CR12-MPED is the functional receptor region of cadherin for Cry1Ab binding and cytotoxicity. This 23.3kDa peptide fragment was fed with Cry1Ab to *M.sexta* larvae. CR12-MPED peptide fragment increased the toxicity of Cry1Ab. CR12-MPED peptide was also shown to enhance the toxicity of Cry1Ac in other lepidopteran insects. It was suggested that when CR12-MPED is fed with Cry1A toxin, CR12-MPED increases the number of binding sites in the microvilli of the insects by binding to the microvilli thereby increasing the probability of Cry1A interaction with receptor thus increasing the Cry1A toxicity (Chen *et al.*, 2007).

Mutations to increase the toxicity of Cry proteins. Rajamohan *et al.* (1996) showed that asparagine which is present at 372nd position of Cry1Ab amino acid sequence, lies in loop 2 of domain II, when substituted with alanine or glycine increased the toxicity of Cry1Ab by 8-fold against *Lymantria dispar*. A triple mutation in Cry1Ab was shown to increase the toxicity by 36 times against *L.dispar*. Substitutions were made at 282nd (located in α-helix 8), 283rd (located in α-helix8) and 372nd (located in loop 2) positions. At 372nd position asparagine was substituted by alanine. At 282nd position alanine was substituted by glycine and at 283rd position leucine was replaced by serine.

Domain swapping to increase the toxicity of Cry toxin. Cry1Ab is moderately toxic to *S. exigua*. Domain III of Cry1Ab when swapped with domain III of Cry1Ca increased the toxicity of Cry1Ab hybrid (Cry1Ab having domain III of Cry1Ca) toxin against *S.exigua* (de Maagd *et al.*, 1996).

Cry1A modified toxins. Cry1A modified toxins have been created by removing the helix α 1 of domain I. Cry modified toxins do not require interaction with cadherin to form oligomers. So they bypass the cadherin receptor and bind directly to the GPI-anchored receptor and insert into lipid micro domain where these GPI-anchored receptors are localised and pores in the apical membrane of midgut cells are created. This induces osmotic shock, bursting of midgut cells and finally leads to the insect death. Cry1A modified toxins are effective against those insects which are resistant to native Cry1A toxins and whose mechanism of resistance is linked to mutation in cadherin gene (Soberon *et al.*, 2009).

11.7.1.5. Deploying second generation toxic genes

Bacterial Genes. There are other insecticidal proteins from bacteria, which are being utilized in developing plant transgenics. Estruch *et al.* (1996) characterized the novel insecticidal proteins produced by certain *Bt* isolates in logarithmic stage of bacterial growth called vegetative insecticidal proteins (VIPs). Sequences encoding for a VIP have been cloned and the protein has been expressed in *E. coli*. Further, VIP has been successfully expressed in monocots and dicot plants.

Photorhabdus luminescens that dwells inside the gut of entomopathogenic nematode belonging to the family Heterorhabditidae favoured the synthesis of high molecular weight protein complexes toxic to insects ranging from

Lepidoptera, Coleoptera to Dictyoptera. Bowen *et al.* (1998) have characterized four insecticidal toxins from the bacterium *P. luminescens* encoded by toxin complex loci *tca*, *tcb*, *tcc*, and *tcd*, representing the second generation of insecticidal transgenes that will complement the novel *Bt* endotoxin in future.

Plant Genes. Plants are known to release localized and systemic signals in response to wounding by phytophagous insects. Systemic induction of resistance implies the production of a signal at the site of primary wounding or infection, the signal being translocated to the other parts of the plant where it induces defence mechanisms. Among the myriad of secondary metabolites, gallic acid and salicylic acid, which are hydroxyl benzoic acids produced by shikimatic acid pathway by plants, are known to play an important role in insect– plant interactions and have considerable practical application in the area of biotechnology of crop protection.

Salicylic acid. This has been known to be a feeding deterrent for the phytophagous soyabean looper, *Pseudaplusia includens*. Survival and adult emergence of *H. armigera* also varied greatly with concentration tested.

Gallic acid. Apart from antibiotic effect, synergistic interactions involving gallic acid have been observed to increase the effect of *Bt* endotoxin on *H. armigera*.

Proteinase inhibitors. Plants have a wide array of defence proteins, including the proteinaceous proteinase inhibitors (PIs) and lectins induced in response to insect attack. PIs inhibit the gut proteinase of the insect which adversely affects the protein digestion in the gut and forces the insect to synthesize alternative proteases to compensate for the inhibited activity. This leads to deficiency of essential amino acids and exhibits physiological stress on the insect, leading to growth retardation. This mechanism of action minimizes the possibility of developing resistance in the insects and reduces crop damage. A direct proof of the protective role of PIs against insect herbivores was provided by Hilder *et al.* (1987), who showed that the transgenic tobacco plants expressing cowpea trypsin inhibitor (T1) were resistant to the tobacco budworm, *H. virescens*.

Lectins. These are proteins having affinity for specific carbohydrate moieties. They bind to glycoprotein in the peritropic matrix lining of the insect midgut to disrupt the digestive processes and nutrient assimilation. A lectin from snowdrop (*Galanthus nivalis*), when expressed in transgenic tobacco and potato, has been found to be toxic to aphids (Hilder *et al.*, 1995). Incorporation of snowdrop lectin at three concentrations of 0.5, 1.0, and 2.0 per cent of total dietary protein in the artificial diet of *Eoreuma loftini* Dyar significantly decreased larval survivability and adult emergence (Setamou *et al.*, 2003). Several plant lectins have been found to have adverse effects on the larval development and survival of different species including the tomato moth (*Lacanobia oleracea*) (Fitches *et al.*, 1997), *H. armigera* (Shukla *et al.*, 2005), *O. nubilalis* and *Diabrotica undecimpunctata* (Czapla and Lang, 1990), *Callosobruchus maculates* and *Zabrotes subfasciatus* (Osborn *et al.*, 1988), and *Aulacorthum solani* (Down *et al.*, 1996).

a-(alpha) amylase inhibitors. The common bean (*Phaseolus vulgaris*) contains a family of seed-related proteins called *a*-amylase inhibitors (A1). A1 forms a complex with certain insect amylases and is supposed to play a role in plant defence against insects. The introduction and expression of the bean alpha A1 in pea confers resistance to bruchid beetles (Shade *et al.*, 1994). For example, bruchids such as *Zabrotes* can feed on plants producing A1 because they possess a serine proteinase able to cleave some kind of A1. It is therefore difficult to evaluate the long-term benefits of the expression of these genes in plants.

Insect chitinase. Chitin is an insoluble structural polysaccharide that occurs in the exoskeleton and gut lining of insects. It is believed to protect the insect against water loss and abrasive agents. Dissolution of chitin by chitinase (either from insects themselves or from fungi) is known to perforate peritrophic matrix and exoskeleton and make insects vulnerable to attack by different pathogens. Expression of cDNA for chitinase obtained from the tobacco hornworm, *Manduca sexta* in tobacco plants offered partial protection against *H. virescens*.

Recombinant strains. A recent study (Naimov *et al.*, 2014) demonstrated increased *S. exigua* toxicity of a novel Cry9Aa5 when produced in recombinant *E. coli*. Gozmanova *et al.* (2014) reported toxicity of Cry9Aa5 expressed by the native 8S4 strain and recombinant Bt51 strains showed no statistically significant difference in LD_{50} values for both preparations.

11.7.2. Management of Resistance to Bacterial Insecticides in Mosquito Populations

Bti presents a low risk for resistance because it produces a diverse mixture of insecticidal toxins that interact with one another synergistically to increase toxicity towards *Ae. aegypti, C. pipiens,* and *An. stephensi* (Crickmore *et al.*, 1995; Poncet *et al.*, 1995). In addition to increasing activity against mosquitoes, these synergies were shown to slow the rate of evolution of resistance. Cyt1A plays the most significant role, since it was demonstrated that resistance evolves to Cry toxins in CytA's absence (Georghiou and Wirth, 1997) and that resistance is suppressed if Cyt1A is combined with the Cry toxins (Wirth *et al.*, 1997). Furthermore, when Cyt1A is combined with a single Cry toxin, the mixture can significantly delay resistance and reduce the level of resistance compared to the resistance that evolves toward either individual toxin (Wirth *et al.*, 2005). An additional benefit is the short residual time that *Bti* has after application that significantly limits the selection pressure exerted on populations. Inevitably some breeding sites escape treatment so that some proportion of the populations escapes exposure. These natural characteristics place *Bti* within 2 categories of resistance management, moderation due to its short residual and the likely presence of refuges due to incomplete coverage, and saturation of the mosquito's defenses by the synergy of its components. These traits are strongly advantageous to avoiding resistance, although there is no guarantee that resistance will never evolve due to the great variability in field conditions and insect populations.

L. sphaericus lacks this inherent capacity to delay resistance. It targets a single receptor in the mosquito larval midgut, and this single-site action places *Bs* at high risk for selecting resistance. However the cases of *Bs* field resistance have revealed variable rates and levels of resistance evolution resulting from the different treatment practices and environmental conditions unique for each location. Although many of the cases of field-resistance show the rapid evolution of *Bs* resistance (Yuang *et al.*, 2000) treatment with *Bs* does not inevitably lead to resistance. This was observed in a program in Recife, Brazil where 2 sites were followed over a 3-year period during which bi-monthly *Bs* treatments occurred (Silva-Filha *et al.*, 2008). The sites included urban breeding sites such as septic tanks and areas along the banks of the Pinheiros River. Along the river, 25 treatments with *Bs* were followed by 5 cycles of treatment with *Bti*. These sites were monitored because of the risk for resistance but no significant changes in susceptibility were detected from either location. In Thailand, field populations of *C. quinquefasciatus* were successfully treated in the field with a mixture of *Bs* and *Bti,* and no control problems were observed after 9 treatments in 9 months, whereas control problems appeared after 9 treatments with *Bs* alone and control failure occurred after 17 treatments (Mulla *et al.*, 2003). Different isolates of *Bs* may vary in their capacity to select for resistance. Selection studies in the laboratory using *Bs* isolate IAB59 revealed slower rates of resistance evolution and lower levels of resistance than observed in parallel selection lines with isolates 2362 and C3-41 (Pei *et al.*, 2002). Resistance to IAB59 was 5.7 and 46-fold after 18 and 12 generations of selection, whereas resistance levels in the same number of generations was > 100,000-fold for C3-41 and 2362 isolates. Although a later report showed that high levels of resistance could eventually evolve to IAB59 (40,000-fold), the process was slower than for the other isolates. IAB59 is known to produce Mtx toxins in addition to the Bin toxins, but as these toxins are primarily present only during vegetative phase, their contribution to strain activity in their native form is limited. Cry48A and Cry49A, were identified in IAB-59 and show activity to susceptible and Bin-resistant larvae (Jones *et al.*, 2007). Their presence may explain the high activity of IAB-59 toward Bin-resistant *Culex* (Jones *et al.*, 2007).

Laboratory studies have shown that the activity of *Bs* can be enhanced by artificially combining it with naturally co-occurring toxins such as the Mtx toxins or with toxins from other bacteria with mosquitocidal activity such as *Bti.* For example, recombinant bacteria expressing Mtx toxins from *Bs* during sporulation synergized *Bs* and Cry toxins against susceptible and resistant mosquitoes (Wirth *et al.*, 2007a). Cyt1A from *Bti* was also shown to suppress high levels of *Bs* resistance (Wirth *et al.*, 2000) and *Bti* Cry toxins can interact synergistically with *Bs* and reduce resistance (Wirth *et al.*, 2004). For example, a combination of *Bs* and *Bti* (5:1) was sufficient to completely suppress high levels of *Bs* resistance. In addition, Cyt1A or *Bti* combined with *Bs*, extended activity to *Aedes aegypti,* an important vector species that is not naturally susceptible to *Bs* (Wirth *et al.*, 2004). These data and others suggest that engineering *Bs* or *Bti* to express additional mosquitocidal toxins may prove an effective strategy for managing insecticide resistance and extending the host range of *Bs*.

Engineering *Bs* or *Bti* to express additional toxins has been investigated and some of the recombinants showed enhanced activity (Federici *et al.*, 2007). However problems with stability have limited the utility of *Bs* recombinants, while *Bti* has proved more amenable to engineering. Although some early recombinants showed no particular improvement in activity relative to the wild-type strains, recent molecular genetic techniques enabled engineering for the expression of high levels of Bin toxins and *Bti*'s own toxin complement in a recombinant *Bti* strain. That strain showed increased activity against both susceptible and resistant mosquitoes (Park *et al.*, 2005). Laboratory selection studies have validated the design of this recombinant as effective for avoiding the evolution of resistance (Wirth *et al.*, 2010). Other researchers have successfully transferred the toxin coding plasmid from *Bti* into *Bs*, and showed that the recombinant strain was able to overcome *Bs* resistance in mosquitoes (Gammon *et al.*, 2006). Stability of the recombinant was not sustainable. Cry4A from *Bti* was cloned and expressed in *Bs* and the recombinant suppressed *Bs* resistance in larvae (Sun *et al.*, 2001). An acrystalliferous strain of *Bti* was engineered to express Bin, Cry11B and Cyt1A, thus incorporating toxins from *Bs, Bti,* and *Bt jegathesan* (Park *et al.*, 2003). That recombinant had greater activity than either *Bs* or *Bti*. *Bt morrisoni* PG-14, which naturally produces the 4 major endotoxins found in *Bti* plus a 144 kDa Cry protein with Lepidopteran activity, was engineered to express Bin (Park *et al.*, 2005). The recombinant showed improved activity against *C. quinquefasciatus* but not against *Ae. aegypti*. *Bt jegathesan* was also engineered to express Bin and Cyt1A, in addition to its normal complement of endotoxins, and a 17-fold improvement in activity against *C. quinquefasciatus* was observed (Park *et al.*, 2003).

Since refuge size requirements can be large and setting aside untreated mosquito breeding sites is counter to the need for high levels of mosquito control to disrupt disease transmission, large untreated areas are not well suited for management of vector borne diseases. However many breeding sites in a treated area are unreachable or not identified, thus some areas escape treatment and serve as *de facto* refuges. In 2010 experiment with the mosquito *Ae. aegypti* in the Cayman Islands, males of a bi-sex RIDL strain – OX513 - were released in a small town on Grand Cayman (Alphey *et al.*, 2010). Monitoring of the wild population, by trapping adults and setting oviposition traps (to count eggs laid by wild females), showed that these RIDL releases were able to suppress the wild population by approximately 80 per cent compared to nontreatment areas during the trial period.

11.7.3. Virulence Management of *Cydia pomonella* Granulovirus

The success and sustainability of CpGV products for management of codling moth (CM) in the future will depend on the implementation of strategies to avoid selection of resistance and to exploit the natural genetic diversity of different CpGV isolates. This diversity is of outstanding importance as it offers not only new tools in CM control but also provides a unique model to study microevolutionary processes in the virus insect interaction between CpGV and CM, as well as of viruses and host organisms in general. A wise application of new CpGV isolates includes that not only a single genotype is applied but that a mix – different isolates at the same time or consecutively - of different CpGV viruses will be considered. In addition, it is important to note that any control strategy – whether chemical or biological– that depends on a single product or a single mode of action is prone to selection of individuals that are not susceptible to this agent. Therefore, it will be of greatest importance for the control of codling moth in organic farming to diversify the control measures and to integrate all available control methods, e.g. mating disruption, but also harvesting and destroying infested apples during the growing season and controlling the overwintering larvae using entomopathogenic nematodes. By applying these techniques it should be possible to sustain the efficacy of one of the most effective biological control agents in organic fruit production.

All CpGV products, which had been registered in Europe, were based on the so-called Mexican (M) isolate, CpGV-M (Tanada, 1964). It was proposed that alternative isolates with differing virulence might be critical in the resistance management based on different CpGV strains (Jehle *et al.*, 2006). In 2006, the novel isolate CpGV-I12 was identified showing a very good efficacy against resistant CM larvae in bioassays (Eberle *et al.*, 2008). This isolate originated from Iran and was tested in the field (Zingg, 2008; Berling *et al.*, 2009). Also in 2006, Andermatt Biocontrol selected a CpGV virus (termed Madex Plus) by subsequent passage of CpGV-M through resistant CM larvae. Madex Plus received registration in Switzerland in December 2007 and might be registered along with other new isolates in other European countries (Zingg, 2008). In addition, several new CpGV isolates, which fully or partly overcome resistance, were isolated. By genome characterization and genome mapping different genome types of CpGV could be identified (Eberle *et al.*, 2009).

11.8. CONCLUSIONS

Insect resistance to synthetic chemical insecticides is largely a consequence of their extensive and indiscriminate use. The microbial control agents provide a welcome alternative to these toxic chemicals. But insects have already shown the capacity to develop resistance to these bioinsecticides as well. Therefore, there is an urgent need to develop resistance management strategies for judicious utilization of microbial control agents in IPM programmes. The dynamics of crop plant-insect pest-entomopathogen interactions need to be understood at the physiological and molecular level in order to develop suitable bioinsecticide resistance management strategies. Suppression of insect immune responses to the attack of pathogens may prove to be rewarding in delaying as well as in combating resistance. The refuge strategy has so far been successful in delaying resistance development especially in cotton bollworms to Bt cotton. But the same has been less successful in Bt corn against the corn rootworm. It is therefore necessary to develop a basket of strategies for resistance management. The resistance management programmes should form an integral component of broader IPM programmes to reduce selection pressure on target pests. This will serve to extend the useful life of these environmentally benign alternatives to synthetic insecticides for sustainable management of pest-insects in future.

REFERENCES

Abdullah M, Sarnthoy O, Tantakom S, Isichaikul S and Chaeychomsri S (2000) Monitoring Insecticide Resistance Development in Beet Armyworm, *Spodoptera exigua* (Hübner) (Lepidoptera: Noctuidae). *Kasetsart Journal (Natural Science)* **34**: 450 – 457.

Abot AR, Moscardi F, Fuxa JR, Sosa-Gómez DR and Richter AR (1996) Development of resistance by *Anticarsia gemmatalis* from Brazil and the United States to a nuclear polyhedrosis virus under laboratory selection pressure. *Biological Control* **7**: 126–130.

Adang MJ, Hua G, Chen J and Abdullah MAF (2005) Peptides for inhibiting insects. United States Patent Application US2005/0283857.

Akhurst RJ, James WL, Bird LJ and Beard C (2003) Resistance to the Cry1Ac delta endotoxin of *Bacillus thurinngiensis* in the cotton bollworm, *Helicoverpa armigera* (Lepidoptera: Noctuidae). *Journal of Economic Entomology* **96**:1290-1299.

Al-Deeb MA, Wilde GE and Blair JM (2003) Effect of *Bt* corn for corn rootworm control on nontarget soil microarthropods and nematodes. *Environmental Entomology* **32**:1164-1170.

Alphey L (2002) Re-engineering the sterile insect technique. *Insect Biochemistry and Molecular Biology* **32**: 1243-1247.

Alphey L and Andreasen MH (2002) Dominant lethality and insect population control. *Molecular and Biochemical Parasitology* **121**: 173-178.

Alphey N, Bonsall MB and Alphey L (2009) Combining pest control and resistance management: synergy of engineered insects with Bt crops. *Journal of Economic Entomology* **102**: 717-732.

Alphey L, Benedict M, Bellini R, Clark GG, Dame DA, Service MW, Dobson SL (2010) Sterile-insect methods for control of mosquito-borne diseases: an analysis. *Vector Borne Zoonotic Disease* ***10***: 295-311.

Alstad DN and Andow DA (1995) Managing the evolution of insect resistance to transgenic plants. *Science* **268**: 1894-1896.

Altre JA, Grafius EJ and Whalon ME (1996) Feeding behavior of CryIIIA-resistant and susceptible Colorado potato beetle (Coleoptera: Chrysomelidae) larvae on *Bacillus thuringiensis tenebrionis* transgenic CryIIIA-treated and untreated potato foliage. *Journal of Economic Entomology* **89**: 311– 317.

Arora R and Shera PS (2014). Biointensive IPM for ecofriendly management of insect pests In: *Souvenir: National Seminar on Reorientation of Agricultural Research to Ensure National Food Security.* Chaudhary Charan Singh Haryana Agricultural University, Hisar, Haryana. January 6-7, 2014, pp. 78-92.

Arora R, Battu GS and Ramakrishnan N (2000) Microbial Pesticides: Current Status and Future Outlook. In: Dhaliwal GS and Singh B (eds), *Pesticides and Environment*. Commonwealth Publishers, New Delhi, pp 344-395.

Asser-Kaiser S, Fritsch E, Undorf-Spahn K, Kienzle J, Eberle K E, Gund N A and Reineke A (2007) Rapid emergence of baculovirus resistance in codling moth due to dominant, sex-linked inheritance. *Science* **317**:1916–1918.

Asser-Kaiser S, Heckel DG and Jehle JA (2010) Sex linkage of CpGV resistance in a heterogeneous field strain of the codling moth *Cydia pomonella* (L.). *Journal of Invertebrate Pathology* **103**: 59-64.

Ballester V, Escriche B, M´ensua JL, Riethmacher GW and Ferr´e J (1994) Lack of cross-resistance to other *Bacillus thuringiensis* crystal proteins in a population of *Plutella xylostella* highly resistant to Cry1Ab. *Biocontrol Science and Technology.* **4**: 437–443

Ballester V, Granero F, Tabashnik BE, Malvar T and Ferre J (1999) Integrative model for binding of *Bacillus thuringiensis* toxins in susceptible and resistant larvae of the diamondback moth (*Plutella xylostella*). *Applied and Environmental Microbiology* **65**: 1413-1419.

Battu GS, Arora R and Dhaliwal GS (2002) Prospects of Baculoviruses in Integrated Pest Management. In: Koul O and Dhaliwal GS (eds), *Microbial Pesticides.*

Taylor & Francis, London and New York, pp 225-247.

Bauer LS (1995) Resistance: a threat to the insecticidal crystal proteins of *Bacillus thuringiensis*. *Florida Entomologist* **78**: 414-443.

Berling M, Rey JB, Ondet SJ, Tallot Y, Soubabere O, Bonhomme A, Sauphanor B and Lopez-Ferber M (2009) Field trials of CpGV virus isolates overcoming resistance to CpGVM. *Virologica Sinica* **24**: 470-477.

Betz FS, Hammond BG and Fuchs RL (2000) Safety and advantages of *Bacillus thuringiensis* protected plants to control insect pests. *Regulatory Toxicology and Pharmacology* **32**: 156-173.

Bolin PC, Hutchison WD and Andow DA (1999) Long-term selection for resistance to *Bacillus thuringiensis* Cry1Ac endotoxin in a Minnesota population of European corn borer (Lepidoptera: Crambidae). *Journal of Economic Entomology* **92**: 1021–1030.

Bonin A, Paris M, Tetreau G, David JP and Despres L (2009) Candidate genes revealed by a genome scan for mosquito resistance to a bacterial insecticide: sequence and gene expression variations. *BMC Genomics* **10:** 551.

Bowen D, Rocheleau TA, Blackburn M, Andreev O, Golubeva E, Bhartia R and Constant RH (1998) Novel insecticidal toxins from bacterium *Photorhabdus luminescens*. *Science* **280**: 2129–2132.

Boyer S, Tilquin M and Ravanel P (2007) Differential sensitivity to *Bacillus thuringiensis* var. *israelensis* and temephos in field mosquito populations of *Ochlerotatus cataphylla* (Diptera: Culicidae): Toward resistance? *Environmental Toxicology and Chemistry* **26**: 157-162.

Boyer S, Paris M, Jego S, Lemperiere G and Ravanel P (2012) Influence of insecticide *Bacillus thuringiensis* subsp. *israelensis* treatments on resistance and enzyme activities in *Aedes rusticus* larvae (Diptera: Culicidae). *Biological Control* **62**: 75-81.

Bravo A, Gomez I, Conde J, Munoz-Garay C, Sanchez J, Miranda R, Zhuang M, Gill SS and Soberon M (2004) Oligomerization triggers binding of a *Bacillus thuringiensis* Cry1Ab pore-forming toxin to aminopeptidase N receptor leading to insertion into membrane microdomains. *Biochimica et Biophysica Acta* **1667**: 38-46.

Briese DT (1986) Insect resistance to baculoviruses. In: Granados RR and Federici BA (eds) *The Biology of Baculoviruses, Vol. 2, Practical Application for Insect Control.* Boca Raton, CRC Press pp. 237-263.

Caccia S, Hernández-Rodríguez CS, Mahon RJ, Downes S, James W, *et al.* (2010) Binding site alteration is responsible for field-isolated resistance to *Bacillus thuringiensis* Cry2A insecticidal proteins in two *Helicoverpa* species. *PLoS One*. **5**: e9975. doi:10.1371/journal.pone.0009975

Campagne P, Kruger M, Pasquet R, Le Ru B and Van den Berg J (2013) Dominant Inheritance of field-evolved resistance to *Bt* corn in *Busseola fusca*. *PLoS One*. **8**(7): e69675. doi:10.1371/journal.pone. 0069675

Carriere Y and Tabashnik BE (2001) Reversing insect adaptation to transgenic insecticidal plants. *Proceedings of the Royal Society B: Biological Sciences* **268**: 1475-1480.

Carriére Y, Crowder DW and Tabashnik BE (2010) Evolutionary ecology of insect adaptation to Bt crops. *Evolutionary Applications* **3**: 561-573.

Castagnola AS and Jurat-Fuentes JL (2012) Bt crops: past and future. In: Sansinenea E (ed.) *Bacillus thuringiensis Biotechnology*. Springer, pp 283- 304.

Catangui MA and Berg RK (2006) Western bean cutworm, *Striacosta albicosta*

(Smith) (Lepidoptera: Noctuidae), as a potential pest of transgenic Cry1Ab *Bacillus thuringiensis* corn hybrids in South Dakota. *Environmental Entomology* **35**: 1439–1452.

Chandrashekar K and Gujar GT (2004) Development and mechanisms of resistance to *Bacillus thuringiensis* endotoxin CrylAc in the American bollworm, *Helicoverpa armigera* (Hubner). *Indian Journal of Experimental Biology* **42**:164-173.

Charles JF and Nielsen-LeRoux C (2000) Mosquitocidal bacterial toxins: diversity, mode of action and resistance phenomena. *Memórias do Instituto Oswaldo Cruz* 95 Suppl., **1**: 201-206.

Chen J, Hua G, Jusat-Fuentes JL, Abdullah MA and Adang MJ (2007) Synergism of *Bacillus thuringiensis* toxins by a fragment of a toxin binding cadherin. *Proceedings of the National Academy of Science, USA.* **104:** 13901-13906.

Cheong H, Dhesi RK and Gill SS (1997) Marginal cross-resistance to mosquitocidal *Bacillus thuringiensis* strains in Cry11A-resistant larvae: presence of Cry11A-like toxins in these strains. *FEMS Microbiology* **153**: 419-424.

Chilcutt CF and Tabashnik BE (1997) Independent and combined effects of *Bacillus thuringiensis* and the parasitoid *Cotesia plutellae* (Hymenoptera: Braconidae) on susceptible and resistant diamondback moth (Lepidoptera: Plutellidae). *Journal of Economic Entomology* **90**: 397-403.

Christoph T (2010) New pest in crop caused by large scale cultivation of Bt corn. In: Breckling B and Verhoeven R (eds) *Implications of GM-Crop Cultivation at Large Spatial Scales. Theorie in der Ökologie* 16. Frankfurt, Peter Lang. 137 pp.

Comins HN (1977) The development of insecticide resistance in the presence of migration. *Journal of Theoretical Biology* **64**: 177-197.

Cory JS and Meyers JH (2003) The ecology and evolution of baculoviruses. *Annual Review of Ecology, Evolution and Systematics.* **34:** 239-272.

Crickmore N, Bone EJ, Williams JA and Ellar DJ (1995) Contribution of the individual components of the δ-endotoxin crystal to the mosquitocidal activity of *Bacillus thuringiensis* subsp. *israelensis*. *FEMS Microbiology Letters* **131**: 249-254.

Curtis CF (1981) Possible methods of inhibiting or reversing the evolution of insecticide resistance in mosquitoes. *Pesticide Science* **12**: 557-564.

Czapla TH and Lang BA (1990) Effect of plant lectins on the larval development of European corn borer (Lepidoptera: Pyralidae) and southern corn rootworm (Coleoptera: Chrysomelidae). *Journal of Economic Entomology* **83:** 2480-2485.

de Maagd RA, Kwa MS, van der Klei H, Yamamoto T, Schipper B, Vlak JM, Stiekema WJ and Bosch D (1996) Domain III substitution in *Bacillus thuringiensis* delta-endotoxin CryIA(b) results in superior toxicity for *Spodoptera exigua* and altered membrane protein recognition. *Applied and Environmental Microbiology* **62**(5): 1537-1543.

Dhurua S and Gujar GT (2011) Field-evolved resistance to *Bt* toxin Cry1Ac in pink bollworm, *Pectinophora gossypiella* (Saunders) (Lepidoptera: Gelechiidae) from India. *Pest Management Science* **67**: 898–903.

Di Fonzo CD and Hammond R (2008) Range expansion of western bean cutworm, *Striacosta albicosta* (Noctuidae), into Michigan and Ohio. *Crop Management*. Online: doi: 10.1094/CM-2008-0519-01-B.

Dorhout DL and Rice ME (2010) Intraguild competition and enhanced survival of western bean cutworm (Lepidoptera:

Noctuidae) on transgenic Cry1Ab (MON810) *Bacillus thuringiensis* corn. *Journal of Economic Entomology* **103**: 54–62. doi: 10.1603/EC09247.

Down RE, Gatehouse AMR, Hamilton WDO and Gatehouse JA (1996) Snowdrop lectin inhibits development and decrease fecundity of the glasshouse potato aphid (*Aulacorthum solani*) when administered *in vitro* and *via* transgenic plants both in laboratory and glasshouse trials. *Journal of Insect Physiology* **42**: 1035-1045.

Downes S, Parker T, Mahon R (2010) Incipient resistance of *Helicoverpa punctigera* to the Cry2Ab Bt toxin in Bollgard II® cotton. *PLOS One* **5**: e12567. doi:10.1371/journal.pone.0012567

Dyck VA, Hendrichs J and Robinson AS (eds.) (2005) *Sterile Insect Technique: Principles and practice in Area-Wide Integrated Pest Management.* Springer, The Netherlands, 801 pp.

Eberle HE, Asser-Kaiser S, Sayed SM, Nguyen HT and Jehle JA (2008). Overcoming the resistance of codling moth against conventional *Cydia pomonella* granulovirus (CpGV-M) by a new isolate CpGV-I12. *Journal of Invertebrate Pathology* **98**: 293-298.

Eberle KE and Jehle JA (2006) Field resistance of codling moth against *Cydia pomonella* granulovirus (CpGV) is autosomal and incompletely dominant inherited. *Journal of Invertebrate Pathology* **93**: 201–206.

Eberle KE, Sayed S, Rezapanah M, Shojai-Estabragh S and Jehle JA (2009) Diversity and evolution of *Cydia pomonella* granulovirus (CpGV). *Journal of General Virology* **90**: 662-671.

Estada U and Ferre J (1994) Binding of insecticidal crystal proteins of *Bacillus thuringiensis* to the midgut brush border of the cabbage looper, *Trichoplusia ni* (Hubner) (Lepidoptera: Noctuidae), and selection for resistance to one of the crystal proteins. *Applied and Environmental Microbiology* **60**: 3840-3846.

Estruch JJ, Warren GW, Mullins MA, Nye GJ, Craig JA and Koziel MG (1996) Vip3A, a novel *Bacillus thuringiensis* vegetative insecticidal protein with a wide spectrum of activities against lepidopteran insects. *Proceedings Natural Academy Science, USA* **93**: 5389-5394.

FAO (2012) Guidelines on prevention and management of pesticide resistance. In: *Intenational Code of Conduct on the Distribution and use of Pesticides.* Food and Agriculture Organization of the United Nations, Rome, Italy, 55 pp.

Fabrick J and Tabashnik B (2007) Binding of *Bacillus thuringiensis* toxin Cry1Ac to multiple sites of cadherin in pink bollworm. *Insect Biochemistry and Molecular Biology* **37**: 97-106.

Fakrudin B, Badariprasad, Prakash SH, Krishnareddy KB and Patil BV (2003) Baseline resistance to Cry1Ac protein in geographic populations of *Helicoverpa armigera* in south Indian Cotton ecosystem. *Current Science* **80**: 1304-1307.

Federici BA and Bauer LS (1998) Cyt1Aa protein of *Bacillus thuringiensis* is toxic to the cottonwood leaf beetle, *Chrysomela scripta*, and suppresses high levels of resistance to Cry3Aa. *Applied and Environmental Microbiology* **64**: 4368–4371.

Federici BA, Park HW and Bideshi DK (2007) Developing recombinant bacteria for control of mosquito larvae. *Journal of the American Mosquito Control Association* **23**(suppl S): 164-175.

Ferre J and Van Rie J (2002). Biochemistry and genetics of insect resistance to *Bacillus thuringiensis. Annual Review of Entomology* **47**: 501-533.

Ferre J, Escriche B, Bel Y and Rie JV (1995) Biochemistry and genetics of insect resistance to *Bacillus thuringiensis* insecticidal crystal proteins *FEMS Microbiology Letters* **132**: 1-7.

Ferr´e J, Real MD, Van Rie J, Jansens S and Peferoen M (1991) Resistance to the *Bacillus thuringiensis* bioinsecticide in a field population of *Plutella xylostella* is due to a change in a midgut membrane receptor. *Proceedings of the National Academy of Sciences, USA* **88**: 5119–5123

Fitches E, Gatehouse AMR and Gatehouse JA (1997) Effects of snowdrop lectins (GNA) delivered via artificial diet and in transgenic plants on the development of tomato moth (*Lacanobia oleracea*) larvae in laboratory and glasshouse trials. *Journal of Insect Physiology* **43:** 727-739.

Forcada C, Alc´acer E, Garcer´a MD, Tato A and Mart´ınez R (1999) Resistance to *Bacillus thuringiensis* Cry1Ac toxin in three strains of *Heliothis virescens*: proteolytic and SEM study of the larval midgut. *Archives of Insect Biochemistry and Physiology* **42**: 51–63

Fritsch E, Undorf-Spahn K, Kienzle J, Zebitz CPW and Huber J (2005). Codling moth granulovirus: First indications of variations in the susceptibility of local codling moth populations. *Nachrichtenbl Deut Pflanzenschutzd* **57**: 29-34.

Fritsch E, Undorf-Spahn K, Kienzle J, Zebitz CPW and Huber J (2006) Codling moth granulovirus: variations in susceptibility of local codling moth populations. In: Boos M (ed) *Proceedings of the 12th International Conference on Cultivation Technique and their Pathological Problem in Organic Fruit Growing.* Weinberg Germany, 31st Jan.-2nd Feb. 2006.

Frutos R, Rang C and Royer M (1999) Managing insect resistance to plants producing *Bacillus thuringiensis* toxins. *Critical Reviews in Biotechnology* **19**: 227-276.

Fuentes G, Dastidar SG, Madhumalar A and Verma CS (2011) Role of protein flexibility in the discovery of new drugs. *Drug Development Research* **72** : 26–35

Gahan LJ, Ma YT, Coble MLM, Gould F, Moar WJ and Heckel DG (2005) Genetic basis of resistance to Cry1Ac and Cry2Aa in *Heliothis virescens* (Lepidoptera: Noctuidae). *Journal of Economic Entomology* **98**: 1357-1368.

Gammon K, Jones GW and Hope SJ (2006) Conjugal transfer of a toxin-coding megaplasmid from *Bacillus thuringiensis* to mosquitocidal strains of *Bacillus sphaericus*. *Applied and Environmental Microbiology* **72**: 1766-1770.

Gassmann AJ, Petzold-Maxwell JL, Keweshan RS, Dunbar MW (2011) Field-Evolved Resistance to Bt Maize by Western Corn Rootworm. *PLOS One*. 6(7): e22629. doi:10.1371/ journal.pone.0022629

Gatehouse JA (2008) Biotechnological prospects for engineering insect resistance in plants. *Plant Physiology* **146:** 881-887.

Georghiou GP (1994) Principles of insecticides resistance management. *Phytoprotection* **75**: 51–59.

Georghiou GP and Wirth MC (1997) Influence of exposure to single versus multiple toxins of *Bacillus thuringiensis* subsp. *israelensis* on development of resistance in the mosquito *Culex quinquefasciatus* (Diptera: Culicidae). *Applied and Environmental Microbiology* **63**: 1095-1101.

Georghiou GP, Malik JI, Wirth M and Sainato K (1992) Characterization of resistance of *Culex quinquefasciatus* to the insecticidal toxins of *Bacillus sphaericus* (strain 2362). University of California, Mosquito Control Research, Annual Report 1992. University of California, Riverside, CA. pp. 34-35.

Glare T, Caradus J, Gelernter W, Jackson T, Keyhani N, Köhl J, Marrone P, Morin L and Stewart A (2012). Have biopesticides come of age? *Trends in Biotechnology* **30**: 250-258.

Goldberg RJ and Tjaden G (1990) Are *B.t.t.* plants really safe to eat. *Biotechnology* **8**: 1011-1014.

Goldman I, Arnold J and Carlton BC (1986) Selection for resistance to *Bacillus thuringiensis* subspecies *israelensis* in field and laboratory populations of the mosquito *Aedes aegypti*. *Journal of Invertebrate Pathology* **47**: 317-324.

Gong P, Epton M, Fu G, Scaife S, Hiscox A, Condon K, Condon G, Morrison N, Kelly D, Dafa'alla T, Coleman P and Alphey L (2005) A dominant lethal genetic system for autocidal control of the Mediterranean fruitfly. *Nature Biotechnology* **23**: 453-456.

González CJ, García M, Hernández-Crespo P, Farinós GP, Ortego F and Castañera P (2013) Resistance to *Bt* maize in *Mythimna unipuncta* (Lepidoptera: Noctuidae) is mediated by alteration in Cry1Ab protein activation. *Insect Biochemistry and Molecular Biology* **43**: 635-643.

Gould F (1998) Sustainability of transgenic insecticidal cultivars: integrating pest genetics and ecology, *Annual Review of Entomology* **43**: 701-726.

Gould F and Anderson A (1991) Effects of *Bacillus thuringiensis* and HD-73 deltaendotoxin on growth, behavior, and fitness of susceptible and toxin-adapted strains of *Heliothis virescens* (Lepidoptera: Noctuidae). *Environmental Entomology* **20**: 30-38.

Gould F, Anderson A, Reynolds A, Bumgarner L and Moor W (1995) Selection and genetic analysis of a *Heliothis virescens* (Lepidoptera : Noctuidae) strain with high levels of resistance to *Bacillus thuringiensis* toxins. *Journal of Economic Entomology*. **92:** 1377-1383.

Gould F, Mart´ınez-Ram´ırez A, Anderson A, Ferr´e J, Silva FJ and Moar WJ (1992) Broad-spectrum resistance to *Bacillus thuringiensis* toxins in *Heliothis virescens*. *Proceedings of the National Academy of Sciences, USA* **89**: 7986–7990.

Gozmanova M, Nedyalkova R, Staykov N, Minkov I and Naimov S (2014) Insecticidal properties of Cry9Aa5 crystal spore preparations. *International Journal of Current Microbiology and Applied Sciences* **3**: 355-358.

Gujar GT, Archana K, Vinay K and Chandrashekar K (2000) Spatial and temporal variation in susceptibility of the American bollworm, *Helicoverpa armigera* (Hubner) to *Bacillus thuringiensis* var. *kurstaki* in India. *Current Science* **78**: 995-1001.

Gulzar A (2010) Studies on resistance and associated fitness costs in Lepidoptera to *Bacillus thuringiensis* toxins. *Ph.D. Thesis*. Imperial College London.

Gulzar A, Sayyed AH, Karimzadeh J and Wright DJ (2012) Lack of cross resistance to non *Bt* insecticides in a mode-1, Cry1Ac-resistant population of *Plutella xylostella* (Lepidoptera: Plutellidae). *Applied Entomology and Phytopathology* **79**: 29-35.

Guo QY, Cai QX, Yan JP, Hu XM, Zheng DS and Yuan ZM (2013) Single nucleotide deletion of cqm1 gene results in the development of resistance to *Bacillus sphaericus* in *Culex quinquefasciatus*. *Journal of Insect Physiology* **59**: 967-973.

Haas-Stapleton EJ, Washburn JO and Volkman LE (2005) *Spodoptera frugiperda* resistance to oral infection by *Autographa californica* multiple nucleopolyhedrovirus linked to aberrant occlusion-derived virus binding in the midgut. *Journal of General Virology* **86**: 1349-1355.

Hama H, Suzuki K and Tanaka H (1992) Inheritance and stability of resistance to *Bacillus thuringiensis* formulations of the diamondback moth, *Plutella xylostella* (Linnaeus) (Lepidoptera: Yponomeutidae). *Applied Entomology and Zoology* **27**: 355–362.

Haseeb M, Liu TX and Jones WA (2004) Effects of selected insecticides on *Cotesia plutellae*, endoparasitoid of *Plutella xylostella*. *BioControl* **49**: 33-46.

Hernández-Martínez P, Navarro-Cerrillo G, Caccia S, de Maagd RA, Moar WJ, Ferre J, Escriche B and Herrero S (2010) Constitutive activation of the midgut response to *Bacillus thuringiensis* in Bt-resistant *Spodoptera exigua*. *PLoS One* **5**: e12795.

Herrero S, Gechev T, Bakker PL, Moar WJ and de Maagd RA (2005) *Bacillus thuringiensis* Cry1Ca-resistant *Spodoptera exigua* lacks expression of one of four aminopeptidase N genes. *BMC Genomic* **6**: 96.

Herrero S, Oppert B and Ferre J (2001) Different mechanisms of resistance to *Bacillus thuringiensis t*oxins in the Indian meal moth. *Applied and Environmental Microbiology* **67**: 1085-1089.

Hilder VA, Gatehouse AMR, Sheerman SE, Barker RF and Boulter D (1987) A novel mechanism of insect resistance engineered into tobacco. *Nature* **330**: 160–163.

Hilder VA, Powell KS, Gatehouse AMR, Gatehouse JA, Gatehouse LN, Shi Y, Hamilton WDO, Merryweather A, Newell CA, Timans JC, Puemans WJ, van Damme E and Boulter D (1995) Expression of snowdrop lectin in transgenic tobacco plants resulted in added protection against aphids. *Transgenic Research* **4**: 18–25

Hodgson E and Gassmann A (2010) Current Challenges in Managing Corn Rootworm. Agriculture and Environment Extension Publications. Book 94. http://lib.dr.iastate.edu/extension_ag_pubs/94.

Hoy MA (1998) Myths, models and mitigation of resistance to pesticides. *Philosophical Transactions of the Royal Society Biological Sciences.* **353**: 1787-1795.

Huang F, Buschman LL, Higgins RA and McGaughey WH (1999) Inheritance of resistance to *Bacillus thuringiensis* toxin (Dipel ES) in the European corn borer. *Science* **284**: 965-967.

Huang F, Higgins RA and Buschman LL (1997) Baseline susceptibility and changes in susceptibility to *Bacillus thuringiensis* subsp. *kurstaki* under selection pressure in European corn borer (Lepidoptera: Pyralidae). *Journal of Economic Entomology* **90**: 1137-1143.

Huang FN, Ghimire MN, Leonard BR, Daves C, Levy R and Baldwin J (2012) Extended monitoring of resistance to *Bacillus thuringiensis* Cry1Ab maize in *Diatraea saccharalis* (Lepidoptera: Crambidae). *GM Crops and Food* **3**:245–254.

Insecticide Resistance Action Committee (2007), Resistance Definition. Retrieved on September 15, 2007. url: http://www.craconline.org/about/resistance

Jahan N and Shahid A (2012) Evaluation of resistance against *Bacillus thuringiensis israelensis* WDG in dengue vector from Lahore, Pakistan. *Pakistan Journal of Zoology* **44**: 945-949.

Jalali SK, Mohan KS, Singh SP, Manjunath TM and Laitha Y (2004) Baseline susceptibility of the old-world bollworm *Helicoverpa armigera* (Hubner) (Lepidoptera: Noctuidae) population from India to *Bacillus thurinngiensis* Cry1Ac insecticidal protein. *Crop Protection* **23**:53-59.

James, Clive. 2013. Global Status of Commercialized Biotech/GM Crops: 2013. *ISAAA Brief* No. 46. ISAAA: Ithaca, NY.

Jaquet F, Hutter R and Luthy P (1987) Specificity of *Bacillus thuringiensis* delta-endotoxin. *Applied and Environmental Microbiology* **53**: 500-504.

Jehle JA, Sayed SM and Wahl-Ermel B (2006). What do we (need to) know about low susceptibility of codling moth against *Cydia pomonella* granulovirus (CpGV)! In: Foko EV (ed.), *Ecofruit - Proceedings to the 12th International conference on cultivation Technique and Phytopathological Problems in Organic Fruit-Growing.* Weinsberg, Germany pp 14-18.

Jones GW, Nielsen-LeRoux C and Yang Y (2007) A new Cry toxin with a unique two-component dependency from *Bacillus sphaericus. Federation of American Societies for Experimental Biology Journal* **21**: 4112-4120.

Kamgang B, Marcombe S, Chandre F, Nchoutpouen E, Nwane P, Etang J, Corbel V and Paupy C (2011) Insecticide susceptibility of *Aedes aegypti* and *Aedes albopictus* in Central Africa. *Parasites and Vectors* **4**: 79.

Khajuria C, Buschman LL, Chen M-S, Siegfried BD and Zhu KY (2011) Identification of a Novel Aminopeptidase P-Like Gene (*OnAPP*) possibly involved in Bt toxicity and resistance in a major corn pest (*Ostrinia nubilalis*). *PLOS One*. **6**: e23983.

Knipling E (1955) Possibilities of insect control or eradication through use of sexually sterile males. *Journal of Economic Entomology* **48**: 459-462.

Kota M, Daniell H, Varma S, Garczynski SF, Gould F and Moar WJ (1999) Overexpression of the *Bacillus thuringiensis* (*Bt*) Cry2Aa2 protein in chloroplasts confers resistance to plants against susceptible and Bt-resistant insects. *Proceedings of the National Academy of Sciences USA* **96**: 1840–1845

Krafsur E (1998) Sterile insect technique for suppressing and eradicating insect populations: 55 years and counting. *Journal of Agricultural Entomology* **15**: 303-317.

Kranthi KR, Dhawad CS, Naidu SR, Mate K, Behere GT, Wadaskar RM and Kranthi S (2006) Inheritance of resistance in Indian *Helicoverpa armigera* (Hübner) to Cry1Ac toxin of *Bacillus thuringiensis*. *Crop Protection* **25**:119–124.

Kranthi KR, Kranthi S and Wanjari RR (2001) Baseline toxicity of CryIAc toxins to *Helicoverpa armigera. International Journal of Pest Management* **45**: 1-5.

Kranthi KR, Kranthi S, Ali S and Banerjee SK (2000) Resistance to Cry1Ac dendrotoxin of *Bacillus thuringiensis* in a laboratory strain of *Helicoverpa armigera* (Hubner). *Current Science* **78**: 1001-1004.

Kruger M, Van Rensburg JBJ and Van den Berg J (2012) Transgenic Bt maize: farmers' perceptions, refuge compliance and reports of stem borer resistance in South Africa. *Journal of Applied Entomology* **136**: 38–50.

Lambert B, Buysse L, Decock C, Jansens S and Piens C (1996) A *Bacillus thuringiensis* insecticidal crystal protein with a high activity against members of the family Noctuidae. *Applied and Environmental Microbiology* **62**: 80–86.

Lee MK, Rajamohan F, Gould F and Dean DH (1995) Resistance to *Bacillus thuringiensis* CryIA δ endotoxins in a laboratory selected *Heliothis virescens* strain is related to receptor alteration. *Applied and Environmental Microbiology* **61**: 3836–3842.

Lee SB (2013) *Toxin binding receptors and the mode of action of Bacillus thuringensis subsp. israelinsis Cry toxins.* Ph.D. Dissertation. University of California. 182 pp.

Levy SM, Falleiros AMF, Moscardi F and Gregório EA (2007) Susceptibility/ resistance of *Anticarsia gemmatalis* larvae to its nucleopolyhedrovirus (AgMNPV): Structural study of the peritrophic membrane. *Journal of Invertebrate Pathology* **96**:183-186.

Li G, Wu K, Gould F, Wang J, Miao J, Gao X and Guo Y (2007) Increasing tolerance to Cry1Ac cotton from cotton bollworm *Helicoverpa armigera*, was confirmed in Bt cotton farming area of China. *Ecological Entomology* **32**: 366–375.

Liang G, Tan W and Guo Y (2000) Study on screening and mode of inheritance to Bt transgenic cotton in *H. armigera. Acta Entomologica Sinica* **43**: 57-62.

Lira J, Beringer J, Burton S, Griffin S, Sheets J, Tan SY, Woosley A, Worden S and Narva KE (2013) Insecticidal activity of *Bacillus thuringiensis* Cry1Bh1 against *Ostrinia nubilalis* (Hübner) (Lepidoptera: Crambidae) and other Lepidopteran pests. *Applied and Environmental Microbiology* **79**: 247590-247597.

Liu YB and Tabashnik BE (1997) Inheritance of resistance to *Bacillus thuringiensis* toxin Cry1C in the diamondback moth. *Applied and Environmental Microbiology* **63**: 2218–2223.

Liu YB, Tabashnik BE and Pusztai-Carey (1996) Field involved resistance to the *Bacillus thuringiensis* toxin Cry1C in diamondback moth (Lepidoptera: Plutellidae). *Journal of Economic Entomology*. **80:** 798-804.

Liu YB, Tabashnik BE, Dennehy TJ, Patin AL and Bartlett AC (1999) Development time and resistance to Bt crops. *Nature* **400**: 519.

Liu YB, Tabashnik BE, Luke M, Baltasar E and Juan F (2000) Binding and toxicity of *Bacillus thuringiensis* protein Cry1C to susceptible and resistant diamondback moth (Lepidoptera: Plutellidae). *Journal of Economic Entomology* **93**: 1-6.

Liu YB, Tabashnik BE, Meyer SK, Carri`ere Y and Bartlett AC (2001) Genetics of pink bollworm resistance to *Bacillus thuringiensis* toxin Cry1Ac. *Journal of Economic Entomology* **94**: 248–252.

Loke SR, Andy-Tan WA, Benjamin S, Lee HL and Sofian-Azirun M (2010) Susceptibility of field-collected *Aedes aegypti* (L.) (Diptera: Culicidae) to *Bacillus thuringiensis israelensis* and temephos. *Tropical Biomedicine* **27**: 493-503.

Loseva O, Ibrahim M, Candas M, Koller CN, Bauer LS and Bulla Jr. LA (2002) Changes in protease activity and Cry3Aa toxin binding in the Colorado potato beetle: implications for insect resistance to *Bacillus thuringiensis* toxins. *Insect Biochemistry and Molecular Biology* **32**: 567-577.

Lu Y, Wu K, Jiang Y, Xia B, Li P, Feng H, Wyckhuys KAG and Guo Y (2010) Mirid Bug Outbreaks in multiple crops correlated with wide-scale adoption of Bt Cotton in China. *Science* **328**: 1151–1154.

Luttrell RG, Wan L and Knighten K (1999) Variation in susceptibility of Noctuid (Lepidoptera) larvae attacking cotton and soybean to purified endotoxin proteins and commercial formulations of *Bacillus thuringiensis. Journal of Economic Entomology* **92**: 21-32.

M¨uller-Cohn J, Chaufaux J, Buisson C, Gilois N, Sanchis V and Lereclus D (1996) *Spodoptera littoralis* (Lepidoptera: Noctuidae) resistance to CryIC and cross resistance to other *Bacillus thuringiensis* crystal toxins. *Journal of Economic Entomology* **89**: 791-797.

Mahon RJ, Downes SJ and James B (2012) Vip3A resistance alleles exist at high levels in Australian targets before release of cotton expressing this toxin. *PLoS One* **7**: e39192

Mallet J and Porter P (1992) Preventing insect adaptation to insect-resistant crops: are seed mixtures or refugia the best strategy? *Proceedings of the Royal Society B: Biological Sciences* **250**: 165-169.

Marec F, Neven L, Robinson A, Vreysen VJM, Goldsmith M, Nagaraju J and Franz G (2005) Development of genetic sexing strains in Lepidoptera: from traditional to transgenic approaches. *Journal of Economic Entomology* **98**: 248-259.

Martinez-Ramirez AC, Gould F and Ferre J (1999) Histopathological effects and growth reduction in a susceptible and a resistant strain of *Heliothis virescens* (Lepidoptera:Noctuidae) caused by sublethal doses of pure Cry1A Crystal proteins from *Bacillus thuringiensis*. *Biocontrol Science and Technology* **9**: 239–246.

Matten SR, Head GP and Quemada H D (2008) How governmental regulation can help or hinder the integration of *Bt* crops into IPM programs. In: Romeis J, Shelton AM and Kennedy GG (eds), *Integration of Insect-Resistant Genetically Modified Crops within IPM Programs* Springer, New York pp. 27– 39.

McGaughey WH (1985) Insect resistance to the biological insecticide *Bacillus thuringiensis*. *Science* **229**: 193-194.

McGaughey WH and Beeman RW (1988) Resistance to *Bacillus thuringiensis* in colonies of Indian meal moth and almond moth (Lepidoptera: Pyralidae). *Journal of Economic Entomology* **81**: 28–33.

McGaughey WH and Johnson DE (1987) Toxicity of different serotypes and toxins of *Bacillus thuringiensis* to resistant and susceptible Indian meal moths (Lepidoptera: Pyralidae). *Journal of Economic Entomology* **80**: 1122–1126.

McGaughey WH and Johnson DE (1992) Indian meal moth (Lepidoptera: Pyralidae) resistance to different strains and mixtures of *Bacillus thuringiensis*. *Journal of Economic Entomology* **85**: 1594–1600.

McGaughey WH and Johnson DE (1994) Influence of crystal protein composition of *Bacillus thuringiensis* strains on cross resistance in Indian meal moths (Lepidoptera: Pyralidae). *Journal of Economic Entomology* **87**: 535-540.

Mendelsohn M, Kough J, Vaituzis Z, Matthews K (2003) Are Bt crops safe? *Nature Biotechnology* **21:** 1003-1009.

Milks ML, Burnstyn I and Myers JH (1998) Influence of larval age on the lethal and sublethal effects of the nucleopolyhedrosis of *Trichoplusia nis* in the cabbage looper. *Biological Control* **12:** 119-126.

Milne RE, Pang ASD and Kaplan H (1995) A protein complex from *Choristoneura fumiferana* gut-juice involved in the precipitation of δ- endotoxin from *Bacillus thuringiensis* subsp. *sotto*. *Insect biochemistry and molecular biology* **25**: 1101-1114.

Mittal P (2005) Laboratory selection to investigate the development of resistance to *Bacillus thuringiensis* var. *israelensis* H-14 in *Culex quinquefasciatus* Say (Diptera: Culicidae). *National Academy Science Letters India* **28**: 281-283.

Moar WJ, Pusztai-Carey M, Faassen HV, Bosch D, Frutos R, Rang C, Luo K and Adang MJ (1995) Development of *Bacillus thuringiensis* CryIC resistance by *Spodoptera exigua* (Hubner) (Lepidoptera: Noctuidae). *Applied and Environmental Microbiology* **61**: 2086-2092.

Monsanto (2000) *The Whole Plant. The Whole Season.* Available at: http://www. monsanto. com/monsanto/agriculture/ current_products/yieldgard_ipcorn/whole _ plant_season.html.

Monsanto (2010) *Cotton in India.* http://www.monsanto.com/monsanto_toda y/for_the_record/india_pink_bollworm.asp.

Mulla MS, Thavara U, Tauratsin A, Chompoori J, Zain M, Su TY (2003) Laboratory and field evaluation of novaluron a new insect growth regulator (IGR) against *Aedes aegypti*. *Journal of Vector Ecology*. **28:** 241-254.

Muller-Cohn J, Chaufax J, Buisson C, Gilois N, Sanchis V and Lereclus D (1996) *Spodoptera littoralis* (Lepidoptera: Noctuidae) resistance to Cry1C and cross resistance to other *Bacillus thuringiensis* crystal toxins. *Journal of Economic Entomology* **89**: 791-797.

Naimov S, Nedyalkova R, Staykov N, Weemen-Hendriks M, Minkov I and de Maagd RA

(2014) A novel Cry9Aa with increased toxicity for *Spodoptera exigua* (Hübner). *Journal of Invertebrate Pathology* **115**: 99-101.

Nair R, Kalia V, Aggarwal KK and Gujar GT (2013) Variation in the cadherin gene sequence of Cry1Ac susceptible and resistant *Helicoverpa armigera* (Lepidoptera: Noctuidae) and the identification of mutant alleles in resistant strains. *Current Science* **104**: 215-222.

Neppl C C (2000) *Management of Resistance to Bacillus Thuringiensis toxins,* The Environmental Studies Program. University of Chicago. Available at *http://camillapede.tripod.com/bapaper.html*

Nielson-LeRoux C, Raghunatha Rao D, rodcharoen JM, Carron A, Mani TR, Hamon S and Mulla MS (2001) Various levels of cross resistance to *Bacillus sphaericus* strains in *Culex pipiens* (Diptera: Culicidae) colonies resistant to *B. sphaericus* strain 2362. *Applied and Environmental Microbiology* **67:** 5049-5054.

O'Rourke PK and Hutchison WD (2000) First report of the western bean cutworm, *Richia albicosta* (Smith) (Lepidoptera: Noctuidae), in Minnesota corn. *Journal of Agricultural and Urban Entomology* **17**: 213–217.

Osborn TC, Alexander DC, Sun SSM, Cardona C and Bliss FA (1988) Insecticidal activity and lectin homology of arcelin seed protein *Science.* **240:** 207-210.

Onstad DW, Mitchell PD, Hurley TM, Lundgren JG, Porter RP, Krupke CH, Spencer JL, Difonzo CD, Baute TS, Hellmich RL, Buschman LL, Hutchison WD and Tooker JF (2011) Seeds of change: corn seed mixtures for resistance management and integrated pest management. *Journal of Economic Entomology* **104**: 343-352.

Oppert B (1999) Protease interactions with *Bacillus thuringiensis* insecticidal toxins. *Archives of Insect Biochemistry and Physiology* **42**: 1-12.

Oppert B, Kramer KJ, Beeman RW, Johnson D and McGaughey WH (1997) Proteinase-mediated insect resistance to *Bacillus thuringiensis* toxins. *Journal of Biological Chemistry* **272**: 23473-23476.

Oppert B, Kramer KJ, Johnson DE, MacIntosh SC and McGaughey WH (1994) Altered protoxin activation by midgut enzymes from a *Bacillus thringiensis* resistant strain of *Plodia interpunctella. Biochemical and Biophysical Research Communications.* **198:** 940-947.

Pang ASD. and Gringorten JL (1998). Degradation of *Bacillus thuringiensis* δ-endotoxin in host insect gut juice. *FEMS Microbiology Letters* **167**: 281-285.

Pardo-Lopez L, Munoz-Garay C, Porta H, Rodriguez-Almazan C, Soberon M and Bravo A (2009) Strategies to improve the insecticidal activity of Cry toxins from *Bacillus thuringiensis. Peptides* **30**: 589-595.

Paris M, Tetreau G, Laurent F, Lelu M, Despres L and David J-P (2011) Persistence of *Bacillus thuringiensis israelensis* (Bti) in the environment induces resistance to multiple Bti toxins in mosquitoes. *Pest Management Science* **67**: 122-128.

Park HW, Bideshi DK and Federici BA (2003) Recombinant strain of *Bacillus thuringiensis* producing Cyt1A, Cry11B, and the *Bacillus sphaericus* binary toxin. *Applied and Environmental Microbiology* **69**: 1331-1334.

Park HW, Bideshi DK, Wirth MC, Johnson JJ, Walton WE and Fedici BA (2005) Recombinant larvicidal bacteria with markedly improved efficacy against culex vectors of West Nile Virus. *American Journal of Tropical Medicine and hygiene.* **75**: 732-738.

Paul A, Harrington L C, Zhang L and Scott J G (2005) Insecticide resistance in *Culex pipi-*

ens from New York. *Journal of American Mosquito Control Association* **21**: 305–309.

Pedigo LP (2002) *Entomology and Pest Management*, 4th edition. Prentice-Hall, Upper Saddle River, NJ, pp 1742.

Pei G, Oliveira CM, Yuan Z, Nielsen-LeRoux C, Silva-Filha MH, Yan J and Regis L (2002) A strain of *Bacillus sphaericus* causes slower development of resistance in *Culex quinquefasciatus*. *Applied and Environmental Microbiology* **68**: 3003–3009.

Perez CJ and Shelton AM (1997) Resistance of *Plutella xylostella* (Lepidoptera: Plutellidae) to *Bacillus thuringiensis* Berliner in Central America. *Journal of Economic Entomology* **90**: 87–93.

Petzold-Maxwell JL, Cibils-Stewart X, French BW and Gassmann AJ (2012) Adaptation by western corn rootworm (Coleoptera: Chrysomelidae) to Bt maize: Inheritance, fitness costs and feeding preference. *Journal of Economic Entomology* **105**: 1407-1418.

Pinnock DE and Hess RT (1977) Electron microscope observations on granulosis virus replication in the fruit tree leaf roller, *Archips argyrospila*: infection of the midgut. *Journal of Invertebrate Pathology* **30**: 354-361.

Poncet S, Bernard C, Dervyn E, Cayley J, Klier A and Ropopart G (1997) Improvement of *Bacillus sphaericus* toxicity against Dipteran larvae by integration via homolgous recombinations of the Cry1A toxin gene from *Bacillus thuringiensis* subsp. *israelensis*. *Applied and Environmental Microbiology*. **63:** 4113-4201.

Poncet S, Delécluse A, Klier A and Rappoport (1995) Evaluation of synergistic interactions among the CryIVA, CryIVB, and CryIVD toxic components of *Bacillus thuringiensis* subsp. *israelensis*. *Journal of Invertebrate Pathology* **66**: 131-135.

Poopathi S and Abidha S (2010) Mosquitocidal bacterial toxins (*Bacillus sphaericus* and *Bacillus thuringiensis* serovar *israelensis*): Mode of action, cytopathological effects and mechanism of resistance. *Journal of Physiology and Pathophysiology* **1**: 22-38.

Rahardja U and Whalon ME (1995) Inheritance of resistance to *Bacillus thuringiensis* subsp. *tenebrionis* CryIIIA δ endotoxin in Colorado potato beetle (Coleoptera: Chrysomelidae). *Journal of Economic Entomology* **88**: 21–26.

Rajagopal R, Sivakumar S, Agrawal N, Malhotra P and Bhatnagar RK (2002) Silencing of midgut aminopeptidase N of *Spodoptera litura* by double-stranded RNA establishes its role as *Bacillus thuringiensis* toxin receptor. *The Journal of Biological Chemistry* **277**: 46849-46851.

Rajamohan F, Alzate O, Cotrill JA, Gould F and Dean DH (1996) Protein engineering of *Bacillus thuringiensis* δ-endotoxin: Mutations at domain II of CryIAb enhance receptor affinity and toxicity toward gypsy moth larvae. *Proceedings of the National Academy of Sciences, USA* **93**: 14338-14343.

Ramachandran S, Buntin GD, All JN, Tabashnik BE and Raymer PL (1998) Survival, development, and oviposition of resistant diamondback moth (Lepidoptera: Plutellidae) on transgenic canola producing a *Bacillus thuringiensis* toxin. *Journal of Economic Entomology* **91**: 1239–1244.

Rao DR, Mani TR, Rajendran R, Joseph AS, Gajanana A and Reuben R (1995) Development of a high-level of resistance to *Bacillus sphaericus* in a field population of *Culex quinquefasciatus* from Kochi, India. *Journal of American Mosquito Control Association* **11**: 1-5.

Raymond B, Sayyed AH and Wright DJ (2005) Genes and environment interact to determine the fitness costs of resistance to *Bacillus thuringiensis*. *Proceedings of Royal Society London, B: Biological Sciences* **272**: 1519–1524.

Raymond B, Sayyed AH and Wright DJ (2007) Host plant and population determine the fitness costs of resistance to *Bacillus thuringeinsis*. *Biology Letters*. **3:** 83-86.

Rendon P, McInnis D, Lance D and Stewart J (2004) Medfly (Diptera:Tephritidae) genetic sexing: large scale field comparison of males-only and bisexual sterile fly releases in Guatemala. *Journal of Economic Entomology* **97**: 1547-1553.

Ribeiro LMS, Teixeira VW, Siqueira HAA, Oliveira ACB, Lemos AJJM and Teixeira AAC (2013) Midgut histopathology of resistant and susceptible *Plutella xylostella* exposed to commercial formulations of *Bacillus thuringiensis*. *Bulletin of Insectology* **66**: 161-171.

Rice ME (2000) *Western bean cutworm hits northwest Iowa*. Integrated Crop Management. IC-484. **22**: 163. Iowa State University Extension, Ames, IA. (www.ipm.iastate.edu/ipm/icm/2000/9-18-2000/wbcw.html)

Rodcharoen J and Mulla MS (1996) Cross-resistance to *Bacillus sphaericus* strains in *Culex quinquefasciatus*. *Journal of American Mosquito Control Association* **12**: 247-250.

Romeis J, Dutton A and Bigler F (2004) *Bacillus thuringiensis* toxin (Cry1Ab) has no direct effect on larvae of the green lacewing *Chrysoperla carnea* (Stephens) (Neuroptera: Chrysopidae). *Journal of Insect Physiology* **50**: 175-183.

Saleh MS, El-Meniawi FA, Kelada NL and Zahran HM (2003) Resistance development in mosquito larvae *Culex pipiens* to the bacterial agent *Bacillus thuringiensis* var. *israelensis*. *Journal of Applied Entomology* **127**: 29-32.

Sauphanor B, Berling M, Toubon JF, Reyes M, Delnatte J, Allemoz P (2006) *Carpocapse des pommes. Cas de résistance au virus de la granulose en vergers biologique. Phytoma-La Défense des Végétaux* **590**: 24–27.

Saxena D and Stotzky G (2001) *Bacillus thuringiensis* (Bt) toxin released from root exudates and biomass of Bt corn has no apparent effect on earthworms, nematodes, protozoa, bacteria, and fungi in soil. *Soil Biology and Biochemistry* **33**: 1225-1230.

Sayyed AH, Haward R, Herrero S, Ferre J and Wright DJ (2000). Genetic and biochemical approach for characterization of resistance to *Bacillus thuringiensis* toxin Cry1Ac in a field population of the diamondback moth, *Plutella xylostella*. *Applied and Environmental Microbiology* **66**: 1509-1516.

Schulze-Bopp S and Jehle JA (2013) Development of a direct test of baculovirus resistance in wild codling moth populations. *Journal of Applied Entomology* **137**: 153–60. doi: 10.1111/j.1439-0418.2012.01733.

Setamou M, Bernal JS, Mirkov T E and Legaspi JC (2003) Effects of snowdrop lectin on mexican rice borer (Lepidoptera: Pyralidae) life history parameters. *Journal of Economic Entomology* **96:** 950-956.

Shade RE, Schroeder HE, Pueyo JJ, Tabe LM, Murdock LL, Higgins TJV, Chrispeels MJ (1994) Transgenic pea seeds expressing amylase inhibitor of the common bean are resistant to bruchid beetles. *Bio Technology* **12**: 793-796.

Shao Z, Cui Y, Yi H, Ji J and Yu Z (1998) Processing of δ-endotoxin of *Bacillus thuringiensis* HD-1 in *Heliothis armigera* midgut juice and the effects of protease inhibitors. *Journal of Invertebrate Pathology* **72**: 73-81.

Shikano I and Cory JS (2014) Genetic resistance to *Bacillus thuringiensis* alters feeding behaviour in the cabbage looper, *Trichoplusia ni*. *PLoS One*. **9**(1): e85709. doi:10.1371/journal.pone.0085709.

Shukla S, Arora R and Sharma HC (2005) Biological activity of soybean trypsin inhibitor and plant lectins against cotton bollworm/legume pod borer, *Helicoverpa armigera*. *Plant Biotechnology* **22:** 1-6.

Siegel JP (2001) The mammalian safety of *Bacillus thuringiensis*- based insecticides. *Journal of Invertebrate Pathology* **77**: 13-21.

Silva-Filha M-H, Regis L, Nielsen-LeRoux C and Charles J-F (1995) Low level resistance to *Bacillus sphaericus* in a field-treated population of *Culex quinquefasciatus* (Diptera: Culicidae). *Journal of Economic Entomology* **88**: 525-530.

Silva-Filha MHNL, Chalegre KDJM and Anastacio DB (2008) *Culex quinquefasciatus* field populations subjected to treatment with *Bacillus sphaericus* did not display high resistance levels. *Biological Control* **44**: 227-234.

Sims SR and Stone TB (1991) Genetic basis of tobacco budworm resistance to an engineered *Pseudomonas fluorescens* expressing the δ endotoxin of *Bacillus thuringiensis kurstaki*. *Journal of Invertebrate Pathology* **57**: 206-210

Soberon M, Gill SS and Bravo A (2009). Signaling versus punching hole: How do *Bacillus thuringiensis* toxins kill insect midgut cells? *Cellular and Molecular Life Sciences* **66**: 1337-1349.

Stone TB, Sims SR and Marrone PG (1989) Selection of tobacco budworm for resistance to a genetically engineered *Pseudomonas fluorescens* containing the δ endotoxin of *Bacillus thuringiensis* subsp. *kurstaki*. *Journal of Invertebrate Pathology* **53**: 228–234.

Su T and Mulla MS (2004). Documentation of high-level *Bacillus sphaericus* 2362 resistance in field populations of *Culex quinquefasciatus* breeding in polluted water in Thailand. *Journal of American Mosquito Control Association* **20**: 405-411.

Sun F, Yuan Z, Li T, Zhang Y, Yu J and Pang Y (2001) Reduction of resistance of *Culex pipiens* larvae to the binary toxin from *Bacillus sphaericus* by coexpression of *cry4Ba* from *Bacillus thuringiensis* subsp. *israelensis* with the binary toxin. *World Journal of Microbiology and Biotechnology* **17**: 385-389.

Suresh R, Buntin GD, All JN, Tabashnik BE, Raymer PL, Adang MJ, Pulliam DA and Stewart, Jr. CN (1998) Survival, development, and oviposition of resistant diamondback moth (Lepidoptera: Plutellidae) on transgenic canola producing a *Bacillus thuringiensis* toxin. *Journal of Economic Entomology* **91**: 1239-1244.

Tabashnik BE (1991) Determining the mode of inheritance of pesticide resistance with backcross experiments. *Journal of Economic Entomology* **84**:703-712.

Tabashnik BE (1994a) Delaying insect adaptation to transgenic plants: seed mixtures and refugia reconsidered. *Proceedings of the Royal Society London, B: Biological Sciences* **255**: 7-12.

Tabashnik BE (1994b) Evolution of resistance to *Bacillus thuringiensis*. *Annual Review of Entomology* **39**: 47-79.

Tabashnik BE (1997) Seeking the root of insect resistance to transgenic plants. *Proceedings of the National Academy of Sciences, USA* **94**: 3488-3490.

Tabashnik BE, Brévault T and Carrière Y (2013) Insect resistance to Bt crops: lessons from the first billion acres. *Nature Biotechnology*. **31**: 510 DOI:10.1038/nbt.2597

Tabashnik BE, Cushing NL, Finson N and Johnson MW (1990) Field development of resistance to *Bacillus thuringiensis* in diamondback moth (Lepidoptera: Plutellidae). *Journal of Economic Entomology* **83**: 1671–1676.

Tabashnik BE, Finson N, Groeters FR, Moar WJ and Johnson MW (1994a) Reversal of resistance to *Bacillus thuringiensis* in *Plutella xylostella*. *Proceedings of the*

National Academy of Sciences, USA **91**: 4120-4124.

Tabashnik BE, Finson N, Johnson MW and Heckel D (1995) Prolonged selection affects stability of resistance to *Bacillus thuringiensis* in diamondback moth (Lepidoptera: Plutellidae). *Journal of Economic Entomology* **88**: 219-224.

Tabashnik BE, Finson N, Johnson MW and Moar WJ (1993) Resistance to toxins from *Bacillus thuringiensis* subsp. *kurstaki* causes minimal cross-resistance to *Bacillus thuringiensis* subsp. *aizawai* in the diamondback moth (Lepidoptera: Plutellidae). *Applied and Environmental Microbiology* **59**: 1332–1335.

Tabashnik BE, Gassmann AJ, Crowder DW and Carriére Y (2008) Insect resistance to Bt crops: Evidence versus theory. *Nature Biotechnology* **26**: 199–202.

Tabashnik BE, Liu YB, de Maagd RA and Dennehy TJ (2000) Cross-resistance of pink bollworm (*Pectinophora gossypiella*) to *Bacillus thuringiensis* toxins. *Applied and Environmental Microbiology* **66**: 4582–4584.

Tabashnik BE, Liu YB, Finson N, Masson L and Heckel DG (1997a) One gene in diamondback moth confers resistance to four *Bacillus thuringiensis* toxins. *Proceedings of the National Academy of Sciences USA* **94**: 1640-1644.

Tabashnik BE, Liu YB, Malvar T, Heckel DG, Masson L, Ballester V, Granero F, Mensua JL and Ferre J (1997b) Global variation in the genetic and biochemical basis of diamondback moth resistance to *Bacillus thuringiensis*. *Proceedings of the National Academy of Sciences, USA* **94**: 12780-12785.

Tabashnik BE, Malvar T, Liu YB, Finson N and Borthakur D (1996) Cross resistance of diamondback moth indicates altered interactions with domain II of *Bacillus thuringiensis* toxins. *Applied and Environmental Microbiology* **62**: 2839–2844.

Tabashnik BE, Sanchez DM, Whalon ME, Hollingworth RM and Carrière Y (2014) Defining terms for proactive management of resistance to Bt crops and pesticides. *Journal of Economic Entomology* **107**: 496-507.

Tabashnik BE, Roush RT, Earle ED and Shelton AM (2000) Resistance to Bt toxins. *Science* **87**: 42.

Tabashnik BE, Van Rensburg JBJ and Carrière Y. (2009) Field-evolved insect resistance to Bt Crops: definition, theory, and data. *Journal of Economic Entomology* **102**: 2011–2025.

Tabashnik BE, Liu YB, Malvar T, Heckel DG, Masson L and Ferre J (1998) Insect resistance to *Bacillus thureingiensis: uniform or diverse? Philosphical Transactions of the Royal Society B London.* **353:** 1751-1756.

Tanada Y (1964) A granulosis virus of the codling moth, *Carpocapsa pomonella* Linnaeus) (Olethreutidae, Lepidoptera). *Journal of Invertebrate Pathology* **8**: 378-380.

Tang JD, Collins HL, Metz TD, Earle ED and Zhao JZ (2001) Greenhouse tests on resistance management of Bt transgenic plants using refuge strategies. *Journal of Economic Entomology* **94**: 240-247.

Tang JD, Collins HL, Roush RT, Metz TD, Earle ED and Shelton AM (1999) Survival, weight gain, and oviposition of resistant and susceptible *Plutella xylostella* (Lepidoptera: Plutellidae) on broccoli expressing Cry1Ac toxin of *Bacillus thuringiensis*. *Journal of Economic Entomology* **92**: 47-55.

Tang JD, Gilboa S, Roush RT and Shelton AM (1997) Inheritance, stability, and lack of-fitness costs of field-selected resistance to *Bacillus thuringiensis* in diamondback moth (Lepdoptera: Plutellidae) from Florida. *Journal of Economic Entomology* 90: 732-741

Tang JD, Shelton AM, Van Rie J, de Roeck S and Moar WJ (1996) Toxicity of *Bacillus thuringiensis* spore and crystal protein to resistant diamondback moth (*Plutella xylostella*). *Applied and Environmental Microbiology* **62**: 564-569.

Teshima R, Watanabe T, Okunuki H, Isuzugawa K, Akiyama H, Onodera H, Imai T, Toyoda M and Sawada J (2002) Effect of subchronic feeding of genetically modified corn (CBH351) on immune system in BN rats and B10A mice. *Shokuhin Eiseigaku Zasshi* **43**: 273-279.

Tetreau G, Bayyareddy K, Jones CM, Stalinski R, Riaz MA, Paris M, David JP, Adang MJ and Despres L (2012) Larval midgut modifications associated with *Bti* resistance in the yellow fever mosquito using proteomic and transcriptomic approaches. *BMC Genomics* **13**: 248.

Tetreau G, Stalinski R, David JP and Laurence D (2013) Monitoring resistance to *Bacillus thuringiensis* subsp. *israelensis* in the field by performing bioassays with each Cry toxin separately. *Memórias do Instituto Oswaldo Cruz* **108**: 894-900.

Tiewsiri K and Wang P (2011) Differential alteration of two aminopeptidases N associated with resistance to *Bacillus thuringiensis* toxin Cry1Ac in cabbage looper. *Proceedings of National Academy of Sciences, USA* **108**: 14037–14042.

U.S. Environmental Protection Agency (2011) *Resistance Terms*. www.entsoc.org/PDF/2013/EPAResistanceTerms-2013.pdf

Undorf-Spahn K, Fritsch E, Huber J, Kienzle J, Zebitz CPW and Jehle JA (2012) High stability and no fitness costs of the resistance of codling moth to *Cydia pomonella* granulovirus (CpGV-M). *Journal of Invertebrate Pathology* **111**: 136-142.

Vaeck M, Reynaerts A, Hoftey H, Jansens S, DeBeuckleer M, Dean C, Zabeau M, Van Montagu M and Leemans J (1987). Transgenic plants protected from insect attack. *Nature* **327**: 33-37.

Van Rensburg JBJ (2007) First report of field resistance by stem borer, *Busseola fusca* (Fuller) to *Bt*-transgenic maize. *South African Journal of Plant and Soil* **24**: 147–151.

Van Rie J, McGaughey WH, Johnson DE, Barnett BD, Van Mellaert H (1990) Mechanism of insect resistance to the microbial insecticide *Bacillus thuringiensis*. *Science* **247**:72–74.

Vasquez MI, Violaris M, Hadjivassilis A and Wirth MC (2009) Susceptibility of *Culex pipiens* (Diptera: Culicidae) field populations in Cyprus to conventional organic insecticides, *Bacillus thuringiensis* subsp. *israelensis* and methoprene. *Journal of Medical Entomology* **46**: 881-887.

Vasquez-Gomez M (1983) *Investigations of the possibility of resistance to Bacillus thuringiensis ser. H-14 in Culex quinquefasciatus through accelerated selection pressure in the laboratory*. Ph.D. Thesis. University of California.

Vélez AM (2013) *Characterization of Resistance to the Cry1F Toxin from* Bacillus thuringiensis *in Resistant Fall Armyworm,* Spodoptera frugiperda *(J. E. Smith) (Lepidoptera: Noctuidae) from Puerto Rico*. PhD dissertation, University of Nebraska-Lincoln, Nebraska, USA.

Vettori C, Paffetti D, Saxena D, Stotzky G and Giannini R (2003) Persistence of toxins and cells of *Bacillus thuringiensis* subsp.. *kurstaki* introduced in sprays to Sardinia soils. *Soil Biology and Biochemistry* **35**: 1635-1642.

Wan P, Huang YX, Wu HH, Huang MS, Cong SB, Tabashnik BE and Wu KM (2012) Increased frequency of pink bollworm resistance to *Bt* toxin Cry1Ac in China. *PLOS One* **7**:e29975.

Wan P, Huang Y, Wu H, Huang M, Cong S and Tabashnik B E (2012) Increased frequency of pink bollworm resistance to *Bt* toxin Cry1Ac in China. *PLoS One*. **7**: e29975.

Wang L, Li XF, Zhang J, Zhao JZ, Wu QJ, Xu B and Zhang YJ (2007) Monitoring of resistance for the diamondback moth to *Bacillus thuringiensis* Cry1Ac and Cry1Ba toxins and a Bt commercial formulation. *Journal of Applied Entomology*. **131**: 441–446.

Wang S, Just DR and Pinstrup-Andersen P (2008) Bt-Cotton and secondary pests. *Journal of Biotechnology* **10**: 113–120.

Whalon ME, Miller DL, Hollingworth RM, Grafius EJ and Miller JR (1993) Selection of a Colorado potato beetle (Coleoptera: Chrysomelidae) strain resistant to *Bacillus thuringiensis. Journal of Economic Entomology* **86**: 226–233.

Williams JL, Ellers KC, Orth RG, Gassmann AJ, Head G, Tabashnik BE and Carrière Y (2011) Fitness cost of resistance to Bt cotton linked with increased gossypol content in pink bollworm larvae. *PLoS One* **6**: e21863.

Wirth MC and Georghiou GP (1997) Cross-resistance among CryIV toxins of *Bacillus thuringiensis* subsp. *israelensis* in *Culex quinquefasciatus* (Diptera: Culicidae). *Journal of Economic Entomology* **90**: 1471-1477.

Wirth MC, Cluse AD and Walton WE (2004) Laboratory selection for resistance to *Bacillus thuringiensis* subsp. *jegathesan* or a component toxin, *Cry*11B, in *Culex quinquefasciatus* (Diptera: Culicidae). *Journal of Medical Entomology* **41**: 435-441.

Wirth MC, Delécluse A and Walton WE (2001) Lack of cross resistance to Cry19A from *Bacillus thuringiensis* subsp. *jegathesan* in *Culex quinquefasciatus* (Diptera: Culicidae) resistant to Cry toxins from *Bacillus thuringiensis* subsp. *israelensis. Applied and Environmental Microbiology* **67**: 1956-1958.

Wirth MC, Federici BA and Walton WE (2000) CytA from *Bacillus thuringiensis* synergizes activity of *Bacillus sphaericus* against *Aedes aegypti* (Diptera: Culicidae). *Applied and Environmental Microbiology* **66**: 1093-1097.

Wirth MC, Georghiou GP and Federici BA (1997) CytA enables CryIV endotoxins of *Bacillus thuringiensis* to overcome high levels of CryIV resistance in the mosquito, *Culex quinquefasciatus. Proceedings of the National Academy of Sciences, USA* **94**: 10536-10540.

Wirth MC, Park HW, Walton WE and Federici BA (2005) CytA of *Bacillus thuringiensis* delays resistance to Cry11A in the mosquito *Culex quinquefasciatus. Applied and Environmental Microbiology* **71**: 185-189.

Wirth MC, Walton WE and Federici BA (2010) Resistance to the *Bacillus sphaericus* Bin toxin is delayed significantly through combination with the mosquitocidal proteins of *Bacillus thuringiensis* subsp. *israelensis. Environmental Microbiology* 1254-1260.

Wirth MC, Yang Y, Walton WE, Federici BA and Berry C (2007a) Mtx toxins synergize *Bacillus sphaericus* and Cry11Aa against susceptible and insecticide-resistant *Culex quinquefasciatus* larvae. *Applied and Environmental Microbiology* **73**: 6066-6071.

Wirth MC, Zaritsky A and Ben Dov E (2007b) Cross-resistance spectra of *Culex quinquefasciatus* resistant to mosquitocidal toxins of *Bacillus thuringiensis* towards recombinant *Escherichia coli* expressing genes from *B. thuringiensis* ssp. *israelensis. Environmental Microbiology* **9**: 1393-1401.

Wood RJ (1981) Strategies for conserving susceptibility to insecticides. *Parasitology* **82**: 69-80.

Wright DJ, Iqbal M, Granero F and Ferr´e J (1997) A change in a single midgut receptor in the diamondback moth (*Plutella xylostella*) is only in part responsible for field resistance to *Bacillus thuringiensis* subsp. *kurstaki* and *B. thuringiensis* subsp. *aizawai. Applied and Environmental Microbiology* **63**: 1814–1819

Xie RY, Zhuang MB, Ross LS, Gomez I, Oltean DI, Bravo A, Soberon M, Gill SS (2005) Single amino acid mutations in the cadherin receptor from *Heliothis virescens* affect its toxin binding ability to Cry1A toxins. *Journal of Biological Chemistry* **280:** 8416–8425.

Xu X and Wu Y (2008) Disruption of Ha_BtR alters binding of *Bacillus thuringiensis* δ-endotoxin Cry1Ac to midgut BBMV's of *Helicoverpa armigera*. *Journal of Invertebrate Pathology*. **97:** 27-32.

Xu XJ, Yu LY and Wu YD (2005) Disruption of a cadherin gene associated with resistance to Cry1Ac δ-endotoxin of *Bacillus thuringiensis* in *Helicoverpa armigera*. *Applied and Environmental Microbiology* **71**: 948–954.

Yenagi BS, Patil VC, Biradar DP and Khadi BM (2011) Refuge cropping systems for *Helicoverpa armigera* (Hubner) resistance management in BT cotton (*Gossypium hirsutum*). *Academic Journal of Entomology* **4**: 102-107.

Yuang Z, Zhang Y, Cai Q and Liu EY(2000) High-level field resistance to *Bacillus sphaericus* C3-41 in *Culex quinquefasciatus* from southern China. *Biocon Sci Technol* **10**: 41-49.

Zago HB, Siqueira HÁ, Pereira EJ, Picanço MC and Barros R (2014) Resistance and behavioural response of *Plutella xylostella* (Lepidoptera: Plutellidae) populations to *Bacillus thuringiensis* formulations. *Pest Management Science* **70**: 488–495.

Zhang HY, Yang CJ, Huang JY and Lu L (2004) Susceptibility of field populations of *Anopheles sinensis* (Diptera: Culicidae) to *Bacillus thuringiensis* subsp. *israelensis*. *Biocontrol Science and Technology* **14**: 321-325.

Zhang XB, Candas M, Griko NB, Taussig R, Bulla LA (2006) A mechanism of cell death involving an adenylyl cyclase/PKA signaling pathway is induced by the Cry1Ab toxin of *Bacillus thuringiensis*. *Proceedings of the National Academy of Sciences USA* **103**: 9897–9902.

Zhang H, Tian W, Zhao J, Jin L, Yang J, Liu C, Yang Y, Wu S, Wu K, Cui J, Tabashnik TE and Wu Y (2012). Diverse genetic basis of field-evolved resistance to Bt cotton in cotton bollworm from China. *Proceedings of the National Academy of Sciences, USA* **109**: 10275– 10280.

Zhang H, Yin W, Zhao J, Jin L, Yang y, Wu S, Tabashnik BE and Wu Y (2011) Early warning of cotton bollworm resistance associated with intensive planting of Bt cotton in China. *PLoS One* 6: e22874.

Zhao JH, Ho P and Azadi H (2011) Benefits of Bt cotton counterbalanced by secondary pests? Perceptions of ecological change in China. *Environmental Monitoring and Assessment* **173:** 985–994.

Zhao JZ, Collins HL, Tang JD, Cao J, Earle ED, Roush RT, Herrero S, Escriche B, Ferre J and Shelton AM (2000) Development and characterization of diamondback moth resistance to transgenic broccoli expressing high levels of Cry1C. *Applied and Environmental Microbiology*. **66:** 3784-3789.

Zingg D (2008) Madex Plus and Madex I12 overcome virus resistance of codling moth. In: Boos M (ed) 13th International Conference on Cultivation Technique and Phytopathological Problems in Organic Fruit-Growing. FÖKO, Weinsberg, Germany.

Biological and Molecular Approaches in Pest Management
Balwinder Singh, Ramesh Arora & S S Gosal (Eds)
Scientific Publishers (India), Jodhpur, 2015, pp 312-328.

Chapter - 12

Biological and Molecular Approaches in Management of Mite Pests

Paramjit Kaur and M B Bhullar
Department of Entomology, Punjab Agricultural University, Ludhiana – 141 001, India
E-mail: paramjitkaur@pau.edu

12.1. INTRODUCTION

Mites are the most diverse representatives of an ancient lineage in phylum Arthropoda, subphylum Chelicerata, class Arachnida, subclass Acari. There are around 48,200 described species of Acari (Halliday *et al.,* 2000), but certainly many more remain to be explored. Interestingly phytophagy in Acari is seen in very few families like Tetranychidae, Tenuipalpidae, Tarsonemidae, Eriophyidae and Tuckerellidae. The Tetranychidae is a moderately large family and comprising over 1,200 described species (Krantz and Walter, 2009), some like the two-spotted spider mite (*Tetranychus urticae* Koch), European red mite (*Panonychus ulmi* (Koch)) are serious pests of crops; among tarsonemids, the yellow mite (*Polyphagotarsonemus latus* (Banks)) and among Eriophyids the purple mite and pink tea mite (*Acaphylla thea* (Watt & Mann), *Calacarus carinatus* (Green)), the coconut perianth mite (*Aceria guerreronis* Keifer) etc. are causing considerable loss in the revenue of the farmers.

The phytophagous mites feed on leaf tissues with needle like chelicerae which causes white specks and patches on the leaves leading to loss of chlorophyll adversely affecting the growth and vitality of the plants. These mites usually feed on the epidermis resulting in blotching, stippling or bronzing, crinkling of leaves, uneven growth of leaf surfaces and sometimes are accompanied by leaf fall. Mineral contents of plant like potassium, iron, magnesium, zinc, and calcium; inorganic compounds like nitrate, nitrite; organic compounds, protein, carbohydrate and chlorophyll are depleted affecting the normal functioning of the cell. Webbing with silken threads is a characteristic feature of two-spotted spider mites. They spin a fine web over the leaves of the plant and many times the entire plant may be covered by this webbing. The heavy webbing and the dust particles deposited on it adversely affect the normal functioning of the leaves. Eriophyids cause different types of damage like rusting, russetting, kill buds, malformations like galls, erineum, blisters on leaves and leaf sheaths (Helle and Sabelis, 1985; Hoy, 2011).

Apart from the direct damage the mites cause to the crops, some also act as vectors. In India, *Aceria cajani* Channabasavanna transmits the virus that causes pigeon pea sterility mosaic disease in *Cajanus cajan* (L.) Millspaugh; *A. tulipae* (Keifer) transmits wheat streak mosaic; *Phyllocoptus fructiphilus* Keifer spreads rose rosette disease and *Aculus schlechtendali* (Nalepa) causes apple rust. In recent years due to indiscriminate usage of pesticides and adoption of several modern agricultu- ral practices, many species of mites which were of less importance have assumed the status of major pests. Favourable host plant and suitable weather are pre-requisites for abundance of spider mites. They attained pest status in many ecosystems with the introduction of sprays of new synthetic pesticides. Management is a serious challenge to humans since there are evidences of development of resistance to acaricides. Hence in recent years, several new molecules of acaricides have been introduced.

The effects of pesticides on predators and host-plant nutrition, the development of resistance to pesticides, and altered management practices have all played a part. Currently, chemical control has to deal with serious control failure in mite populations, since the evolution of pesticide resistance in phytophagous mites is very common (Hoy, 2011). Consequently, different chemical molecules are being currently developed to face this phenomenon, mainly in spider mites. However, the low level of immigration of susceptible individuals and the rapid reproductive rate associated to these mite groups have made it difficult to manage population in crops.

Mites cause yield loss to the tune of 10-80 per cent in a variety of crops – mango (50-80%), litchi (30-40%), rice (20-25%), red gram (15-30%), sugarcane (20-30%), cotton (20-30%), tea (15-20%), ber (15-30%), copra loss in coconut (10 – 30%) (Ramaraju *et al.*, 2005) and vegetables (10-15%) (Chauhan and Dahiya, 2003). Besides above, total failure of crops due to mite infestation has been reported in okra, brinjal, beans and ornamental crops. In Punjab, crop loss in brinjal due to two-spotted spider mite, *T. urticae* was found to be 18.14 per cent (Anonymous, 2012).

12.2. INTEGRATED MITE MANAGEMENT

Due to the economic losses caused by mite pests, management tactics need to be established to keep population levels under the economic threshold of infestation. This practice should be based on integrated mite management including spraying chemical products, using biological control agents and / or resistant varieties and other control measures. Integrated mite control is based on an understanding of the necessary interrelatedness of pesticides and natural enemies. Integration of a biocontrol agent into most agricultural systems will not be successful unless the natural enemy can survive the pesticides used in that crop system due to disruption of trophic relationships. Selectivity may be achieved by a pesticide's specific activity towards a narrow range of taxa, or by limiting pesticide exposure of a biocontrol agent in its natural environment while killing its host or prey. Knowledge of insecticide and acaricide selectivity to beneficial arthropods is important to their utility in integrated mite management programs. The integrated control methods are integrated in such a way that pest populations are prevented from causing economic damage. Better timing, selection and application of pesticides reduce the amount of chemical residues on crop plants at harvest. The use of biological control agents and less pesticide reduces the chances of pests developing resistance to pesticides. To overcome pesticide resistance, it is essential that eco-friendly approaches are explored, which would be safer to the environment and are also acceptable to the markets, which demand low level of pesticide residues. Various strategies to be adopted include cultural, mechanical, physical, biological, chemical and most recent is the use of biotechnology in management of mite pests.

12.2.1 Biological Control

Biological control is an environmentally safe, cost effective and energy efficient pest control, either on its own or as a component of integrated pest management. Destruction of predators by the pesticide sprays has been shown in many instances to cause increase of pest mites. Natural enemies of mites are many and varied and are classified into three groups namely pathogens, micro predators (predatory mites), and macro predators (predatory insects) (Gerson and Smiley, 1990). Different natural enemies with their characteristics and possible role in managing mite pests are discussed.

12.2.1.1. Microbial Control Agents

The study of pathogens of mites has not been as intense or as complete as those of insects. Lipa (1971) was one of the first to highlight the importance of mite pathogens and his review was one of the first to summarize research on this subject. However, as compared with insect pathogens, diseases of mites have not attracted enough attention that they deserve. Nonetheless, accessibility to advanced and sophisticated equipment as well as availability of molecular techniques in recent times, has given a boost to the subject. Many review papers have been published during the last decade on mite pathogens (Chandler *et al.*, 2000; Van der Geest *et al.*, 2000; Mietkiewski *et al.*, 2000 and Van der Geest, 2004). The different microbial agents which proved to be effective for mite management are discussed below.

Fungal pathogens: Fungi are among the most frequent noticed groups of mite pathogens. The most encountered fungal pathogens of mites are the Entomophthorales. *Entomophthora* genus is worldwide in distribution and attacks mites as well as insects. *Entomophthora* infections have been commonly reported on citrus red mite, *Panonychus citri* (McGregor). Infected mites become sluggish and darken in colour. Epizootics of *Hirsutella* occur when the correct interactions between the fungus, host, temperature, and humidity occur. Citrus rust mite, *Phyllocoptruta oleivora* (Ashmead) is an economic pest in USA and *Hirsutella thompsonii* Fisher is a key factor in its natural control. The fungus is the most important natural regulator of citrus rust mite, where it causes regular epizootics resulting in the suppression of the mite populations. The epizootics last for 3-4 weeks and the inoculum persists in the environment to suppress the fresh mite populations for several months. Warm, wet weather and high densities of the mite population are favourable to the development of the epizootics (McCoy, 1981). Mycoacraicide, *H. thompsonii* was commercially produced by Abbott Laboratories under the name of Mycar in 1976 and was recommended for the control of both spider mites and eriophyid mites on coconut and in green houses (McCoy, 1996).

The genus *Neozygites* has the deleterious effect on tetranychids, which was first noticed by Fisher (1951), who reported an adult mortality ranging from 32-95 per cent in populations of citrus red mite, *P. citri* from Florida, USA. Though many fungal pathogens are known to be infective to tetranychid mites, *Neozygites floridana* Weiser & Muma is the most specific pathogen (Chandler *et al.*, 2000) and it has been observed on several species of spider mites on various agricultural crops (van der Geest *et al.*, 2000). The Banks grass mite, *Oligonychus prataensis* (Banks), a pest of field corn and sorghum, and the two-spotted spider mite, *T. urticae* can be decimated in the Western Great Plains of USA by *Neozygites adjarica* (Tsintsadze & Vartapetov) (Dick and Buschman, 1995). Unlike *Hirsutella*, which can be readily mass produced and formulated as products, *Neozygites* species have so far not been cultured on artificial media on a large scale because of their obligatory nature. However, there is a possibility of using cadavers of mites carrying the fungal inoculum as a product. The major hurdle in the mass production of *Neozygites* has recently beer overcome with the development of culture media for *N. floridana* from *T. urticae* (Leite *et al.*, 2000) and for *Neozygites parvispora* (MacLeod and Carl) from thrips (Grundschober *et al.*, 1998).

In addition to the mite-specific fungi, other established insect pathogens like *Beauveria bassiana, B. brongniartii, Verticillium lecanii* and *Metarhizium anisopliae* are known to have good potential as mycoacricides. Forty isolates of different entomopathogenic fungi from six genera were assessed against *T. urticae* on tomato in UK with interesting results (Chandler *et al.*, 2005). *Paecilomyces eriophyes P. farinosus, P. fumosoroseus* and *P. terricola* are also known to be effective to Acari (Chandler *et al.*, 2000). Species of *Sporothrix* are also known to be pathogenic to mites (Sreerama Kumar *et al.*, 2004) in addition to their entomopathogenic nature. *H. thompsonii* is the most important bioregulator of the coconut mite in the southern states. Based on intensive research on many other fungal pathogens of *A. guerreronis*, NBAII (National Bureau of Agriculturally Important Insects, Bangalore, India) developed a mycoacricide (Mycohit) with *H. thompsonii* strain MF (Ag) 5 (ITCC 4962; IMI 385470) as its active ingredient (Sreerama Kumar and Singh, 2000 and 2001). During 2002-04, four new formulations, *viz.* Mycohit LG10, Mycohit LG20, Mycohit OS and Mycohit GP, all based on *H. thompsonii* were developed for field evaluation in different states. Under field conditions, the formulated fungus was found to be capable of bringing down the mite population up to 90%, resulting in considerable reduction in pre-harvest nut damage (Sreerama Kumar, 2009). A mycelia alone variant (Mycohit M) was developed during 2006-07, which when applied after tank-mixing separately with three selected adjuvants brought down the post-treatment population of the coconut mite by 85.6-97.1%. Application of the fungus in combination with one of adjuvants (glycerol) resulted in a tolerable mean nut damage grade of 2.0 during the pre-harvest stage, compared with an acute score of 4.0 in control palms (Sreerama Kumar and Singh, 2008).

Viral pathogens: Viruses associated with mites are necessarily pathogenic because several plant viruses are known to be transmitted by mites. No polyhedrosis virus has yet been reported in mites. But particles of non-occluded virus in tissues of the European red mite, *P. ulmi,* and citrus red mite, *P. citri* have been reported. They are bacilliform virions related to the Baculoviridae. The *P. citri* virus occurs in natural mite populations and at times reduces the population level below the economic threshold. But under field conditions, sunlight and high temperatures result in rapid loss of viral infectivity in aqueous solutions of virus. Muma (1955) reported the first indication of the occurrence of a virus disease in citrus red mite in Florida, USA. Diseased mites showed symptoms of diarrhoea and when dead, were often stuck to the leaf surface by a smear of black resinous material that was excreted through the anus. Diseased mites can easily be recognised by the darker red colour of the immature mites, but adult mites do not show a difference in colour. The most conspicuous symptoms, however, is the presence of birefringent inclusions inside the midgut of the mite with a radiating crystalline structure. The virus causes only natural epizootics in dense populations.

Bacterial pathogens: The few bacteria that have been associated with diseased mites appear to be isolated instances of infection. Trial with the spore crystal complex of *Bacillus thuringiensis* did not produce infection in the few tests conducted with spider mites; however, the *B. thuringiensis* exotoxin was shown to have a detrimental effect on *Tetranychus telarius* Linn., and the citrus red mite, *P. ulmi*. Recently, *Pseudomonas putida* biotype B was found to have a strong efficacy in causing mortality in *T. urticae* (Aksoy *et al.,* 2008), but there is no field evidence.

12.2.1.2. Major insect predators of mites

Several species of insects prey upon tetranychid mites. They belong to orders Coleoptera, Thysanoptera, Hemiptera, Diptera, Neuroptera and Dermaptera. The degree of adaptation to prey upon mites as well as their efficacy in manipulating mite populations varies with predator species and environmental conditions.

***Stethorus* spp. (Coleoptera : Coccinellidae):** About 60 species of *Stethorus* have been described and fourty per cent of theses known species have been reported to attack spider mites of economic importance. They attack all stages of mites. The larvae of *Stethorus* suck the prey, liquefying the body contents by extra oral digestion. The adults chew and eat the whole mite and prefer large mobile prey. They also need a minimum level of mite population to colonize infested crops successfully, and alternative mite hosts in the vicinity for their survival. Biddinger *et al.* (2009) have reviewed the use of Stethorini worldwide for the biological control of mites.

Thysanoptera: Thrips belonging to families Phlaeothripidae, Aeolothripidae and Thripidae have been reported in literature as predaceous on spider mites in different agro ecosystems. From India, *Scolothrips indicus* (Priesner) and *S. sexmaculatus* (Pergande) have been reported as acariphagous occurring and feeding on spider mites on different plant species. *S. sexmaculatus* appears to be much more specialized on tetranychid mites and the silk threads of the mites are utilized by the predator to trace mites. Eggs are completely consumed but other instars of the mite are only partially sucked. This behaviour increases the number of prey killed and so improves the predation efficiency. *Scolothrips rhagebianus* and *S. asura* are also very common predators in South India (Poorani, 2010)

Heteroptera: Bugs belonging to families Anthocoridae, Miridae, Nabidae and Lygaeidae are facultative predators of mites. *Anthocoris nemorum* (L.), *Orius laevigatus* (Fieber) and *Blephoridopterous angulatus* (Fallen) have been reported feeding on different spider mites. Kaur *et al.* (2012) reported the possibility of controlling two-spotted spider mite by releasing of 7-days old anthocorid bug at the level of 1:50 under laboratory conditions. Therefore, this can be explored further as biological control agent in combination with other control measures to control two-spotted spider mite.

Neuroptera: Chrysopids are mainly aphid predators, but being general feeders their prey also includes mites. *Chrysoperla carnea* Stephens is reported as general predator of spider mites. Chrysopids have better searching capacity than *Stethorus*. They find their prey at extremely low mite density. The predatory potential of *Chrysoperla arabica zastrowi* was found to be more as compared to anthocorid bug, *Blaptostethus pallescens* for *T. urticae*. The percent consumption of two-spotted spider mite by *Chrysoperla* and anthocorid bug ranged from 89.66 to 98.33 and 81.65 to 85.32, respectively (Anonymous, 2012).

12.2.1.3. Mite predators of mites

Among the mite predators, phytoseiids are most promising since they have shorter life cycle than other groups of predators and they can be mass produced. More than 2280 species are known from the world and about 189 species from India (Gowda, 2009). Predaceous mites belong to family Phytoseiidae, Stigmaeidae, Bdellidae, Cunaxidae, Tydeidae, Anystidae, Cheyletidae and Erthryaeidae (Yadav *et al.*, 2003). Some of the important phytoseiid species are *Amblyseius (Neoseiulus) longispinosus* (Evans), *A. tetranychivorus* (Gupta), *A. ovalis* (Evans), *A. largoensis* (Muma), *A. finlandicus* (Oudemans). *Neoseiulus longispinosus* (Evans) is capable of suppressing the spider mite population on rose under polyhouse conditions (Mallik *et al.*, 1998). Phytoseiid mites have been more widely included in biological control programs, due to their capacity for surviving and reproducing on other arthropods. Additionally, some phytoseiid mites have shown to be resistant or less susceptible to chemical compounds commonly used to control pest mites in commercial crops, thus making them suitable for their use in integrated programs. The potential of phytoseiid mites as predators has been reported in many areas.

Phytoseiids are found in all parts of world. *Phytoseiulus persimilis* Athias-Henriot is an excellent predator of spider mites in humid greenhouses with moderate temperatures. It does not perform best on dense low growing plants and is used extensively on field strawberries in California for control of spider mite. Biological control of the two-spotted spider mite in commercial cucumber fields was achieved by using *P. persimilis* at a rate of 20-25 individuals/bit. Release of *P.*

persimilis at 5, 15 and 45 predators/plant reduced *T. urticae* on roses in five weeks after their release (Carvalho and Vis, 1996). *A. cucumeris* (Oudemans) lives on low growing plants and feeds actively on eggs of clover mite, *Bryobia praetiosa* Koch, but it is an unimportant predator of brown mite, *B. rubrioculus* (Scheuton), which inhabits trees. *Typhlodromus occidentalis* Nesbitt is a more general predator and has wide scope. *N. longispinosus* (Evans) has a wide distribution and better ability to adapt to warm temperature inside green houses under south Indian conditions. It is a promising predator, which can suppress the spider mite population both in laboratory and field conditions when released at appropriate predator : prey ratios (Mallik *et al.*, 1998 and Onkarappa, 1999). The results of Fraulo and Liburd (2007) indicated that one release of *N. californicus* (McGregor) is able to sustain control of two-spotted spider mite in strawberry throughout growing season if released when the population of pest mite is low early in the season in USA. Potential of *N. longispinosus* to control spider mites on rose and carnation in polyhouses has been assessed. The annual cost of management of *T. urticae* using *N. longispinosus* was Rs. 74,908 per hectare as compared to Rs. 1,60,607 per hectare, where acaricides were used (Rajashekarappa, 2000 and 2010). Studies show that more the predators released faster is the rate at which the spider mites are eliminated. If the predators are released at 1:50 predator to prey ratio, the time for reduction in population of the spider mite is lesser than the time required when they are released at 1:200 or 1:300 ratios. It eliminated *T. urticae* completely, 5 weeks after release at 1:50 ratio in polycarbonate house (Jayasinghe, 2008). On the bases of experiments conducted in greenhouse, Oliveira *et al.* (2009) concluded that *Phytoseiulus macropilis* (Banks) can locate and reduce two-spotted spider mite population on strawberry plants under greenhouse conditions.

Kasap (2011) examined the efficacy of the predatory mite, *T. athiasae* Porath and Swirski as a biological control agent of the citrus red mite, *P. citri* on seedlings of citrus cultivars at 1:10, 1:20 and 1:40 (predator : prey) release ratios under greenhouse conditions. At predator : prey ratios of 1:10, significant reductions on *P. citri* populations were observed one week after release of *T. athiasae*, and populations remained at low levels thereafter. He therefore, recommended a release ratio of 1:10 for effective control of *P. citri* in greenhouses on seedlings of citrus. Kaur and Zalom (2013) observed the predatory preference of *N. californicus* towards nymphs as compared to adults of *T. urticae* and *Eotetranychus lewisi* (McGregor). The results showed that *N. californicus* appeared to prefer *E. lewisi* as compared to *T. urticae.* The predatory efficiency of *N. californicus* for both the prey mites makes it a useful predator in an IPM system for the control of both of these mite pests of strawberry.

Stigmaeidae family are red to yellow, ovoid or elongate mites whose dorsum is covered with several plates. A few species prey on scale insects or parasitize flies. *Agistemus* spp. and *Zetzellia* spp are common predators of eriophyid, tetranychid and tenuipalpid mites. *Agistemus longisetus* Gonzalez feeds on eggs and active stages of spider mites infesting apple trees. *A. exsertus* Gonzalez and *Zetzellia mali* (Ewing) have been reported to readily feed and control *Brevipalpus phoenicis* (Geijskes) on tea in Indonesia.

In family Bdellidae, *Bdella depressa* Ewing is a voracious feeder on various spider mites. *Cunaxa capreolus* (Berlese) belonging to family Cunaxidae is reported to feed on *Eutetranychus orientalis* (Klein). *Cunaxa oliveri* (Schruff) was observed to feed on gall mite, *Calepitrimerus vitis* (Canestrini) on grape vines in Germany. In family Tydeidae, *Homeopronematus anconai* (Baker) is reported as an excellent predator of tomato russet mite, *Aculus lycopersici* (Massee). *Anystis agilis* (Banks) belonging to family Anystidae has been reported feeding on mite pests and other small insects infesting different crops. In family Cheyletidae, *Cheyletus eruditus* (Schrank) commonly occurs in stored food, where it feeds on pest mites, especially *Acarus* spp. *Balaustium putmani* Smiley (Family - Erythraeidae) feeds on many arthropods and prefers eggs of *P. ulmi* (Koch).

Considering requirements for effective predation of phytophagous mites, an effective predator should have shorter development period than their prey, high reproductive potential, capacity to consume more prey and survive at low prey populations, effective searching capacity at low densities, same microhabitat preference as that of prey, seasonal cycle corresponding with that of prey, ability to tolerate weather extremes and ability to tolerate pesticides. None of the micro as well as macro predators meets all the requirements but each possesses some characters of predation. However, many phytoseiids have shorter cycle than their prey, equivalent reproductive potentials of prey, good searching capacity and ability to survive on relatively fewer preys. So most of the phytoseiids are effective predators and hold promise in management of phytophagous mites.

12.2.2. Molecular Approaches

DNA-based methods have revolutionized many fields of biology, and nowadays are currently used in an increasingly number of taxa to address a wide diversity of questions. With the advent of the PCR in 1987, the ability to amplify numerous copies of a gene or genomic region of interest opened a world of possibilities not only in terms of identification of organisms, genes and genotypes, but also, data obtained through DNA based analyses can aid to address ecological and evolutionary questions. The application of molecular markers to study mites has recently yielded new insights into their populations structure and taxonomic relationships. Genetic markers reflect differences in DNA sequences and are used to provide raw information to make estimates of genetic diversity at all taxonomic levels revealing otherwise unobtainable information of the relationships among taxonomical units (Ben-Ali *et al.*, 2000; Navajas and Fenton 2000; Behura 2006). The utility of molecular data to provide solid taxonomic criteria is now well established, being widely used to accurately differentiate between species (Caterino *et al.*, 2000; Tautz *et al.*, 2003). For population studies, molecular information allows estimating gene flow, identifying haplotypes and lineages or predicting migration and colonization history (Sunnucks 2000; Garant and Kruuk, 2005).

Grbic *et al.* (2011) completely sequenced and annotated spider mite, *T. urticae* genome which is representing the first complete chelicerate genome. At 90 megabases, *T. urticae* has the smallest sequenced arthropod genome. The spider mite genome shows unique changes in hormonal environment and organization of the Hox complex as compared to other arthropods and also reveals evolutionary innovation of silk production. They found strong signatures of phylogeny and detoxification in gene families associated with feeding on different plants which shows how this pest responds to a changing host environment. Thus, *T. urticae* genome offered new insights into arthropod evolution and plant-herbivore interactions, and provides unique opportunities for developing novel plant protection strategies.

12.2.2.1. Molecular taxonomic Studies

One significant application of molecular methods is for species identification. Navajas and Boursot (2003) showed that a nuclear marker, the ribosomal ITS2 region, was diagnostic to separate two closely related but well defined spider mite species, *T. urticae* and *T. turkestani* Ugarov & Nikolskii. As for any taxa, the data on eriophyoid systematics obtained by DNA data would be a more efficient tool if combined with classical systematic information. Eriophyoidea systematics based exclusively on morphological characters may present several limitations. Because of the considerable reduction and simplification in the body plan of eriophyoids, the structures that can be used for eriophyoid morphologically-based systematics are scarce compared to most of the other mites. It is now widely accepted that molecular tools can help to clarify the systematics of taxa presenting such kind of misleading situations. Several examples combining data on nucleotide variation to more traditional morphological features help in establishing reliable criteria to determine species in the Eriophyoidea.

Grapevine eriophyoid mites—the bud mite and the blister mite, *Colomerus vitis* (Pagenstecher), and the rust mite, *Calepitrimerus vitis* Nalepa are well documented pests. The bud mite and blister mite, while morphologically identical are thought to represent two strains of one species based on the type of damage they cause to grapes. The identity of these mites has recently been investigated using molecular markers (Carew *et al.*, 2004). Two types of markers—PCR-RFLP of the ITS-1 and microsatellites—were used to gain insights into the biology and population structure. Results suggested that the bud and blister mite are different entities and should then be treated as separate agricultural problems with distinct control strategies. Importantly, infestations of vineyards with the blister mite will not result in outbreaks of the bud mite. Tsagkarakou *et al.* (1996) identified the presence of *Wolbachia* in a strain of *T. urticae* based on amplification and sequencing of part of the 16S rDNA and *ftsZ* genes. They found that for both 16 S and *ftsZ* genes the *Wolbachia* carried by *T. urticae* clustered together with *Wolbachia* found in other arthropods. As we know *Wolbachia,* a maternally inherited bacterium, is involved in alternations of arthropod sexuality, so we can use this microorganism for reproductive incompatibilities in mites, which will give a classical control measure by breaking their life cycles.

For more applied issues, information provided through molecular studies can significantly contribute to define management strategies for eriophyoid pests. A significant pest control concern is the use of eriophyoid mites as biological control agents of weeds or other invasive plants. The knowledge of the genetic intraspecific variability of a biological control agent is the basis to establish efficient biological control strategies. The intraspecific variability is also relevant information on species that are vectors of plant pathogens especially when the pathosystem management should be based on the host genetic resistance. Diverse vector biotypes can present different levels of resistance to host plant genes involved in mite resistance. In addition, among the phytophagous mites, eriophyoids are being increasingly recognized for their potential to become invasive species (Hong *et al.*, 2006). It is necessary to adopt preventive measures to avoid introductions and fast dissemination of eriophyoid pests in new geographical areas.

A series of studies on species identification, phylogeny and intraspecific variability in the *Cecidophyopsis* genus, have been conducted since 1995 (Fenton *et al.*, 1995, 1996, 1997, 2000). This group includes mite species known to occur on twelve plant species of the genus *Ribes* and several of them are serious agricultural pests. Some species feed in young buds, causing galls resulting in the sterility of the flower buds and leading to considerable crop losses (De Lillo and Duso, 1996). In addition, at least one species, *Cecidophyopsis ribis* (Westwood), transmits the agent of the reversion disease, the most important virus disease of blackcurrants, *Ribes nigrum* L., worldwide (Oldfield and Proeseler, 1996). A PCR multiplex technique based on species-specific differences of the ITS-1 sequences, was developed to identify *Cecidophyopsis* mites (Kumar *et al.*, 1999). Anderson and Trueman (2000) identified *Varroa destructor* and *Varroa jacobsoni* as different species based on molecular analysis of both the species.

An important issue that molecular tools can help to address is testing hypotheses of the occurrence of cryptic species. The eriophyid *Abacarus hystrix* (Nalepa) has long been considered as a generalist species, having been reported from grasses of at least 30 genera, it was considered as unlikely evolving host specialization. Some studies however aimed to discriminate between three populations of *A. hystrix* from quackgrass, ryegrass and smooth brome. Using sequences of the mitochondrial COI gene and the nuclear D2 region of the 28S rDNA showed that populations from different grasses form distinct clades supporting the hypothesis of *A. hystrix* as being a complex of species and not a single generalist eriophyid taxon (Skoracka and Dabert, 2010). Likewise, several authors have suggested the occurrence of cryptic species within the wheat curl mite, *Aceria tosichella* (Keifer), through morphological variation, although this could never be conclusively demonstrated. Using the mitochondrial 16S rRNA gene and two nuclear—ITS 1 and ANT—regions, it was shown that this taxon from Australia consists of at least two separate lineages that may represent putatively distinct species (Carew *et al.*, 2009). Thus, molecular information while helping to detect and define the occurrence of cryptic species has not been concomitantly used by classical taxonomists to guide detailed morphological or morphometric studies in the search for diagnostic characters to be used in the description of new taxa.

12.2.2.2. Molecular approach for mite management

A valuable application of using genetic data to assist pest management is to investigate specialisation of mites to their host plants, which in some cases have uncovered host races. Although poorly used in eriophyoids, the approach has been used to investigate some economically important issues. The eriophyid mite *A. cajani* (Channabasavanna) is the vector of the agent of pigeonpea sterility mosaic disease (PSMD), a very damaging virus like disease in the Indian subcontinent (Ghanekar *et al.*, 1992). When this disease occurs early in the season, yield losses can reach over 90%. Integrated management of PSMD includes the development of resistant cultivars. However, pigeonpea genotypes resistance was found to be location specific. It is possible that the breakdown in PSMD resistance at various locations is due to the occurrence of different *Aceria* species or biotypes of *A. cajani*. Aiming to test this hypothesis, the variation of *A. cajani* was assessed using nucleotide sequences variation and patterns of restriction enzymes in the ITS region and associated rDNA genes (Kumar *et al.*, 2001). Results strongly suggested that *A. cajani* on pigeonpea across the Indian subcontinent constitutes a single species. It could be concluded that no other *Aceria* species and probably no *A. cajani* biotypes that differ in their vectoring ability are involved in the transmission of the agent of PSMD. It seems most likely therefore that this variation in resistance is due to the occurrence of strains of the PSMD agent and host interaction with these strains.

The tremendous economic impact caused by invasive species has received increasing attention and motivated much research aimed at understanding invasive processes (Grbic *et al.*, 2007 and Facon *et al.*, 2008) and help to establish control measures (Fagan *et al.*, 2002). One valuable approach is the study of sources and introduction routes of invasive arthropods facilitated by the use of molecular markers (Darling and Blum, 2007). Molecular information has been used to trace the invasion routes of plant mites, including the eriophyoid coconut mite, *A. guerreronis*. This species has recently spread and rapidly established in important coconut production areas worldwide, being considered as a highly destructive invasive species. The mite has not been recorded in the Indo-Pacific region, the area of origin of the coconut, suggesting that it has infested coconut only recently. In a recent phylogeography study conducted to investigate the geographical origin, ancestral host associations, and colonization history of the mite (Navia *et al.*, 2005), DNA sequences variation from one mitochondrial (16S) and one nuclear region (ITS) were used to investigate mites obtained from samples of 29 populations originating from the Americas, Africa and the Indo-ocean region, the three continents where the mite has been reported. The authors suggested that the mite originated from the Americas and not from the ancestral region of coconut in South East Asia. Thus, colonization of coconut by this mite is a recent event, perhaps facilitated by modern transportation of coconut or propagation material potentially from a palm different from coconut.

The study prompts a reassessment of efforts using quarantine measures by knowing movements of the pest. Saylor *et al.* (2008) determined the genetic diversity of panicle rice mite (*Stenotarsonemus spinki* Smiely) in rice growing areas of Arkansas, USA and concluded that all mitochondrial cytochrome oxidase subunit I (COI) from USA populations have been identical supporting the hypothesis that these infestations are very recent invasions into these research greenhouses, likely occurring around the same time period since they have not had a chance to diverge yet.

Bowman and Hoy (2012) studied the phytoseiid populations imported from Mauritius for evaluation for a classical biological control program in Florida, USA. They morphologically identified this population as *A. largoensis* Muma, a species associated with the red palm mite in South Florida. They used RAPD-PCR markers in addition to Bayesian and sequence divergence analysis of the *12S* rRNA sequence for confirmation of both the population. The results showed that the Mauritius and South Florida populations were genetically different but they were not sure whether these are species or population difference is unknown. Niu *et al.* (2012) explored novel candidate genes potentially involved in the pesticide detoxification of *P. citri* and also constructed digital gene expression libraries of short term trancriptome responses of *P. citri* to pesticides, which resulted in the identification of 120 unique transcripts potentially associated with pesticide detoxification.

12.3 CONCLUSIONS

Plant feeding mites have become a menace in virtually all the major economically important crops throughout India. Biological control seems to have considerable potential as an alternative to the conventional acaricides in the light of many of the proven chemicals already in the pipeline for banning. It is now imperative for acarologists to closely monitor the natural regulatory role of pathogens in populations of phytophagous mites in various ecosystems, eventually leading to the development of bioacaricides. The studies on molecular approach in class Acari is in the nascent stage. Various markers provide a powerful tool for studies of intraspecific variations of closely related species. Various molecular techniques provide unique opportunities for developing novel plant protection strategies for management of phytophagous mites.

REFERENCES

Aksoy HM, Ozman-Sullivan SK, Ocal H, Celik N and Sullivan GT (2008) The effects of *Pseudomonas putida* biotype B on *Tetranychus urticae* (Acari: Tetranychidae). *Experimental and Applied Acarology* **46**: 223-230.

Anderson DL and Trueman JW (2000) *Varroa jacobsoni* (Acari: Varroidae) is more than one species. *Experimental and Applied Acarology* **24**: 165-189.

Anonymous (2012) *Annual Progress Report.* All India Network Project on Agricultural Acarology, XII Group meeting. Thrissur, India.

Behura SK (2006) Molecular marker systems in insects: current trends and future avenues. *Molecular Ecology* **15**: 3087–3113.

Ben-Ali Z, Boursot P, Said K, Lagnel J, Chatti N and Navajas M (2000) Comparison of ribosomal ITS regions among *Androctonus* spp. scorpions (Scorpionida: Buthidae) from Tunisia. *Journal of Medical Entomology* **37**: 787–790.

Biddinger DJ, Weber DC and Hull LA (2009) Coccinellidae as predators of mites: Stethorini in biological control. *Biological Control* **51**: 268-283.

Bowman HM and Hoy M (2012) Molecular discrimination of phytoseiids associated with the red palm mite, *Raoiella indica* (Acari : Tenuipalpidae) from Mauritius and South Florida. *Experimental and Applied Acarology* **57**: 395-407.

Carew M, Schiffer M, Umina P, Weeks A and Hoffmann A (2009) Molecular markers indicate that the wheat curl mite, *Aceria tosichella* Keifer may represent a species complex in Australia. *Bulletin of Entomological Research* **99**: 479-486.

Carew ME, Goodisman MAD and Hoffmann AA (2004) Species status and population genetic structure of grapevine eriophyoid mites. *Entomologia Experimentalis et Applicata* **111**: 87–96.

Carvalho FS and Vis RD (1996) Biological control of *Tetranychus urticae* Koch (Acari : Tetranychidae) with *Phyoseiulus persimils* Athias-Henriot (Acari : Phytoseiidae) on roses in the Sabana of Bogota. *Revista Colombiana de Entomologia* **22** : 153-157.

Caterino MS, Cho S and Sperling FAH (2000) The current state of insect molecular systematics: a thriving tower of Babel. *Annual Review of Entomology* **45**: 1–54.

Chandler D, Davidson G and Jacobson RJ (2005) Laboratory and glasshouse evaluation of entomopathogenic fungi against the two-spotted spider mite, *Tetranychus urticae* (Acari : Tetranychidae) on tomato, *Lycopersicon esculentum. Biocontrol Science and Technology* **15**: 37-54.

Chandler D, Davidson G, Pell JK, Ball BV, Shaw K and Sunderland KD (2000) Fungal Biocontrol of Acari. *Biocontrol Science and Technology* **10**: 357-384.

Chauhan R and Dahiya B (2003) Phytophagous mites, an over view. In: Yadav PR, Chauhan R, Putatunda BN and Chhillar (eds), *Mites, their Identification and Management*. ICAR Centre of Advanced Studies, CCS Haryana Agricultural University, Hisar, pp. 28-31.

Darling JA and Blum MJ (2007) DNA-based methods for monitoring invasive species: a review and prospectus. *Biological Invasions* **9**: 751–765.

De Lillo E and Duso C (1996) Currants and berries. In: Lindquist EE, Sabelis MW, Bruin J (eds.) *Eriophyoid mites—their biology, natural enemies and control*. Elsevier, Amsterdam, pp 583–591.

Dick GL and Buschman LL (1995) Seasonal occurrence of fungal pathogen, *Neozygites adjarica* (Entomophthorales : Neozygitaceae) infescting Banks grass mites, *Oligonychus prataensis,* and two-spotted spider mite, *Tetranychus urticae* (Acari : Tetranychidae), in field corn. *Journal of the Kansas Entomological Society* **68**: 425-436.

Facon B, Pointier JP, Jarne P, Sarda V and David P (2008) High genetic variance in life-history strategies within invasive populations by way of multiple introductions. *Current Biology* **18**: 363–367.

Fagan W, Lewis M, Neubert M and van den Driessche P (2002) Invasion theory and biological control. *Ecology Letters* **5**: 148–157.

Fenton B, Birch ANE, Malloch G, Lanham PG and Brennan RM (2000) Gall mite molecular phylogeny and its relationship to the evolution of plant host specificity. *Experimental and Applied Acarology* **24**: 831–841.

Fenton B, Jones AT, Malloch G and Thomas WP (1996) Molecular ecology of some *Cecidophyopsis* mites (Acari: Eriophyidae) on *Ribes* species and evidence for their natural cross colonisation of blackcurrant (*R. nigrum*). *Annals of Applied Biology* **128**: 405–414.

Fenton B, Malloch G, Jones AT, Amrine JWJ, Gordon SC, Hara AS, McGavin WJ and Birch ANE (1995) Species identification of *Cecidophyopsis* mites (Acari: Eriophyidae) from different *Ribes* species and countries using molecular genetics. *Molecular Ecology* **4**: 383–387.

Fenton B, Malloch G and Moxey E (1997) Analysis of eriophyid rDNA internal transcribed spacer sequences reveals variable simple sequence repeats. *Insect Molecular Biology* **6**: 23–32.

Fisher FE (1951) An Entomophthora attacking citrus red mite. *Florida Entomologist* **34**: 83-88.

Fraulo A and Liburd OE (2007) Biological control of twospotted spider mite,

Tetranychus urticae, with predatory mite, *Neoseiulus californicus*, in strawberries. *Experimental and Applied Acarology* **43**: 109–119.

Garant D and Kruuk L (2005) How to use molecular marker data to measure evolutionary parameters in wild populations. *Molecular Ecology* **14**: 1843–1859.

Gerson U and Smiley RL (1990) *Acarine Biocontrol Agents: An Illustrated Key and Manual.* Chapman and Hall, London.

Ghanekar AM, Sheila VK, Benival SPS, Reddy MV and Nene YL (1992) Sterility mosaic of pigeonpea. In: Singh US, Mukhopadhyay AN, Kumar J, Chaube HS (eds.) *Plant diseases of international importance*, vol 1. Prentice Hall, New Jersey, pp 415–428.

Gowda CC (2009) Fauna of phytoseiid mites (Acari : Phytoseiidae) associated with plants in southern Karnataka. Ph.D. thesis, UAS, Bangalore, pp 208.

Grbic M, Khila A, Lee KZ, Bjelica A, Grbic V Whistlecraft J, Verdon L, Navajas M and Nagy L (2007) Mite model: *Tetranychus urticae*, a candidate for chelicerate model organism. *BioEssays* **29**: 489–496.

Grbic M, Leeuwen TV, Clark RM, Rombauts S, Rouze P, Crbic V, Osborne EJ, Dermauw W, Ngoc PCT, Ortego F, Crespo PH, Diaz I, Martinez M, Navajas M, Sucena E, Magalhaes S, Nagy L, Pace RM, Djuranovic S, Smagghe G, Iga M, Christiaens O, Veerstra JA, Ewer J, Villalobes RM, Hutter JL, Hudson SD, Velez M, Yi SV, Zeng J, dsSilva AP, Roch F, Cazaux M, Navarro M, Zhurov V, Acevedo G, Bjelica A, Fawcett JA, Bonnet E, Martens C, Baele G, Wissler L, Rodriguez AS, Tirry L, Blais C, Demeestere K, Henz SR, Gregory TR, Mathieu J, Verdon L, Farinelli L, Schmutz J, Lindquist E, Feyereisen R and Peer YV De (2011) The genome of *Tetranychus urticae* reveals herbivorous pest adaptations. *Nature* **479**: 487-492.

Grundschober A, Tuor U and Aebi M (1998) In vitro cultivation and sporulation of *Neozygites parvispora* (Zygomycetes : Entomophthorales). *Systematic and Applied Microbilogy* **21**: 461-469.

Halliday RB, Oconnor BM and Baker AS (2000) Global diversity of mites. In : Raven P H and Williams T (ed) *Nature and Human Society: the Quest for a Sustainable World : Proceeding of the 1997 forum on biodiversity*. Board on Biology, National Research Council. National Academy Press, pp. 192-212.

Helle W and Sabelis MW (1985) *Spider Mites: Their Biology, Natural Enemies and Control*. Vol. 1 Part A. Elsevier, Amesterdam, 406 pp.

Hong XY, Wang DS and Zhang ZQ (2006) Distribution and damage of recent invasive eriophyoid mites (Acari: Eriophyoidea) in mainland China. *International Journal of Acarology* **32**: 227–240.

Hoy MA (2011) Agricultural Acarology: An introduction to Integrated mite management. CRC Press, FL.

Jayasinghe GJ (2008) Studies on ecology and biological control of the two-spotted spider mite, Tetranychus urticae Koch (Acari : Tetranychidae) infesting tomato. Ph.D. Thesis, UAS, Bangalore, pp 152.

Kasap I (2011) Biological control of the citrus red mite *Panonychus citri* by the predator mite *Typhlodromus athiasae* on two citrus cultivars under greenhouse conditions. *Biological Control* **56**: 327–332.

Kaur P, Bhullar MB and Kaur R (2012) Predatory potential of an anthocorid predator, *Blaptostethus pallescens* Poppius against two-spotted spider mite, *Tetranychus urticae* Koch on brinjal. In : *Proceedings of 2nd International Conference on Agrochemicals Protecting Crops, Health and Natural Environment : Role of Chemistry for Sustainable Agriculture from February 15-18, 2012 at IARI, New Delhi, India.* 232 pp.

Kaur P and Zalom FG (2013) Preference of the predatory mite, *Neoseiulus californicus* (McGregor) for *Tetranychus urticae* Koch and *Eotetranychus lewisi* (McGregor) on strawberry at different temperatures. In : *Proceedings 4th International Confe-rence on Insect Science"* GKVK, UAS, Bangalore Feb 14-17, 2013. pp. 99.

Krantz GW and Walter DE (2009) *A Manual of Acarology*. Pp. 298, Third edition. Texas Tech University Press.

Kumar L, Fenton B and Jones AT (1999) Identification of *Cecidophyopsis* mites (Acari: Eriophyidae) based on variable simple sequence repeats of ribosomal DNA internal transcribed spacer-1 sequences via multiplex PCR. *Insect Molecular Biology* **8**: 347–357.

Kumar PL, Fenton B, Duncan GH, Jones AT, Sreenivasulu P and Reddy DVR (2001) Assessment of variation in *Aceria cajani* using analysis of rDNA ITS regions and scanning electron microscopy: implications for the variability observed in host plant resistance to pigeonpea sterility mosaic disease. *Annals of Applied Biology* **139**: 61–73.

Leite LG, Smith L, De Moraes GJ and Robert DW (2000) In vitro production of hyphal bodies of the mite pathogenic fungus *Neozygites floridana*. *Mycologia* **92**: 201-207.

Lipa JJ (1971) Microbial control of mites and ticks. In : Burges HD and Hussey NW. (eds) *Microbial Control of Insects and Mites*, Academic Press, New York, pp. 357-373.

Mallik B, Onkarappa S and Harish Kumar M (1998) Management of spider mites, *Tetranychus urticae* Koch on rose using phytoseiid predators, *Amblyseius longispinosus* (Evans) in polyhouse. *Pest Management Horticulture Ecosystem* **4**: 46-48.

McCoy CW (1981) Pest Control by the fungus *Hirsutella thompsonii*. In: Burges H D (ed), *Microbial Control of Pests and Plant Diseases*. Academic Press, London, UK, pp. 499-512.

McCoy CW (1996) Pathogens of eriophyid mites. In : Lindquist EE, Sabelis MW and Bruin J (eds), *Eriophyid Mites- Their Biology, Natural Enemies and Control.* Elsevier Science B V, Amsterdam, The Netherlands, pp. 481-490.

Mietkiewski R, Balazy S and Tkaczuk C (2000) Mycopathogens of mites in Poland – a review. *Biocontrol Science and Technology* **10**: 459-465.

Muma MH (1955) Factors contributing to the natural control of citrus insects and mites in Florida. *Journal of Economic Entomology* **48**: 432-438.

Navajas M and Boursot P (2003) Nuclear ribosomal DNA monophyly Versus mitochondrial DNA polyphyly in two closely related mite species: the influence of life history and molecular drive. In : *Proceedings of the Royal Society of London B (Suppl)* **270**: 124–127.

Navajas M and Fenton B (2000) The application of molecular markers in the study of diversity in acarology: a review. *Experimental and Applied Acarology* **24**: 751–774.

Navia D, Moraes GJD, Roderick G and Navajas M (2005) The invasive coconut mite *Aceria guerreronis* (Acari: Eriophyidae): origin and invasion sources inferred from mitochondrial (16S) and nuclear (ITS) sequences. *Bulletin of Entomological Research* **95**: 505–516.

Niu JN, Dou W, Ding TB, Shen GM, Zhang K, Smagghe G and Wang JJ (2012) Transcriptome analysis of citrus red mite, *Panonychus citri* and its gene expression by exposure to insecticide/acaricide. *Insect Molecular Biology* **21**: 422-436.

Oldfield GN and Proeseler G (1996) Eriophyoid mites as vectors of plant pathogens. In: Lindquist EE, Sabelis MW, Bruin J (eds) *Eriophyoid Mites—their Biology, Natural*

Enemies and Control, Elsevier, Amsterdam, pp 259–275.

Oliveira H, Fadini MAM, Venzon M, Rezende D, Rezende F and Pallini A (2009) Evaluation of the predatory mite *Phytoseiulus macropilis* (Acari: Phytoseiidae) as a biological control agent of the two-spotted spider mite on strawberry plants under greenhouse conditions. *Experimental and Applied Acarology* **47**: 275–283.

Onkarappa S (1999) Management of two-spotted spider mite, *Tetranychus urticae* Koch (Acari : Tetranychidae) on rose. Ph.D. Thesis. UAS, Bangalore, pp 252.

Poorani J (2010) Insect natural enemies of mites with special reference to coccinellids and thrips. In : Srinivasa N, Mallik B and Chinnemade G (eds) *Advances in Agricultural Acarology*, pp 158-163.

Rajashekarappa K (2000) Biological suppression of spider mite, *Tetranychus urticae* Koch using phytoseiid predator, *Neoseiulus longispinosus* Evans on carnation in polyhouse conditions. M.Sc. Thesis, UAS, Bangalore, pp. 89.

Rajashekarappa K (2010) Management of *Tetranychus urticae* Koch (Acari : Tetranychidae) on rose and carnation in polyhouse using phytoseiid predator, *Neoseiulus longispinosus* (Evans) (Acari : Phytoseiidae). Ph.D. Thesis, UAS, Bangalore.

Ramaraju K, Palaniswamy S, Annakkodi P, Varadarajan MK, Muthukumar M and Bhaskaran V (2005) Impact of coconut eriophyid mite, *Aceria guerreronis* K. (Acari: Eriophyidae) on the yield parameters of coconut. *Indian Coconut Journal* **35**: 12-15.

Saylor RJ, Dowling APG and Cartwright RD (2008) Molecular detection and genetic characterization of the panicle rice mite, *Stenotarsonemus spinki. AAEs Research Series* **571**: 121-126.

Skoracka A and Dabert M (2010) The cereal rust mite *Abacarus hystrix* (Acari: Eriophyoidea) is a complex of species: evidence from mitochondrial and nuclear DNA sequences. *Bulletin of Entomological Research* **100**: 263-272.

Sreerama Kumar P (2009) *Hirsutella thompsonii* as a mycoacaricide for *Aceria guerreronis* on coconut in India: research, development, and other aspects. In : Sabelis M W and Bruin J (eds), *Trends in Acarology : Proceedings of the Twelfth International Congress*. Springer Science and Business Media B V, Amsterdam, The Netherlands, pp. 441-444.

Sreerama Kumar P and Singh L (2008) Enabling mycelial application of *Hirsutella thompsonii* for managing the coconut mite. *Experimental and Applied Acarology* **46**: 169-182.

Sreerama Kumar P and Singh SP (2000) *Hirsutella thompsonii :* The best biological control option for the management of the coconut mite in India. *Indian Coconut Journal* **31**: 11-17.

Sreerama Kumar P and Singh SP (2001) Mycoacaricides : The emerging class of biopesticides. In : *Proceedings of the IIIrd Asia Pacific Crop Protection Conference-2001,* Pesticides Manufacturers and Formulators Association of India, Habitat Centre, New Delhi, 6-7 September 2001, pp. 60-67.

Sreerama Kumar P, Singh SP and Anuroop CP (2004) Investigations on *Sporothrix* fungorum de Hoog and de Vries, a newly recorded pathogen of *Aceria guerreronis* Keifer, the coconut eriophyid mite. *Journal of Biological Control* **18**: 13-20.

Sunnucks P (2000) Efficient genetic markers for population biology. *Trends in Ecology and Evolution* **15**: 199–203.

Tautz D, Arctander P, Minelli A, Thomas RH and Vogler AP (2003) A plea for DNA taxonomy. *Trends in Ecology and Evolution* **18**: 70–74.

Tsagkarakou A, Guillemaud T, Rousset F and Navajas M (1996) Molecular identification of a *Wolbachia* endosymbiont in a *Tetranychus urticae* strain (Acari : Tetranychidae). *Insect Molecular Biology* **5**: 217-221.

Van der Geest LPS (2004) Fungi against phytophagous mites. *Entomologische Berichten* **64**: 146-156.

Van der Geest LPS, Elliot SL, Breeuwer JAJ and Beerling EAM (2000) Diseases of mites. *Experimental and Applied Acarology* **24**: 497-560.

Yadav PR, Chauhan R, Putatunda BN and Chillar BS (2003) *Mites, their Identification and Management.* ICAR Centre for Advanced Studies, Department of Entomology, CCSHAU, Hisar, 215 pp.

Biological and Molecular Approaches in Pest Management
Balwinder Singh, Ramesh Arora & S S Gosal (Eds)
Scientific Publishers (India), Jodhpur, 2015, pp 329-343.

Chapter - 13

Biosystematics, Molecular Characterization and Management of Shoot and Fruit Borer *Conogethes* Spp. (Crambidae: Lepidoptera)

A K Chakravarthy*, P R Shashank, B. Doddabasappa, S B Kandakoor and Chandrashekharaiah
Department of Entomology, University of Agricultural Sciences, Bangalore-560065, India
*E-mail: chakravarthyakshay@gmail.com

13.1. INTRODUCTION

The genus *Conogethes* is a large and taxonomically complex taxon. Until now ten named species of genus *Conogethes* have been known from Eastern Palaearctic and Indo-Australian regions (Meyrick, 1884). The fruit and shoot borer *C.* (=*Dichocrocis*) *punctiferalis* Guenee commonly called as yellow peach moth, is an important pest incurring severe crop losses as the larvae directly inflict damage to reproductive or economically important plant parts. The pest mainly occurs in tropical and subtropical countries (Pena *et al.*, 2002) and is distributed in Asia and Australia. Larvae of this crambid moth are typically polyphagous pests attacking more than 120 wild and cultivated diversified plants, *viz.*, peach, chestnut, durian, citrus, papaya, castor, cardamom, ginger, eggplant and maize (Sekiguchi, 1974). The type locality of *C. punctiferalis* is India. So, many closely allied species may be included. However, taxonomic revision of them has been neglected for a long time. Species belonging to genus *Conogethes* are receiving increasing attention because of expanding host range, geographical occupancy and complexity involved in species identifications. As such the pest, an internal tissue borer is difficult to manage in fruit orchards and plantations. This insect group is undergoing speciation, genomic changes or evolving into new taxon. So biosystematics, molecular characterization and management aspects of *Conogethes* spp. are dealt with in this chapter.

13.2. BIOSYSTEMATICS

Genus *Conogethes* is a large, taxonomically complex taxon belonging to superfamily Pyraloidea, family Crambidae, subfamily Pyraustinae. Earlier references put *Conogethes* under Pyralidae. Maes (1998) demonstrated that differences in structure of tympanal organ or ears that is present at base of abdomen called the *praecinctorium*, which joins two tympanic membranes in the Crambidae, is absent in the Pyralidae.

Since publication of Sir George Hampson's Fauna volumes in 1896 in the 'Fauna of British India' series, considerable changes in the taxonomy of this group have taken place. The species *punctiferalis* was placed in genus *Conogethes* by Meyrick (1884), although moved to *Dichocrocis* after that. It has been placed back in *Conogethes* by Munroe (1989). Shaffer *et al.* (1996) placed it in *Conogethes* as a revised combination. The Natural History Museum (London) card index is yet to be updated (10/10/2012), so we considered GlobIZ (Global Information System on Pyralidae) as a better source. Hence, retention of *Conogethes* for the crambid moth which is now followed by entomologists world over. Hampson in his 'Fauna of British India' reported 20 species of *Dichocrocis* mainly based on the wing venation and arrangements of black spots on wings. Subsequently, few preliminary studies have been done by lepidopterists in India and abroad on this genus. Five species of *Dichocrocis* were identified, namely *D. evaxalis* Walker, *D. punctiferalis* Walker, *D. nigrilnealis* Walker, *D. plutusalis* Walker and *D. surusalis* Walker from light trap collections in Kerala, India (Mathew & Menon, 1984). Azam and Ali (1965) studied the morphology of larva of *D. punctiferalis* with special reference to chaetotaxy. The genitalial morphology of *D. punctiferalis* and *D. plutusalis* was studied and main modification in the morphology of valvae, uncus, saccus, phallus, bursae and ductus in these species were elucidated (Mathew & Menon, 1989). Until now, ten named species of genus *Conogethes* Meyrick (1884) have been known from Eastern Palaearctic and Indo-Australian regions (Shaffer *et al.*, 1996). Inoue & Yamanaka (2006) redescribed *C. punctiferalis* along with two new (*C. parvipunctalis and C. pinicolalis*), closely allied species from Eastern Palaearctic and Oriental regions. In mid-1980s Chakravarthy (unpublished observations) found differences in morphology of *Conogethes* moths reared on castor and cardamom (*Elettaria cardamomum* Maton) in Western Ghats of Karnataka. The male genitalia also differed between the two populations.

Molecular taxonomy is giving an additive support for species identification with traditional taxonomy. Now-a-days DNA barcoding is a major tool for identificat-ion of species. Seventeen species (Table 1) from six countries, *viz*. Australia, Papua New Guinea, Cambodia, China, Indonesia and Nepal are barcoded for genus *Conogethes* and deposited in the Barcode of Life Data Systems (BOLD) to date.

Unfortunately, there is no single species from India that is barcoded. There is a need for taxonomic revision of this genus with the help of molecular taxonomy. The species listed in Hampson's fauna differ from those that are barcoded. Recently Shashank (2012) barcoded *Conogethes* moths reared on castor and cardamom from different geographical locations of India.

Table 1. *Conogethes* species with records on BOLD

S No.	Species	Speci-mens	Sequ-ences	Barcodes (COI of >500bp)
1	*Conogethes diminutiva*	2	2	1
2	*Conogethes ersealis*	3	3	3
3	*Conogethes evaxalis*	9	8	8
4	*Conogethes haemactalis*	3	3	3
5	*Conogethes nr. diminutiva*	1	1	1
6	*Conogethes nr. haemactalis*	3	3	3
7	*Conogethes parvipunctalis*	1	1	1
8	*Conogethes pluto*	12	12	12
9	*Conogethes punctiferalis*	67	26	26
10	*Conogethes punctiferalis PS2*	1	0	0
11	*Conogethes semifascialis*	11	8	8
12	*Conogethes* sp.	1	0	0
13	*Conogethes* sp.ANIC 1	2	1	1
14	*Conogethes* sp. ANIC 2	3	3	3
15	*Conogethes* sp.ANIC 3	1	0	0
16	*Conogethes* sp. *complex*	4	3	3
17	*Conogethes tharsalea*	5	4	4

Source: http://www.boldsystems.org/views/speciessummary.php

13.3. MOLECULAR CHARACTERIZATION

Mitochondrial genes are often chosen for evolutionary studies and barcoding as they have a number of positive characteristics. The protein coding genes are the most frequently sequenced mitochondrial genes for evolutionary studies and phylogenetic analysis.

The CBOL established the "All-Leps Barcodes of Life" project because the Lepidoptera is the second most diverse order of insects. There are about 180,000 known species, and it is likely that there are another 300,000 species awaiting description. The initiative involves campaigns on three geographic scales: Global (Geometridae, Saturniidae and Sphingidae), Continental (North America and Australia) and Regional (Great Smoky Mountains National Park, USA and Area de Conservacion, Guanacaste) (Bravo *et al.*, 2008). Until now, 632,006 lepidopteran species are with barcodes; of them 4,443 are Crambidae (International Barcode of Life, 2012). Hebert *et al.* (2004) studied the morphological and DNA barcoding of *Astraptes fulgerator* Walch widely distributed neotropical skipper butterfly (Lepidoptera: Hesperiidae) in north western Costa Rica with museum specimens. They showed that *A. fulgerator* is a complex of at least 10 species in Costa Rica. Largely sympatric, these taxa have mostly different caterpillar food plants, mostly distinctive caterpillars, and different ecosystem preferences but only subtly differing adults with no genitalic divergence.

Barcodes were generated on *Conogethes* from specimens collected on castor and cardamom from 6 and 8 locations, respectively. The locations were chosen including all the spots in a cultivated zone of the crops. The DNA barcodes were appended to an existing dataset from BOLD for *Conogethes* species. The entire data set on *Conogethes* included 115 DNA signatures categorized into four clades and these further have been classified into more than 50 clusters. The *Conogethes* on castor and cardamom belonged to two distinct clades, moths were distinguished based on the host plants. With specimen matching 91 per cent of the standard signature in BOLD, with signature corresponding to *C. punctiferalis*. The *Conogethes* specimen on cardamom is a new signature, this DNA signature did not match with any signatures in BOLD deposited earlier. So this indicated that the moths reared on cardamom possibly belong to a new species. This needs to be further confirmed. The mean within species divergence for *C. Punctiferalis* was 2.102 per cent. However, the pair wise divergence between *C. Punctiferalis* and the moths reared on cardamom was >5 per cent (Shashank, 2012) (Figure 1).

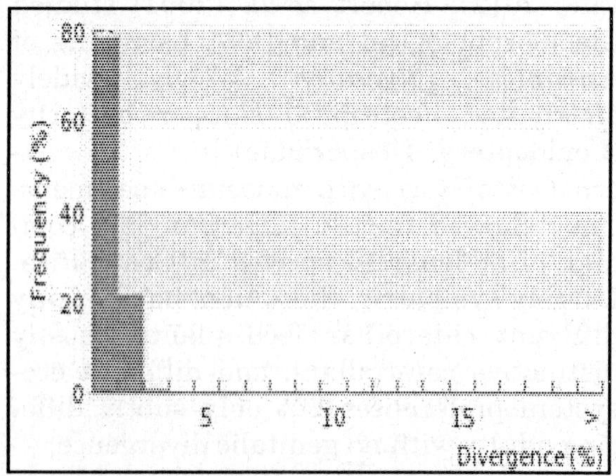

Fig. 1. Divergence in barcodes within *Conogethes* species (Shashank, 2012)

The family Crambidae, subfamily Pyraustinae, has 4585 species with barcodes. The genus *Conogethes* has 138 barcode sequences and seventeen species. Armstrong (2010) compared DNA barcoding of different populations of *Conogethes* and revealed that Australian and Asian specimen form separate clades divergent by ~6 per cent. The barcode data successfully distinguished *C. punctiferalis* and *C. pluto*, but unexpectedly revealed divergence between the Asian and Australian populations. Morphologically these were determined to be the same species, and distinct from other closely related species found on the east coast of Australia such as *C. haemactalis* Walker, *C. semifascialis* Walker, or *C. tharsalea* Walker.

The neighbour joining tree (NJ tree) was constructed based on all the 33 DNA barcodes using BOLD analysis tool. Based on NJ tree two clades were recognized: those clades which includes all the Conogethes individuals breeding on cardamom werenamed as *Conogethes* sp. while from castor were named as *Conogethes punctiferalis* (Figure 2).

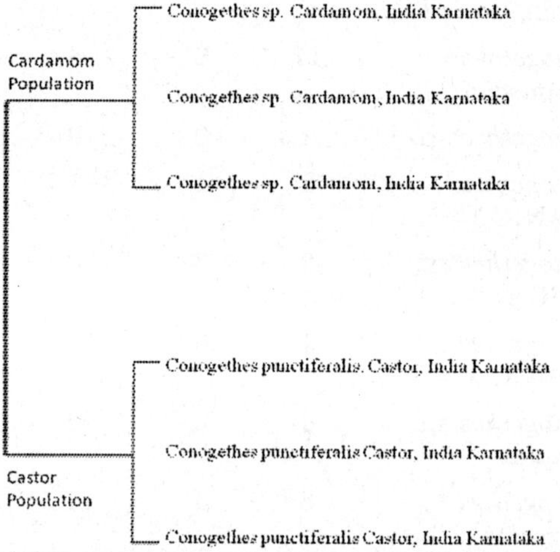

Fig. 2 Neighbour joining analysis of COI DNA barcode sequences for *Conogethes* species breeding on castor and cardamom (Shashank, 2012)

The sequence details are analyzed carefully, submitted to NCBI for Gen Bank Accessions and subsequently uploaded to COBOL Barcoding facility for formally obtaining the Barcodes. All generated sequences, together with photographs and collection details, have been deposited at the Barcode of Life Database (BOLD; www.boldsystems.org) under the project code. DNA barcoding analyses were done through the online interface of the BOLD website. The taxon identification tree was based on the Kimura 2-parameter distance model (Kimura, 1980), with the filter set to sequences with length >100 base pairs, and all codon positions included. A sequence distance summary was provided through the online interface of BOLD, with only sequences longer than 425 base pairs included in order to ensure reliable sequence distance estimates (Shashank, 2012). Preliminary morphological, biological and barcode analysis of *Conogethes* moths reared on castor, cardamom, turmeric, ginger, guava and cocoa revealed two distinct clusters. Moths on castor, guava and cocoa fall in one cluster while moths on cardamom, ginger and turmeric into another. Further experiments in this direction are in progress.

13.4. *Conogethes pinicolalis* INOUE AND YAMANAKA, 2006 : AN EVOLVED SPECIES

In Japan, researchers for convenience identified two different types, in absence of precise identification, the fruit tree type and the conifer type from angiosperms and Pinaceae gymnosperms, respectively. Later it was reported that the adults from coniferous plant, Japanese Cedar, *Cryptomeria japonica* (L. f.) D. Don were of the fruit tree type (Sekiguchi, 1974). Thus the fruit tree type was renamed as Fruit-feeding type (FFT) and the conifer type, the Pinaceae-feeding type (PFT) (Konno *et al.,* 1981). Subsequently, Honda and Mitsuhashi (1989) studied the morphological differences between the fruit-feeding and pinaceae-feeding types of *C. punctiferalis*. Nine quantitative characters were morphometrically evaluated on the male geni-talia and female ovipositor. Male moths of the two types were easily distinguishable by the angular and linear characters on the valve and tegument. Differences were also observed between mandible, labrum, epipharynx and pinacular of larvae, and cremaster of pupae. Based on these morphological differences in adults, larvae and pupae, the authors suggested pinaceae-feeding type of the yellow peach moth as a segregated species of the genus *Conogethes*. Until 2006 *C. pinicolalis* was included under *C. punctiferalis*. Now *C. pinicolalis* represents a new species of *Conogethes* and shares similar but not the same wing patterns with the other species. The present species could be identified by the following characters: the second segment of labial palp almost black; hind tibia and hind tarsus with large tufts of fuscous scales (Inoue and Yamanaka, 2006).

13.5. PHEROMONES

Sex pheromone was detected in Japan in extracts of the abdominal tips of females of the fruit-feeding type of *C. punctiferalis*. It was identified by gas liquid chromoatography, mass spectrometry, ozonolysis and electro antennography as (E)-10-hexadecenal. The synthetic compound was attractive to males (10 ng being equivalent to 1 live virgin female) in the laboratory, but in the field, traps baited with this compound trapped only a few males. However, four times as many males were trapped with a mixture of (E) - and (Z)-10 hexadecenal in 9:1. Males of the type of *C. punctiferalis* feeding on the leaves of Pinaceae were also caught in traps baited with the mixture at 9:1 and 8:2 (Konno *et al.*, 1982). It was therefore, concluded that (Z)-10-hexadecenal was a minor component of the sex pheromone of *C. punctiferalis*.

Konno (1986) investigated the [illegible]nship between daily changes in the sex pheromone quantity and calling behaviour in females of *C. punctiferalis* in the laboratory at 23°C, 70-80% RH and LD15:9 hrs. He recorded that quantity of the sex pheromone, (E)-10-hexadecenal, in the pheromone gland increased when the lights were turned off, reached a maximum after 5 hrs and then decreased. On the other hand, calling started from 5 hrs after light-off, reached a maximum at 7.5 has after light-off and then decreased. Nakano *et al.* (2012) detected ultrasound produced by male *C. punctiferalis* for mating. Males developed mesothoracic tymbal organs for generating ultrasonic clicks in mating. Mori *et al.* (1990) studied the synthesis and biological evaluation of two sex-pheromone mimics ((E) - and (Z)-Tetr-adecenyl formate) for *C. punctiferalis*. A 10:1 mixture of (E) -and (Z)-tetradecenyl formate was found as attractive as the natural pheromone for adults of the crambid.

Chakravarthy and Thyagaraj (1997) studied activity of seven pheromone compounds ((Z)-9-hexadecenyl acetate, (Z)-7-tetradecenal,(E)-11-tetradecenal, (Z)-11-tetradecenal, (Z)-11-hexadecenal, (E)-10-hexadecenal and (Z)-10-hexadecenal) against *C. punctiferalis*. Tests were carried out in laboratory and cardamom fields in Mudigere and Sakleshpur, Karnataka, India during September-December between 1985-93. In laboratory, males responded to (E)-10-hexadecenal, (Z)-10-hexadecenaland (Z)-11-hexadecenal at 1000 ng. A blend of (E)-10-hexadecenal and (Z)-10-hexadecenal at 9:1 had maximum attractancy. Laboratory studies on the mating behaviour showed that an airborne sex pheromone was released from the calling female. Males were attracted to virgin female extract in laboratory tests suggesting that the extract contained pheromone components for attraction. (E)-10-hexadecenal and (Z)-10-hexadecenal at 1000 ng concentration when tested separately, and a blend of these two compounds at 9:1 attracted maximum number of male moths in the laboratory. In field trials male moths were attracted to traps baited with (E)-10-hexadecenal and (Z)-10-hexadecenal at 9:1 in cardamom plantations (Chakravarthy and Thyagaraj, 1998).

The sex pheromone components of *Conogethes* in Japan, Korea, China and India showed variations (http://www.pherobase.com). *Conogethes* represents a cryptic species complex (Konno *et al.*, 1980). The non-polar components of female body wax and pheromone gland extracts of yellow peach moth synergistically enhanced male behavioral responses from close to pheromone resources in wind tunnel experiments in Japan (Xiao *et al.*, 2011).

13.6. INSECT-HOST PLANT INTERACTIONS

Studies on the biology of *Conogethes* on castor and cardamom revealed that the insect could complete the life cycle on the three types of the castor and the cardamom. The insect required longer period (in days) of time to complete the life cycle on cardamom than on castor i.e., on an average 34.5 days on castor compared to 39 days on cardamom (Doddabasappa, 2012). Egg incubation period (in days) was the least (2.67 days) on spineless spike castor as also the larval period (26.68 days) when compared with compact spiny and spiny loose castor types. The trend in the egg incubation, larval and pupal periods was consistent across three castor types. *Conogethes* larvae preferred compact spiny castor type for feeding over other two types. *Conogethes* larvae and pupae gained more weight on compact spiny castor type than on the spineless and spiny loose castor types. This is obvious because compact spiny type was more suitable to *Conogethes* than the other two castor types.

When *Conogethes* reared on cardamom were implanted on castor, larvae suffered cent per cent mortality. Besides, larvae completed the period early probably because of physiological stress throughout by host plant shift (11-12 days). When castor larvae were on castor itself, larvae took 14-16 days.

When *Conogethes* reared on castor were implanted on cardamom, larvae suffered cent per cent mortality. Larvae completed the period early probably because of physiological stress brought out by host plant shift (13-14 days). When cardamom larvae were implanted on cardamom, larvae took 20-22 days.

13.7. BIOECOLOGY

Krishnamurthy *et al.* (1989) reported that *Conogethes* completed one life cycle on cardamom in 25-40 days with five generations a year at Mudigere, Chikmangalur, Karnataka at 26°C and 70% RH.

Eggs of *C. punctiferalis* are round, light yellow, 0.63 x 0.41 mm. After incubation of 6 to 7 days egg turn dark brown with a dark head (Bilapate, 1978; Jarvis, 1914 and Thyagaraj, 2003). The phenology of host plant influences the size, growth and development of eggs (Bilapate, 1978; Jacob, 1981 and Twine, 1971). There was a significant difference in per cent egg hatching from 65.0±0.76 to 90.5±1.38 and incubation period from 4.19±0.80 to 9.35±1.05 days under varied temperature and relative humidity (Thyagaraj, 2003 and Wang and Cai, 1997). Temperature and relative humidity play an important role in egg hatching (Rajan, 1965). The egg characters under ranging temperature and relative humidity conditions have been extensively studied in laboratory (Kalshoven, 1929 and Thyagaraj, 2003).

First instar (neonate) larva bored the pseudostem or capsule. On pseudostem, larva bored at the base of leaf axis and entered inside the cardamom shoot tissue (Jacob, 1981). The excreta plugged at the entry hole on the shoot indicated larval boring (Thyagaraj, 2003). Larva fed on the shoot was light greenish. While those on capsules were dull yellow (Bilapate & Talati, 1977 and Thyagaraj, 2003). Larva remains inside the pseudostem till pupation. Therefore, it is difficult to study the larval instars directly. Dyar's law was applied to record number of larval instars and there were five larval instars (Asante *et al.*, 1991; Dyar, 1890 and Thyagaraj, 2003). Observations revealed that there was no major difference among the instars excepting for head and body size. All larval instars were active and when disturbed, tried to fall down with a fine silken thread (Bilapate and Talati, 1978; Kondo and Miyahara, 1930; Kodoi and Kavida, 1990; Thyagaraj, 2003; Twine, 1971 and Young and Shaw, 1962).

The duration of the larval stage varied from 20-23 days in August-September at 21-35°C to 22-26 days in

October -January at 14-28°C and larvae were found in the field until February. (Bilapate, 1977; Jacob, 1981; Kondo and Miyahara, 1930; Twine, 1971; Wang and Cai, 1997 and Xi *et al.*, 1996). The larval period varied from 12.55±2.00 to 19.59±5.50 days and the per cent survival varied from 49.6±0.18 to 92.8±1.39. A temperature of 28.0±1.0°C and 80.0±5.0% RH were most favorable for larval development (Thyagaraj, 2003).

Pupation of *Conogethes* took place in cocoons inside or between the capsules and the pupal stage lasted for 7-10 days (Bilapate and Talati, 1978; Gour and Sriramulu, 1992). According to Wu (1995) 86.5 per cent larvae of *C. punctiferalis* pupated in leaf axils and in ears of maize and 13.5 per cent pupated in stalks.

There was a clear difference in size, shape and weight of male and female pupae. Female pupae were larger (17.81x6.29 mm with 0.127 gm in weight). Male pupa measured 14.30 x 4.26 mm with 0.108 gm in weight, shorter, slightly narrower and the genital opening was located in the posterior region of the ninth abdominal segment and flanked by a pair of pads. Significant difference was also noticed between the sexes in terms of length and diameter (Thyagaraj, 2003).

Bilapate and Talati (1978) revealed that the males survived for 14.00 ± 3.80 days as compared with 15.80 ± 2.50 days for females. The ratio of males to females of the progeny ranged between 1:1 and 1:2. Kuang *et al.* (2009) reported that in China adults emerged mainly between 22.00-08:00 h. The emergence rate was 92.22 per cent. The dates corresponding to the emergence of 16, 50 and 84 per cent of the total number of moths were in 8 May, 11 May and 15 May, respectively. The longevity of adults was significantly affected by adult nutritional conditions. Adults fed on water, 10 per cent honey solution or 10 per cent sugar and vinegar liquid survived significantly longer than unfed advents. *C. Punctiferalis* males have hair pencils at the abdominal end and thus can be differentiated from the other sex. Mating occured only in the dark after 7.30 PM (Stanley *et al.* 2009).

Hybridization experiments with *Conogethes* moths reared on castor and cardamom were conducted under laboratory conditions. The mating experiments were conducted in single pair/cage and multiple pairs (4 pairs/cage). In cages where multiple pairs were enclosed, higher mating success was achieved. When the moths were reared on 'shifted' plants, no mating occurred although the moths attempted to mate several times (Shashank, 2012), the sexes were reproductively isolated.

Life cycle

Life cycle of the insect varied with the changed environmental conditions (Abraham, 1965 and Bilapate and Talati, 1978). Under laboratory conditions (28.0 ± 1.0°C and 80.0 ± 5.0% RH), the incubation period lasted for a mean of 5.3±0.49 days. Larval period 17.62±4.88 days, pupal period 8.81 ± 0.69 days and the mean male and female longevity worked out to be 14.26 ± 3.29 and 15.29 ± 3.39 days, respectively in *Conogethes* species. The total life cycle from egg to adult emergence required a mean of 31.75 ± 10.16 days (Pruthi, 1944 and Young and Shaw, 1962). Under field conditions in cardamom plantations the incubation lasted 8.51 ± 0.65 days, larval period 25.49 ± 4.76 days, pupal period 9.55±1.12 days and the total life cycle from egg to adult emergence was 43.63±11.23 days. There was 8-10 days difference in the total number of days in a particular generation.

Shoot and fruit borer, *C. punctiferalis* occurred throughout the year on Cardamom in Western Ghats in South India. Two peaks in the population were noticed in a year, i.e. one during April - May and the other during November – December (Thyagaraj, 2003). The population coincided with the period of less or no rainfall, i.e. during pre and post monsoon periods (Ballard, 1927; Ono, 1937 and Thyagaraj, 2003). Temperature and rainfall influenced greatly the growth and development of life stages (Moralesranous & Cote, 1992; Rao, 1992 and Shanuowr *et al.*, 1993).

13.8. FEEDING BEHAVIOR

Larvae of *C. punctiferalis* exhibit diversified feeding habit; feeding on underground rhizome/root to cones/needles in pine to fruiting bodies from ground to fruit trees. On cardamom, moths lay eggs singly on the top of the leaf axis of young pseudostems, rarely two larvae are found in a pseudostem. This kind of egg laying habit probably is to avoid larval competition for food within the same pseudostem (Thyagaraj, 2003). The young caterpillar bores at the base of the pedicel and later at the base (0.3m above ground level) of the seedlings in the nursery and at the nodal region of the grown-up suckers, feeding the central tissue and tunneled the shoot causing a dead heart. The live caterpillar indicates its presence by faecal matter (excreta) from the entry hole (Kalshoven, 1922; Clausen, 1927; Anonymous, 1944 and Thyagaraj, 2003). Same larva feeds on capsules from July to November (Anonymous, 1918; Thyagaraj, 2003). Second instar larvae bored in to the capsule, fed on the immature seeds leaving empty capsule and then moved to another capsule or psuedostem to soil for pupation (Kalshoven, 1929; Smith, 1937). Occasionally larvae bored in to panicles also. Such panicles dried up without bearing capsules.

13.9. ALTERNATE HOST PLANTS

Review of literature on the host plants of *C. punctiferalis* clearly indicated that for moths 31 plants served as alternate host plants. This showed that this insect has got a wide range of host plants. Some of the plants were tested in the laboratory to know the preference of host plants. Cardamom was the most preferred followed by *Hedycium* spp., *Alpinia* spp and *Ammomum* spp. (Thyagaraj, 2003). *Gossypium hirsutum* L., *Zea mays* L, *Sorghum bicolor* (L.) Moench ssp. *bicolor*, *Psidium guajava* L., *Curcuma longa* L., *Zingiber officinale* Roscoe, *Citrus* L., *Mangifera indica* L. and *Punica granatum* L. also plants served as host plants for this pest (Gurmey, 1918; Tryon, 1920; Flether, 1922; Ballard, 1924; Clausen, 1927; Veitch, 1931; Hutson, 1937; Narasimhaswamy, 1937; Sloan, 1945; Twine, 1971; Bilapate and Talati, 1977; Cai and Mu, 1993; Lu *et al.*, 1995; Wu, 1995; Konno and Shishido, 1996; Peter, 1996; Chakravarthy *et al.*, 1997 and Park *et al.*, 1998). *Theobroma cacao* L. and *Vitis vinifera* L. were recorded as host plants (Mohanan and Kumar, 1976; Gour and Sriramulu, 1992; Ram, 1997). *Ricinus communis* L. plant served as another major host plant (Flether, 1920 and Chakravarthy *et al.*, 1997). *Diospyros virginiana* L. is also attacked by this pest (Ono, 1937 and Tomomatsu *et al.*, 1995).

This has been also recorded on yellow peach as major pest (Konno *et al.*, 1980; Kodoi and Kaneda, 1990; Kimura and

Honda, 1999) in Japan and other countries.

13.10. CROP LOSS

The shoot damage in cardamom due to *Conogethes* infestation varied from five to ten per cent. In terms of capsule yield loss (dry weight basis), it varied from 6.79 to 9.18 per cent. Therefore, combining together the loss was estimated to be more than 20 per cent every year (Kapadia, 1996 and Thyagaraj, 2003). Studies on castor in Salem, Tamil Nadu, South India revealed 10.80 to 26.70 per cent capsule damage (Suganthy, 2011). Kapadia (1996) estimated 42.30 per cent crop loss in castor in India. Fifty per cent grape yield reduction was recorded on grapes due to *C. punctiferalis* attack (Ram *et al.*, 1997).

13.11. MANAGEMENT

C. punctiferalis has a wide range of host plants. Since many plants serve as alternate hosts, the pest has adapted well in agro-ecosystems. Some of the cultural practices like increased planting distance reduced the pest damage (Sharma *et al.*, 1992). Reducing the nitrogenous fertilizer levels also minimized the pest damage to a great extent (Chakravarthy and Thyagaraj, 1999). Clean cultivation in cardamom plantations can reduce the borer damage (Sharma, 1992). Mechanical methods also checked the pest to the extent of 5-10 per cent (Butani, 1978; Tomomatsu *et al.,* 1995; Chakravarthy *et al.,* 1997 and Boo, 1998) in crops like castor and pomegranate. Intercropping is an important tool for *C. punctiferalis* management. In semi arid tropics in India introducing cluster bean, cowpea, blackgram and/or groundnut as intercrops in castor (1:2 proportions) reduced *C. punctiferalis* infestation and build up of natural enemies of crop pests has been recorded (Rao *et al.*, 2012). Some of the predators and parasites were also identified and can be exploited in the management of this pest (Clausen, 1927; Abe and Sanari, 1992 and Anandraj and Peter, 1996). There are proven insecticides that can be used (Rama, 1980; He, 1997; Chakravarthy *et al.,* 1997 and Thyagaraj, 2003).

An integrated approach embracing thrashing (removing dried leaves, suckers, panicle etc. during February-March), and the timely spraying of monocrotophos 36 SL @ 15 ml /10 liter of water (March-April) followed by spraying phosalone 30 EC @ 20 ml/ 10 liter of water twice at an interval of 25-30 days after each spray could be adopted for economical management of the pest (Krishnamurthy *et al.,* 1989; Devasahayam, 1998 and Thyagaraj, 2003) in cardamom plantations.

13.12. CONCLUSION

The genus *Conogethes* is a large, taxonomically complex taxon. Till date 17 species have been identified by DNA barcodes and deposited in BOLD (http://www.boldsystems.org). Integration of several approaches like conventional taxonomy, DNA barcode, electrophysiological and behavioral responses of adult and larvae, host plant relationship patterns, hybridization experiments, biochemical analysis and analysis of and responses to pheromone components of *Conogethes* reared on castor and cardamom confirmed that the populations belong to two different species. Differences in sex pheromone system are mostly decisive for reproductive isolation in allied species. Bioecology and management of *Conogethes* have been well studied. This pest occurs throughout the year, but with one or two peaks. Peak population of the pest generally coincides with the period of rainfall. Temperature and rainfall greatly influences the growth and development of life stages. More than 31 alternate host plants for the borer have been recorded. This pest damages cardamom crop to an extent of 20 per cent every year. Since cardamom is cultivated in evergreen tropical forest tract and is an export oriented crop, it needs safer management practices to avoid toxic residual chemical compounds in its end products. Intercropping, mixed cropping, use of pheromone traps and timely harvests and destruction of affected plant parts can substantially contribute to borer suppression.

REFERENCES

Abe Y and Sanari G (1992) Notes on the peach moth *Conogethes punctiferalis* attacking cynipid and aphid galls. *Japanese Journal of Entomology* **60**(1): 108.

Abraham P (1965) The cardamom in India. ICAR *Farm Bulletin* **37**: 6-8.

Anandraj M and Peter KV (1996) *Biological Control in Spices.* Indian Institute of Spice Research, Calicut, India. 52pp.

Anonymous (1918) Cultivation of castor. *Tropical Agriculturist* **5**: 297-301.

Anonymous (1944) Annual Report of Dept. of Agriculture and stock for the year, Brisbane, *Queensland Agricultural Journal* **3**: 88.

Armstrong K (2010) DNA barcoding: a new module in New Zealand's plant biosecurity diagnostic toolbox. *Bulletin OEPP/ EPPO* **40**: 91-100.

Asante SK, Danthanarayana W and Heatwol H (1991) Bionomics and population growth static of aptirousvirginoparae of wooly aphid, *Eriosoma lanigereum*, at constant temperature. *Entomological Experimentalis et Applicata* **60**: 261-270.

Azam KM and Ali MH (1965) Morphology of larva of *Dichocrocis punctiferalis* Guen. the castor shoot and capsule borer with special reference to chaetotaxy (Lepidoptera: Pyralvidae). *Indian Journal of Entomology* **27**(4): 423-431.

Ballard E (1924) Entomological Note. *Queens-Land Agricultural Journal* **11**: 31.

Ballard E (1927) Some insects associated with cotton in Papua and in the mandated territory of New Guinea. *Bulletin of Entomological Research* **27**: 295-300.

Bilapate GG and Talati GM (1978) Some studies on bionomics of castor short and fruit borer *Dichocrocis punctiferalis* from Gujarat. *Journal of the Maharashtra Agricultural Universities* **3**(1): 75-76.

Bilapate GG and Talati GM (1977) Effect of different food plants on the larval and pupal development of *Dichocrocis punctiferalis. Madras agricultural Journal* **64**(1): 53-54.

Boo KS (1998) Variation in sex pheromone composition of a few selected lepidopteron, *Journal of Asia Pacific Entomology* **1**(1): 17-23.

Bravo JP, Silva JLC, Munhoz REF and MA Fernandez (2008) DNA barcode information for the sugar cane moth borer *Diatraea saccharalis. Genetics and Molecular Research* 7(3): 741-748.

Butani DK (1978) Insect pests of fruit crops and their control. *Pesticides* **12** (8): 53-59.

Cai RX and Mu ZL (1993) Trap trials for *Conogethes punctiferalis* with sex pheromones in citrus orchards. *China Citrus* **22** (1): 33.

Chakravarthy AK and Thyagaraj NE (1997) Response of cardamom shoot and fruit borer to different pheromone compounds. *Insect Environment* **2**(4): 127-128.

Chakravarthy AK and Thyagaraj NE (1998) Evaluation of selected synthetic sex pheromones of the cardamom shoot and fruit borer, *C. punctiferalis* in Karnataka.

Pest management in Horticulture Ecosystem **4** (2): 78-82.

Chakravarthy AK, Shivanna S, Chandrappa L and Naik L (1997) Insect pests on castor in Karnataka. *Insect Environment* **2** (4): 135.

Chakravarthy AK and Thyagaraj N E (1999) Effect of different NPK fertilizer levels on the incidence of cardamom pests, *Insect Environment*, **4** (4): 139-140.

Clausen CP (1927) The citrus insects in Japan. *Technical Bulletin of USDA* **15**: 15.

Devasahayam S, Koya MA, Reddy PP, Kumar WKK and Varghese A (1998) IPM in spices, Challenges for the future. Advances in IPM for Horticultural crops: Environmental Complications and Thrust, Bangalore. India Oct. 15-17, 1997, 157-64

Doddabassappa B (2012) Factors associated with resistance mechanism with shoot and fruit borer *Conogethes punctiferalis* Guenee (Lepidoptera, Crambidae) population infesting castro and cardamom. PhD thesis submitted to UAS, GKVK, Bangalore.

Dyar HG (1890) The number of molts of Lepidopterus larvae. *Psyche* **5**: 420-22.

Flether TB (1922) Report of the Imperial Entomologist. Scientific reports of the Imperial Institute of Agricultural Research. Pusa., pp. 51-67.

Gour TB and Sriramulu M (1992) Grapevine, *Vitis vinifera* -A new host of castor shoot and capsule borer *Conogethes punctiferalis*. *Tropical Pest Management* **38**(4): 459.

Gurmey WB (1918) The insect pests of maize. *Agriculture Gazetteer Bydney* **9**: 641-50

Hampson (1863) The Fauna of British India, Cyclone. Moths, Vol. IV. pp. 305-12.

He RQ (1997) Study on the control of yellow peach moth in china. *South China Fruits* **26** (4): 33.

Hebert PDN, Penton EH, Burns JM, Janzen DH and Hallwachs W (2004) Ten species in one: DNA barcoding reveals cryptic species in the neotropical skipper butterfly *Astraptes fulgerator*. *Proceedings of the National Academy of Sciences* **101**(41): 14812–14817.

Honda H and Mitsuhashi W (1989) Morphological and morphometrical differences between the fruit- and Pinaceae Feeding type of yellow peach moth, *Conogethes punctiferalis* (Guenee) (Lepidoptera: Pyralidae). *Journal of Applied Entomology and Zoology* **24** (1): 1-10.

http://www.boldsystems.org/views/speciessummary.php

Hutson JC (1937) Report on the work of the Entomological Division. *Report of The Director of Agriculture Cyclone* pp. 22-28.

Inoue H and Yamanaka H (2006) Redescription of *Conogethes punctiferalis* (Guenee) and description of two new closely allied species from Eastern Palaearctic and Oriental regions (Pyralidae, Pyraustinae). *Tinea* **19**(2): 80-91.

Jacob SA (1981) Biology of *Conogethes punctiferalis* on turmeric. *Journal of Plantation Crops* **9** (2): 119-23.

Jarvis E (1914) A new fruit boring caterpillar of Banana occurring at Tweed keeds. *Queens Land Agricultural Journal* **5**:280-84.

Jung JK, Han KS, Choi KS and Boo KS (2000) Sex pheromone composition for field-trapping of *Dichocrocis punctiferalis* (Lepidoptera: Pyralidae) males. *Korean Journal of Applied Entomology* **39**:105-10.

Kalshoven LGE (1922) Notes on Forest Zoology for the Dutch East India. *Tectona* **15**: 944-59.

Kalshoven LGE (1929) Injuries, diseases and pests of Teak forests in Java. *Tectona* **21**: 593-23.

Kapadia MN (1996) Estimation of loses due to pod borer in oil seed crops. *Journal of Oilseeds Research* **13**(1): 139-40.

Kimura M (1980) A simple method for estimating evolutionary rate of base substitutions through comparative studies of nucleotide sequences. *Journal of Molecular Evolution* **16**: 111-120.

Kimura T and Honda N (1999) Identification and possible functions of the hair pencil scent of the yellow peach moth *Conogethes punctiferalis* in Japan. *Applied Entomology and Zoology* **34** (1): 147-53.

Kodoi M and Kaneda M (1990) Development of yellow peach mouth *Conogethes punctiferalis*on apple fruit. *Research Bulletin of the Plant Protection Service, Japan* **26**: 61-63.

Kondo T and Miyahara T (1930) Fruit borer and a summary of their life histories in Knangtuny, *China Journal of Plant Protection* **17**: 85-94.

Konno Y and Shishido T (1996) Glutathione – dependent o-alkyl and oral conjunctions for dicapthon and fertikothine in several insects. *Journal of Pesticide Science* **21** (4): 430-33.

Konno Y (1986) Time-lag between sex pheromone content and the calling behavior in the yellow peach moth, *Conogethes punctiferalis* (Guenee) (Lepidoptera: Pyralidae). *Applied Entomology and Zoology* **21**: 622- 624.

Konno Y, Arai K, Sekiguchi K and Matsumoto Y (1982) (E)-10-hexadecenal, a sex pheromone component of the yellow peach moth, *Dichocrocis punctiferalis* Guenée (Lepidoptera: Pyralidae). *Applied Entomology and Zoology* **17**: 207-217.

Konno Y, Honda N and Matsumoto Y (1980) Observation on the mating behaviour and biology for the six pheromone of yellow peach moth, *Dichocrocis punctiferalis. Applied Entomology and Zoology* **15**(3) 321-322.

Konno Y, Honda N and Matsumoto Y (1981) Mechanism of reproductive isolation between the fruit feeding and pin ace feeding type of yellow peach moth, *Dichocrocis punctiferalis, Japanese Journal of Applied Entomology and Zoology* **25**(4): 253-258.

Krishnamurthy K, Khan MM, Avadhani KK, Venkatesh J, Siddaramaiah L, Chakravarthy AK and Gurumurthy BR (1989) Three decades of cardamom Research, Regional Research Station (1958-1988). *Station Technical Bulletin* **2**: 44-68.

Kuang MH, Liu SW, Ji BZ, Gao JY, Gao YG and Wang GX (2009) Investigation on the overwintering status and bionomics of needle feeding type of *Conogethes punctiferalis. Chinese Bulletin of Entomology* **46**(4): 569-573.

Lu Zx, Yang ZF, Wang GY and BO WL (1995) The niche of the corn and peach borers and interspecific competition between them in corn. *Acta Agriculture Zhajiangensis* **7** (1): 31-34.

Maes KVN (1998) Pyraloidea: Crambidae, Pyralidae. Pp. 87–98, 426-435. – *In:* Polaszek, A., *African Cereal Stem Borers: Economic Importance, Taxonomy, Natural Enemies and Control.* – CAB International, Oxon & New York, Pp 74..

Mathew G and Menon MGR (1984) The pyralid fauna (Lepidoptera: Pyralidae) of Kerala (India). *Journal of Entomological Research* **8**(1): 5-13.

Mathew G and Menon MGR (1989) Identification of some Indian Pyraustinae (Lepidoptera: Pyraustidae). *Journal of Entomological Research* **13**(1): 76-89.

Meyrick E (1884) On the classification of the Australian Pyralidina. *Tractions of the Entomological Society of London* **1884**: 277-350.

Mishra SC and Teotia TPS (1965) Studies on the biology of the castor shoot and capsule borer. *Indian oil seeds Journal* **9**(3): 184-87.

Mohanan RC and Kumar RH (1976) The castor capsule borer *Dichocrocis punctiferalis* as a pest of cocoa is India. *Agricultural Research Journal of Kerala* **14** (1): 79-80.

Moralesranous JA and Cote JR (1992) Rate of increase and adult longevity of *Catolocus grandis* in the laboratory at four temperature. *Environmental Entomology* **21**: 620-27.

Mori K, Watanabe H, Fujiwhara M and Kuwachau (1990) (E) – and (Z). Tetra decenyl format, potent sex pheromone mimics against the yellow peach moth. *Liebigs Annalen der Chemie* **12**: 1257-59

Munroe E (1989) Changes in classification and names of Hawaiian Pyralidae since the publication of Insects of Hawaii, Volume 8, by E. C. Zimmerman (1958). *Bishop Museum Occasional Papers* **29**: 199-212.

Nakano R, Takanashi T, Ihara F, Mishiro, K, Toyama M. and Ishikawa Y (2012) Ultrasonic courtship song of the yellow peach moth, *Conogethes punctiferalis* (Lepidoptera: Crambidae). *Applied Entomology and Zoology* **47**(2): 87-93.

Narasimha Swamy RL (1937) Staminodes in *Elettaria cardamomum*. Maton. *Current Science* **5**(8): 427.

Ono K (1937) On the insect pests of persimmon. *Insect world* **41** (5): 195-199.

Park JD, Lec SG, Kim CS and Byun RK (1998) The bionomics of the oak nut weevil, *Mehorisur sulus* and the insect pests of the corer in Korea. *Journal of Forest Science* **5**: 151-156.

Patel MM, Naik MM and Vyas HN (1988) Field evaluation of certain insecticides against castor capsule borer.*Agricultural Science Digest* **8**(3): 122-24.

Pena JE, Nadel H, Pereira MB and Smith D (2002) Pollinators and pests of *Annona* Sp. *(in) Tropical Fruit Pests and pollinators: Biology, Economic Importance, Natural Enemies and Control*, 208 pp. Pena J, Sharp J, Wysoki (Eds.) CABI Publication, UK.

Peter KV (1996) Major spices, In: *50 Years of Crop Science Research in India*. CAR, pp. 771-796.

Pruthi HS (1944) Report of the Imperial Entomologist. Abriay. *Scientific reports of the Agricultural Research Institute, Pusa*. pp 64-71.

Rajan SV (1965) Protect your cardamom against insect pests. *Indian Farming* **14**(11): 21-24.

Ram KT, Hugar PS and Somasekhar Patil BV (1997) Record of *Conogethes punctiferalis* (Guenee.) on grapes. *Pest Management in Horticultural Ecosystem* **3**: 37-39.

Rama CSV (1980) Plant protection in plantation crops through the soil. *Journal of Plantation Crops* **8** (1): 1-12.

Rao AS (1992) Preliminary studies on the seasonal occurrence of insect pests a soap nut (*Sapindus* sp.). *Indian Forester* **118** (6): 432-37.

Rao MS, Rao CAR, Srinivas K, Pratibha G, Sekhar SMV, Vani GS and Venkateswarlu B (2012) Intercropping for management of insect pests of castor in the semi-arid tropics of India. *Journal of Insect Science*, USA.

Sekiguchi K (1974) Morphology, biology and control of the yellow peach moth, *Dichocrocis punctiferalis* Guenée (Lepidoptera: Pyralidae). *Bulletin of Ibaraki Horticulture Experiment Station* Special Issue. p. 89.

Shaffer M, Nielsen ES and Horak M (1996) Pyralidae.In Nielsen, E. S.; E. D. Edwards and T. V. Rangsi, Checklist of the Lepidoptera of Australia. *Monograph Australia Lepidoptera*, 4:164-199.

Shanuowr TG, Guttierray AP and Wightman A (1993) Effect of temperature on development rate, fecundity and longevity of groundnut leaf minor, *Aproaerema modicella* (Lepidoptera: Gelechiidae) in India. *Bulletin of Entomological Research* **83**: 413-19.

Sharma ML, Shukla A and Reddy RK (1995) Field testing of caster cultivar for yield potential and resistance against shoot and capsule borer *Conogethes punctiferalis. Crop Research* **10**(1): 54-58.

Sharma YR, Devasahayam S and Anandraj M (1992) Black Pepper and cardamom: problems and prospects. *Proceedings of National Seminar on Black Pepper and Cardamom.* May, 17-18 1992, Calicut, Kerala, India.

Shashank PR (2012) Biosystematics and pheromone components of *Conogethes punctiferalis* Guenee (Lepidoptera: Crambidae) with special reference to populations infesting castor (*Ricinus communis* L.) and cardamom (*Elettaria cardamomum* Moton). Ph. D. thesis submitted to UAS, GKVK, Bangalore.

Sloan WJS (1945) Insect pests of grain sorghum. *Queensland Agricultural Journal* **61** (4): 221-39.

Smith, J. H., 1937. Insects and allied pests of the Papua. *Queensland Agricultural Journal* **48**: 553-57.

Stanley J, Chandrasekaran S and Preetha G (2009) *Conogethes punctiferalis* (Lepidoptera: Pyralidae) its biology and field parasitization. *Indian Journal of Agricultural Science* **79**(11): 906–909.

Suganthy M (2011) Field screening of promising castor cultures against capsule borer, *Conogethes punctiferalis* and leafhopper, *Empoasca flavescens. Madras Agricultural Journal* **98** (4/6): 178-179.

Thyagaraj NE (2003) Integrated Management of some important cardamom pests of hill region of Karnataka, South India. Ph. D. Thesis, Dr. B. R. Ambedkar University, Uttar Pradesh, India.

Tomomatsu S, Sakaguchi T, Ogino T, Hiramatsu T and Kaukami F (1995) Methyl bromide fumigator for quarantine control of persimmon fruit moth and yellow Peach moth on Japanese persimmon. *Research Bulletin of the Plant Protection Service* **31**: 67-73.

Tryon H (1920) Report of the Entomologist, Vegetable pathologist, *Queensland Annual Report of Department of Agriculture* **19** (16): 49-63.

Twine PH (1971) Insect pests of maize. *Queens Land Agricultural Journal* **97** (10): 523-31.

Veitch R (1931) Report of the chief Entomologist, Annual Report of Department of Agriculture. Queens Land Agricultural Journal pp. 43-46.

Wang YX and Cai RT (1997) The emergence and control of *Dichocrosis punctiferalis* for younai plum variety. *South China* Fruits, pp. 26-30.

Wu LM (1995) A study on the pupation pattern of three pyralid caterpillars helps is maize field. *Plant Protection* **21** (3):24-26.

Xi K, Xu FY and Dai YS (1996) Studies on the Chinese chestnut cone life table. *Journal of Joangsu Forestry Science and Techonology* **23** (4): 6-10.

Xiao Wei Honda and Matsuyama S (2011) Monoenyl hydrocarbons in female body wax of the yellow peach moth as synergists of aldehyde pheromone components. *Applied Entomology and Zoology* **46**(2): 239-246.

Young HY and Shaw KY (1962) Preliminary studies on the peach borer *Dichocrocis punctiferalis. Entomologica Sinica* **1**(2): 109-134.

Biological and Molecular Approaches in Pest Management
Balwinder Singh, Ramesh Arora & S S Gosal (Eds)
Scientific Publishers (India), Jodhpur, 2015, pp 344-372.

Chapter - 14

Molecular Techniques as Precision Diagnostics for Diseases and Mites of Honey Bees

Pardeep K Chhuneja[1*] and Kuldeep Singh[2]
1 Department of Entomology, Punjab Agricultural University, Ludhiana – 141 004, India
2. School of Agricultural Biotechnology, Punjab Agricultural University, Ludhiana – 141 004, India
*E-mail: pkchhuneja@pau.edu

14.1. INTRODUCTION

Honey bees are highly beneficial insects which impart us several useful hive products including honey, bees wax, pollen, propolis, royal jelly and bee venom. Though the honey bees are best known for the honey they produce, their economic role in nature is to pollinate flowering plants and assure setting of seed or fruit. Worldwide, 90 per cent food supply is contributed by 82 commodities assigned to plant species and bees are pollinators of 63 (i.e. 70%) of these plant species and are the most important known pollinators of 39 (48%) of these species. In USA, about 130 agricultural plant species have been reported to be pollinated by bees with the annual pollination value of 5.7 billion US dollars. In USA, during 2004, due to severe attack of *Varroa* mite on *Apis mellifera* L. colonies, annual almond production worth 15 billion US dollars was threatened. The annual benefit of honey bee pollination in Canada is estimated to be about 10 times (443 million Canadian dollars) the value of honey and bees wax produced (49 million Canadian dollars) and annually, 47,000 colonies are rented out for planned pollination of crops. In UK, 39 food crops are pollinated by insects, the majority being honey bees and bumble bees. Annual value of insect pollination of 13 major field crops and two glasshouse crops has been estimated to be £ 202 million of which contribution due to honey bees is £ 137.8 million. In European Union, annual value of 30 most important insect pollinated crops is estimated to be £ 5 billion of which £ 4.3 billion was attributed to bees and honey bees. Optimizations of honey bee pollination using bee attractants increased fruit size in pears resulting in $ 400-1,055/ ha increase in farm revenue; 40 per cent increase in cranberry yield with $ 8,804/ ha increase in revenue and 7 per cent increase in blue berry yield with $ 986/ ha increase in revenue.

Beekeeping is a very fascinating occupation. It supplements income generation of the rural people. It is being increasingly realized that bees are an economical input for promoting sustainable and eco-friendly agriculture and for enhancing crop productivity. Honey bees also play a vital role in sustaining plants bio-diversity resulting in environmental stability. Ample yield increases of crops have been recorded in India (apple 25-55%, apricot 48%, citrus 22%, blue currant 34.7-41.8%, peach 2.6 times, guava 2.69 times, plum 2.5-3.7 times, litchi 2-3 times; onion seed 195.7%; radish pods 24.51% and radish seed 79.17-86%). National Bee Board has, however, reported exponential increases in fruit yields (apple 180-6,950%, pear 240-6,014%, cherries 56-1,000%, strawberry 17.4-91.9%, litchi 4,538-10,246%, orange 47-900%, citrus 7-33.3%, guava 70-140%, papaya 5-10%). At the global level, 33 fruit crops and 15 vegetables have been identified which need bee pollination for increase in their yields. In India, 12 crops including eight fruit crops (almond, apple, citrus, coconut, guava, grapes, mango, papaya) are most dependent on honey bee pollination.

Based on the requirement of 2-3 colonies / ha depending upon the acreage under these crops in India, it has been estimated that 73.7 million honey bee colonies are required for optimizing crop pollination. This would have resulted in the additional production of these crops worth Rs.30 billion annually. However, the number of hive bee colonies in the country is too less i.e. only 1.9 million necessitating 71.8 million additional hive honey bee colonies to obtain optimum pollination of these crops. As per the latest estimates by National Bee Board of India, the country needs 150-200 million honey bee colonies for pollination services in various crops in the country, which includes 31.92 million for fruit pollination and 34.90 million for vegetables pollination.

Beekeeping industry has a great employment generation potential providing direct self employment to rural and forest based population and to rural educated youths in collecting, processing and marketing of bee-products, which may go up to 30 million for optimum number of honey bee colonies in India (Chhuneja, 2012; Chhuneja & Singh, 2013).

At the global level, about 50 million bee colonies, mostly, *A. mellifera*, are being maintained and the world production of honey is estimated at about 1.4 million tonnes. Fifteen countries of the world account for 90 per cent of the global honey production. China is the only Asian country for this group producing nearly 0.25 million tonnes of honey; and is the biggest exporter of hive products. There are about 0.7 million *A. mellifera* and 0.30 million *A. cerana* colonies in China with a plan to increase the number to 50 million in future.

14.2. BEEKEEPING IN INDIA

All the known four traditional honey bee species, viz. *Apis dorsata* Fab. (Rock bee/ Dammer bee/ *Dumna*), *Apis florea* Fab. (Small bee/ Little bee/ *Chhoti makhi*), *Apis cerana* Fab. (=*A. indica, A. cerana indica*) (Asian bee/ Indian bee/ *Pahadi makkhi*) and *Apis mellifera* Linn. (European bee/ Western bee) of the world are presently found in India. The first two species are wild ones (cannot be hived) and have strong absconding instinct, ferocious (more so the *A. dorsata* which may lead to fatal results) and usually do not ripen their honey and thus have little export potential. *A. cerana* and *A. mellifera* are the hive honey bee species which produce multiple parallel combs in their colonies and are docile and can be hived in artificial movable frame hives. Till 1975, only *A. cerana* was hived in India. The various problems in our earlier beekeeping with *A. cerana*, forced the Indian apiculture experts to make efforts to introduce Western/ European honey bee, *A. mellifera* into India.

14.3. *A. MELLIFERA* INTRODUCTION BY PAU: SAFEGUARD AGAINST INTRODUCTION OF DISEASES AND ENEMIES

Punjab Agricultural University success-fully introduced *A. mellifera* into India which is docile, less prone to swarming, with negligible absconding instinct and suitable and productive even in hot plains of the country, besides relatively less prone to wax-moth attack. Within a little more than three decades of release of *A. mellifera* into fields, the Punjab has made strides with this exotic species that the State beekeepers, at present, are producing more than 14,000 tonnes of honey and the beekeeping has ushered in 'Sweet Revolution' in the State which is exporting its produce to about 48 countries. The apicultural activities were estimated to contribute Rs. 4,750 million towards State's economy. Punjab has emerged as hub of beekeeping activity in India. After the Punjab, Bihar in a very short time has also emerged as another leading State after adoption of *A. mellifera* in 1989. Himachal Pradesh, Jammu & Kashmir, Uttarakhand, Orissa, Andhra Pradesh, Karnataka, Assam and Kerala are some other states where beekeeping, particularly with *A. mellifera,* has flouri-shed. According to an estimate based on bee floral acreages of the Punjab, the state has a potential of sustaining one million honey bee colonies (Chhuneja, 2012; Chhuneja, 2013; Chhuneja & Singh, 2013).

After the successful introduction of *A. mellifera* into India by the PAU, it was feared by a group of scientists that this bee species might spread bee diseases already known elsewhere in the world to the native *A. cerana* and may put our earlier beekeeping based on this latter species in danger (Deodikar, 1971 a,b). However, it is worth mentioning that foreseeing the risk of introduction of new bee diseases, the Punjab Agricultural University (PAU) had followed a unique technique of 'inter-specific queen introdu-ction' in which the exotic queens from disease free stock were legally imported and were introduced into the queenless colonies of *A. cerana,* and all the importa-tions went through the country's quaran-tine (Atwal, 1964; Atwal & Sharma, 1967, 1968).

14.4. BEE DISEASES AND MITES

At the global level, hive bees have been reported to be infected by various diseases which are inflicted by bacteria, viruses, fungi, protozoa, etc. and by various astigmatic and mesostigmatic mites. The bacterial diseases include chiefly American foulbrood, caused by Gram positive sporulating *Paenibacillus larvae larvae* (White) Heyndricx *et al.* and European foulbrood caused by the bacterium *Melissococcus plutonius* (White)

Truper & De'Clari. The other minor bacterial diseases include Half-moon disorder caused by *Bacillus coagulans,* and other caused by *Spiroplasma* spp., *Bifidobacterium asteroids* and *Bifidobacterium indicum.* Viral diseases mainly include sacbrood, thai sacbrood, chronic bee paralysis and clustering disease while some minor diseases include Araknsas viral disease, slow bee paralysis, and those inflicted by satellite virus, Kashmir virus, X and Y virus, filamentous virus, etc. Fungal diseases chiefly include two – chalk brood caused by *Ascosphaera apis* (Maassen ex Claussen) Olive & Spiltoir and stone brood caused by *Aspergillus flavus* Link. Among the other diseases, two are important: Nosema caused by the microsporidian *Nosema apis* Zander and *Nosema ceranae* Fries and the amoeba disease caused by *Malpighamoeba mellificae* Prell.

Important mites of concern include ectoparasitic species *Varroa jacobsoni* Oudemans, *Varroa destructor* Anderson & Trueman and *Tropilaelaps clareae* Delfinado & Baker and the endoparasitic *Acarapis woodi* Rennie (Chhuneja, 2013).

14.5. OCCURRENCE OF BEE DISEASES AND MITES IN INDIA

American foulbrood, the most deadly bee disease of the world, was reported from India during 1961 on *A. cerana* from Nainital (Singh, 1961) i.e. before the initiatives of the PAU towards the introduction of *A. mellifera* in the country. Another bacterial foulbrood disease i.e. European foulbrood was first reported on *A. cerana* during 1971-74 from Maharashtra and Karnataka (Diwan *et al.,* 1971; Kshirsagar & Godbole, 1974; Bailey, 1974; Bailey, 1983) i.e. before even the *A. mellifera* was released by the PAU into the fields. On *A. mellifera,* however, the disease in the country was first noticed in 1998 and in the Punjab in 1999-2K (Anonymous, 1999; Singh & Garg, 2000; Chhuneja *et al.,* 2001, 2002). Thai sacbrood, the viral disease, outbreak occurred during 1979-86 on *A. cerana* in the northern India which *inter-alia* compelled I.C.A.R. to give its decision to disseminate *A. mellifera* to even other states of the country as *A. mellifera* showed immunity to this disease. In the southern peninsula, its serious outbreak was reported on *A. cerana* in early nineties of the last century (Channa Basavanna, 1992). This disease is widely distributed and occasionally observed in *A. cerana* colonies almost throughout the country. Sacbrood disease was first reported in India from *A. melifera* colonies in 1998 from Himachal Pradesh (Chandel *et al.,* 1999) followed by from Punjab in 2000 (Chhuneja *et al.,* 2001) and is now widely spread, though in very low intensity in the country (Chhuneja, 2013). Even clustering disease, another viral disease, was found on *A. cerana* in India even before *A. mellifera* was introduced in the country (Mutto, 1956; Singh, 1979; Shah & Shah, 1979; Kshirsagar *et al.,* 1975). It is of rare occurrence on *A. mellifera* in the country. Nosema was found on *A. cerana* in 1974 (Kshirsagar *et al.,* 1974) and the first report of this disease on *A. mellifera* in the country appeared much later (Rahman & Rahman, 1994). Ectoparasitic mite, *T. clareae* was reported earlier on *A. dorsata* in 1968 (Bhardwaj, 1968) which shifted to the exotic bee, *A. mellifera* during 1969-71 in the country (Atwal, 1969; Atwal & Goyal, 1971). Endoparasitic mite, *A. woodi* was also reported much earlier (in 1957) on *A. cerana* (Singh, 1957; Singh & Adlakha, 1958, 1960). On *A. mellifera*, it was reported during 1967 (Atwal, 1967) and subsequently there had been several reports (Adlakha, 1976; Singh & Verma, 1981; Gupta, 1986; Mattu *et al.*, 1994). Another ectoparasitic mite, *V. jacobsoni* has been reported both from *A. cerana* and *A. mellifera* in India. It was reported in India on *A. cerana* during 1966 (Phadke *et al.*, 1966), i.e. much before the *A. mellifera* was released by the PAU into the fields and its first occurrence on *A. mellifera* in India was reported only during 1988 (Kumar *et al.*, 1988). Further, *V. jacobsoni* haplotypes present in India did not multiply on *A. mellifera*. The Korean haplotype of *V. destructor* which is avirulent on *A. cerana* and highly virulent on *A. mellifera,* was reported for the first time on *A. mellifera* from the Punjab during 2004 (Gatoria *et al.,* 2004) but it has now spread very fast in almost whole of the northern India (Chhuneja *et al.,* 2005). *V. destructor* and European foulbrood and Sacbrood are introduced problems which got into India during the introduction of untested and unevaluated races and so-called hybrids of *A. mellifera* into our country against the country's quarantine since the latter half of the last decade of the last century (Chhuneja, 2013).

14.6. DIAGNOSIS OF BEE DISEASES AND PESTS

Bee diseases and mites are some current very serious challenges thrusted on beekeepers, brought into India through import of bee material against quarantine laws. Bee disease and mite diagnostics is the major concern in the country due to lack of which the wrong identifications/ diagnosis are leading to injudicious use and misuse of chemicals in beekeeping industry leading to several honey marketing problems and restriction on honey consignments for export.

Previously, the diagnosis of bee diseases and mites was based on clinical / field symptoms and further confirmation through microscopic techniques. Due to certain limitations in diagnosis / identification like in mixed kind of symptoms and morphologically quite related microorganisms, strong need has been felt for development of highly precise and quick diagnostics (Chen *et al.*, 2005). Further due to wide variation with respect to virulence among different isolates / haplotypes, only the biochemical / molecular diagnostics would be relied upon.

Clinical diagnoses which are based on field / *in-vivo* symptoms, require long expertise, are not always reliable encounter ambiguity and may even need supplementation of microscopy. Microscopic diagnoses include direct identification of the causative agent under microscope and hanging drop / Brownian movement or are based on Gram reaction. Through microscopy, it is also not possible to differentiate closely related species / strains / haplotypes.

Biochemical techniques involve various characterization tests including Carbohydrate test, Nitrate reduction test, Catalase test, Indole test, Voges-Proskauer test etc. (Lochhead, 1937; Bailey, 1956; Bitner, 1971; Haynes, 1972; Hitchcock & Wilson, 1973; Hornitzky & Karlovskis, 1989; Hornitzky & Wilson, 1989; Bailey & Ball, 1991; Allen & Ball, 1993; Kumar, 2003; Gatoria *et al.,* 2006).

Several immunological/ serological diagnostic techniques including radio immunoassay, immuno osmoelectrophoresis, immune serum electron microscopy, immunodiffusion, immunofluoresence and ELISA have been developed as diagnostic tools for bee diseases and mites (Anderson,

1984; Pinnock & Featherstone, 1984; Allen *et al.*, 1986; Ragsdale & Furgala, 1987; Ragsdale & Kjer, 1989; Olsen *et al.*, 1990). Though these involve low application cost, are simple faster and sensitive, require few reactants offer repeatability and may sometimes help to differentiate strains, yet, the serological methods have the drawback of limited availability of antisera; besides, raising of antisera is time laborious (Benjeddou *et al.*, 2001). Moreover, these techniques including ELISA may sometimes fail to differentiate among the closely related viral and other microorganismic strains and even among some mite haplotypes.

14.7. MOLECULAR TECHNIQUES AS DIAGNOSTICS

PCR/RT-PCR have been developed for most of the bee pathogens. Different viruses (CBPV, ABPV, BQCV, etc.) are being diagnosed using RT-PCR (Benjeddou *et al.*, 2001; Grabensteiner *et al.*, 2001; Chen *et al.*, 2005; Berényi *et al.*, 2006). Lately, M-RT-PCR has been standardized for simultaneous detection of several pathogens in the case of mixed infections in honey bees (Topley *et al.*, 2005). Bacterial diseases (AFB, EFB) and the strains of their pathogens, fungal and protozoan diseases have been diagnosed by molecular techniques. PCR techniques and molecular markers (RAPD, RFLP) were successful in diagnosing bee mites and their haplotypes and also closely related *Paenibacillus* species/ subspecies (Tangjingjai *et al.*, 2003; Alippi *et al.*, 2002).

14.7.1 Polymerase Chain Reaction

The Polymerase Chain Reaction (PCR) is an extremely versatile technique for copying DNA. PCR allows a single DNA sequence to be copied (millions of times), or altered in predetermined ways (Mullis, 1990). PCR can also be used to determine whether a particular DNA fragment is found in a cDNA library. PCR has many variations, like reverse transcription PCR (RT-PCR) for amplification of RNA, and, more recently, Real-Time PCR (QPCR) which allows for quantitative measurement of DNA or RNA molecules. PCR is most important because most of the fungal, bacterial and viral diseases are correctly and precisely diagnosed with the help of PCR. PCR amplifies a single or few copies of a piece of DNA across several orders of magnitude, generating thousands to millions of copies of a particular DNA sequence (Sambrook & Russell, 2001). PCR based methods are suitable where enough amount of template DNA of the pathogen available.

14.7.2 Multiplex Polymerase Chain Reaction

It is a modification of PCR consisting of multiple primer sets within a single PCR mixture to produce amplicons of varying sizes that are specific to different DNA sequences and mainly used for genotyping applications, detection of pathogens and for microsatellite analyses (Hayden *et al.*, 2008). Many advantages of multiplex PCR include precise detection and identification of several honey bee viruses in a single step reaction. It saves costs and time in comparison to individual PCR assays. It enables simultaneous amplification of many targets of interest in one reaction by using more than one pair of primers.

14.7.3 Real-Time Polymerase Chain Reaction

It is also called quantitative real time polymerase chain reaction (Q-PCR/qPCR/

qrt-PCR) or kinetic polymerase chain reaction (KPCR) which enables both detection and quantification of one or more DNA samples. It is used under situations where the pathogen DNA contributes only an insignificant fraction of the bee sample DNA as in case of viral pathogens or in case of unapparent infections. The procedure follows the general principle of polymerase chain reaction and amplified DNA is detected as the reaction progresses in real time (Joyce, 2002).

RT-PCR is combined with reverse transcription to quantify messenger RNA and non-coding RNA in cells or tissues. Real-time RT-PCR provides a method in which the amplicons can be visualized as the amplification progresses using a fluorescent reporter molecule. There are three major kinds of fluorescent reporters used in real time RT-PCR, which are general non-specific DNA Binding Dyes such as SYBR Green I, TaqMan Probes and Molecular Beacons (including Scorpions). There are several advantages of RT-PCR like amplification can be monitored real-time, not influenced by non-specific amplification, no post-PCR processing of products, ultra-rapid cycling (as fast as 25 minutes), wider dynamic range of up to 10^{10}-fold, requirement of 1,000-fold less RNA than conventional assays, most specific, sensitive and reproducible and not much more expensive than conventional PCR. Detection of honey bee viruses are carried out by sequencing the PCR products RT-PCR is commonly used in studying the genomes of viruses whose genomes are composed of RNA.

14.8. MOLECULAR MARKERS

Molecular marker is a piece of the DNA molecule that is associated with a certain trait specific to the organism. It helps to determine the existence of specific DNA segments/ regions/ sequences/ genes that are either associated with or control important traits in different insect species. Markers help us to determine the location (map) of genes that control important traits. A number of different types of molecular markers have been exploited for molecular analysis of various insect pathogens. The methodology to develop these is derived from those earlier developed for plants but modified for application to small amount of DNA obtained from insects. The molecular analysis or screening involves investigating whole or part of genomic DNA (nuclear, nDNA and mtDNA). The purified DNA is analyzed after it is digested prior to being probed to detect differences (RFLP) or used as such in techniques like AFLPs, sequencing and SSRs which utilize the polymerase chain reaction (PCR) to amplify regions of interest (Hunt & Page, 1995; Solignac *et al.*, 2004; Weinstock *et al.*, 2006).

Types of Molecular Markers

Several types of molecular markers which have been developed and used in plants are restriction fragment length polymorphism (RFLP), sequence tagged sites (STS), expressed sequence tags (ESTs), simple sequence repeats (SSRs) or microsatellites, randomly amplified polymorphic DNA (RAPDs), sequence characterized amplified regions (SCARs) and amplified fragment length polymorphic (AFLP) markers. In addition several other variations of these markers have been developed (Table 1). Catherine and Hugh (2004) have reviewed different types of molecular markers and their applications in beneficial insects. Vignal *et al.* (2002) have reviewed the role of SNPs and other molecular markers in animal genetics.

Table 1. An over view of the molecular markers

Marker (1)	Abbrev. (2)	Description (3)
Restriction fragment length polymorphism	RFLP	The occurrence of variation in the length of DNA fragments that are produced after cleavage with a restriction enzyme. There are differences in DNA lengths because of the presence or absence of recognition site(s) for that particular restriction enzyme.
Variable number tandem repeat or minisatellites	VNTR	A short DNA sequence that is present as tandem repeats [motifs of 10-100 bases] and in highly variable copy number.
Simple sequence length polymorphism	SSLP	The sequence of unique DNA that flanks microsatellites [i.e. tandemly repeated motifs of 1-6 bases] is used to construct oligonucleotide primers. Variation is based on size differences of the microsatellite. **SSLP, SSR, STR, STMS and microsatellite are synonyms**
Simple sequence repeats	SSR	
Short tandem repeats	STR	
Sequence tagged microsatellite site	STMS	
Inter-simple sequence repeat.	ISSR	The stretches of unique DNA in between the SSRs are amplified. SSRs such as [GATA] 4 are used to prime PCR on genomic DNA Tetranucleotide repeats perform better than di-or tri-nucleotide repeats
Single primer amplification reaction.	SPAR	
Cleaved amplified polymorphic sequences	CAPS	PCR amplified DNA that is not yet polymorphic is digested with restriction endonucleases to reveal polymorphism.
Derived cleaved amplified polymorphic sequences.	dCAPS	A restriction enzyme recognition site that includes the SNP is introduced into the PCR product by a primer containing one or more mismatches to template DNA.The resulting PCR product is subjected to restriction enzyme digestion to determine the presence or absence of the SNP.
Allele –specific Polymerase chain reaction.	AS-PCR	Refers to amplification of specific alleles or DNA sequence variants, at the same locus. Specificity is achieved by designing one or both PCR primers so that they partially overlap the site of sequence difference between the amplified alleles. Only a template with exactly the same sequence as the target DNA will be amplified.
PCR-amplification of specific alleles.	PASA	
Allele-specific oligonucleotide	ASO	
Allele-specific associated primers	ASAP	

Contd...

(1)	(2)	(3)
Single-nucleotide amplified polymorphisms	SNAP	Modified AS-PCR procedure for assaying SNPs. Adding an extra mismatch, coupled with the natural [SNP] mismatch at the 3'end, produces a dramatic reduction in PCR product yield of non-specific allele, but has a relatively minor effect on the amplification of the specific allele.
Single-strand conformational polymorphism.	SSCP	DNA fragments [200-800 bp] amplified by PCR using specific 20-25bp primers, followed by gel-electrophoresis of single strand DNA to detect nucleotide sequence variation. The method is based on the dependence of the electrophoretic mobility of single strand DNA on the secondary structure [conformation] of the molecule, which changes significantly in case of mutation. SSCP provides a method to detect nucleotide variation without the need to sequence DNA samples.
Random amplified polymorphic DNA	RAPD	A PCR product obtained from genomic DNA using a single or a combination of typically 10-mer oligonucleotides. Variation is based on the position and orientation of primer-annealing sites and the interval they span.
Arbitrary primed Polymerase chain reaction	AP-PCR	Related to RAPD; uses longer arbitrary primers than RAPD does.
DNA amplification fingerprinting	DAF	Related to RAPD; uses 5-8 bp primers to generate more fragments.
Multiple arbitrary Amplicon profiling	MAAF	Collective term for techniques using single arbitrary primers.
Sequence characterized amplified region	SCAR	DNA fragments amplified by PCR using specific 15-30bp primers designed from nucleotide sequences established in cloned RAPD fragments. By using longer PCR primers, SCAR does not face the problem of poor reproducibility.
Amplified fragment length polymorphism	AFLP	AFLPs are DNA fragments [80-500bp] obtained from endonuclease restriction, followed by ligation of oligonucleotide adaptors to the fragments and selective amplification by PCR.The PCR-primers consist of a core sequence [part of the adaptor], a restriction enzyme – specific sequence and 1-3 selective nucleotides. The AFLP fragments are separated by gel-electrophoresis and generally scored as a dominant marker.
Selective amplification of microsatellite polymorphic loci	SAMPL	The template is identical to the AFLP template .The rare cutter primer is replaced by $[CA]_{22}$ $[TA]_{22}$ or $[GT]_{22}$ $[TA]_{2}$
Diversity Array	DArT	Diversity array technology DArT ™ is a high throughput DNA polymorphism analysis method that combines miniaturization of microarrays and restriction based PCR. It is capable of providing comprehensive genome coverage even in organisms without any DNA sequence information (Jaccoud *et al.*, 2001).

A. Southern Hybridization based Markers

i). Restriction Fragment Length Polymorphism (RFLP): These are single or low copy DNA fragments and are simply inherited. These probes could be genomic clones, cDNA clones, or even cloned genes. The RFLP markers show co-dominance and are highly reliable in linkage analysis and breeding. Their detection is based on radio labelling, require large quantities of DNA, are labour intensive and relatively expensive and hazardous. Hence, their large-scale use in practical plant breeding may be restricted.

B. Polymerase Chain Reaction (PCR) based Markers

i). Sequence Tagged Sites (STS): The RFLP probes, linked to desirable traits can be converted to polymerase chain reaction (PCR) based markers. In this the RFLP probes are end-sequenced and complementary primers are synthesized. These primers (generally 20 mers) are then used for amplifying specific genomic sequences using PCR. For example, STS markers have been developed for RFLP markers linked with bacterial blight resistance genes *xa5, xa13* and *Xa21*, powdery mildew and stem rust resistance gene in barley etc. One major limitation of these markers is the reduced polymorphism than the corresponding RFLP marker.

ii). Expressed Sequence Tags (EST): These markers are developed by end sequencing of random cDNA clones. The cDNA markers are first mapped as RFLP markers and then partially sequenced to convert them into PCR based markers. Thus, these are like STS markers. These can be used for synteny mapping and cloning of specific genes. Most of these could be functional genes. A large number of EST markers have been identified in honey bee (Whitfield *et al.*, 2002)

iii). Simple Sequence Repeats (SSRs): Also called as microsatellites, these are short tandem repeats dispersed throughout the genome. These are generally di-to-tetra- nucleotide repeats and are highly hyper variable. These are flanked with unique sequences that are highly conserved. The flanking unique sequences are analyzed and their complementary primers synthesized. These can thus be assayed with PCR and act as codominant markers. Referred to as simple sequence length polymorphism (SSLP), allelic differences are usually a result of variable number of repeat units. These are highly polymorphic. Their major limitation is the cost involved in its development. Microsatellite markers have been developed for honey bee.

iv). Randomly Amplified Polymorphic DNAs (RAPDs): Williams *et al.* (1990) originally developed the technique. In this arbitrary decamer sequences are used as primers for amplification. These markers are dominant markers because the polymorphism is due to presence or absence of a particular amplified fragment. One major advantage of these markers is that this does not need any prior sequence information. These markers have been used for constructing linkage maps in several species and also for tagging genes of economic importance. One major limitation of these markers is lack of repeatability in certain cases.

v). Sequence Characterized Amplified Regions (SCARs): These markers overcome the limitation of RAPDs. In this, the RAPD fragments that are linked to a gene of interest are cloned and their termini sequenced. Based on the terminal sequences, longer primers (20 mers) are designed. These SCAR primers lead to a more specific amplification of particular locus. These are similar to STS markers in construction and application. The presence or absence of the band indicates variation in sequences. The SCAR markers thus are dominant markers. These however, can be converted to codominant markers in certain cases by digesting the amplified fragment with tetra cutting restriction enzymes. The RAPD primers linked to gene of importance have been converted to SCAR markers in several cases.

vi). Amplified Fragment Length Polymorphism (AFLP): This technique was developed in Vos *et al.* (1995). In this technique restriction fragments generated by a frequent (4 base) and a rare (6-base) cutter are anchored with oligonucleotide adapters of a few oligonucleotide bases. This method generates a large number of restriction fragments facilitating the detection of polymorphism. The number of DNA fragments amplified can be controlled, by choosing different base numbers and composition of nucleotides in the adapters. This technique is more reliable since stringent reaction conditions are used for primer annealing. This technique thus shows an ingenious combination of RFLP and PCR techniques and is extremely useful in detection of polymorphism even between closely related genotypes. AFLP maps have been constructed in several species and integrated into already existing RFLP maps.

III. Single Nucleotide Polymorphism (SNPs): The ultimate level of variation between any two individuals of the same species or of different species is at base pair level. The large scale genome sequencing projects has resulted in specific amplification and sequencing of individual genes of genome segments. Comparison of base pair sequences in these regions has led to identification of a large number of variations at base pair level (Henikoff & Comai, 2003). These markers have the highest level of polymorphism.

14.9. STEPS IN MOLECULAR DIAGNOSIS

The method for diagnosing using molecular technique through PCR, in general, essentially involve the following steps (Singh *et al.*, 2009).

14.9.1 Sample Collection

The disease suspected bee stages are collected and stored immediately at -80°C before using for DNA/RNA extraction.

14.9.2 RNA/ DNA Extraction

Individual samples in Eppendorf tubes are homogenized in extraction buffers. DNA/RNA is isolated using simple protocols of standard molecular biology kits which provide easy methods for isolation of high quality DNA/ RNA with additional ability/ ease of handling large number of samples in short times. Extracted nucleic acid is stored at subzero temperatures until used to avoid their self degradation.

14.9.3 Designing Pathogen Specific Primers/ Probes

The success of PCR based diagnostics strictly depend upon specificity of short oligomers (primers) that are designed based upon the pathogen specific nucleotide sequence. Ample information is available on the primer sequences of bee pathogens that hold specificity for these organisms.

14.9.4 PCR-Amplification

For this, a mixture of the total DNA from the diseased sample (template DNA), four deoxyribonucleotides, primers and Taq polymerase enzyme in a buffered reaction mixture are exposed to programmed time-temperature cycles in a PCR machine or thermocycler to allow 25-30 continuous amplification cycles. This results in million fold amplification of pathogen specific DNA regions to provide enough amount of DNA to be sufficient for further analysis. The amplified product is analyzed on agarose gels for its expected size. This method clearly differentiates amongst different pathogen types, the amplified DNA after restriction with suitable restriction enzymes followed by separation of restricted fragments helps to differentiate amongst even the closely related pathogen strains.

This method is directly applicable to isolated DNA from all bacterial, fungal and protozoan pathogens. However, for differentiating amongst bee viral pathogens (RNA viruses), a modified version of the method includes additional step to synthesize a complementary DNA (cDNA) through a 'Reverse transcription reaction (RT reaction) that involves activity of another enzymes 'Reverse transcriptase'. In this, whole of the reaction is carried out as in the case of DNA amplification except that PCR reaction mixture also includes Reverse transcriptase enzymes, besides a modified PCR amplification programme.

14.10. DIAGNOSTICS DEVELOPMENTS FOR BEE PATHOGENIC TAXA

The use of various molecular techniques used as diagnostics for bee diseases and mites are tabulated (Table 2) and discussed below.

Table 2. Molecular techniques for identification of different bee pathogenic taxa

Bee Diseases	**Head starters**	**Technique**	**Reference**
(1)	**(2)**	**(3)**	**(4)**
a. Bacterial			
American Foulbrood	16S rRNA	PCR	Govan *et al.*, 1999; Dobbelaere *et al.*, 2001
	16S rDNA	PCR and RFLP	Alippi *et al.*, 2002; Antunez *et al.*, 2009
		PCR	Alippi *et al.*, 2004; Neuendorf *et al.*, 2004
European Foulbrood	16S rRNA	PCR	Govan *et al.*, 1998
	DNA	PCR	Djordjevic *et al.*, 1998
b. Fungi			
Chalk brood	rDNA	PCR	Horintzky, 2001; Murray *et al.*, 2005; Rehner & Evans, 2009
c. Microsporidian			
Nosema disease	rRNA	PCR and RFLP	Robert, 2001
	RNA	RT-PCR	Bakonyi *et al.*, 2002
	DNA	PCR	Webster *et al.*, 2004; Chen *et al.*, 2009; Cornman *et al.* 2009
	16S rRNA	Duplex PCR	Martin-Hernandez *et al.*, 2007
	RNAi		Paldi *et al.*, 2010
		RT-PCR	Hamiduzzaman *et al.*, 2010
d. Viruses			
Acute Bee Paralysis Virus	RNA	RT-PCR	Govan *et al.*, 2000; Benjeddou *et al.*, 2001; Bakonyi *et al.*, 2002; Davison *et al.*, 2003; Topley *et al.*, 2005; Berényi *et al.*, 2006

Contd...

(1)	(2)	(3)	(4)
Black Queen Cell Virus		RT-PCR	Leat *et al.*, 2000; Benjeddou *et al.*, 2001; Davison *et al.*, 2003; Berényi *et al.*, 2006
Cloudy Wing Virus		RT-PCR	Benjeddou *et al.*, 2001
Chronic Bee Paralysis Virus		RT-PCR	Ribière *et al.*, 2002; Berényi *et al.*, 2006
Deformed Wing Virus		RT-PCR	Benjeddou *et al.*, 2001; Chen *et al.*, 2004b; Tentcheva *et al.*, 2004; Chen *et al.*, 2005a; Yue & Genersch, 2005; Berényi *el al.*, 2006; Lanzi *et al.*, 2006; Fievet *et al.*, 2006
Kashmir Bee Virus	RNA	RT-PCR	Stoltz *et al.*, 1995; Hung & Shimanuki, 1999; Hung *et al.*, 1999; Benjeddou *et al.*, 2001; Evans, 2001; de Miranda *et al.*, 2004; Topley *et al.*, 2005; Berényi *et al.*, 2006
Sacbrood virus		RT-PCR	Ghosh *et al.*, 1999; Topley *et al.*, 2005; Benjeddou *et al.*, 2001; Grabensteiner *et al.*, 2001; Berényi *et al.*, 2006
Acute Bee Paralysis Virus, Sacbrood virus, Black Queen Cell Virus		M-RT-PCR	Topley *et al.*, 2005
e. Acarine			
Varroa destructor, Varroa jacobsoni	mt DNA	RAPD	Kraus & Hunt, 1995; Anderson & Fuchs, 1998; Anderson & Trueman, 2000
		PCR and RFLP	Zhou *et al.*, 2004; Cornman *et al.*, 2010
Tropilaelaps clareae			
T. koenigerun	Nuclear rRNA	PCR and RFLP, RAPD	Tangjingjai *et al.*, 2003

Table 3. Primers used for AFB diagnosis

Direction	Sequence	PCR-product size	Specificity level
Forward	5'-AAG-TCG-AGC-GGA-CCT-TGT-GTT-TC-3'	973 bp	Species
Reverse	5'-TCT-ATC-TCA-AAA-CCG-GTC-AGA-GG-3'		
Forward	5'-CTT-GTG-TTT-CTT-TCG-GGA-GAC-GCC-A-3'	1106 bp	Species
Reverse	5'-TCT-TAG-AGT-GCC-CAC-CTC-TGC-G-3'		
Forward	5'-CGA-GCG-GAC-CTT-GTG-TTT-CC-3'	700 bp	Sub species
Reverse	5'-TCA-GTT-ATA-GGC-CAG-AAA-GC-3'		

14.10.1 Bacterial Bee Diseases

American Foulbrood (*Paenibacillus larvae larvae*): Govan *et al.* (1999) used specific primer made from 16S rRNA region of *P. l. larvae* causing AFB and amplified with DNA of five bacteria previously associated with honey bees. The gel showed a band just below 1090 bp at 973 bp for the test bacterium while no amplified product was visualized with DNA of other bacteria which ensured the specificity of the primer for the pathogenic test bacterium. Later, Dobbelerae *et al.* (2001) amplified the primer made from16S rRNA regions of *P. larvae* subsp. *larvae* (pathogenic) and *P. larvae* subsp. *pulvifaciens* (antagonist) and reported even the specific primers not to differentiate between the two subspecies, with 99.44 per cent similarity. So the primer could detect only upto genus and species level but not at subspecies level. However, Alippi *et al.* (2002), using a restriction enzyme *Hae*III for the PCR amplified product, could differentiate between the subspecies but it was not possible with the *Msp*I restriction enzyme.

The detection of *P. l.* larvae DNA by purifying the bacterial DNA and amplifying its 16 S rRNA gene has been described for practically diagnosing AFB (Thai Agricultural Standard, 2008), stating that microscopic examination of smears from larvae with no clinical signs is far less sensitive at detecting spores in colonies compared with PCR-based methods and PCR-based methods detect spores in colonies that never develop clinical signs of AFB. Primers for use for the detection have been given Table 3.

PCR method has also been described to detect. *P. l. larvae* in winter beehive wax debris to create a fast and sensitive method of detecting bacterium from beehive debris based that does not require long-lasting cultivation steps. Various methods of extracting spores from beehive debris were compared: the original method of extraction of spores into toluene, and alternative spore extraction methods into Tween 80, into water, into isopropanol and into 95 per cent ethanol. Isolation of DNA from various spore extractions was evaluated too. Best results were provided by isolation of DNA using the QIAamp DNA Mini Kit, without heat treatment. DNA of spores was detected by PCR from 0.25 g of bees wax debris, with the detected titer of 10^5 in 1g according to the cultivation tests (Ryba *et al.,* 2009).

In addition to honey samples, this real-time PCR-based strategy may be also applied to confirm AFB diagnosis in honey bee brood and to screen other apiary supplies and products (bees, pollen, wax), thus broadening the control of AFB spreading. A rapid, selective, with low detection limit, sensitive and specific method to detect and quantify vegetative cells and spores of *P. larvae* is now available (Martínez *et al.,* 2010). A real-time PCR that allowed selective identification and quantification of *P. larvae* 16S rRNA sequence was developed. Using standard samples quantified by flow cytometry, detection limits of 37·5 vegetative cells ml^{-1} and 10 spores ml^{-1} were determined. Compared to spread plate method, this real-time PCR-based strategy allowed, in only 2 h, the detection of *P. larvae* in contaminated honeys. No false-positive results were obtained. Moreover, its detection limit was 100 times lower than that of the culture method (2 *vs* 200 spores g^{-1} of honey).

European Foulbrood (*Melissococcus plutonius*): The specific primer from 16S rRNA region of *M. plutonius* (causal organism of EFB) amplified with its DNA (Govan *et al.*, 1998) showed the presence of 812 bp band for diseased larvae which was specific to *M. plutonius*.

OIE (2010) has recommended and described the details of the use of hemi-nested PCR method for detecting *M. plutonius* and described Primers as given below:

Name	Sequence	PCR-product size
EFB-F	5'-GAA-GAG-GAG-TTA-AAA-GGC-GC-3'	812 bp
EFB-R	5'-TTA-TCT-CTA-AGG-CGT-TCA-AAG-G-3'	
MP1	5'-CTT-TGA-ACG-CCT-TAG-AGA-3'	486 bp
MP2	5'-ATC-ATC-TGT-CCC-ACC-TTA-3'	
MP3	5'-TTA-ACC-TCG-CGG-TCT-TGC-GTC-TCT-C-3'	276 bp

14.10.2 Viral Bee Diseases

ABPV has a single positive strand RNA genome consisting of 9470 nucleotides excluding the poly (A) tail (Govan *et al.*, 2000). BQCV has a single positive strand RNA genome consisting of 8550 nucleotides excluding the poly (A) tail (Leat *et al.*, 2000). SBV has a single positive strand RNA genome consisting of 8832 nucleotides (Ghosh *et al.*, 1999).

ABPV specific primer amplified with RNA of ABPV, BQCV, CWV and KBV found a single band (900 bp) specific only to ABPV while amplified products from RNA of other viruses were not seen (Benjeddou *et al.*, 2001). Further, BQCV specific primer was amplified with RNA of different viruses but it visualized only single band specific to BQCV (700 bp) while no amplified product was found in the other lanes. Even RNA from ABPV and BQCV infected honey bees amplified with their specific primer only. RT-PCR used to detect a variety of RNA viruses including the picorna-like insect viruses. This technique is reliable, specific, and sensitive (Benjeddou *et al.*, 2001). A reverse transcriptase PCR (RT-PCR) assay was developed for the detection of acute bee paralysis virus (ABPV) and black queen cell virus (BQCV), two honey bee viruses. Complete genome sequences were used to design unique PCR primers within a 1-kb region from the 3′ end of both genomes to amplify a fragment of 900 bp from ABPV and 700 bp from BQCV. The combined guanidinium thiocyanate and silica membrane method was used to extract total RNA from samples of healthy and laboratory-infected bee pupae. In a blind test, RT-PCR successfully identified the samples containing ABPV and BQCV. Sensitivities were approximately 1,600 genome equivalents of purified ABPV and 130 genome equivalents of BQCV. Oligonucleotides for detection of ABPV and BQCV were designed within a 1-kb region measured from the 3′ ends of their respective genomes (ABPV accession number, AF150629 (Govan *et al.*, 2000) and BQCV accession number, AF183905 (Leat *et al.*, 2000). The ABPV primers were ABPV1 (5′-TTA TGT GTC CAG AGA CTG TAT CCA I-3′) and ABPV2 (5′-GCT CCT ATT GCT CGG TTT TTC GGT I-3′), corresponding to nucleotides 8460 to 8484 and 9336 to 9360, respectively. These primers amplified a 900-bp fragment. The BQCV primers were BQCV1 (5′-TGG TCA GCT CCC ACT ACC TTA AAC I-3′) and BQCV2 (5′-GCA ACA AGA AGA AAC GTA AAC CAC I-3′), corresponding to nucleotides 7850 to 7874 and 8526 to 8550, respectively. These primers amplified a 700-bp fragment. Since different strains of ABPV and BQCV may vary slightly in genome sequence, deoxyinosine residues were incorporated into the 3′ end of each primer to ensure that it would anneal to the template in case of a strain-specific mismatch at this position.

Chen *et al.* (2005a) detected DWV on different stages of honey bees and also in *V. destructor*. RNA taken from apparently healthy adult bees, the bees with wing deformities, pupae, larvae, eggs, adult drones and *V. destructor,* was amplified with DWV specific primer and the product showed a 702 bp band in all bee stages, as positive indication for infection, and also in *V. destructor* which was established to serve as carrier for the virus.

DNA of SBV infected honey bee larvae was amplified by Grabensteiner *et al.* (2001) from five different regions with five different primers and used sequenced amplified product for phylogenetic studies. This study revealed the existence of at least three distinct genotypes of SBV; Nepal 1, 2 and 4 strains were close to one Indian strain and these would infect *A. cerana* but Nepal 3 strain was closer to Central Europe strain which infect *A. mellifera*. It depicted the occurrence of two strains within Nepal while the third in South Africa which differed from earlier two strains on nucleotide level and also striking divergence at amino acid level, thereby proving at least three distinct strains of SBV.

Bakonyi *et al.* (2002) compared the sensitivities of AGID and RT-PCR for ABPV; AGID was capable of detecting the infection only up to 16 times dilution while RT-PCR could detect it up to 10,000 times dilution. Benjeddou *et al.* (2001) tested the sensitivity of RT-PCR for ABPV and BQCV; RT-PCR could detect ABPV up to 10^{-5} and BQCV up to 10^{-6} dilutions. Ribière *et al.* (2002) proved the sensitivity of RT-PCR to detect CBPV even up to 10^{-8} dilution. Berényi *et al.* (2006) reported RT-PCR to detect ABPV, BQCV and SBV up to 10^{-6} and CBPV, DWV and KBV up to 10^{-7} dilutions.

Real field infections were also tried to be diagnosed with different diagnostics. Ribière *et al.* (2002) compared EI, AGID and RT-PCR for CBPV infection in healthy H1, unapparently infectious H2 and diseased H3 colonies (Table 4). Only RT-PCR could detect infection in H2; in general while the first two methods could detect infection upto 10^{-3} times dilution but could not detect further dilution which could be detected only by RT-PCR. Its superiority in detecting CBPV infected apiaries and hives was also established (Table 5).

Topley *et al.* (2005) established the utility of M-RT-PCR for simultaneous detection of multiple viral bee infections by amplifying RNA from infected pupae with their specific primers.

Six most important viruses in beekeeping include sacbrood virus (SBV), chronic bee paralysis virus (CBPV), black queen cell virus (BQCV), deformed wing virus (DWV), acute bee paralysis virus (ABPV), and Kashmir bee virus (KBV). Due to difficulties with the classical diagnostic methods in bee virology (i.e., absence of bee tissue cultures, dependence on the season to obtain pupae for experimental infection, or requirement of type-specific serum for agar gel immunodiffusion tests), scientific interest

Table 4. Comparison of EI, AGID and RT-PCR for CBPV detection in different hives

Hive (symptoms)	CBPV status	Samples tested	Results of test			Sample No.
			EI	AGID	RT-PCR	
H1 (-)	Non infected	H1 bees	ND	-	-	1
		H1 bees inoculated with 10^{-1} H1 bee extract	-	-	-	2
H2(-)	Inapparent infection	H2 bees	ND	-	+	3
		H2 bees inoculated with 10-3 H2 bee extract	+	+	+	4
		H2 bees inoculated with 10-4 H2 bee extract	-	-	+	5
		H2 bees inoculated with water	-	-	+	6
H3 (+)	Clinical disease	H3 bees	ND	+	+	7
		H1 bees inoculated with 10^{-8} H3 bee extract	+	+	+	8
		H1 bees inoculated with purified virus	+	+	+	9

EI – Experimental Infection; ND – Not determined

Table 5. Detection of CBPV in field using AGID and RT-PCR

Tests	Apiaries		Hives	
	With clinical symptoms and losses	Without clinical symptoms and losses	With clinical symptoms and losses	Without clinical symptoms and losses
AGID-RT-PCR	9	0	24	0
AGID-RT-PCR	0	2	1	6
AGID-RT-PCR	0	1	0	2
Total	9	3	25	8

has turned towards molecular techniques. Since complete or at least partial genome sequence information on the aforementioned viruses is available in gene bank databases, diagnostic methods based on reverse-transcription PCR (RT-PCR) have been developed for the detection of viral RNA in honey bee samples (Bakonyi *et al.*, 2002; Tentcheva *et al.*, 2004). The presence of bee viruses was demonstrated by amplification of virus-specific nucleic acid by employing RT-PCR methods. Oligonucleotide primers were designed, based on the genome sequences of SBV,

CBPV, BQCV, DWV, ABPV, and KBV by using the Primer Designer 4 program for Windows 95, version 5.20 (Scientific & Educational Software) (Berényi *et al.*, 2006). The sequences, orientations, and locations of the six primer pairs are shown in Table 6. The RT-PCR assays were able to detect viral RNAs at dilutions of up to 10^{-6} (ABPV, BQCV, and SBV) and 10^{-7} (CBPV, DWV, and KBV). The samples containing *V. destructor* were mostly infected with ABPV, BQCV and DWV. *N. apis*, together with viruses, was found in 12 per cent of the samples; usually ABPV, BQCV, SBV, and DWV were associated with *Nosema* infestation. *Malpighamoeba mellificae* was identified in only one sample, which was also positive for SBV, BQCV, and DWV.

Table 6. Sequences, orientations and locations of primer pairs

Primer name	Primer sequence	Primer position on the genome	Length of amplified product
ABPV23f	GTC CTA TCT TGG AAT ACT AC	7928-7947	618
ABPV24r	AAG GYT TAG GTT CTA CTA CT	8527-8546	
BQCV 3f	AGT AGT TGC GAT GTA CTT CC	252-277	472
BQCV 4r	CTT AGT CTT ACT CGC CTT AC	710-729	
CBPV 111f	TGT CGA ACT GAG GAT CTT AC	111-130	315
CBPV 426r	GAC CTG ATT AAC GAC GTT AG	407-426	
DWV 2345f	ATT GTG CCA GAT TGG ACT AC	2345-2364	434
DWV 2779r	AGT TGC AAT GGA GGA TAC AG	2760-2997	
KBV 5406f	GAT GAA CGT CGA CCT ATT GA	5406-5425	395
KBV 5800r	TGT GGG TTG GCT ATG AGT CA	5781-5800	
SBV 1f	ACC AAC CGA TTC CTC AGT AG	221-240	487
SBV 2r	CCT TGG AAC TCT GCT GTG TA	689-708	

Table 7. Primers used for identification of various bee viruses

Virus	Primers	Product size (bp)	References
ABPV	ABPV-F: (5-ttatgtgtccagagactgtatcca-3')	900	Benjeddou *et al.* (2001)
	ABPV-R: (5'-gctcctattgctcggtttttcggt-3')		
CBPV	CBPV-F: (5'-agttgtcatggttaacaggatacgag-3')	455	Ribiere *et al.* (2002)
	CBPV-R: (5'-tctaatcttagcacgaaagccgag-3')		
BQCV	BQCV-F: (5'-tggtcagctcccactaccttaaac-3')	700	Benjeddou *et al.* (2001)
	BQCV-R: (5'-gcaacaagaagaaacgtaaaccac-3')		
DWV	DWV-F: (5'-atcagcgcttagtggaggaa-3')	702	Chen *et al.* (2004b)
	DWV-R: (5'-tcgacaattttcggacatca-3')		
KBV	KBV-F: ('5-gatgaacgtcgacctattga-3')	415	Stoltz *et al.* (1995)
	KBV-R: (5'-tgtgggttggctatgagtca-3')		
SBV	SBV-F: (5'-gctgaggtaggatctttgcgt-3')	824	Chen *et al.* (2004c)
	SBV-R: (5'-tcatcatcttcaccatccga-3')		

Detection of Multiple Viruses in Queens by RT-PCR: The development of molecular technologies, however, has provided a powerful tool for specific, sensitive, and rapid identification of bee viruses. The availability of complete genome sequences of five viruses, acute bee paralysis virus (ABPV; GenBank Accession No. AF150629), black queen cell virus (BQCV; GenBank Accession No. AF183905), deformed wing virus (DWV; GenBank Accession No. NC004830), Kashmir bee virus (KBV; GenBank Accession No. NC004807), and sacbrood virus (SBV; GenBank Accession No. AF092924) and partial genome sequences of chronic bee paralysis virus (CBPV; GenBank Accession No. AF461061) makes the molecular detection and characterization of these viruses possible (Bakonyi *et al.*, 2002; Benjeddou *et al.*, 2001; Chen *et al.*, 2004c, 2005b; Evans, 2001; Genersch, 2005; Grabenstiner *et al.*, 2001; Hung *et al.*, 1996; Ribiere *et al.*, 2002; Stoltz *et al.*, 1995; Tentcheva *et al.*, 2004). Utilizing RT-PCR methods, viruses have been detected in different life stages of bees as well as in the parasitic mite *Varroa destructor* (Chen *et al.*, 2004a, 2005a). Despite the increasing knowledge of bee virus infections, there is little information regarding the virus status of the honey bee queens. Utilizing RT-PCR methods, viruses are detected in different life stages of queen bee. Twenty-nine queens from honey bee colonies maintained in Beltsville, MD and Sapelo Island, GA were used in this study. The nucleotide sequences of PCR products (Table 7) were analyzed and compared with sequences published at the Gen Bank, National Center for Biotechnology Information, NIH. This study demonstrates that adult queen bees can harbor multiple viruses. Sequence analyses indicated that RT-PCR bands representing BQCV, CPBV, DWV, KBV, and SBV were virus specific. Among the queens, 86 per cent were positive for BQCV, 14 per cent for CBPV, 100 per cent for DWV, 21 per cent for KBV, and 62 per cent for SBV. The results also indicated that queens could be co-infected with multiple viruses, simultaneously. Of the 29 queens examined, 93 per cent had multiple virus infections of two, three, four or five viruses.

14.10.3. Fungal Diseases

Chalk Brood (*Aschosphaera apis*): Identification of individual *Ascosphaera* species based on morphological features has been difficult due to a lack of distinguishing characteristics. Most identifications are based on the size and shape of the ascomata, spore balls and conidia. Unfortunately, much overlap occurs in the size of these structures, and some *Ascosphaera* species will not produce sexual structures in vitro. James and Skinner (2005) reported a quick and reliable diagnostic method for identifying *Ascosphaera* infections using PCR markers that employ genus-specific primers for *Ascosphaera*. Using these methods, species identifications can be performed directly on bees, including asymptomatic individuals. Furthermore, the PCR markers can detect co-infections of multiple *Ascosphaera* species in a single host. A marker was identified for *A. apis*. The *A. apis* species-specific primers were used for the 5.8s ribosomal DNA and the internal transcribed spacers (ITS 1 and ITS 2) using the program Primer-BLAST. The *A. apis*-specific DNA sequence of 442 bp was amplified, cloned successfully for diagnostic molecular marker Quick and reliable diagnostic method. The technique eliminated the need for culturing samples used to detect these latent or inapparent infections, and could be used to process a large number of field collected bee larvae.

14.10.4. Microsporidian Diseases

Nosemosis: The diagnosis of *Nosema* disease has been traditionally done by detecting spores of *Nosema* spp. through microscopic analyses. However, with the recent finding that both *Nosema ceranae* and *Nosema apis* affect *A. mellifera*, molecular techniques are required that can reliably differentiate between these different species of microsporidia, because the spores of the two *Nosema* species cannot be reliably distinguished by their morphology (Fries *et al.*, 2006). In addition, microscopic analyses are not as sensitive at detecting low levels of *Nosema* infection as molecular methods, such as PCR, can be. Moreover, microscopic examination of *Nosema* spores is costly, laborious and time-consuming. Nosemosis is caused by *N. apis* and *N. ceranae* (Fries). A simpler method of detecting *N. apis* and *N. ceranae* was developed by Martin-Hernández *et al.* (2007). They created a duplex PCR-based method that simultaneously amplified the 16S rRNA regions of *N. apis* or *N. ceranae* using a single reaction with two pairs of primers allowing for parallel detection of both microsporidians. More recently, a real time duplex PCR assay was developed amplifying the same 16S rRNA regions that also allowed for quantification of *N. apis* and *N. ceranae* in a single reaction (Bourgeois *et al.*, 2010). However, real-time PCR thermocyclers and techniques are not readily available in many laboratories and are more costly than standard PCR thermocyclers and techniques. An improved multiplex-PCR method for the differentiation and quantification of *Nosema* spp. in honey bees in a single reaction was developed by Hamiduzzaman *et al.* (2010). It combines the duplex PCR-based method of Martin-Hernández *et al.* (2007) with the relative RT-PCR approach of Dean *et al.* (2002), which includes a house-keeping gene of the host in each reaction for relative quantification based on PCR band intensity. This semi-quantitative method made it possible to estimate *Nosema* infection levels using a standard PCR thermocycler. Newly improved method has a number of advantages compared to previously published, standard PCR-based assays and makes it possible to both detect and quantify *Nosema* spores in honey bee samples. Triplex PCR-based method was developed to detect *N. ceranae* and *N. apis*, while amplifying the honey bee RpS5 gene as a control at the same time. By comparison, the duplex PCR detection method of Martin-Hernández *et al.* (2007) did not include a honey bee control gene. The triplex PCR assay was highly sensitive in detecting the presence of very low numbers of *Nosema* spores. Bands were clearly visible even at levels equivalent to only 100 spores/bee, and faint bands were also detectable at levels equivalent to10 spores/bee based on singleplex PCR. When compared with a microscopic analysis, this is at least 500 times more sensitive in detecting *Nosema* spores, which makes it highly useful for *Nosema* diagnosis in the early stages of the disease. Detection level is also more sensitive than those reported previously for *Nosema* spp. when standard PCR assays were used (Klee *et al.*, 2007; Martin-Hernández *et al.*, 2007). However, real-time PCR may be even more sensitive as Bourgeois *et al.* (2010) claimed potential detection levels as low as one spore/bee.

14.10.5. Differentiation and Identification of Acarines

Tangjingjai *et al.* (2003) proved the usefulness of polymorphism of ITS (Internal Transcribed Spacer) in differentiating between morphologically close species of brood mites, *Tropilaelaps clareae* and *Tropilaelaps koenigerum*. The two species could be differentiated by using restriction enzyme *Mse*I for digesting ITS region of *T. clareae* and *T. koenigerum* which resulted in bands at separate positions.

In *Varroa destructor*, differentiation among genotypes was established by using RFLP (Zhou *et al.*, 2004) with two restriction enzymes (*Sac* I, *Xho* I) for digesting mitochondrial DNA. It was observed that in one sample, mitochondrial region was digested with only *Xho* I (sample # 1-8 & 10) and one sample (# 9) with both the restriction enzymes. It proved that samples 1-8 and 10 were Korean haplotype and sample 9 the China or China 2 haplotype.

14.11. ADVANTAGES AND DISADVANTAGES OF MOLECULAR TECHNIQUES

Advantages of molecular techniques include detection of even the latent infections, offer very high degree of sensitivity, reliability, reproducibility and above all no prior DNA sequence information is required.

For molecular analyses, very large quantities of DNA are required especially for restriction digestion. Molecular analysis cannot be done *in situ* and is highly sensitive to reaction conditions; single base changes are not easy to detect.

14.12. CONCLUSIONS

Molecular techniques are highly efficient in diagnosing / identifying bee diseases and mites and produce more desirable effects as compared to biochemical techniques. Molecular markers can be used for differentiating subspecies and strains, and using M-RT-PCR, several viruses can be detected simultaneously. Molecular techniques being very fast and highly sensitive would be more reliable as precise diagnostic tools.However, there is need to develop strain specific genetic markers for diagno-sing bee diseases and mites that should be helpful in differentiating sub-species / strains, etc. of the pathogenic microorga-nisms and mites.

REFERENCES

Adlakha RL (1976) Acarine disease of adult honey bees in India. *American Bee Journal* **16** (7): 324-344.

Alippi AM, Lopez AC and Aguilar OM (2002) Differentiation of *Paenibacillus larvae* subsp. *larvae*, the cause of American foulbrood of honey bees, by using PCR and restriction fragment analysis of genes encoding 16S rRNA. *Applied and Environmental Microbiology* **68** (7): 3655-3660.

Alippi AM, Lopez AC and Aguilar OM (2004) A PCR-based method that permits specific detection of *Paenibacillus larvae* subsp. *larvae,* the cause of American Foulbrood of honey bees, at the subspecies level. *Letters in Applied Microbiology* **39** (1): 25-33.

Allen M and Ball BV (1996) The incidence and world distribution of honey bee viruses. *Bee World* **77**: 141-162.

Allen MF, Ball BV, White RF and Antoniw JF (1986) The detection of acute paralysis virus in *Varroa jacobsoni* by the use of a simple indirect ELISA. *Journal of Apicultural Research* **25**: 100-105.

Anderson DL (1984) A comparison of serological techniques for detecting and identifying honey bee viruses. *Journal of Invertebrate Pathology* **44**: 233-243.

Anderson DL and Trueman JWH (2000) *Varroa jocobsoni* is more than one species. *Experimental Applied Acarology* **24**: 165-189.

Anonymous (1999) *Annual Report : 1998-99 (Solan Centre), All India Coordinated Research Project on Honey Bee Research and Training*, presented at Annual Workshop of the Coordinated Project, held at Assam Agricultural University, Jorhat (India), June 8-9, 1999.

Antunez K, Martin-Hernandez R, Prieto L, Meana A, Zunino P and Higes M (2009) Immune-suppression in the honey bee (*Apis meliifera*) following infection by *Nosema ceranae* (Microsporidia) *Environ*-mental Microbiology* **11**: 2284-2290.

Atwal AS (1964) Introduction di api regine di razza italiana in alveari di Apis indica. 1' Apicoltore di Italia, August 1964, pp. 100-101.

Atwal AS (1967) Acarine Disease Problem of Honey Bees in the Punjab. *Technical Report of College of Agriculture, PAU, Ludhiana, India*, 39 pp.

Atwal AS (1969) Serious infestations of *Apis mellifera* L. bee by the ectoparasitic mite, *Tropilaelaps clareae. XXII International Beekeeping Congress,* Summary **22** : 108.

Atwal AS and Goyal NP (1971) Infestation of honey bee colonies with *Tropilaelaps* and its control. *Journal of Apicultural Research* **10** (3):137-142.

Atwal AS and Sharma OP (1967) Introduction of *Apis mellifera* queens into dequeened colonies of *Apis indica* F. *XXI International Beekeeping Congress, Beltsvile, U.S.A.*, pp. 70-71.

Atwal AS and Sharma OP (1968) Introduction of *Apis mellifera* queens into *Apis indica* colonies and the associated behaviour of the two species. *Indian Bee Journal* **30** (2): 41-56.

Bailey L (1956). Etiology of European Foul Brood; a disease of the larval honey bee. *Nature* **178**: 1130.

Bailey L (1974) An unusual type of *Streptococcus pluton* from the Eastern hive bee *Apis cerana indica* F. *Journal of Invertebrate Pathology* **23** : 246-247.

Bailey, L. 1983. *Melissococcus pluton,* the cause of European foulbrood of honey bees (*Apis* spp.). *Journal of Applied Bacteriology* **55** : 65-69.

Bailey L and Ball BV (1991). *Honey Bee Pathology*, pp. 35-52, Academic Press Limited, London, UK.

Bakonyi T, Farkas R, Szendroi A, Dober-Kovacs M and Rusvai M (2002) Detection of acute bee paralysis virus by RT-PCR in honey bee and *Varroa destructor* field samples: rapid screening of representative Hungarian apiaries. *Apidologie* **33**: 63-74.

Benjeddou M, Leat N, Allsopp M and Davison S (2001). Detection of Acute Bee Paralysis virus and black queen cell virus from honey bees by reverse transcriptase PCR. *Applied and Environmental Microbiology* **67**(5): 2384-2387.

Berényi O, Bakonyi T, Derakhshifar I, Köglberger H and Nowotny N (2006) Occurrence of six honey bee viruses in diseased Austrian apiaries. *Applied and Environmental Microbiology* **72**(4): 2414-2420.

Bhardwaj RK (1968) A new record of the mite, *Tropilaelaps clareae*: a parasite of honey bee in South East Asia. *Bee World* **49** (3): 115.

Bitner AR (1971) Nitrite test as an aid in the identification of *Bacillus larvae* grown in yeast-starch agar. *American Bee Journal* **28**: 385-391.

Bourgeois AL, Rinderer TE, Beaman LD and Danka RG (2010) Genetic detection and quantification of Nosema apis and N. Ceranae in the honey bee. *Journal of Invertebrate Pathology* **103**: 53–58

Catherine, MacDonald and Hugh Loxdale (2004) Molecular markers to study population structure and dynamics in beneficial insects (predators and parasitoids). *International Journal of Pest Management* **50:** 215-224.

Chandel YS, Arun Kumar and Ball BV (1999) Sac brood disease in Italian honey bee, *Apis mellifera* L., in Himachal Pradesh, India. *Pest Management and Economic Zoology* **7** (2) : 181-182.

Channa Basavanna GP (1992) Thai sac brood disease in *Apis cerana* in South India. *IUSSI Newsletter, Bangalore* **6**: 10-11.

Chen YP, Pettis JS, Evans JD, Kramer M and Feldlaufer MF (2004a) Molecular evidence for transmission of Kashmir bee virus in honey bee colonies by ectoparasitic mite, *Varroa destructor*. *Apidologie* **35**: 441–448.

Chen YP, Smith B, Collins AM, Pettis S and Feldlaufer MF (2004b) Detection of deformed wing virus infection in honey bees, *Apis mellifera* L., in the United States. *American Bee Journal* **144**: 557–559.

Chen YP, Zhao Y, Hammond J, Hsu HT, Evans JD and Feldlaufer MF (2004c) Multiple virus infections in the honey bee and genome divergence of honey bee viruses. *Journal of Invertebrate Pathology* **87**: 84–93.

Chen YP, Higgins JA and Feldlaufer MF (2005a) Quantitative real-time reverse transcription-PCR analysis of deformed wing virus infection in the honeybee (*Apis mellifera* L.). *Applied and Environmental Microbiology* **71**: 436-441.

Chen YP, Pettis JS and Feldlaufer MF (2005b) Detection of multiple viruses in queens of the honey bee *Apis mellifera* L. *Journal of Invertebrate Pathology* **90**: 118–121.

Chen YP, Evans JD, Murphy C, Gutell R, Zuker M, Gundensen-Rindal D and Pettis JS (2009) Morphological molecular and phylogenetic characterization of *Nosema ceranae,* a microsporidian parasite isolated from the European honey bee, *Apis mellifera. Journal of Eukaryotic Microbiology* **56:** 142-147.

Chhuneja PK (2012) Apiculture – a viable rural enterprise, pp. 491-554. **In:** Singh, R. and Mangraj, S. (eds), *Entrepreneurship Development through Agro-industrial Enterprises – Principle & Practice.* LAP Lambert Academic Publishing GmbH Co. KG, Germany, 601 pp.

Chhuneja PK (2013) Honey bee diseases and mites in India – First occurrences and inter-specific spread, pp. 734-749. **In**: Dhawan, A.K.; Singh, B.; Brar Bhullar, M. and Arora, R. (2013), *Integrated Pest Management,* Scientific Publishers (India), Jodhpur, India, 749 pp.

Chhuneja PK, Anderson DL and Gatoria GS (2005) Causes of occurrence and fast spread of Korean haplotype of *Varroa destructor* Anderson & Trueman in *Apis mellifera* Linnaeus colonies in northern India and its management. *Indian Bee Journal* **67** (3 & 4) : 193-198.

Chhuneja PK, Garcha S and Gatoria GS (2001) Bee diseases - Impending threat and challenge to Indian beekeeping. *Indian Bee Journal* **63** (3 & 4): 59-63.

Chhuneja PK, Gatoria GS, Aulakh R, Singh J and Garcha S (2002) Surveillance of *Apis mellifera* Linn. colonies for bee diseases and enemies in the Punjab (India). **In** : *'6th Asian Apicultural Association (AAA) International Conference and World Apiexpo, Bangalore, India*, pp. 133.

Chhuneja PK and Singh J (2013) Apiculture, pp. 15-36. **In**: *Agripreneurship – An Opportunity for Rural Empowerment*. Punjab Agriculture Management and Extension Training Institute, Ludhiana, India, 281 pp.

Cornman RS, Schatz MC, Johnston JS, Chen YP, Pettis J, Hunt G, Bourgeois L, Elsik C, Anderson D, Crozinger CM and Evans JD (2010) Genomic survey of the ectoparasitic mite *Varroa destructor,* a major pest of the honey bee *Apis mellifera*. *BMC Genomics* **11:** 602. http://dx.doi.org/10.1186/1471-2164-11-602.

Cornmann RS, Street C, Chen Y, Zhao Y, Schatz M, Salzberg S, Egholm M, Hutchinson S, Pettis JP, Lipkin WI and Evans JD (2009) Genomic analyses of the microsporidian *Nosema ceranae,* an emerging parasite of honey bees. *PLoS Pathogens* **5** e1000466. doi:10.1371/journal.ppat.1000466.

Davison S, Leat N and Benjeddou M (2003) Development of molecular tools for honey bee virus research: the South African contribution. *African Journal of Biotechnology* **2** (12): 698-713.

De Miranda JR, Drebot M, Tyler S, Shen M, Cameron CE, Stoltz DB and Camazine SM (2004) Complete nucleotide sequence of Kashmir bee virus and comparison with acute bee paralysis virus. *Journal of General Virology* **85**: 2263-2270.

Dean JD, Goodwin PH and Hsiang T (2002) Comparison of relative RT-PCR and Northern blot analyses to measure expression of ß-1,3-glucanase in *Nicotiana benthamiana* infected with *Colletotrichum destructivum*. *Plant Molecular Biology Report* **20**: 347–356.

Deodikar GB (1971a) Dangers from introduction of exotic pathogenic diseases through import of useful insects like bees, silkworms or lac insects. *Indian Bee Journal* **33** : 1-2.

Deodikar GB (1971b) Code of conduct for conservation of bees and beekeeping. *Indian Bee Journal* **33** : 14-22.

Diwan VV, Kshirsagar KK, Rao AVR, Raghunath D, Bhambure CS and Godbole SH (1971) Occurrence of a new bacterial disease of Indian honey bee *Apis indica* F. *Current Science* **40** : 196-197.

Djoodjevic SP, Noone K, Smith L and Hornitzky MAZ (1998) Development of a heminested PCR assay for the specific detection of *Melissococcus pluton. Journal of Apicultural Research* **37**: 165-174.

Dobbelaere W, De Graaf DC, Peeters JE and Jacobs FJ (2001) Development of fast and reliable diagnostic method for American foulbrood disease (*Paenibacillus larvae* subsp. *larvae*) using a 16S rRNA gene based PCR. *Apidologie* **32**: 363-370.

Evans JD (2001) Genetic evidence for co infection of honey bees by acute bee paralysis and Kashmir bee viruses. *Journal of Invertebrate Pathology* **78**: 189–193.

Fievet J, Tentcheva D, Gauthier L, de Miranda JR, Cousserans F, Colin ME and Bergoin M (2006) Localization of deformed wing virus infection in queen and drone *Apis mellifera* L. *Virology Journal* **3**: 16.

Fries I, Martin R, Meana A, Garcia-Palencia P and Higes M (2006) Natural infections of Nosema ceranae in European honey bees. *Journal of Apicultural Research* **45**: 230–233.

Gatoria GS, Chhuneja P K, Singh J, Anderson DL and Blossom (2005) *Varroa destructor* Anderson & Truman - a new species of mite on *Apis mellifera* Linn. in India. Paper presented at *'First Congress on Insect Science'* organized by Indian Society for the Advancement of Insect Science, at PAU, Ludhiana, December 15-17, 2005. Summary published in Proceedings, p. 218-219.

Gatoria GS, Chhuneja PK, Singh Y and Blossom (2006) Annual Report : All India

Coordinated Project on Honey Bee Research and Training. Presented at Kanakakunnu Palace, Thiruvananthapuram, Jan. 27, 2006.

Gatoria GS, Chhuneja PK, Singh J and Blossom (2004) *Varroa* - a potential threat to Indian Apiculture. *Indian Bee Journal* **66** (3-4): 112-122.

Genersch E (2005) Development of a rapid and sensitive RT-PCR method for the detection of deformed wing virus, a pathogen of the honeybee (*Apis mellifera*). *Veterinary Journal* **169**: 121–123.

Ghosh RC, Ball BV, Wilcocks MM and Carter MJ (1999) The nucleotide sequence of sacbrood virus of the honey bee: an insect picorna-like virus. *Journal of General Virology* **80**: 1541-1549.

Govan VA, Leat N, Allsop M and Davison S (2000) Analysis of the complete genome sequence of acute bee paralysis virus shows that it belongs to the novel group of insect-infecting RNA viruses. *Virology* **277**: 457–463.

Govan VA, Allsopp MH and Davison S (1999) A PCR detection method for rapid identification of *Paenibacillus larvae. Applied Environmental Microbiology* **65**: 2243-2245.

Govan VA, Brözel V, Allsopp MH and Davison S (1998) A PCR detection method for rapid identification of *Melissococcus pluton* in honey bee larvae. *Applied and Environmental Microbiology* **64** (5): 1983-1988.

Govan VA, Leat N, Allsopp M and Davison S (2000) Analysis of the complete genome sequence of acute bee paralysis virus shows that it belongs to the novel group of insect infecting RNA viruses. *Virology* **277**: 457-463.

Grabensteiner E, Ritter W, Carter MJ, Davison S, Pechhacker H, Kolodziejek J, Boecking O, Derakshifar I, Moosbeckholer R, Licek E and Nowotny N (2001) Sacbrood virus of the honey bee (*Apis mellifera*): rapid identification and phylogenetic analysis using reverse transcription – PCR. *Clinical and Diagnostic Laboratory Immunology* **8**: 93-104.

Gupta M (1986) Occurrence of acarine Disease in *Apis mellifera* L. in Hisar. *National Symposium on New Dimensions in Parasitology*, Allahabad, India, pp. 37-38.

Hamiduzzaman Mollah Md, Guzman-Novoa Ernesto and Goodwin Paul H (2010) A multiplex PCR assay to diagnose and quantify *Nosema* infections in honey bees (*Apis mellifera*). *Journal of Invertebrate Pathology* **105** (2): 151-155.

Hayden MJ, Nguyen TM, Waterman A and Chalmers KJ (2008) Multiplex-ready PCR: a new method for multiplexed SSR and SNP genotyping. *BMC Genomics* **9**: 80. doi: 10.1186/1471-2164-9-80.

Haynes WC (1972) The catalase test: An aid in the identification of *Bacillus larvae. American Bee Journal* **112**: 130-131.

Henikoff S and Comai L (2003) Single-nucleotide mutations for plant functional genomics. *Annual Review of Plant Physiology and Plant Molecular Biology* **54:** 375-401.

Hitchcock JD and Wilson WT (1973) Pathogenicity to honey bees of a strain of *Bacillus larvae* that does not reduce nitrate. *Journal of Economic Entomology* **66**: 901-902.

Hornitzky M (2001) Literature review of Chalkbrood – a fungal disease of honey bees. Rural Industries Research and Development Corporation. http://www.rirdc.gov.au

Hornitzky MAZ and Karlovskis S (1989) A culture technique for the detection of *Bacillus* larvae in honey bees. *Journal of Apicultural Research* **28**:118-120.

Hornitzky MAZ and Wilson SC (1989) A system for the diagnosis of the major bacterial brood diseases of honey bees.

Journal of Apicultural Research **28**:191-195.

Hung ACF, Ball BV, Adams JR, Shimanuki and Knox DA (1996) A scientific note on the detection of American strain of acute paralysis virus and Kashmir bee virus in dead bees in one US honey bee (*Apis mellifera* L) colony. *Apidologie* **27**: 55–56.

Hung ACF and Shimanuki H (1999) A scientific note on the detection of KBV in individual honey bees and *Varroa jacobsoni* mites. *Apidologie* **30** (4) : 353-354.

Hung ACF, Adams JR and Shimanuki H (1999) Bee parasitic mite syndrome (II): the role of *Varroa* mite and viruses. *American Bee Journal* **135** (10) : 702-704.

Hunt Greg J and Page Robert E Jr (1995) Linkage map of the honey bee, *Apis mellifera*, based on RAPD markers. *Genetics* **139**: 1371-1382.

Jaccoud D, Peng K, Feinstein D and Kilian A (2001) Diversity arrays: a solid state technology for sequence information independent genotyping. *Nucleic Acids Research* **29:** e25.

James RR and Skinner JS (2005) PCR diagnostic methods for *Ascosphaera* infections in bees. *Journal of Invertebrate Pathology* **90** (2): 98-153.

Joyce C (2002) Quantitative RT-PCR. A review of current methodologies. *Methods of Molecular Biology* **193**: 83–92. doi: 10.1385/1-59259-283-X:083.

Klee J, Besana A M, Genersch E, Gisder S, Nanetti A, Tam D Q, Chinh T X, Puerta F, Ruz JM, Kryger P, Message D, Hatjina F, Korpela S, Fries I and Paxton RJ (2007) Widespread dispersal of the microsporidian Nosema ceranae, an emergent pathogen of the western honey bee, Apis mellifera. *Journal of Invertebrate Pathology* **96**: 1–10.

Kraus B and Hunt G (1995) Differentiation of *Varroa jacobsoni* oud populations by random amplification of polymorphic DNA (RAPD). *Apidologie* **26**: 283-290.

Kshirsagar KK and Godbole SH (1974) Indian *Streptococcus pluton. Indian Bee Journal* **36** (1/4) : 39.

Kshirsagar KK, Chauhan RM and Singh YK (1974) Detection of *Nosema apis* Zanders in *Apis cerana indica* Fab. in Uttar Pradesh. *Indian bee Journal* **36**: 23-24.

Kshirsagar KK, Mahindre DV, Salvi SR and Mittal MC (1975) Occurrence of a viral disease in Indian honey bee *Apis cerana indica* F. – a preliminary report. *Indian bee Journal* **37**: 19-20.

Kumar J, Gupta JK, Dogra GS (1988) Discovery of the ectoparasitic mite, *Varroa jacobsoni,* on *Apis mellifera* in Himachal Pradesh, India. *American Bee Journal* **128** (2): 124.

Kumar V (2003) *Surveillance of bacterial diseases of Apis mellifera Linn. in various agroclimatic zones of the Punjab.* M.Sc. Thesis, Department of Entomology, Punjab Agricultural University, Ludhiana, India.

Lanzi G, de Miranda JR, Boniotti MB, Cameron CE, Lavazza A, Capucci L, Camazine SM and Rossi C (2006) Molecular and biological characterization of deformed wing virus of honeybees (*Apis mellifera* L.). *Journal of Virology* **80**(10): 4998-5009.

Leat N, Ball B, Govan V and Davison S (2000) Analysis of the complete genome sequence of black queen cell virus, a picorna-like virus of honey bees. *Journal of General Virology* **81**: 2111-2119.

Lochhead AG (1937) The nitrate reduction test and its significance in the detection of *Bacillus larvae. Canadian Journal Research* **15**: 79-86.

Martínez J, Simon V, Gonzalez B and Conget P (2010) A real-time PCR-based strategy for the detection of *Paenibacillus larvae* vegetative cells and spores to improve the diagnosis and the screening of American foulbrood. *Letters in Applied Microbiology* **50**: 603–610.

Martin-Hernández R, Meana A, Prieto L, Salvador AM, Garrido-bailon E and Higes M (2007) Outcome of colonization of *Apis mellifera* by Nosema ceranae. *Applied Environmental Microbiology* **73**: 6331–6338.

Mattu VK, Sharma N and Chaudhary DK (1994) Some biological observations on predatory wasps and parasitic mites infesting *Apis* species in Himachal Pradesh, India. **In** *Proc. 2nd Conference of Asian Apicultural Associatiuon, Yogakarta, Indonesia, July 25-29, 1994.*

Mullis K. (1990) The unusual origin of the polymerase chain reaction. *Scientific American* **262** (4): 56–61, 64–65. doi:10.1038/scientificamerican0490-56.

Murray KD, Aronstein, KA and Jones, WA (2005) A molecular diagnostic method for selected *Ascosphaera* sp. Using PCR amplification of internal transcribed spacer region of rRNA. *Journal of Apicultural Research* **44**(2) : 61-64

Mutto RN (1956) Bee diseases in Indian honey bee. *Indian Bee Journal* **18** : 14-18.

Neuendorf S, Hedtke K, Tangen G and Genersch E (2004) Biochemical characterization of different genotypes of *Paenibacillus larvae* subsp. *larvae*, a honeybee bacterial pathogen. *Microbiology* **150**: 2381-2390.

OIE (2010) European foulbrood of honey bees, Chapter 2.2.3. **In:** Manual of Diagnostic Tests and Vaccines for Terrestrial Animals 2010. The Animal Health and Production Compendium. World Organization for Animal Health.

Olsen PE, Grant GA, Nelson DL and Rice WA (1990) Detection of American foulbrood disease of the honey bee, using a monoclonal antibody specific to *Bacillus larvae* in an enzyme linked immunosorbant assay. *Canadian journal of Microbiology* **36**: 732-735.

Paldi N, Glick E, Oliva M, Ziberberg Y, Aubin L, Pettis JS, Chen YP and Evans JD (2010) Effective gene silencing of a microsporidian parasite associated with honey bee (*Apis mellifera*) colony declines. *Applied and Environmental Microbiology* **76:** 5960-5964.

Phadke KG, Bisht DS and Sinha RBD (1966) Occurrence of the mite *Varroa jacobsoni* Oudemans in the brood cells of the honey bee, *Apis indica* F. *Indian Journal of Entomology* **28**: 411-412.

Pinnock DE and Featherstone NE (1984) Detection and quantification of Melissococcus pluton infection in honey bee colonies by mean of enzyme linked immunosorbant assay. *Journal of Apicultural Research* **23**: 168-170.

Ragsdale DW and Furgala B (1987) A serological approach to the detection of *Acarapis woodi* parasitism in honey bees using an enzyme linked immunosorbant assay. *Apidologie* **18** : 1-9.

Ragsdale DW and Kjer KM (1989) Diagnosis of Tracheal mite (*Acarapis woodi* Rennie) parasitism of honey bees using a monoclonal based enzyme linked immunosorbent assay. *American Bee Journal* **129**: 550-553.

Rahman A and Rahman S (1994) Diagnosis and control of Nosema disease of *Apis mellifera* L. in Assam. **In :** Veeresh, G.K. and Viraktamath, C.A. (eds), *Honey Bee Diseases, their Diagnosis and Management,* Oxford and IBH publishers, New Delhi, India.

Rehner SA and Evans JD (2009) Microsatellite loci for the fungus *Ascosphaera apis*: cause of honey been chalkbrood disase. *Molecular Ecology Resources* **9:** 855-858.

Ribiere M, Triboulot C, Mathieu L, Aurieres C, Faucon JP, Pepin M (2002) Molecular diagnosis of chronic bee paralysis virus infection. *Apidologie* **33**: 339–351.

Robert N (2001) Nosema disease in honey bees – Genetic variation and control, Rural Industries Research and Development Corporation. http://www.rirdc.gov.au.

Ryba S, Titera D, Haklova M and Stopka P (2009) A PCR method of detecting American Foulbrood (*Paenibacillus larvae*) in winter beehive wax debris. *Veterinary Microbiology* **139** (1-2):193-196.

Sambrook Joseph and Russel David W (2001) In vitro Amplification of DNA by the Polymerase Chain Reaction. **In:** *Molecular Cloning: A Laboratory Manual* (3rd ed.). Cold Spring Harbor, New York (USA).

Shah FA and Shah TA (1979) More on Apis iridescent virus. *Indian Bee Journal* **41** (3/4): 24-25.

Singh Kuldeep, Chhuneja P, Vikal Y and Kaur S (2009) *Techniques in Plant Molecular Biology: A Laboratory Manual.* Punjab Agricultural University, Ludhiana, India, 170 pp.

Singh Rajpal and Garg R (2000) *Streptococcus pluton* White in Paonta valley of Himachal Pradesh. *Insect Environment* **6** (2): 64.

Singh S (1957) Acarine disease in the Indian bee (*Apis indica* F.). *Indian Bee Journal* **19**: 27-28.

Singh S (1961) Appearance of American foulbrood disease in Indian honey bee (*Apis indica* Fab.) *Indian Bee Journal* **23** : 46-50.

Singh S and Adlakha RL (1958) Acarine disease of adult honey bee. *Indian Bee Journal* **20**: 64-8.

Singh S and Adlakha RL (1960) Survey of acarine disease of adult honey bees in Shimla hills. *Indian Bee Journal* **22** : 3-6.

Singh Y (1979) Iridescent virus in the Indian honey bee (*Apis cerana indica* F.). *American Bee Journal* **119** : 398.

Singh Y and Verma SK (1981) Occurrence of acarine disease in *Apis cerana indica* in Uttar Pradesh (India). *Indian Bee Journal* **43** (2): 37-39.

Solignac M, Vautrin D, Baudry E, Mougel F, Loiseau A and Cornuet J-M (2004) A microsatellite-based linkage map of the honey bee, *Apis mellifera L. Genetics* **167**: 253-262.

Stoltz D, Shen XR, Boggis C and Sisson G (1995) Molecular diagnosis of Kashmir bee virus infection. *Journal of Apicultural Research* **34**: 153–160.

Tangjingjai W, Verakalasa P, Sittipraneed S, Klingbunga S and Lekprayoon C (2003) Genetic differences between *Tropilaelaps clareae* and *Tropilaelaps koenigerum* in Thailand based on ITS and RAPD analyses. *Apidologie* **34**: 513-524.

Tentcheva D, Gauthier L, Jouve S, Canabady-Rochelle L, Dainat B, Cousserants F, Colin ME, Ball VB and Bergoin M (2004) Polymerase chain reaction detection of deformed wing virus (DWV) in *Apis mellifera* and *Varroa destructor. Apidologie* **35**: 431–439.

Thai Agricultural Standard (2008) Diagnosis of American Foulbrood in Bee. TAS 10351-2007. National Bureau of Agricultural Commodity and Food Standards, Ministry of Agriculture and Cooperatives, Thailand.

Topley E, Davison S, Leat N and Benjeddou M (2005) Detection of three honey bee viruses simultaneously by a single multiplex reverse transcriptase PCR. *African Journal of Biotechnology* 4(7): 763-767.

Vignal Alain Milan, Denis; Sancristobal, Magali and Eggen, André (2002) A review on SNP and other types of molecular markers and their use in animal genetics. *Genetics, Selection and Evolution* **34**: 275-305

Vos Pieter, Hogers Rene, Bleeker Marjo, Reijans Martin, Lee Theo van de, Hornes

Miranda, Friters Adrie, Pot Jerina, Paleman Johan, Kuiper Martin and Zabeau Marc (1995) AFLP: a new technique for DNA fingerprinting. *Nucleic Acids Research* **23** (21): 4407-4414.

Webster TC, Pomper KW, Hunt G, Thacker EM and Jones SC (2004) *Nosema apis* infection in worker and Queen *Apis mellifera*. *Apidologie* **35**(1): 49-54.

Weinstock GM, Robinson GE, Gibbs RA, Weinstock GM and Weinstock GM (2006) Insights into social insects from the genome of the honeybee *Apis mellifera*. *Nature* **443**: 931-949.

Whitfield CW, Band MR, Bonaldo MF, Kumar CG, Liu L, Pardinas JR, Robertson HM, Soares MB and Robinson GE (2002) Annotated expressed sequence tags and cDNA microarrays for studies of brain and behavior in the honey bee. *Genome Research* **12**(4):555-566.

Williams JGK, Kubelik AR, Livak KJ, Rafalski JA and Tingey SV (1990) DNA polymorphisms amplified by arbitrary primers are useful as genetic markers. *Nucleic Acids Research* **18**: 6531-6535.

Yue C and Genersch E (2005) RT-PCR analysis of deformed wing virus in honey bees (*Apis mellifera*) and mites (*Varroa destructor*). *Journal of General Virology* **86**: 3419-3924.

Zhou T, Anderson DL, Huang ZY, Huang S, Yao J, Ken T and Zhang Q (2004) Identification of varroa mites (Acari: Varroidae) infesting *Apis cerana* and *Apis mellifera* in China. *Apidologie* **35**: 645-654.

Biological and Molecular Approaches in Pest Management
Balwinder Singh, Ramesh Arora & S S Gosal (Eds)
Scientific Publishers (India), Jodhpur, 2015, pp 373-429.

Chapter - 15

Biointensive Integrated Pest Management for Sustainable Agriculture

P S Shera* and Ramesh Arora
Department of Entomology, Punjab Agricultural University, Ludhiana-141004
*E-mail: psshera@pau.edu

15.1. INTRODUCTION

Sustainable agriculture is an integrated system of plant and animal production practices having a site-specific application that will, over the long term satisfy human food and fiber needs; enhance environmental quality and the natural resource base upon which the agricultural economy depends; make the most efficient use of nonrenewable resources and on-farm resources and integrate, where appropriate, natural biological cycles and controls; sustain the economic viability of farm operations; and enhance the quality of life for farmers and society as a whole (FAO, 1990). It primarily focuses on ensuring improved agricultural production by utilizing optimum natural resources with minimal adverse impact on environment and also livelihood of communities.

The losses due to biotic and abiotic stresses are the major constraints in ensuring sustainable production in many countries. The average crop loss resulting from animals (mostly insects), diseases, viruses and weeds at global level has been reported as 32.1 per cent in major crops, i.e. cotton, maize, potatoes, rice, soybean and wheat (Oerke, 2006). The efforts to avoid these crop losses due to insect pests have predominantly relied on the use of chemical insecticides, which have undoubtetly played a pivotal role in increasing the agricultural production worldwide. This approach is under increasing pressure around the world, due to high levels of pesticide resistance in many pest species, pesticide residues, environmental toxicity, fishery losses, ground water and surface water contamination, depletion of rhizosphere microflora, food safety hazards and human health concerns, etc (Singh, 2012).

Therefore, there is an urgent need to curtail and ultimately eliminate the use of chemical pesticides in integrated pest management (IPM) programmes. It is in this context that biointensive integrated pest management (BIPM) is considered the desirable path to sustainability in agriculture. The principles of BIPM are essential for sustainability of agricultural production systems and its implementation is essential for profitable production of safe and wholesome food supply while minimizing hazards of environmental pollution that could threaten entire ecosystems.

Biointensive IPM incorporates ecological and economic factors into agricultural system design and decision making, and addresses public concerns about environment quality and food safety. The benefits of biointensive IPM can include reduced chemical input costs, reduced on-farm and off-farm environmental impacts, and more effective and sustainable pest management. An ecology-based IPM has the potential of decreasing inputs of fuel, machinery, and synthetic chemicals—all of which are energy intensive and increasingly costly in terms of financial and environmental impact. Benbrook (1996) defined Biointensive IPM as "A systems approach to pest management based on an understanding of pest ecology. It begins with steps to accurately diagnose the nature and source of pest problems, and then relies on a range of preventive tactics and biological controls to keep pest populations within acceptable limits. Reduced-risk pesticides are used if other tactics have not been adequately effective, as a last resort, and with care to minimize risks". The primary goal of biointensive IPM is to provide guidelines and options for the effective management of pests in an ecological context. The flexibility and environmental compatibility of a biointensive IPM strategy make it useful in all types of cropping systems (Dufour, 2001).

15.2. STRATEGIES IN BIOINTENSIVE IPM

The theory and principles supporting IPM have been developed over the last six decades or more. However, there is a need to shift the IPM paradigm from focusing on pesticide management to a system approach relying primarily on biological knowledge of pests and on their interaction with the crops and other components of the cropping system. An important difference between conventional and biointensive IPM is that the emphasis of the latter is on proactive measures to redesign the agricultural ecosystem to the disadvantage of a pest and to the advantage of its natural enemy complex. At the same time, biointensive IPM shares many of the same components as conventional IPM, including monitoring, use of economic thresholds, record keeping, and planning. IPM options may be considered proactive or reactive. Proactive options, such as cultural control, host plant resistance, and creation of habitat for beneficial organisms, permanently lower the carrying capacity of the farm for the pest. The second set of options is more reactive. This simply means that the grower responds to a situation, such as an economically damaging population of pests, with some type of short term suppressive action. Reactive methods generally include inundative releases of biological controls, mechanical and physical controls, and reduced risk pesticides (Dufour, 2001). This suggests that bio-intensive IPM programmes should represent a sustainable approach to manage pests utilizing cultural control, host plant resistance, biocontrol (predators, parasitoids or pathogens), biorationals, biopesticides and molecular approaches to ensure favourable economic, ecological and sociological consequences.

15.2.1. Physical, Cultural and Mechanical Control

Physical control methods involve the modification of the physical environment of insects in such a way that they no longer pose a threat to the agricultural crop. This can be achieved by generating stress levels ranging from agitation to death or by using devices such as physical barriers that protect produce or plants from infestation. Many physical control methods target an ensemble of physiological and behavioral processes, whereas chemical methods have well-defined and limited modes of action. Physical controls can be classified as passive (trenches, fences, organic mulch, particle films, inert dusts, and oils), active (mechanical, polishing, pneumatic, impact, and thermal), and miscellaneous (cold storage, heated air, flaming, and hot-water immersion). The passive methods do not require additional input after establishment to be effective over a given period (Vincent *et al.*, 2003). The efficacy of active methods depends on continued input over the period of control. The level of control achieved is related to the amount and intensity of the input. Effective physical control methods protect plants during the entire season from emergence to postharvest. However, postharvest conditions are better suited to physical control methods because the environment is rather confined, the material is of high economic value, and the use of insecticides is usually inappropriate or even unlawful. Physical control methods, such as cold, heat, and ionizing radiation, are used extensively as post-harvest quarantine treatments where disinfestation of a given pest at a predetermined level of control must be achieved (Hallman, 2001).

Cultural practices refer to that broad set of management techniques or options which may be manipulated by agricultural producers to achieve their crop production goals, or the manipulation of the environment to improve crop production. Cultural control on the other hand, is the deliberate alteration of the production system, either the cropping system itself or specific crop production practices, to reduce pest populations or avoid pest injury to crops. The functional mechanisms include: impediments to pest colonization of the crop; the creation of adverse biotic conditions that reduce survival of individuals or populations of the pest; modifications of the crop in such a way that pest infestation results in reduced injury to the crop and enhancement of natural enemies by manipulating the environment (Ferror, 2013).

Cultural practices are among the oldest techniques used for pest suppression, and many of the preventative practices used in conventional and organic farming today have their roots in traditional agriculture. It represents an important component in the development of a coherent, holistic approach to pest management and warrants a great deal more attention than it is currently receiving. Many of the methods interfere with the pests' ability to colonize a crop, promoting dispersal, reducing reproduction or survival. With the awareness of environmental problems, exploitation of different cultural and mechanical practices, viz. planting time, plant geometry, tillage, crop sanitation, fertilizer as well as water management, intercropping, crop rotation, manipulation of carryover sources, barrier/trap crops, etc. have been advocated as a vital approach to curb pest population, as these are simple and easy to follow. There has been a tendency for a greater emphasis to be placed on limited evaluations of treatments rather than on experiments to evaluate processes and as a result, cultural control is often considered to include a mismatch of techniques that can have no cohesive relations. However, over recent years more has been done to develop an integrated philosophy based on a greater knowledge of the host, crop/pest biology and their underlying ecological principles, providing a framework for a more integrated approach in the future (Dent *et al.*, 2000).

15.2.2. Host Plant Resistance

The production of crop plants with heritable arthropod resistant traits has been recognized for more than a century years as a sound approach to crop protection (Smith, 2004). The earliest documentation on host plant resistance dates back to the earlier days of applied entomology and plant pathology (Painter, 1951). An outstanding early success in utilizing host plant resistance in pest management was the control of the grape phylloxera *Daktulasphaira vitifoliae* (Fitch) in France. Balachowsky (1951) described the "brutal appearance" of this North American pest species in France in 1861 and its spread to the vineyards of European and Mediterranean countries. The pest was successfully controlled by 1890 when the distinguished 19th century Entomologist Charles Valentine Riley imported North American rootstocks to France and French vineyards were reconstituted using grafts of the susceptible European grapevine scions of *Vitis vinifera* L. on resistant North American rootstocks particularly of *V. labrusca* L. In India, the early work of Hussain and Lal (1940) led to evolving hairy cotton varieties resistant to jassid and by 1943 resistant varieties such as Punjab 4F, LSS and 289 F/43 covered extensive areas where jassid had posed a serious threat. Classical studies over the past five decades have elucidated genetic basis of resistance in major crop plants.

The breeding of arthropod-resistant plants was formalized after the rediscovery of Mendel's law of heredity in 1900 and blossomed as a field of research in the first half of the twentieth century with the work of Painter (1951). Over the past 60 years, breeding crops for pest resistance has accelerated through the efforts of public and private sector plant breeders and entomologists in several countries, including agricultural research centers. These efforts led to spectacular successes in developing arthropod-resistant crops during the green revolution in Southeast Asia during the 1960s. Hundreds of insect resistant crop varieties have been developed worldwide and are grown extensively for increasing and stabilizing the crop productivity (Dhaliwal and Singh, 2004).

The annual value of arthropod rsistance genes currently deployed in global agriculture is greater than US$2 billion (Smith and Clement, 2012). A classic example is the development of the rice cultivar IR36, which is resistant to multiple insect pests. This cultivar, developed at the International Rice Research Institute in the 1970s and cropped on 11 million ha by the early 1980s, provided an annual income increase of approximately $1 billion to Asian rice farmers. The economic value of genetic resistance in wheat to all major worldwide arthropod pests amounts to just over $250 million per year. Wheat cultivars resistant to Hessian fly provided a 120-fold greater return on investment than pesticides. Insect resistant cultivars of alfalfa, corn and beet returned $320 for each $ invested in research (Smith, 2005).

Some of the major limitations in adopting host plant resistance for pest management are lack of high level of resistance against target pests in the germplasm, long time required to develop resistant crop varieties and prevalence of a pest complex at any given time against which multiple pest resistance may be lacking. In addition, breakdown of resistance due to development of virulent biotypes and secondary pest outbreaks have also restricted the wide scale adoption of resistant varieties (Bentur, 2004). With the advent and use of molecular tools over the past 30 years, the field of plant resistance to arthropods has been transformed into a new era, offering enormous opportunities for continued development of new crop cultivars with genes for durable arthropod resistance (Smith and Clement, 2012). Some of the recent developments in host plant resistance are discussed herein.

15.2.2.1. Marker Assisted Selection. In traditional plant breeding, phenotypic evaluation determines the initial identification of a source of resistance or progeny from crosses made between resistant and susceptible parents. With marker-assisted/aided selection (MAS) individual plants can be selected based on their genotype. They can be more effectively selected, which may enable certain traits to be 'fast-tracked' resulting in quicker line development and variety release. New agricultural research involving genomics and MAS presents the potential to improve the efficiency of plant breeding by allowing for the transfer of specific genomic regions of interest and accelerating the recovery of the elite parent background. Genomic sequencing in major crops has made it possible to identify and map precisely a number of genes through linkage to DNA markers. MAS can be used for monitoring the presence or absence of these genes in breeding populations and can be combined with conventional breeding approaches. Marker-assisted backcross breeding has been used to effectively integrate major genes or quantitative trait loci with large effect into widely grown varieties. Pyramiding different resistance genes using MAS provides opportunities to breeders to develop broad-spectrum resistance for diseases and insects from the fine mapped position of the genes (Smith, 2004).

Molecular markers have been identified that are linked both to single major genes for resistance and to groups of loci controlling the expression of quantitative resistance, known as quantitative trait loci (QTLs) (Knapp, 1994). More than 100 monogenic and polygenic arthropod plant resistance gene loci and their linked markers have been characterized by molecular mapping, and several are in use to track resistance via MAS in breeding lines for deployment in cultivars. Close linkages between insect resistant genes and molecular markers is useful in transferring genes following marker aided selection. More than 30 major genes conferring resistance against 11 insect species of 6 crops and fruit plants have been tagged and mapped (Yencho *et al.*, 2000). Seven of the gall midge resistance genes have been tagged and mapped. Gm1 has been tagged and mapped on rice chromosome 9 and flanking markers

RM219 and RM 444 are able to detect the gene in wide range of germplasm (Biradar *et al.*, 2004). Using wheat microsatelite primers, Russian aphid resistance genes Dn1, Dn2, Dn5 and Dnx have been mapped to wheat chromosome 7 DS (Liu *et al.*, 2002). Mapping and marker assisted selection of brown planthopper (BPH) resistance gene *bph2* in rice would be very beneficial for BPH resistance improvement. AFPL analysis of *bph2* has been mapped on the long arm of chromosome 12 within 1.0cM region (Sun *et al.*, 2006).

There are a few comparisons of the efficiency of MAS of resistant plants with phenotypic selection. MAS of genes for Russian wheat aphid, *Diuraphis noxia* Mordvilko for resistance in wheat and cereal cyst nematode, *Heterodera avenae* Woll for resistance in barley can be accomplished approximately 30 times faster and approximately 75 per cent more cheaply per evaluation compared to plant phenotypes. Over the past decade, QTL mapping has been used to characterize antixenic/ antibiotic resistance, and, to a lesser extent, plant tolerance, thereby increasing the possibility to develop arthropod-resistant germplasm with which to examine the individual effects of specific QTL. Such knowledge about specific QTL provides the potential to broaden the genetic bases of arthropod plant defense and to develop more durable resistance (Smith and Clement, 2012). Molecular genetic approaches have successfully elucidated QTL linked with the majority of the DIMBOA [2,4-dihydroxy-7-methoxy-2H-1,4-benzoxazin-3 (4H)-one] metabolic pathway *bx* genes (Butron *et al.*, 2010), and QTL mapping has identified the *p1* locus as a key to maysin biosynthesis (Zhang *et al.*, 2003). In the past, it has been difficult to pyramid multiple resistance genes because they generally show the same phenotype, necessitating a progeny test to determine which plants possess more than one gene. With linked DNA markers, the number of resistance genes in any plant can be easily determined. The incorporation of quantitative resistance controlled by QTLs offers another promising strategy to develop durable resistance (Sandhu and Arora, 2013).

15.2.2.2. Induced Resistance. Numerous agricultural and wild plants demonstrate induced responses to insect damage. Induced responses represent any change in host suitability for the pest that occurs after herbivore attack. Many studies have documented negative effects of induced responses on herbivore preference or performance (Karban and Baldwin, 1997). Elicitors are natural or synthetic compounds that initiate plant induced responses to herbivory when applied to foliage or roots. Plant reactions to both insect and disease attack may include hypersensitive cell death, as in the case of *Meu1.2* activation of DR genes and the redirection of normal cell-maintenance genes to plant defense. In DR-gene activation, plants produce elicitors that activate plant gene expression and the synthesis of volatile and non-volatile allelochemicals (Smith, 2004).

Jasmonic acid is found in many plant species and is involved in regulating diverse plant functions, including plant resistance and senescence. It is produced by the plant after caterpillar damage and results in increased production of compounds involved in resistance. In tomatoes, application of jasmonic acid results in induction of proteinase inhibitors and polyphenol oxidase and in a decrease in the preference, performance, and abundance of many common herbivores in the field, including *Frankliniella occidentalis* Pergrande, *Spodoptera exigua* (Hübner) and *Trichoplusia ni* Hübner, *Epitrix hirtipennis* Melsheimer, and *Macrosiphum euphorbiae* Thomas, and *Myzus persicae* Sulzer (Thaler *et al.*, 1996). Pathways involving jasmonates (JA) and methyl jasmonate (MeJA), ethylene (ET) and salicylic acid (SA) induce plant defenses after insect attack, and some types of induced resistance are elicited by unknown and as yet unexplained types of elicitors. Methyl JA-treated wheat plants produce increased amounts of the defensive compound hydroxamic acid and sustain reduced phloem ingestion by the aphid, *Rhopalosiphum padi* (L.), compared with control plants (Slesak *et al.*, 2001). It has been established that jasmonate and its pure derivative, jasmonic acid can induce the emission of volatiles similar to those induced by herbivory which fed on the plant. Nathan *et al.* (2009) indicate that JA application induces systemic defenses in rice that have a direct negative impact on *Nilaparvata lugens* Stal survivorship. At high concentration (5 mM) several abnormalities possibly related to defective moulting, were observed along with malformed eggs.

15.2.3. Biological Control Agents

Biological control or biocontrol refers to the use of living organisms to suppress the population density or impact of specific pest organism, making it less abundant or less damaging than it would otherwise be. Biocontrol is present in all ecosystems, both natural and human made and is always active. The result of natural biological control is that the earth is green and the plants can produce sufficient biomass to sustain other forms of life. Natural (biological) control is the reduction of pest organisms that occurs "for free" since the evolution of the first ecosystem some 500 million years ago. Natural control is constantly active in all world terrestrial ecosystems on 89.5 million kms^2. Most of the potential arthropod pests (95%, nearly 10000 arthropod species) are under natural control; all other control methods used today are targeted at the remaining 5000 arthropod pest species. This ecosystem function of natural biological control is estimated to have an annual minimum value of US$400 billion per year, which is an enormous amount compared to the only 8.5 billion US$ annually spent on insecticides (Van Lenteren, 2008). This is why biointensive IPM seeks to strengthen the role of biological control agents for management of insect pests.

The three basic strategies for the exploitation of biocontrol agents in insect pest management include importation or classical biological control, augmentation and conservation. Classical biological control involves intentional introduction of exotic usually co-evolved biocontrol agents to control a pest in an area where it has been accidentally or speciously introduced. Conservation biological control refers to modification of the environment or existing practices to protect and enhance natural enemies or other organisms to reduce the effects of pests. Augmentation involves two approaches, viz. inoculative and inundative releases, designed to increase the population of biocontrol agents by mass multiplication and periodic releases. These approaches have different applications depending on the adaptability of natural enemy and environment in which they must operate (Eilenberg *et al.*, 2001). Classical biological control is applied on 350 million hectares, which is about 8 per cent of land under culture, and has very high benefit-cost ratios of 20-500: 1. More than 5,000 introductions of about 2,000 species of exotic arthropod agents for control of arthropod pests in 196 countries or islands have been made during the past 120 years. However, augmentative, commercial biological control is utilized only on 16 million hectares, which is 0.4 per cent of land under culture, and has a benefit-cost ratio of 2-5: 1, which is similar to or better than chemical pest control (Van Lenteren, 2008). More than 150 species of natural enemies (parasitoids, predators and pathogens) are currrently commercially available (Van Lenteren *et al.*, 2006).

A diverse range of natural enemies including vertebrates (birds, reptiles, amphibians, fishes and small mammals), nematodes, microorganisms and arthropods serve as agents of biocontrol. These natural enemies may function as predators, parasitoids or entomopathogens.

15.2.3.1. Predators. Predators are the species that kill and eat other living organisms for development, sustenance and reproduction. Insects are eaten by other insects, spiders, fishes, birds and small mammals. Intelligent manipulation of predator complexes for biological control in cropping systems requires knowledge of predator taxonomy and biology, specificity and rates of predation (Van Driesche *et al.*, 2008). Birds and arthropods are the most important predators of insects in the agroecosystem and are discussed here briefly.

(A) Avian Predators: There are approximately 10,000 identified species of birds and more than 50 per cent of these are predominantly insectivorous and another 24 per cent consume invertebrates at least occasionally (Sekercioglu, 2006). Natural pest control services provided by birds and other vertebrate predators are poorly understood, under appreciated and poorly utilized. But increasing evidence from a range of ecosystems suggests that insectivorous birds are important in controlling the populations, behaviour and evolution of their invertebrate prey (Van Bael *et al.*, 2008; Mooney *et al.*, 2010; Triplett *et al.*, 2012). Whelan *et al.* (2008) conducted a meta analysis of a large number of studies on top-down effects of birds on herbivorous insects. In agricultural ecosystems, all the studies showed a decline in pest numbers, and plant damage (wherever recorded) was also reduced. The key aspect in need of further study is the extent to which trophic cascades involving avian predators have measurable economic benefits in terms of increased plant growth or agricultural production. Perhaps the most exciting example comes from Dutch apple orchards. Mols and Visser (2002) reported that emplacement of nesting boxes to attract great tits (*Parus major*) reduced caterpillars and fruit damage and increased fruit yield by an astonishing 66 per cent from 4.7 to 7.8 kg apples per tree. Triplett *et al.* (2012) have reviewed the studies reporting economic benefits from control of insect pests by avian predators in agricultural landscapes (Table 1). Providing resources for insectivorous birds is an important management option in biointensive IPM. The feasibility and effectiveness of habitat manipulations that boost either the population of avian predators (e.g. deploying nest boxes) or their effectiveness as biological control agents (e.g. providing perches for foraging) need to be thoroughly evaluated (Wenny *et al.*, 2011), so that these can be gainfully exploited in IPM programmes.

Table 1. Selected examples of economic benefits from control of insect pests by avian predators in agricultural landscapes

Crop/stock	Avian species	Cost/benefit estimate
(1)	(2)	(3)
Corn	Red-winged blackbird (*Agelaius phoeniceus*)	Correction between red-winged blackbird density and predation on northern corn rootworm (*Diabrotica longicornis*) populations
Coffee	Various	Abundance of insects increased with bird exclusion (Particularly flatid planthopper (*Petrusa epilepsies*); less so coffee leaf miner (*Leucoptera coffeella*)
Field bean	House crow (*Corvus splendens*), House sparrow (*Passer domesticus*), Golden oriole (*Oriolus oriolus*), Black drongo (*Dicrurus adsimilis*), common mynah (*Acridotheres tristis*) plus other insectivorous birds	Seed yield was positively correlated with bird species richness. Birds assumed to control pod borer of field bean
Coffee	Various	Birds reduced large arthropods by 64-80%. Increase in damage to plants when birds were excluded
Coffee	Various	Bird control of coffee berry borer (*H. hampei*) valued at US$310/ha
Coffee	Various	Birds reduced coffee berry borer infestations, increasing quantity of saleable fruits
Apple	Great tit (*Parus major*)	Birds reduced caterpillar abundance and increased fruit yield from 4.7 to 7.8 kg of apples per tree
Coffee	Various	Birds reduced density of simulated lepidopteran larvae in crop with high floristic diversity of surrounding forest
Apple	Pine siskin (*Carduelis pinus*)	Density of winter moth (*Operophtera brumata*) larvae reduced where birds had access to the crop
Apple	Blue tit (*Parus caeruleus*), great tit, marsh tit (*Peoecile palustris*), willow tit (*Poecile montanus*)	Where birds had access, codling moth (*Cydia pomonella*) larvae were reduced by 94.7, 94.9 and 95.6% in separate trials

contd....

(1)	(2)	(3)
Corn	Various	Density of cutworms (*Agrotis* spp.) and weevils (*Sphenophorus* spp.) were reduced where birds had access to crop. Results for aphids and European corn borer (*Ostrinia nubilalis*) not significant, but tended toward a reduction where birds had access to crop
Oil palm	Various	The exclusion of insectivorous birds from oil palm seedlings resulted in an increase in herbivory rate of 0.047% per day. This amounts to 28% foliage damage over the effective life of an oil palm leaf.
Coffee	Buff-throated Foliage Gleaner (*Automolus ochrolaemus*) Rufous-breated Wren (*Pheugopedius rutilus*) Rufous-capped Warbler (*Basileuterus rufifrons*) White-tailed Emerald (*Elvira chionura*) Yellow Warbler (*Setophaga petechia*)	The natural control of coffee pests by birds prevented US$75-US$310 ha/year damage in Costa Rica

Source: Modified after Triplett *et al.* (2012), Karp *et al.* (2013)

B. Arthropod Predators: Predaceous insects of potential use in biocontrol are found in Dermaptera, Mantodea, Hemiptera, Thysanoptera, Coleoptera, Hymenoptera and Diptera with Hemiptera, Coleoptera, Hymenoptera and Diptera being most important (Table 2). More than 30 families of insects are predaceous and of these, the Anthocoridae, Nabidae, Reduviidae, Geocoridae, Carabidae, Coccinellidae, Nitidulidae, Staphylinidae, Chrysopidae, Formicidae, Cecidomyiidae and Syrphidae are commonly important in crops. Important insect predators include lady beetles, ground beetles, rove beetles, flower bugs and other predatory true bugs, lacewings, and hover flies. Spiders and some families of mites are also predators of insects and mite species, and other arthropods.

The coccinellid beetle *Cryptolaemus montrouzieri* Mulsant has provided spectacular control of heavy infestations of sucking pests, especially mealybugs (Mani 1990; Mani and Krishnamoorthy 2008) and some soft scales (Kumar and Prakasam 1984; Mani and Krishnamoorthy 1990). The predator has the potential to be exploited as an efficient biological control agent against citrus mealybug, *Planococcus citri* Risso (Singh 1978; Afifi *et al.*, 2010), pink mealybug, *Maconellicoccus hirsutus* (Green) (Reddy and Narayan 1986; Mani and Thontadarya 1988; Gautam *et al.*, 1998; Kairo *et al.*, 2000; Moses *et al.*, 2002) and cotton mealybug, *Phenacoccus solenopsis* Tinsley (Kaur and Virk, 2011; Rashid *et al.*, 2012).

Approximately 90 worldwide species of the tribe Stethorini Dobzhansky (genera

Stethorus Weise and *Parasthethorus* Pang and Mao) are the only specialist mite predators in the Coccinellidae. Various species of Stethorini have received considerable attention over the last five decades because of their potential as biological control agents of spider mites. At least 12 species of Stethorini have been imported into the US for this purpose (Gordon, 1985) and many more have been purposefully redistributed throughout the world. Chazeau (1985) reported that 40 per cent of the 68 species attacked spider mites of economic importance. Based on 40-year case history, *Stethorus punctum* (LeConte) has been found to be key component of biological mite control in Pennsylvania apple orchards (Biddinger *et al.*, 2009).

Genetic improvement of arthropod predators has been successfully carried out to enhance their efficacy in pest management (Arora and Shera, 2014). A strain of *Chrysoperla zastrowi arabica* Henry *et al.* (PTS-8) having tolerance to endosulfan, acephate and fenvalerate has been developed by National Bureau of Agricultural Insect Resources (NBAII), Bangalore. PTS-8 acquires higher activity of detoxifying enzymes (esterase and glutathione-S-transferase activity) as against susceptible population. It has been found to be effective against sucking pests of cotton under pesticide sprayed conditions (Venkatesan *et al.*, 2011).

15.2.3.2. Parasitoids. Insects that parasitize other insects are known as parasitoids. Six orders of insects, including 86 families have been reported to contain parasitoids. These orders include the Coleoptera, Hymenoptera, Lepidoptera, Neuroptera and Strepsiptera. However, the most important of these by far are the Hymenoptera and Diptera in that order (Table 2). The most significant groups of parasitoids in the Hymenoptera include small parasitic wasps in the Ichneumonoidea (Ichneumonidae, Braconidae) and Chalcidoidea (Encyrtidae, Aphelinidae and Trichogrammatidae being the most important).

Table 2. List of commercially produced parasitoids and predators

Class /order/insect (1)	Family (2)	Main target pest(s) (3)
INSECTA		
Coleoptera		
Adalia bipunctata	Coccinellidae	Aphididae
Aleochara bilineata	Staphylinidae	*Delia antiqua, Delia radicum (= Delia brassicae)*
Atheta coriaria	Staphylinidae	Thripidae (thrips), Sciaridae (fungus gnats), Ephydridae (shore flies)
Chilocorus baileyii	Coccinellidae	Diaspididae
Chilocorus bipustulatus	Coccinellidae	Diaspididae, Coccidae (*Saissetia oleae*)
Chilocorus nigrita	Coccinellidae	Diaspididae, Asterolecaniidae
Coccinella septempunctata	Coccinellidae	Aphididae
Cryptolaemus montrouzieri	Coccinellidae	*Planococcus citri*

Contd....

(1)	(2)	(3)
Delphastus catalinae	Coccinellidae	Aleyrodidae (*Trialeurodes vaporariorum, Bemisia tabaci*)
Rhyzobius lophanthae	Coccinellidae	Diaspididae (*Pseudolacaspis pentagona*), *Quadraspidiotus perniciosus, Chrysomphalus dictyospermi, Parlatoria blanchardi*
Rodolia cardinalis	Coccinellidae	*Icerya purchasi*
Scymnus rubromaculatus	Coccinellidae	Aphididae
Stethorus punctillum	Coccinellidae	*Panonychus ulmi*
Diptera		
Aphidoletes aphidimyza	Cecidomyiidae	Aphididae (*Aphis gossypii, Myzus persicae, Macrosiphum* sp., *Aulacorthum* sp.)
Episyrphus balteatus	Syrphidae	Aphididae
Feltiella acarisuga	Cecidomyiidae	*Tetranychus urticae, T. cinnabarinus*
Hemiptera		
Anthocoris nemoralis	Anthocoridae	Psyllidae (orchards)
Anthocoris nemorum	Anthocoridae	*Cacopsylla pyri*, thrips
Macrolophus pygmaeus	Miridae	Aleyrodidae
Orius albidipennis	Anthocoridae	Thrips
Orius laevigatus	Anthocoridae	Thripidae (*Frankliniella occidentalis, Thrips tabaci*)
Orius majusculus	Anthocoridae	Thripidae (*Frankliniella occidentalis, Thrips tabaci*)
Picromerus bidens	Pentatomidae	Lepidoptera
Podisus maculiventris	Pentatomidae	Lepidoptera, *Leptinotarsa decemlineata*
Hymenoptera		
Anagrus atomus	Mymaridae	Cicadellidae
Anagyrus fusciventris	Encyrtidae	Pseudococcidae
Anagyrus pseudococci	Encyrtidae	Pseudococcidae
Aphelinus abdominalis	Aphelinidae	Aphididae (*Macrosiphum euphorbiae, Aulacorthum solani*)
Aphidius colemani	Braconidae	Aphididae (*Aphis gossypii, Myzus persicae, M. nicotianae*)
Aphidius ervi	Braconidae	*Aulacorthum solani, Macrosiphum euphorbiae*
Aphidius matricariae	Braconidae	*Myzus persicae*

Contd...

(1)	(2)	(3)
Aphytis diaspidis	Aphelinidae	Diaspididae, *Quadraspidiotus perniciosus, Pseudaulacaspis pentagona*
Aphytis holoxanthus	Aphelinidae	Diaspididae
Aphytis lingnanensis	Aphelinidae	*Aonidiella aurantii, Chrysomphalus dictyospermi*
Aphytis melinus	Aphelinidae	*Aonidiella aurantii*
Aprostocetus hagenowii	Eulophidae	Blattidae (*Periplaneta* spp.)
Bracon hebetor	Braconidae	Lepidoptera (on stored products)
Coccophagus lycimnia	Aphelinidae	Coccidae
Coccophagus rusti	Aphelinidae	Coccidae
Coccophagus scutellaris	Aphelinidae	Coccidae
Comperiella bifasciata	Encyrtidae	Diaspididae (*Chrysomphalus aonidum, Aonidiella aurantii*)
Cotesia marginiventris	Braconidae	Lepidoptera (Noctuidae)
Dacnusa sibirica	Braconidae	Agromyzidae (*Liriomyza* spp.)
Diglyphus isaea	Eulophidae	Agromyzidae (*Liriomyza* spp.)
Encarsia citrina	Aphelinidae	Diaspididae
Encarsia formosa	Aphelinidae	Aleyrodidae (*Trialeurodes vaporariorum, Bemisia tabaci*)
Encyrtus aurantii	Encyrtidae	Coccidae
Encyrtus infelix	Encyrtidae	Coccidae
Eretmocerus eremicus	Aphelinidae	*Bemisia tabaci*
Eretmocerus mundus	Aphelinidae	*Bemisia tabaci*
Gyranusoidea litura	Encyrtidae	*Pseudococcus longispinus*
Leptomastidea abnormis	Encyrtidae	Pseudococcidae
Leptomastix dactylopii	Encyrtidae	*Planococcus citri*
Leptomastix epona	Encyrtidae	Pseudococcidae, especially *Pseudococcus viburni*
Metaphycus flavus	Encyrtidae	Coccidae, *Saissetia oleae, Coccus hesperidum*
Metaphycus helvolus	Encyrtidae	Coccidae (*Saissetia oleae Coccus hesperidum*)
Metaphycus lounsburyi	Encyrtidae	Coccidae (*Saissetia oleae*)
Metaphycus swirskii	Encyrtidae	Coccidae
Microterys nietneri	Encyrtidae	Coccidae (*Saissetia oleae*)
Nasonia vitripennis	Pteromalidae	Stable flies, especially Musca domestica
Opius pallipes	Braconidae	*Liriomyza bryoniae*

Contd...

(1)	(2)	(3)
Praon volucre	Braconidae	Aphididae
Pseudaphycus maculipennis	Encyrtidae	Pseudococcidae
Scutellista caerulea	Pteromalidae	Coccidae (*Saissetia oleae, S. coffeae, Ceroplastes rusci*)
Tetracnemoidea peregrina	Encyrtidae	Pseudococcidae
Tetracnemoidea brevicornis	Encyrtidae	Pseudococcidae
Thripobius javae	Eulophidae	Thysanoptera (*Heliothrips* spp.)
Trichogramma chilonis	Trichogrammatidae	Lepidoptera
Trichogramma japonicum	Trichogrammatidae	Lepidoptera
Trichogramma brassicae	Trichogrammatidae	Lepidoptera (*Ostrinia nubilalis*)
Trichogramma cacoeciae	Trichogrammatidae	Lepidoptera
Trichogramma dendrolimi	Trichogrammatidae	Lepidoptera
Trichogramma evanescens	Trichogrammatidae	Lepidoptera (including on stored products)
Trichogramma pintoi	Trichogrammatidae	Mainly Lepidoptera species: *Ostrinia nubilalis, Helicoverpa armigera, Cydia nigricana, Cydia pomonella, Cydia funebrana, Plutella xylostella, Mamestra brassicae, Lacanobia oleracea*
Neuroptera		
Chrysoperla carnea	Chrysopidae	Aphididae etc.
Thysanoptera		
Franklinothrips megalops	Aeolothripidae	Thysanoptera
Franklinothrips vespiformis	Aeolothripidae	Thysanoptera
Karnyothrips melaleucus	Phlaeothripidae	Coccidae, Diaspididae (*Howardia biclavis*)
ARACHNIDA		
Acarina		
Amblyseius barkeri	Phytoseiidae	Thysanoptera (*Thrips tabaci, Frankliniella occidentalis*), Tarsonemidae
Amblyseius degenerans	Phytoseiidae	Thysanoptera
Amblyseius swirskii	Phytoseiidae	*Bemisia tabaci, Trialeurodes vaporariorum, Frankliniella occidentalis*
Cheyletus eruditus	Cheyletidae	Storage mites, spider (foliage) mites
Hypoaspis aculeifer	Laelapidae	Sciaridae, *Rhizoglyphus echinopus*
Metaseiulus occidentalis	Phytoseiidae	Tetranychidae
Neoseiulus californicus	Phytoseiidae	Tetranychidae
Neoseiulus cucumeris	Phytoseiidae	Thysanoptera (*T. tabaci, F. occidentalis*)
Phytoseiulus persimilis	Phytoseiidae	Tetranychidae (*Tetranychus urticae*)

Contd...

(1)	(2)	(3)
Stratiolaelaps miles	Laelapidae	Sciaridae, *Rhizoglyphus echinopus*
Typhlodromus pyri	Phytoseiidae	*Panonychus ulmi, Tetranychus urticae, Eriophyes vitis, Epitrimerus vitis*

Source: Van Lenteren (2012)

The subfamilies of Ichneumonidae have been grouped on the basis of type of hosts, *viz.* endoparasitoids of lepidopteran larvae (subfamilies Banchinae, Ophioninae, Porizontinae); endoparasitoids of lepidopteran pupae (Ichneumoninae); ectoparasitoids of larvae or pupae of diverse orders in plant tissue (subfamily Pimplinae) as well as of exposed larvae of Lepidoptera and sawflies (subfamily Typhoninae). Parasitic wasps belonging to family Braconidae have been extensively utilized as biological control agents as endoparasitoids of aphids (subfamily Aphidiinae); endoparasitoids of larvae of Lepidopterans, and Coleopterans (subfamilies Meteorinae, Blacinae, Microgasterinae, Rogadinae); egg-larval and larval endoparasitoids of Lepidoptera (subfamily Cheloninae); egg-larval and larval endoparasitoids of Cyclorrhaphous Dipterans (subfamilies Alysiinae and Opiinae); and endoparasitoids of Lepidopteran and Coleopteran larvae in concealed places (subfamilies Braconidae and Doryctinae) (Van Driesche *et al.*, 2008).

The important genera in family Aphelinidae include *Aphelinus*, *Aphytis*, *Encarsia* and *Eretmocerus*. Members of this family are important parasitoids of scales, mealybugs, aphids, psyllids and whiteflies. Among these, *Encarsia Formosa* Gahan and *Eretmocerus eremicus* Rose & Zolnerowich have been mass reared and extensively used against whiteflies in greenhouse crops. The important genera in the the family Encyrtidae include Ap*oanagyrus*, *Anagyrus*, *Comperiella*, *Ooencyrtus* and *Hunterellus*. They parasitize mealybugs, scales and eggs or larvae of various Blattaria, Coleoptera, Diptera, Lepidoptera, Hymenoptera, Neuroptera and Orthoptera.

The trichogrammatids which parasitize eggs of lepidopteran pests are the most widely utilized biocontrol agents. About ten *Trichogramma* spp. are mass reared extensively for augmentative release against Lepidopteran pests in corn, sugarcane, cotton, vegetables and other crops. The superior strains of *Trichogramma chilonis* Ishii, an important egg parasitoid of lepidopterans have been developed by artificial selection for adaptation to high as well as low temperature regime at NBAIR, Bangalore. The strain adapted to high temperature regime can be utilized in temperatures above 35^{0}C. High temperature tolerant strain of *T. chilonis* is useful against various insect pests of sugarcane, cotton and vegetable crops during hot months (Singh, 2003). This strain has also proved superior in reducing the incidence of sugarcane stalk borer, *Chilo auricillius* (Dudegeon) and early shoot borer, *Chilo infuscatellus* (Snellen) as against local strain under Punjab conditions (Singh *et al.*, 2007).

An endosulfan-resistant strain of *T. chilonis* termed as "endogram" has been developed by sequentially exposing adult parasitoids to various concentrations of endosulfan (0.004 to 0.09%) for 341 generations. The tolerant strain was found to parasitize 56 per cent of *Helicoverpa armigera* (Hubner) eggs immediately after insecticide spray as compared to 3 per cent by susceptible strain (Jalali *et al.*, 2006) under net house on potted cotton plants. Endogram strain has been further improved for multiple resistance to monocrotophos and fenvalerate to obtain multiple insecticide resistance strain through selection process for over 72 generations. Field efficacy of this resistant strain has shown higher parasitism in *H. armigera* eggs on tomato by 260.4 per cent as compared to susceptible strain under sprayed conditions (Jalali and Venkatesan, 2011). A strain of *T. chilonis* has been developed for tolerance to multiple traits known as MITT (multiple-insecticide and temperature tolerant) strain for effective control of the pest even in harsh climatic conditions and under high insecticide pressure in different economically important crops. This strain has showed tolerance to three major groups of insecticides which include endosulfan (organochlorine), monocrotophos (organophosphate) and fenvalerate (synthetic pyrethroid) and to high temperature (32–38°C) through selection for 81 generations. Increase in parasitization from 35 to 90–95 per cent and reduction in mortality from 100 to 57–70 per cent after 6 hours of constant exposure to three insecticides and high temperature has been reported (Kumar *et al.*, 2008). In Punjab, 11 to 13 releases of this multiple insecticide resistant strain of *T. chilonis* @ 1,50,000 per ha at weekly interval during July to October alongwith 6 to 9 insecticide applications have been reported to be effective for the control of cotton bollworms (Brar *et al.*, 2007).

An aphelinid endoparasitoid, *Encarsia flavoscutellum* Zehntner brought from Assam has been well established in sugarcane growing areas of Sankeshwar and other neighboring districts of Northern Karnataka and is considered as a case study in biological control of the sugarcane woolly aphid (SWA), *Ceratovacuna laniger* Zehntner (Mallapur *et al.*, 2008). NBAIR has launched a nationwide programme on classical biological control of the papaya mealybug, *Paracoccus marginatus* Williams & Granara de Willink using natural enemies. Three species of exotic parasitoids namely *Acerophagous papayae*, *Anagyrus loecki* and *Pseudleptomastix mexicana* specific to papaya mealybug imported with the help of USDA-APHIS have been mass produced and distributed to Krishi Vigyana Kendras, CIPM centres, entomologists of SAU's of Tamil Nadu, Kerala, Karnataka, Andhra Pradesh, Maharashtra and Tripura as well as horticultural universities of Karnataka and Andhra Pradesh. Intensive training has been imparted to 90 entomologists/ subject matter specialists on the mass production, field release and conservation of the parasitoids. The parasitoid *A. papayae* has now multiplied in enormous numbers and established well in the farmers fields in Tamil Nadu, Karnataka and Maharashtra. The parasitoids collected from these fields are now being distributed to fields of other farmers. Wherever the parasitoids are now seen, the papaya mealybug attack is reduced significantly. It is anticipated that within the next six to 12 months, the pest problem will be solved in a sustainable and eco-friendly manner on a long term (NBAIR, Bangalore).

In the Diptera, members of thirteen families, i.e. Acroceridae, Bombyliidae, Cecidomyiidae, Cryptochetidae, Conopidae, Calliphoridae, Nemestrinidae, Phoridae, Pipunculidae, Pyrgotidae, Sciomyzidae, Sarcophagidae, and Tachinidae are parasitic to arthropods but the Tachinidae stand out as the most significant parasitoids (Van Driesche *et al.*, 2008).

15.2.3.3. Entomopathogens. Like other organisms, insects also suffer from diseases caused by bacteria, viruses, fungi, microsporidia, rickettsia and nematodes. So far over 3000 species of microorganisms have been reported to cause diseases in insects but many more remain undiscovered or unidentified. Biopesticides based on living microbes and their bioactive compounds have been researched and promoted as replacements for synthetic pesticides for many years (Battu and Arora, 1997; Arora *et al.*, 2000; Arora and Battu, 2001; Rosell *et al.*, 2008). Some of these microbial insecticides are registered with the Central Insecticide Board and Registration Committee in India for the control of different insect pests (http://www.cibric.nic.in) and are commercially applied for management of several important insect pests (Table 3).

The most widely exploited biopesticide in pest management is the crystalliferous spore forming soil and phylloplane-inhabiting bacterium, *Bacillus thuringiensis* (Berlina). Even though the bacterium was isolated in 1901 and the first commercial product 'Sporeine' was produced in France in 1938, but it was not until 1970 that it emerged as a successful biopesticides with the discovery of the potent strain HD1 of *B. thuringiensis* var *kurstaki* (Sanchis, 2011). Hundreds of Bt toxins each with a unique activity spectrum have been discovered since then and details of these are available at the Bt toxin Nomenclature website (http://www.lifesci.sussex.ac.uk. home/ Neil_crickmore/ Bt).

Expansion in the use of Bt in the 1970s and 1980s was largely due to the development of methods of large scale fermentation, and increased efficiency in production and quality control, so that formulations with high activity could be developed. Bt products are widely used for management of insect pests in forestry, in organic cultivation and on high value horticultural crops . The major limitation of Bt is the rapid photo-degradability of foliar sprays on the crop. To overcome this problem the Bt delta endotoxin genes have been inserted into a number of crop plants (refer to section 15.2.5. A of this chapter for details). Other entomopathogens including baculoviruses, fungi and nematodes, etc. have so far been able to capture only niche markets in specialized areas. This is because the microbial control agents have a rather narrow spectrum of activity, are slow acting (especially as compared to chemical pesticides) and are easily inactivated on the foliar surface (Sanahiya *et al.,* 2011). The microbial control agents (MCAs) are discussed in detail in chapter 6 of this volume.

Recent advances in molecular biology have provided the necessary tools for genetic improvement of entomopathogens to overcome their major limitations. There have been significant improvements in mass production technology for all the major groups of entomopathogens, during the last 2-3 decades. However, further refinements in production and formulation technology are required to make these products competitive with chemical pesticides.

The environmental and ecological problems caused by extensive and intensive usage of chemical pesticides have forced the society to minimize the usage of these toxicants. Therefore, the chemical insecticide market is slowly shrinking. In contrast, microbial control agents have shown annual growth rate of more than 10 per cent during the last decade due to their safety, specificity and self-perpetuating action (Glare *et al.,* 2012). There is a need to intensify research on entomopathogens in all the important areas: search for new microbial species/ strains causing epizootics in insect pests in nature; studies on tritrophic/ multitrophic interaction involving host plants insect-pests (hosts)-pathogens; molecular studies for understanding the mechanism of action and for increasing host range, enhancing the speed of kill and improving photostability, development of economical methods for mass production alongwith improvements in formulation and application technology (Vega and Kaya, 2012). With advances in these areas, the microbial control agents are likely to become a major component of biointensive IPM programmes in the coming decades.

Table 3. Microbial insecticides used in India

Insecticides	Taxus	Products	Targets
Bacillus thuringiensis subsp. *israelensis*	Bacterium	Tacibio, Technar	Lepidopteran pests
Bacillus thuringiensis subsp. *kurstaki*	Bacterium	Bio-Dart, Biolep Halt, Taciobio-Btk	Lepidopteran pests
Beauveria bassiana	Fungus	Myco-Jaal, Biosoft, ATEC beauveria, Larvo-Guard, Biorin, Biolarvex Biogrubex, Biowonder, Veera, Phalada 101B, Bioguard, Bio-power	Coffee berry borer, diamondbackmoth, thrips, grasshoppers, whiteflies, aphids, codling moth
Metarhizium Anisopliae	Fungus	ABTEC, Verticillium, Meta-Guard, Biomet, Biomagic, Meta, Biomet, Sun Agro meta, Bio-Magic	Coleoptera and lepidoptera, termites, mosquitoes, leafhoppers, beetles, grubs
Paecilomyces fumosoroseus	Fungus	Nemato-Guard, Priority	Whitefly
Paecilomyces lilacinus	Fungus	Yorker, ABTEC, Paecilomyces, Paecil, Pacihit, ROM biomite, Bio-Nematon	Whitefly
Verticillium lecanii	Fungus	Verisoft, ABTEC, Verticillium, Vert-Guard, Bioline, Biosappex, Versatile, Ecocil, phalada 107V, Biovert Rich, ROM verlac, ROM Gurbkill, Aun Agro Verti, Bio-Catch	Whitefly, coffee green bug, homopteran pests
Helicoverpa armigera nucleopolyherdrosis virus	Virus	Helicide, Virin-H, helocide, Biovirus-H, Helicop, Heligard	*Helicoverpa armigera*
Spodoptera litura nucleopolyherdrosis virus	Virus	Spodocide, Spodoterin, Spodicide, Biovirus-S	*Spodoptera litura*

Source: Rabindra and Grzywacz (2010).

15.2.4. Biorationals

Biorationals or reduced risk pesticides are synthetic or natural compounds that affect those targets within the insect body which are unique to insects e.g. integument, endocrine and communication systems etc. Most of the biorational pesticides are preferable to conventional ones because of their specificity to target pests, their effectiveness at low rates and their non-persistent characteristics in the environment.

15.2.4.1. Insect Growth Regulators. The chemicals that affect the growth and development of insects are called insect growth regulators (IGRs), a name in keeping with the familiar plant growth regulators (PGRs) that influence plant growth and physiology. IGRs are known to be extremely safe due to their biodegradable nature, selectivity in action and have mode of action different from the conventional insecticides which make these compounds less prone to development of cross-resistance. The growth regulators that are of interest for pest control include insect hormones (moulting and juvenile hormones) and their analogues and chitin synthesis inhibitors (CSIs).

Williams (1967) was first to raise the possibility of use of juvenile hormones (JHs) as the 'Third Generation Pesticides' which led to the synthesis of a large number of analogues termed JHAs. However, natural JHs *per se* could not be of much use in IPM programmes given to their instability in UV light and rapid metabolism in insects coupled with difficulties in their synthesis The effectiveness of JHAs, Methoprene, Hydroprene, Fenoxycarb and Pyriproxyfen has been demonstrated for the effective management of many insect species of agricultural and medical importance (Singh and Suri, 2013).

Moulting hormones (MHs) represented by ecdysones are secreted by prothoracic glands and are responsible for moulting, growth and maturation of insects. The exogenous application of moulting hormone analogues (MHAs) or ecdysone agonists results in hormonal imbalance since these cannot be metabolized or excreted rapidly and consequently precocious metamorphosis is ensued. Rohm and Haas Company in 1983 serendipitously discovered the first bisacylhydrazine ecdysteroid agonist (Hsu, 1991). Soon after, a simple and more potent MH analogue, RH-5849 was synthesized that targeted many lepidopteron, coleopteran and dipteran insect pests (Aller and Ramsay, 1988). It was superseded by a group of some cost effective MHAs. The best known representatives of this group comprise the chemical compounds tebufenozide, methoxyfenozide and halofenozide (Singh and Suri, 2013).

Chitin syntheses inhibitors (CSIs) interfere with the production of chitin, the structural polysaccharide found in insect cuticle, and thus affect the integrity of insect exoskeleton. They act by inhibiting the enzyme chitin synthetase required for polymerization of N-acetylglucosamine in the biosynthesis of chitin during the process of moulting of insects. The first chitin synthesis inhibitor introduced into the market as a novel insecticide was a benzoylphenylurea named diflubenzuron, which has been reported to be quite effective against tobacco caterpillar, *Spodoptera litura* (Fabricius) (Acharya and Bhargava, 2007; Arakawa, 2008), Colorado potato beetle, *Leptinotarsa decemlineata* Say (Karimzadeh *et al.*, 2007) and brinjal fruit and shoot borer,

Leucinodes orbonalis Guenee (Jena *et al.*, 2006). Novaluron acts by both ingestion and contact and is found to be highly effective against bollworms on cotton (Vadodaria *et al.*, 2000; Udikeri *et al.*, 2004), pod borer *H. armigera* on pigeon pea (Siddegowda *et al.*, 2004), *L. orbonalis* on brinjal (Sawant *et al.*, 2004) and diamondback moth, *Plutella xylostella* L. on cabbage and cauliflower (Bishnoi, 1999). Sublethal concentrations of lufenuron affected various morphological, biological and physiological parameters of *P. xylostella* (Josan and Singh, 2000). It has also been found to be very effective against leafminer, *Chromatomyia horticola* (Goureau) (Saito, 2004), *L. decemlineata* (Karimzadeh *et al.*, 2007), *H. armigera*, (Vadodaria *et al.*, 2000), cotton bollworms (Udikeri *et al.*, 2004) and cotton leafworm, *Spodoptera littoralis* (Sammour *et al.*, 2008). Buprofezin, acts by inhibiting cuticle deposition. It also suppresses egg laying in female adults with inhibition of prostoglandin synthesis and has effects on levels of hormones associated with moulting in nymphs. It has been used effectively against whitefly, *Bemisia tabaci* Gennadius on cotton (Gogi *et al.*, 2006), planthopper, *N. lugens* on rice (Wang *et al.*, 2008), scale insects (Yarom *et al.*, 1988) and mealybug, *Maconellicoccus hirsutus* Green on grapes (Balikai, 2002; Muthukrishnan *et al.*, 2005).

15.2.4.2. Semiochemicals. The chemicals that trigger various behavioral responses and mediate the interactions between the organisms are called "Semiochemicals". A complex web of chemically mediated conversation exists in the four trophic level system. Plants produce chemicals that inhibit the growth and development of other plants; attract, repel, or stimulate feeding and oviposition by herbivores (insects) and either attract or stimulate the activities of parasitic and predacious insects. Herbivores produce chemicals that influence the other members of the same species, other herbivores, parasitic and predacious insects and even hyperparasitoids. Semiochemicals have powerful biological activity and are often highly specific in nature. Thus they offer a great potential as effective and environmentally sound pest management tools. Semiochemicals are subdivided into allelochemicals and pheromones depending on whether the interactions are interspecific or intraspecific, respectively.

(a) *Pheromones*: Pheromones have been extensively employed in IPM for monitoring, mass trapping and for mating disruption. These uses take advantage of sex pheromones on which a vast majority of insect pests rely to mediate reproduction (Flint and Doanne, 2013). The most widespread and successful applications of sex pheromones concern their use in detection and population monitoring. One of the first widely used monitoring systems that include an action threshold based on trap captures was established for pea moth, *Cydia nigricana* Fab. (Wall *et al.*, 1987). Since pheromone lures are inexpensive and usually reliable, they facilitate the IPM concept, which relies on frequent scouting of target species for planning of control measures and evaluating their efficacy. Hundreds of pheromone compounds have been identified, most of them in Lepidoptera, but also in other insect orders, particularly beetles and flies (El-Sayed, 2008).

Control of insect populations with pheromones is achieved by two principle techniques, mating disruption and mass annihilation. Mating disruption causes disorientation and communication disruption between the sexes, and thus delays, reduces, or prevents fertilization of females. Mass annihilation, by mass trapping or attract and kill, relies on attraction of one or both sexes to a lure, in combination with a large-capacity trap or an insecticide-impregnated target. Mass trapping, without the use of insecticides, has led to a 50 per cent and higher increase in marketable fruit, which has been attributed to the combined effects of mass trapping of brinjal fruit and shoot borer and enhanced impact of natural enemies. Additionally, secondary pests, such as mites and whitefly, were reduced in the pheromone plots (Cork *et al.*, 2005).

(b) *Allelochemicals*: Allelochemicals are those chemicals that mediate interspecific interactions and are distinguished from pheromones, which mediate intraspecific interactions. There are numerous avenues for use of allelochemicals in IPM programs either directly against the insect pests or through enhancement of natural enemies. The chemically mediated host or prey selection behaviours of entomophagous insects attacking herbivorous insect pests, document the vital role of alleochemicals and the effectiveness of the entomophagous insects can be enhanced in biological control programs through behavioural manipulation (Virk and Shera, 2013).

Plants play an important role in the host or prey selection behaviour of entomophagous insect attacking herbivorous insects. Herbivorous hosts may be attacked by particular entomophagous insect on some plant species but not on other. This variation in choice of plants is often due to differences in allelochemicals produced by various plants. Many of the toxic or repellent plant allelochemicals have been found to be attractive or stimulating to parasitoids. Numerous examples are available of natural enemies attacking host of prey in one microhabitat but at much lower rates or not at all in other. Parasitization of *Helicoverpa zea* (Boddie) eggs by *Trichogramma pretiosum* Riley was relatively high in plots of tomato but almost non-existent in adjacent plots of corn. It was revealed that presence of volatile stimuli from tomato plants in corn fields resulted in increased rates of parasitization by *T. pretiosum* (Whitman and Nordlund, 1994). The aphid parasitoid *Diaeretiella rapae* (McIntosh), prefers host insects feeding on *Brassicae* and is attracted by isothiocynates that are typical for these plants (Baer *et al.*, 2004). Bukoviszky *et al.* (2005) reported that odours from crucifers, i.e. Brussels sprout (*Brassica oleracea* L. Gemmifera), white mustard (*Sinapis alba* L.), a feral (*Brassica oleracea* L.), and a non-crucifer malting barley (*Hordeum vulgare* L.) species were more attractive to *Diadegma semiclausum* Hellen, a parasitoid of the diamondback moth, *P. xylostella* than clean air. However, parasitoids preferred the volatile blend from the three cruciferous species over that of malting barley.

The presence of specific herbivore induced plant volatiles may also serve as reliable cues for the presence of a specific host species or host stage (Hilker and Mcneil, 2008). Volatiles may be released by the plant in response to herbivore damage, i.e herbivore induced synomones. Most plants emit relatively few volatiles until their tissues are damaged and damaged tissue causes the release of these stimuli. Egg parasitoids may orientate using oviposition-induced plants cues while larval parasitoids can exploit cues induced by larval feeding. Interestingly, some egg parasitoids also respond to plant cues induced by larval feeding (Lou *et al.*, 2005; Moraes *et al.*, 2005). When a chewing herbivore such as caterpillar bites into leaf, the contents of plant cells and vessels are directly exposed to the air, where rapid volatilization occurs. Not only are existing plant chemicals released, but also entirely new volatile substances can be produced when plant enzymes and protoplasm come into contact with oxygen. Hence, with every caterpillar bite, mini-clouds of plant volatiles are released. These volatile perfumes drift downward, where they stimulate searching carnivores. They not only alert natural enemies of the nearby presence of feeding herbivores, but also serve as chemical road maps for entomophagous arthropods that follow the scent upwind by chemoanemotactic orientation. Many studies have shown that damaged plants are much more attractive to natural enemies than undamaged plants (Whitman and Nordlund, 1994). The females of *Cotesia plutellae* (Kurdjumov), a primary solitary larval endoparasitoid and a specialist of *P. xylostella* are primarily attracted by green leaf volatiles produced by wound left by herbivores or mechanical injury (Liu and Jiang, 2003).

15.2.4.3. *Push-Pull strategy.* Push-pull strategies involve the behavioral manipulation of insect pests and their natural enemies via the integration of stimuli that act to make the protected resource unattractive or unsuitable to the pests (push) while luring them toward an attractive source (pull) from where the pests are subsequently removed. The push and pull components are generally nontoxic. Therefore, the strategies are usually integrated with methods for population reduction, preferably biological control. Push-pull strategies maximize efficacy of behavior manipulating stimuli through the additive and synergistic effect of integrating their use. By orchestrating a predictable distribution of pests, efficiency of population-reducing components can also be increased. The strategy is a useful tool for integrated pest management programs reducing pesticide input (Khan *et al.*, 1997 and Cook *et al.*, 2007). Thousands of farmers in East and South Africa are using push-pull strategies to protect maize and sorghum from stemborers. The strategies involve the combined use of intercrops and trap crops, using plants that are appropriate for the farmers and that also exploit natural enemies. Stem borers are repelled from the crops by repellent nonhost intercrops, particularly molasses grass, silverleaf desmodium or greenleaf desmodium (push), and are concentrated on attractive trap plants, primarily Napier grass or Sudan grass (pull). Molasses grass, when intercropped with maize, not only reduced stem borer infestation, but also increased parasitism by *Cotesia sesamiae* Cameron (Khan *et al.*, 2011). Push-pull strategies has also been effectively demonstrated against *Helicoverpa* in cotton, colorado potato beetle in potato, pea leaf weevil in beans, pollen beetle in oilseed rape, onion maggot on onions, thrips on chrysanthemum, bark beetles on conifers and also for veterinary and medical pests (Cook *et al.*, 2007).

15.2.4.4. Botanicals. The term "botanical pesticides" refers to those compounds or materials derived from plants (excluding synthetic analogues) that demonstrate useful toxicity to pests involved in agriculture, animal husbandry, and human health problems. The plant material is used either directly (e.g., powdered bark) or as a source of extracts for use in formulations that are often combined with other pesticides, synergists, and formulants to increase efficacy and speed of action. At present there are four major types of botanical products used for insect control (pyrethrum, rotenone, neem, and essential oils), along with three others in limited use (ryania, nicotine, and sabadilla). Botanicals like neem, bel, senwar, pyrethrum, tobacco, karanj, mahua and sweet flag etc. have already attained the status of potential pesticides of plant origin to be used in IPM of crop field insects as well as in storage ecosystems (Arora and Dhaliwal, 1994; Dhaliwal and Arora, 2001; Aggarwal, 2013). According to Purohit and Vyas (2004) as many as 2121 plant species are reported to be useful in pest management, out of which 1005 species of plants exhibited insecticidal properties, 384 with antifeedant properties, 297 with repellent properties, 27 with attractant properties and 31 with growth inhibiting properties have been identified.

15.2.4.5. Microorganism-Derived Biorationals. Several important classes of pesticides such as spinosyns, avermectins, polynactins and pyrrole insecticides have been developed, mostly from soil microorganisms by fermentation process and have proved useful as reduced risk pesticides for management of insect pests.

Spinosyns: The spinosyns, derived by the fermentation of soil actinomycetes, *Saccharopolyspora spinosa* Mertz & Yao are very effective against pests of crops, ornamentals, forestry, greenhouse and household. The most effective insecticide of this group is spinosad which is a mixture of Spinosyns A and D (Kirst *et al.*, 1992). It has shown activity against many insect pests of agricultural importance belonging to orders Lepidoptera, Coleoptera, Diptera, Hymenoptera and Thysanoptera.

Avermectins: The avermectins are a group of macrocyclic lactones isolated from fermentation of the soil microorganism *Streptomyces avermitilis*. Abamectin developed for agricultural use is a mixture of about 80% avermectin B1a and 20% avermectin B1b (Fisher and Mrozik, 1989). It acts specifically on phytophagous mites, but also on agromizid leafminers, ants, cockroaches, and selected lepidopteran pests (Lasota and Dybas, 1991). It is considerably less toxic to beneficial insects such as honeybees, parasitoids, and predators. The search for new avermectin derivatives resulted in the development of emamectin benzoate (Proclaim) by Merck (New Jersey, USA) and milbemectin (Milbeknock) by Sankyo (Tokyo, Japan). Emamectin acts specifically on lepidopteran pests (Horowitz and Ishaaya, 2004). Milbemectin acts on a wide range of mites such as the two-spotted spider mite, *Tetranychus urticae*, *T. cinnabarinus*, and the citrus red mite *Panonychus citri* (Sankyo, 1997). It is a mixture of milbemycin A3 and milbemycin A4, both of which are metabolites of *Streptomyces hygroscopicus* (Barrett *et al.*, 1985).

Pyrrole insecticides: Pyrrole insecticides have been derived from natural product, dioxapyrrolomycin, isolated from a strain of *Streptomyces*. Among these, Chlorfenapyr has been developed commercially owing to its broad spectrum activity against many species belonging to orders Lepidoptera, Coleoptera, Thysanoptera and Acarina. It was first registered in the USA for control of termites, cockroaches and nuisance ants under the trade name Phantom®. The label was recently expanded to include food and feed mills, food handling areas, restaurants, and other areas where food is handled and stored, and stored-product insects have been added to the label for the USA.

15.2.4.6. Miscellaneous Biorationals. Several novel insecticides acting on selective biochemical sites have been developed during the past two decades.

Pymetrozine : It is a novel azomethin pyridine insecticide affecting the nerve that controls the salivary pump of some sucking pests, causing irreversible cessation of feeding, followed by starvat and death (Kayzer *et al.*, 1994). The compound is a powerful toxicant against aphids, whiteflies, and planthoppers (Fuog *et al.*, 1998; Zhang, 2011).

Oxadizines: Indoxacarb acts by inhibiting sodium ion entry into nerve cells, resulting in the paralysis and death of target insect pests (Lapied *et al.*, 2001). It has shown activity against lepidopteran pests as well as on certain homopterans and coleopterans in various agricultural systems (Harder *et al.*, 1996; Wing *et al.*, 2000). Its efficacy has been demonstrated against important insect pests such as *Heliothis* and *Helicoverpa* spp., *Spodoptera* sp., *Trichoplusia* sp., *Lygus* sp., *Empoaska* sp., and the Colorado potato beetle, *L. decemlineata* (Harder *et al.*, 1997; Hammes *et al.*, 1998; Michaelides and Irving, 1998; Sullivan *et al.*, 1999; Arora and Singh, 2011).

Phenyl pyrazoles: Fipronil is a phenylpyrazole insecticide developed by Rhone Poulenc in 1987. It acts by blocking γ-aminobutyric acid (GABA) gated chloride channels (Bloomquist, 2001). It acts on a diversity of insect species, such as the diamondback moth (*Plutella xylostella*), *Spodoptera* and *Heliothis* species, Colorado potato beetle, and against household pests, such as termites and cockroaches (Grant *et al.*, 1990; Hamon *et al.*, 1996).

Tetronic Acid Derivatives: Spiromesifen is an insecticide from the new class of spirocyclic tetronic acids that effectively acts against whiteflies and mites (Bretschneider *et al.* 2003; Nauen *et al.*, 2003, 2005; Nauen and Konanz, 2005). It acts as an inhibitor of acetyl-CoA-carboxylase, a lipid metabolism enzyme resulting in decrease of total lipids. This compound has been introduced into several countries during the last few years and is becoming a major compound in whitefly and mite resistance management programs.

Spirodiclofen, another tetronic acid derivative is a broad spectrum acaricide acting via lipid biosynthesis inhibition (LBI) with no cross resistance to currently available acaricides and with additional insecticidal properties. It has been found effective against mites in apple and pear orchards and shows good selectivity against natural enemies (Maeyer and Geerinck, 2009).

Tetramic Acid Derivative: Spirotetramat belonging to the chemical class of ketoenols is a tetramic acid derivative. It interferes with lipid biosynthesis, leading to death of juveniles within two to ten days after application (Palumbo, 2007). It has shown effectiveness against a wide spectrum of sucking insects including aphids (Combs and Reissig, 2008), thrips (Alston *et al.*, 2008), psyllids (Fischer, 2008; Jaworska *et al.*, 2012), whitefly (Kumar *et al.*, 2009) and scale insects (Mckenna *et al.*, 2013), etc.

Diamide: Flubendiamide belongs to the phthalic acid diamide chemical class is a ryanodine receptor agonist, which activates ryanodine-sensitive intracellular calcium release channels in insect neurons. It has larvicidal activity by disrupting calcium ion balance, which causes rapid cessation of feeding. The insecticide has been found to be selective in action against wide range of lepidopteran insect pests such as *H. armigera* (Ameta and Bunker, 2007), *P. xyllostella* (Ameta and Bunker, 2007) and brinjal fruit and shoot borer, *L. orbonalis* (Jagginavar *et al.*, 2009; Latif *et al.*, 2009),

Chlorantraniliprole (Rynəxypyr) also belongs to the anthranilic diamides, which activate the insect ryanodine receptors affecting calcium release during muscle contraction. Insects treated with chlorantraniliprole exhibit rapid feeding cessation, lethargy, regurgitation, muscle paralysis, and ultimately death (Lahm *et al.*, 2007). Chlorantraniliprole has been found to be effective against numerous lepidopteran pests of agricultural importance such as brinjal shoot and fruit borer, *L. orbonalis* (Misra, 2011; Rajavel *et al.*, 2011), *H. zea* on tomato (Kuhar *et al.*, 2010), bollworms and *S. litura* on cotton (Anonymous, 2013), yellow stem borer, *Scirpophaga incertulas* (Walker) and leaffolder *Cnaphalocrocis medinalis* (Guenee) on rice (Suri, 2013).

15.2.5. Biotechnological Approaches

15.2.5.1. Transgenic Crops

The development of transgenic insect resistant plants is the most dramatic and obvious application of biotechnology in IPM (Sharma, 2009; 2012). Remarkable progress based on recombinant DNA technology has been made over the past last three decades for developing crops with novel genes for resistance to insects, plant pathogens and herbicides. Since the commercialization of biotech crops in 1996, the area under genetically engineered crops has increased from 1.7 million hectares to 175 million hectares in 2012. An unprecedented more than 100-fold increase in hectarage between 1996 and 2012 makes biotech crops the fastest adopted crop technology in the history of modern agriculture, being grown by over 17.5 million farmers in 19 developing countries and 8 industrial countries of the world (James, 2013).

Development of insect resistant transgenic crops has focused mainly on two approaches, viz. integration of bacterial genes encoding for the production of toxic proteins, especially from soil inhabiting spore forming bacterium, *Bacillus thuringiensis* Berliner and integration of plant derived genes encoding for enzyme inhibitors, plant lectins and secondary plant metabolites. However, to date, *Bt* genes have subjugated the commercial scene very influentially for the last 15 years, at least in cotton and corn crops.

Bacillus thuringiensis endotoxin

A total of 20 Bt genes, e.g., *cry1Ac*, *cry2Ab2*, *cry1Ab*, *cry1F*, *cry1Fa2*, *cry1C*, *Vip3A*(a), *Vip3Aa20*, *cry9C*, *cry2Ae*, *cry1A*, *cry1A.105*, *mocry1F* for lepidopteran resistance and *cry34Ab1*, *cry35Ab1*, *cry3Bb1*, *cry3A* and *mcry3A* for coleopteran resistance have been expressed in several crop plants, and are being commercially deployed worldwide in different genetically modified crops, e.g., cotton, maize, rice, soybean, brinjal, poplar, tomato and potato, alone or stacked with herbicide resistance genes (www.isaaa.org). Among insect resistant events, the maize event MON810 (49 approvals in 23 countries + EU-27 countries) has received the most number of approvals followed by maize event NK603 (49 approvals in 22 countries + EU-27 countries), maize event Bt11 (45 approvals in 21 countries + EU-27 countries), maize event TC1507 (45 approvals in 20 countries + EU), maize event MON89034 (36 approvals in 19 countries + EU), cotton event MON531 (36 approvals in 17 countries + EU), herbicide tolerant and insect resistant maize event MON88017 (35 approvals in 19 countries + EU), and insect resistant cotton event MON1445 (34 approvals in 15 countries + EU) (James, 2013).

Cotton: Genetically engineered *Bt* cotton is the only crop which has been commercialized in India. Bt cotton was first approved for commercial cultivation by the Genetic Engineering Approval Committee (GEAC), Ministry of Environment and Forests, Government of India in Central and South zones in 2002 and in North zone in 2005 (Karihaloo and Kumar, 2009). Since its commercialization during 2002, the number of events, Bt cotton hybrids and companies marketing approved hybrids have all increased significantly in India. So far, six transgene events, viz. *cry1Ac* (MON531), modified *cry1Ac* (Event 1), *cry1Ab/cry1Ac* (GFM event), *cry1Ac* (BNLA-601), *cry1C* (MLS99124) and *cry1Ac+cry2Ab* (MON 15985) have been approved by GEAC in India. The commercial exploitation of *Bt* cotton in India has significantly established its positive impact by reducing pesticide load which in turn has led to reduction in cost of cultivation and environmental pollution. Besides this, social and economical status of farmers has been uplifted due to yield advantage, net profit and higher economic returns. The global study of benefits generated by biotech crops conducted by Brookes and Barfoot (2013), estimates that India enhanced farm income from Bt cotton by US$ 9.4 billion in the period 2002 to 2010 and US$ 2.5 billion in 2010 alone. Besides, it has halved insecticide consumption, contributed to the doubling of yield and transformed India from a cotton importer to a major exporter.

Maize: Transgenic maize expressing *cry1Ab* gene from *B. thuringiensis* ssp. *kurstaki* strain HD-1 was commercialized in 1996 against European corn borer, *Ostrinia nubilalis* Hubner. The dominant cry1Ab events are Monstanto YieldGard maize (event MON810), Syngenta's BT11sold under NK-brand as YieldGard, and Syngenta's event 176, known as Knockout (Benedict and Ring, 2004). Maize with *cry1F* gene (Dow'z Herculex I) for stemborers (Shelton and Bellinder, 2007) and *cry3Bb* gene (YieldGard VT) for the control of maize rootworm complex has been developed (Sharma *et al.*, 2004). Maize plants transformed with Bt genes have also been found to be effective against stalk borer, corn earworm and fall armyworm.

The first ever eight gene, stacked GMO variety Smartstax maize, provides broadest spectrum of above and below-ground insect resistance, as well as herbicide tolerance. It has been developed by Monsanto and Dow AgroSciences. Smartstax is a combined trait maize product containing events MON 89034, MON 88017, DAS 59122-7 and TC 1507 of the 8 'stacked' transgenes that Smartstax contains, six of these genes generate different forms of insect killing Bt toxins: *cry1A.105*, *cry2Ab2* (both under license from Monsanto), and *cry1F* (Dow) for above ground insects; *cry3Bb1* (Monsanto), and *cry34Ab1*, *cry35Ab1* (both Dow) for underground insects. The two remaining transgenes are resistant to certain herbicides: Glyphosate (traded as Roundup Ready by Monsanto) and Glufosinate (traded as Liberty Link by Dow AgroSciences, under licence from Bayer) (www.epa.gov).

Rice: Chinese Research Institutes in collaboration with IRRI, Manila have transformed rice by inserting *cry1Ab* gene alone and fusion *cry1Ab/cry1Ac* into conventional rice varieties and hybrids against *Chilo suppressalis* (Walker), *Cnaphalocrocis medinalis* (Guenee) and *Scirpophaga incertulas* (Walker) (Tu *et al.*, 2000). A synthetic Cry1Ab toxin gene transformed rice line KMD1 showed cent per cent mortality of newly hatched or third instar larvae of eight lepidopteran rice pest species, *i.e.*, *C. suppressalis*, *S. incertulas*, *C. medinalis*, *Herpitogramma licarisalis* (Walker), *Sesamia inferens* (Walker), *Naranga anescens* Moore, *Mycalesis gotama* Moore and *Parnara guttata* Bremer & Grey (Shu *et al.*, 2000). Several rice lines expressing insecticidal genes such as *cry1Aa, cry1Ab, cry1Ac, cry1Ab/Ac, cry1C, cry2A* etc. have been developed and tested for their efficacy and biosafety in China and elsewhere in the world including India (Chen *et al.*, 2011).

Potato: A modified *cry3A* gene has been expressed in potato plants with resistance to Colorado potato beetle (Perlak *et al.,* 1993). Transgenic potato having *cry1A*(b) gene and truncated gene *cry1A(b)6* resulted in less damage to leaves by potato tuber moth (Jansens *et al.,* 1995). A hybrid *B. thuringiensis* gene has been constructed using a synthetic and truncated *cry1Ba* gene as the scaffold for inserting part of *cry2a* gene encoding domain II (Naimov *et al.,* 2003). Transgenic potato plants expressing this hybrid toxin were resistant to several insect pests, including both Coleoptera (Colorado potato beetle) and Lepidoptera (potato tuber moth and European corn borer). As the target receptor recognition of this hybrid protein is expected to be different from Cry proteins currently in use to control these pests, this strategy provides new opportunities for resistance management studies involving multiple transgenes in crops. Transgenic Indian potato cultivar Kufri Badshah expressing synthetic, modified *cry1Ab* gene has been developed against potato tuber moth in India. Insect bioassay of stored transgenic tubers resulted in significant retardation and mortality in neonate tuber moth larvae (Kumar *et al.*, 2010).

Tomato: The expression of *cry1Ac* gene in tomato has been found highly effective against *H. armigera* (Mandaokar *et al.*, 2000). The transgenic tomato plants expressing *cry1Ab* showed significantly lower foliar and fruit damage by *H. armigera* than non-transgenic control plants (Kumar and Kumar, 2004). Transgenic tomato (cv. *Money maker*) over expressing *Bt* (cry*2Ab*) gene has been produced using *Agrobacterium*-mediated transformation method. Molecular and biochemical analysis confirmed the expression and integration of the transgene into tomato genome. Obvious effects of Cry 2Ab were judged by the mortality of the *H. armigera* and the potato tuber moth, *Phthorimaea operculella* (Zeller) when fed on *Bt* tomato (Saker *et al.*, 2011).

Brinjal: Bt brinjal has been developed by Mahyco in 2000 by inserting *cry1Ac* gene providing resistance to brinjal fruit and shoot borer, *Leucinodes orbonalis*. Since then, it has undergone rigorous scientific evaluation to assess its food safety and environmental effects. In 2006, GEAC granted approval for large scale trials of Bt brinjal. In October 2009, a landmark decision was made by India's Genetic Engineering Approval Committee (GEAC), to recommend the commercial release of Bt Brinjal (Eggplant/Aubergine), which is now pending, subject to final clearance by the government of India (Chaudhary and Gaur, 2009).

The other crops in which *Bt* genes have been inserted and are under different stages of deployment for producing insect resistant transgenic crops in India include rice, sorghum, tomato, cabbage, cauliflower, sugarcane, chickpea and pigeonpea (Gujar and Dhillon, 2011).

Plant Defensive Proteins

Plants have developed defense systems to combat various pathogens throughout their life cycle, from the seed stage until senescence. There are several embryonic defense mechanisms detected, including the production of plant lectins and pathogen-related proteins, in response to pathogen or insect attack (Ye *et al.*, 2001; Guiderdoni *et al.*, 2002). Largely, plant transformation involving plant genes has focused on proteas inhibitors, which disrupt amino acid metabolism, α-amylase inhibitors, which target carbohydrate metabolism, lectins, which cause agglutination and cell aggregation, and enzymes such as chitinase, which target insect exoskeleton.

Proteinase inhibitors: Insects have proteases in their gut which are enzymes helping in digestion of protein. Disruption of amino acid metabolism by inhibition of protein digestion has been a key target for use in insect control (Hilder *et al.*, 1992). Many insects, particularly Lepidoptera, depend on serine proteases (trypsin, chymotrypsin and elastase endopoteases) as their primary protein digestive enzymes. A comprehensive system of classification has been proposed for facilitating the exchange, storage and retrieval of information about this group of proteins (Rawlings *et al.*, 2004). Leo *et al.*, (2002) developed plant –PIs database to facilitate retrieval of information on plant protease inhibitors and related genes (Table 4). However, common polyphagous crop pests are well adapted to deal with a range of different inhibitors, which have only limited effects on fitness as a result. A range of strategies have been attempted to improve effectiveness of proteinase inhibitors as antimetabolites towards insects, including selection for inhibitory activity against insect digestive enzymes, mutagenesis for novel inhibitory activity, and engineering inhibitors with multiple functions. Proteinase inhibitor genes have only been used in transgenic crops in combination with other insecticidal genes. In Chinese genetically engineered cotton varieties which express Bt toxins as an insecticidal protein against lepidopteran larvae, the CpTi (cowpea trypsin inhibitor) gene has been employed as a second transgene to improve protection (Gatehouse, 2011).

Serine protease inhibitors: The first gene of plant origin to be used in transgenic crop protection was isolated from cowpea encoding double headed trypsin inhibitor (CpTi). CpTi is a small polypeptide belonging to the Bowman-Birk type of double-headed serine protease inhibitors (Gatehouse and Boulter, 1983) and is effective against Lepidoptera, Coleoptera, and Orthoptera, based on *in vitro* assays and *in vivo* artificial diet bioassays (Lawrence and Koundal, 2002).

***α*-Amylase inhibitors :** Alpha amylase inhibitors are toxic to insects through interference in the digestion of dietary carbohydrates. They have been shown to be causative factors in the resistance of specific varieties of legumes to coleopteran seed weevils. The bean (*Phaseolus vulgaris*) *α*-amylase inhibitor gene was expressed in seeds of transgenic garden pea (*Pisum sativum*) and other grain legu-mes, using a strong seed-specific promoter (Shade *et al.*, 1994). The resulting seeds contained up to 3 per cent of the foreign protein and were resistant to stored product pests, such as larvae of bruchid beetles, and field pests, such as larvae of the pea weevil *Bruchus pisorum* (Morton *et al.*, 2000). This strategy is directed toward coleopteran seed herbivores, which have a neutral or acidic gut pH, so the inhibitor is not inactivated, and in which starch digestion, not protein digestion, is nutritionally limiting.

Table 4. Plant proteinase inhibitor family in plant-PIs

Plant protease inhibitor family	**Plant-PIs code**
Bowmann-Birk Serine Proteinase Inhibitor	SBI
Cereal Trypsin / Á-Amylase Inhibitors	BRI
Cysteine Proteinase Inhibitors	CYS
Metallo Carboxypeptidase Inhibitors	MCI
Mustard Trypsin Inhibitors	MSI
Potato Type I Inhibitors	PI1
Potato Type II Proteinase Inhibitors	PI2
Serpin	SP1
Soybean Trypsin Inhibitors (kunitz)	KNI
Squash Inhibitors	SQI

Source: Bhattacharjee *et al.* (2012)

Lectins: Lectins are carbohydrate binding proteins that are abundant in seeds and storage tissues of some plant species. Potential exploitation of lectin genes to confer insect resistance in transgenic plants has targeted hemipteran plant pests. Lectins purified from snowdrop (*Galanthus nivalis*-GNA) or garlic are toxic to insects but not to mammals. Expression of the snowdrop lectin (GNA) in transgenic rice plants using constitutive or phloem-specific promoters gave plants that were partially resistant to *N. lugens* and other hemipteran pests. Reductions of up to 50 per cent in survival were observed, with reduced feeding, development, and fertility of survivors (Rao *et al.*, 1998; Foissac *et al.*, 2000). A successful transformation of rice variety Chaitanya with a snow drop lectin gene, GNA, has conferred resistance against brown planthopper, whitebacked plant-hopper and green leafhopper (Naga-dhara *et al.*, 2003). Similar partial resistance to hemipterans has also been obtained by expression of a lectin from garlic (*Allium sativum*) leaves (ASA-L) in transgenic rice (Saha *et al.*, 2006a) and a variety of other transgenic plant species. The transgenic rice plants expressing ASA-L were shown to decrease transmission of Rice tungro virus by its insect vector, presumably as a result of decreased feeding by the pest (Saha *et al.*, 2006b).

Other novel genes: Transgenic expression of certain enzymes and animal proteins has been suggested as another alternative to Bt genes. One of the most important enzyme is chitinase, which is an important component of insect integu-ment. A chitinase from *Serratia marcesens* has been shown to act synergistically with Bt toxin against *S. littoralis* (Rigev *et al.*, 1996). Cholesterol oxidase from *Streptomyces* is highly toxic to cotton boll weevil (*Anthonomus grandis*) (Purcell *et al.*, 1993). Polyphenol oxidase can significantly reduce the protein quality of an insect diet, and thus reduce the growth of larvae feeding on that diet (Felton *et al.*, 1992).

Photorhabdus luminescens, the well-investigated bacterial species associated with nematodes of *Heterorhabditis* species contains a number of potentially insecticidal components (Ffrench-Constant, 2007). The encoding gene *tcdA* was cloned and assembled into expression constructs, containing 5′ and 3′ untranslated region sequences from a tobacco *osmotin* gene to improve expression levels of mRNA and protein in transgenic plants. Expression of toxin A at levels >0.07% of total soluble protein in leaves of transgenic *Arabidopsis thaliana* plants (Liu *et al.*, 2003) gave almost complete protection against larvae of the lepidopteran tobacco hornworm, *Manduca sexta*.

Avidin, aglycoprotein found in chicken egg white, sequesters the vitamin biotin. It has a strong insecticidal effect on many insects, although susceptibility varies widely between different insect species apparently based on biotin requirements. Expression of avidin in transgenic maize initially aimed to produce the protein as a high-value product, but maize seed containing more than 0.1 per cent avidin (of total protein) was fully resistant to larvae of three different coleopteran storage pests (Kramer *et al.*, 2000).

15.2.5.2. RNA interference

RNA interference (RNAi) is a gene silencing technology involving down regulation of the expression of specific genes. It has been widely used for genetic research in insects and has opened new vistas for new generation insect resistant transgenic crops through dsRNA directed against suitable insect target genes. RNAi is a post-transcriptional gene silencing mechanism that is initiated by the introduction of double-stranded RNA (dsRNA) into a cell (Fire *et al.*, 1998). The selective and robust effect of RNAi on gene expression makes it a valuable research tool, both in cell culture and in living organisms (refer to chapter 4 of this volume for details).

15.2.5.3. Genetically Modified Insects

The use of genetically modified insects represents a new innovative tool to address the insect borne disease of humans and crop insect pest losses. However despite rapid advances in the subject area, there is no widely accepted regulatory or biosafety framework that provides guidance on all aspects, although some of these are currently in development (Fontes, 2009; Benedict *et al.*, 2010; WHO, 2009). There are several strategies in the field when considering GM insects which may be classified by propensity of the genetic trait to establish or spread, although it should be noted that these are not necessarily alternatives, as self-limiting strategies can be aimed at population replacement, or self-sustaining ones aimed at suppression, e.g. self-limiting and self-sustaining (including non-GM methods, such as *Wolbachia*) or by desired outcome, e.g. population suppression and population replacement (conversion of the insects to a less harmful form) or by the method by which the heritable modification is achieved (classical genetics, recombinant DNA methods and foreign or novel DNA sequence, paratransgenesis) (Beech *et al.*, 2012). The first integration of exogenous DNA into the genome of an insect, termed 'germline transformation', was described in 1982 (Rubin & Spradling, 1982) based on *P* element-based transposon vectors in *Drosophila melanogaster*.

Insect transgenesis has allowed for deeper understanding of the biology of insects and human disease vectors (Wimmer, 2003) and has provided an important tool for the development of new strategies to control insect pests (Franz & Robinson, 2011). A number of insect species spanning the orders Diptera, Lepidoptera, Coleoptera and Hymenoptera have been transformed by using different transponsable elements, viz. piggyBac, Hermes, Mariner, Mino (Beach *et al.*, 2012).

There is a range of uses for GM insects from evolutionary biology and basic laboratory research to the use of insects as factories for production of proteins, and also for pest control in both agriculture and public health areas. Some current research activities with GM insects are listed in Table 5. A transgenic strain of pink bollworm (*Pectinophora gossypiella*), transformed only with a transgenic fluorescent marker (the red fluorescent protein, DsRed2), was developed to provide the existing sterile insect technique (SIT) programme in south western USA with a more reliable marker of released (and irradiated) moths (Simmons *et al.*, 2011). This strain has undergone a series of cage and open-field trials to assess its suitability for SIT. In 2006-2008, in the world's first open-field trials of a trans-genic insect, the strain was radiation-sterilised and released in large numbers (over 15 million in 2008) over cotton crops in Arizona, USA, for performance assessment and programmatic demonstration. Medfly transgenic sexing strains have undergone contained greenhouse trials to assess mating performance of males (Morrison *et al.*, 2009; Schetelig *et al.*, 2009). Similar sexing strains of olive fly were recently shown to suppress a wild-type population of olive flies in large cages (Ant *et al.*, 2011).

Recently, commerical release of genetically modified mosquito OX513A has been approved by The National Technical Commission for Biosecuirty (CTNBio) in Brazil for the control of *Aedes aegyptii* that spread dengue fever. OX513A, developed by British Biotechnology frim Oxitec Ltd, is described as the first GM insect to be considered safe for commercial use in Brazil. The Oxitec mosquito is a strain of wild species that contains two additional genes. Male mosquitoes (which do not bite) are released in order to mate with wild females. One gene makes the mosquito dependent on the antibiotic tetracycline, which will prevent their offspring from reproducing. The second gene makes the GM insect glow under certain light sources for identification and monitoring in the field. The successive releases of Oxitec males have been shown to reduce substaintially the wild population of dengue mosquitoes (www.oxitec.com). Although mosquitoes are not agricultural pests, the technology can also be adapted for use against insect vectors of crop pathogens as well as directly against crop pests.

15.2.5.4. Paratransgenesis

Symbiotic microbes gave a precious contribution to such a success, playing crucial roles in the biology and life cycle of most insect's species, affecting nutrition, development, reproduction, immunity, defense against natural enemies, speciation (Moran, 2006) and pest status (Hosokawa *et al.*, 2007). In depth knowledge of the molecular mechanisms underlying insect-microbes interactions and their phenotypic manifestation in the host has led to the development of

Table 5. Some examples of GM insects, potential uses and current status

Insect Species	Potential use of modified organism	Current status
Pink Bollworm	Improvements to sterile insect technique	Open field and programmatic scale trials
Mediterranean fruit fly	Protein production	Laboratory research
	Improvements to the sterile insect technique	Laboratory research and confined field trials
Olive fruit fly	Development of the sterile insect technique	Laboratory research and contained field trials
Cabbage looper	Production of recombinant antibodies	Laboratory research
	Developmental biology	Laboratory research
Silkworms	Protein production	Laboratory research and potential contained commercial production
Honey bee	Insecticide resistance	Laboratory research
	Comparative genomics	Laboratory research (transient expression only)
Yellow fever/dengue mosquito (*Aedes*)	Control of vector transmitting dengue	Open field trials
Aedes fluviatilis	Inhibition of malaria parasite	Laboratory research
Anopheles albimanus	Inhibition of malaria parasite	Laboratory research
Anopheles gambiae	Inhibition of malaria parasite	Laboratory research

Source: Beech *et al.* (2012)

methods for insect control. Two approaches have particular potential: the disruption of microbial symbionts required by insect pests, and manipulation of microorganisms with major impacts on insect traits contributing to their pest status (e.g. capacity to vector diseases, natural enemy resistance). The complete elimination of endosymbionts *in vitro* is possible using antibiotics; nevertheless, the use of antibiotics for field pest management is not a viable option, and finding new active compounds against endosymbionts will be required (Douglas, 2007). An alternative strategy is the manipulation of insect-fitness traits through symbionts known as paratransgenesis. It is a genetic manipulation of host's symbiotic microorganisms that resides within the host. This novel approach has been evaluated to eliminate the pathogen from vector population and to manage insect pests also. The majority of related research has been conducted in human disease vectors (Aksoy *et al.*, 2008; Favia *et al.*, 2008; McMeniman *et al.*, 2009; Hurwitz *et al.*, 2011), where the objective is to reduce the competence of the vector for human diseases, although there have been some agricultural applications for pests of citrus, grapevines and sugarcane

(Miller *et al.*, 2006, Pittman *et al.*, 2008, Gai *et al.*, 2009).

The first and successful example of this technique used blood sucking reduviid bug *Rhodnius prolixus* (Stal), an important vector of *Trypanosoma cruzi*, the causative agent of Chagas' disease in humans. The endosymbiont of this vector, *Rhodococcus rhodnii* has been transformed to express anti-parasite molecules that are lethal to *T. cruzi* (Beard *et al.*, 2001). Against plant pathogens, this approach has been evaluated to counteract the threat of Pierce's disease of Grape caused by xylem limited bacterium *Xylella fastidiosa* which is transmitted by glassy-winged sharpshooter *Homalodisca coagulata* (Say). Initially, this paratransgenic approach was focused to block the transmission of pathogens vectored by insects but now has been widened to manage insect pests also by targeting obligatory symbionts necessary for survival and development of insect. A prototype has been developed for paratransgenic control of termites based on genetically modified yeast that express lytic peptides inside termite gut thereby killing gut protozoa responsible for digestion of cellulose and subsequently, death of the termites (Hussender and Grace, 2005). Symbiotic flora are widespread in insects and can be exploited to deliver foreign genes to disrupt the pathogen transmission by insects as well as targeting obligatory symbionts which are necessary for survival and development of the insects via a Trojan-horse approach. The production and ultimate release of paratransgenic insects is potentially very promising but also worthy of much careful consideration with respect to ecological risks, human health and safety concerns. Moreover, the level of knowledge about the relationships between microorganisms associated with arthropods (symbionts) and microorganisms in the environment is also limited.

15.2.5.5. Genetic Improvement of Biocontrol Agents

The main reason for the limited success achieved with augmentative biological control is that biocontrol agents suffer from a number of important limitations, viz. narrow spectrum of activity, susceptibility to adverse environmental stress and slow speed of kill, etc. Recent advances in molecular biology techniques have offered opportunities to enter into a novel realm in overcoming these limitations through genetic improvement of biocontrol agents. It has broadened available techniques for genetic manipulation for diverse traits in species of interest (Atkinson *et al.* 2001; Kramer 2004; Arora and Shera, 2014).

Genetic improvement involves directed purposeful genetic alterations to enhance the efficacy of natural enemies for biological control. This may be achieved by conventional approaches as well as through recombinant DNA techniques. The conventional methods include strain selection, serial passage through hosts, mutation, conjugation, transduction, selective breeding, hybridization, etc. whereas genetic engineering approaches comprise indirect gene transfer methods (vector mediated) by using *Agrobacterium tumefaciens* as vector and direct gene transfer (vectorless methods) such as microprojectile bombardment with DNA or biolistics, direct DNA transfer into isolated protoplasts, electroporation, microinjection, etc. However, genetic improvement of arthropod natural enemies has received little attention because of the concern that breeding and prolonged rearing under

Table 6. Commercial genetically engineered *Bt* products for pest control

***Bt* strain**	**Trade Name**	**Company**	**Target insects**
Kurstaki recipient; *aizawai* donor	Condor, Cutglass (transconjugant)	Ecogen	Lepidoptera
Aizawai recipient; *kurstaki* donor	Agree, Design, Turex (transconjugant)	Thermo Trilogy	Lepidoptera (*Bt* resistant *Plutella xylostella*)
Bt kurstaki ED 7841	CRYMAX	Ecogen	Lepidoptera
Bt kurstaki ED7826	Leptinox (recombinant)	Ecogen	Lepidoptera
Bt kurstaki EQ7673	Raven (recombinant)	Ecogen	Lepidoptera, Coleptera
δ-endotoxin encapsulated in *P. fluroscens*	MVO, MATCH, M-Trak (Cell cap®)	Mycogen	Lepidoptera, Coleoptera
Bt israelensis	Acrobe	Cynamid	Diptera
Bt israelensis	Skeetal	Novo-Nordisk	Mosquito
Bt israelensis	Teknar, Teknar HPD	Zoecon	Black flies, mosquito
Bt israelensis	Vectobac G, Gnatrol, Bactimos	Abbott	Mosquitoes, blackflies
Bt israelensis 187, CS-8	MieJueLing Preparation	Huazhong Agri. University	Diptera

Source: Modified after Sharma (2009)

artificial condition would necessarily result in laboratory adapted strains that would perform poorly in the field. Entomophagous insects may be improved for climatic tolerance, sex ratio, host finding ability, host preference, increased host range, increased pesticide resistance, etc. (Sharma, 2009).

The main objectives in genetically altering microbes are to increase host range, virulence and persistence. The *cry* genes from *Bacillus thuringiensis* have been cloned and expressed in wide variety of organisms (baculoviruses to cyanobacteria) as well in plants in attempts to improve their delivery and efficacy against insect pests. Several commercial products and novel strains of *Bt* have been developed through biotechnological intervention (Table 6). Apart from *B. thuringiensis*, binary toxin from different *B. sphaericus* strains has been expressed in different hosts like *Escherichia coli*, non or low toxic *B. sphaericus* and crystal minus *Bt israelensis* as well as in *Caulobacter cresentus* or cynobacteria *Anabaena* sp (Xu *et al.*, 2000).

Insect viruses especially baculoviruses are mostly specific viruses which can replicate only in hosts. The recombinant DNA technology has its current applications in inserting foreign genes into insect baculoviruses and achieving their rapid and efficient expression in the recipient host systems (Table 7). Candidate genes for hyperexpression in the baculoviruses include those encoding insect specific enzyme genes (juvenile hormone esterase gene); hormone genes (eclosion hormone gene; diuretic hormone

Table 7. Genetically engineered baculoviruses for insect pest management

Insect species	NPV	Gene/Toxin	Impact on biological activity	References
Autographa californica	AcNPV	Scorpion, *Androctonus australis* toxin gene (*AaIT*)	Falling off infected *H. virescens* larvae 5 to 11 hours before death and unable to climb plant to continue feeding; Less foliage consumption	Hoover *et al.* (1995)
	AcMNPV	*AaIT*	Reduces time to kill (25-40%) of *S. eridania, T. ni* and *H. virescens*	Stewart *et al.* (1991); McCutchen *et al.* (1991); Carbonell *et al.* (1988)
	AcNPV	Enhancin gene from *Trichoplusia ni* granulovirus	Increase in infection by 21-folds in *S. exigua*	Hayakawa *et al.* (2000)
	AcMNPV	*Agelenopsis perta* (*Mu-Aga-IV*) *Anemonia sulcata* (*AsII*) *Stydactyla* (*sh1*)	Increase in effectiveness to kill of *S. frugiperda* and *T. ni*	Prikhodko *et al.* (1996)
	AcMNPV	Juvenile esterase from *Heliothis virescens*	Increase in speed of kill by 20 %	Hammock *et al.* (1990); Bonning *et al.* (1999)
	AcMNPV	Viral gene (*egt*) - deletion	Increase in speed of kill, reduced feeding	O'Reilly and Miller (1991)
	AcNPV	Mite toxin	Increase in speed of kill	Tomalski and Miller (1991)
	AcNPV	Scorpion *Leiurus quinquestriatas hebraeus* toxin gene (*LqhIT1, LqhiT2*)	Improvement in speed of kill (32 %); Decrease in median survival time (34%) in *H. virescens*;	Gershburg *et al.* (1998) Harrison and Bonning (2000)
	AcNPV	Spider *Diguetia canities* toxin gene (*DTX9.2*)	Induced paralysis and improved speed of kill	Hughes *et al.* (1997)
Rachiplusia ou	RoMNPV	Scorpion *L. quinquestriatas hebraeus* toxin gene (*LqhiT2*)	Improvement in speed of kill (40 %) in *O. nubilalis* and *H. zea*	Harrison and Bonning (2000)
Bombyx mori	BmNPV	*AaIT*	Increase in biological activity	Maeda *et al.* (1991)
	BmNPV	Diuretic hormone	Haemolymph of the infected larvae was decreased by 30 per cent with an increased mortality	Maeda (1989)
Heliothis zea	HzNPV	*AaIT*	Mortality at faster rate	Treacy *et al.*, (2000)

gene) and insect specific foreign toxic genes (scorpion venom toxin genes; predatory mite toxin gene; predatory spider toxin gene; parasitic wasp venom gene and Bt δ-endotoxin genes) (Bonning *et al.,* 2003; Hu *et al.,* 2003). It is worthwhile to mention that genetic engineering of BCAs is potentially very promising and have led to the development of more effective entomopathogens with desired pathogenicity, virulence, broad host range and persistence, etc. providing a valuable tool for sustainable pest management. The potential environment risks associated with the release of these genetically manipulated bioagents need to be addressed at all four trophic levels if we are to really exploit these biotechnological interventions in improving the biocontrol agents for sustainable pest management.

15.3. AREA WIDE - IPM

Area-wide integrated pest management (AW-IPM) focuses on the preventive management of pest populations throughout the ecosystem. It seeks to treat all habitats of the pest population so that none produces migrants to re-establish significant infestations in areas of concern. In contrast, the conventional strategy focuses narrowly on defending the valued entity (crop, livestock, people, buildings, etc.) from direct attack by pests. Suppression across a broad area may result in reduced reinfestation by migration from nearby unmanaged areas, and the pest management tactics employed may be more effective, particularly ecologically based tactics, when applied areawide (Elliot *et al.*, 2008). AW-IPM also has ancient roots in coping with locust plagues and vector-borne diseases (Klassen, 2000). By systematic use of quarantines, Venice and Milan in 1347–1350 contained bubonic plague transmitted by the oriental rat flea *Xenopsylla cheopis* (Rothschild), and this approach was gradually adopted throughout Europe.

Two conceptual mathematical models has been developed for estimating the minimum area that needs to be considered to successfully apply a series of control tactics according to the AWPM approach against insect pests. The prototype model creates basis for a decision-support tool to assess the minimum dimensions of an intervention area required for the establishment of a pest-free area. For the development of the model, two main situations have been considered: (1) the control area is fixed in size (fixed-area model) and there is no advancing pest control front, and (2) the control area is expanding according to the "Rolling-carpet principle" (Barclay *et al.*, 2011).

Hendrichs *et al.* (2005) describe the basic spatial elements of an AW-IPM program. The first is the core area, in which the aim is to reduce (in case of a suppression strategy) or eliminate the pest species. The core area may contain the actual resource of value, but in other cases, removal of the pest from the core area may simply have a strategic value by protecting crops situated elsewhere or by protecting humans or livestock against disease vectors (in case of a containment or a prevention strategy). The second is a buffer zone that borders the core area on one or more sides and within which control methods attempt to kill the target insects within that zone, including those that enter the zone from outside. In the case of the Fixed-area model, there is a core area to be protected and a buffer zone on all sides of the core area. This model was followed in Chile for fruit fly eradication (Gonzalez and Tronsco, 2007). In the

Rolling-carpet model, there is a buffer on only one side and pest free zones on the other sides. The width of the buffer zone is central to determine the minimum area of an AWPM program, since it defines the smallest possible program. Both these models consist of two components such as a biological component (i.e., dispersal) and an economic component (break-even analysis). The dispersal part describes the movement of the insects across the buffer zone and will determine the width of the buffer zone. The economic component of the model will, given a certain width of the buffer zone determined by the dispersal part, allow a calculation of costs and revenues of the control program and will determine the break-even size of the core area at which control costs equal revenues. The Rolling-carpet model extends the Fixed-area model by introducing a temporal element to the model, that is, the success of the control program permits the core area to be extended regularly when the buffer zone moves onwards (Hendrichs *et al.*, 2005).

AW-IPM came into a degree of prominence in the late 19th and early 20th centuries with the development of classical biological control (1888), the use of pest-resistant plants (such as the grafting of all European grapes on phylloxera-resistant American rootstocks in the 1870s), campaigns to eradicate the Gypsy moth *Lymantria dispar* (L.) in Massachusetts (1890–1901) and cattle ticks *Boophilus* spp. from the Southern USA (1906–1943), the suppression of yellow fever and malaria vectors in Havana (1898) and in the Panama Canal Zone (1904), and the organization of mosquito-abatement districts, first in West Africa and then in several continents (1900).

Several eradication programmes have been highly successful using areawide concepts as an integral part of the programme. The late Edward F. Knipling, an ARS pioneer in insect control, was a strong proponent of the areawide IPM concept. One of his major achievements was development of the sterile-male release technique, which eliminated screwworm and other insect infestations in many parts of the world. The major requirement for successful application of SIT is the ability to deliver to the field population, over large areas, large numbers of sexually active genetically sterile males.

SIT is species-specific and environmentally benign, and increases in effectiveness as the size of the target population declines. The technologies used to rear, sterilize and distribute the males might all have a negative impact on final field fitness, indicating that larger numbers of insects must be released than those predicted by simple models (Benedict and Robinson, 2003). The paradigm of this approach has been the successful elimination of the New World screwworm, *Cochliomyia hominivorax* (Coquerel), from the southern states of the USA, Mexico and all of Central America (Knipling, 1979; Baumhover, 2002). It was eradicated from the island of Curacao in 1954, Florida in 1958-59, entire USA during 1962. A permanent barrier of sterile insects is maintained over eastern Panama to protect NWS- free areas of USA, Mexico, C. America and Panama. Since 1991, the screwworm also has been eliminated from Belize (1994), Guatemala (1994), El Salvador (1995) and Honduras (1995) (USDA-APHIS, 1998).

SIT has also been successfully used in fruitfly, *Dacus cucurbitae* eradication programme undertaken extensively in Japan from 1972 onwards and the *Bactocera dorsalis* suppression programme in Taiwan from 1975 onwards. On the other hand Mediterranean fruitfly, *Ceratitis capitata* eradication programme was a huge success in Mexico and other Central American Countries. Large scale programmes are in progress in many countries for onion fly, *Delia antiqua*; European cockchafer, *Melolontha melolontha*; European cherry fruitfly, *Rhagoletis cerasi*, codling moth, *Cydia pomonella* and grape moth, *Eupoecilia ambiguella*. However, the complexity in implemen-tation puts these techniques beyond the reach of individual farmers and the programmes have to be planned, organized and implemented by govern-ment agencies. Perhaps owing to this reason, hardly any SIT programmes have been undertaken in Asian countries where farms are very small and large scale centralized pest control strategies are difficult to implement (Dhaliwal and Arora, 2006)

In 1999, the US Department of Agri-culture (USDA), Agricultural Research Service (ARS) initiated the Hawaii Fruit Fly AW-PM programme to suppress fruit flies below economic thresholds thereby reducing the use of organophosphate insecticides (Vargas *et al.*, 2003). The programme included developing and inte-grating biologically based pest technology into a comprehensive IPM package, i.e. field sanitation, application of protein bait sprays, male and female annihilation with male lures and other attractants, sterile insect releases and conservation or release of beneficial parasitoids (Vargas *et al.*, 2008).

Cassava mealybug, *Phenacoccus manihoti* used to be impediment in Cassava crop in African continent. In 1973, Cassava in Central Africa was found to be attacked by the Cassava mealybug, *Phenacoccus manihoti* (Matile-Ferrero). The attack of this insect pest was so profound that it created starvation for 200 million people for whom cassava had become a staple crop. A team led by Dr. Hans Herren of the International Institute for Tropical Agriculture (IITA) successfully implemented the largest classical biological control programme in history. The area wide aerial application of mass reared parasitoid, *Apoanagyrus lopezi* (DeSantis) brought Cassava mealybug under control (Klassen 2005).

Areawide programmes using phero-mones as a single control tactic have been conducted to disrupt lepidopteran pests or to trap-out bark beetles over large areas (Huber *et al.*, 1979; Schneider, 1989). Areawide programmes using augmen-tation or release of natural enemies, insect pathogens, sterile males, insecticides, etc. have been implemented for control of pests, including codling moth, cotton bollworm, tobacco budworm, fruit flies, parasitic flies, boll weevil, corn rootworm, fire ants and whiteflies. A few models have addressed areawide crop- and pest-specific parameters of growth, dispersal, host resistance and predators or parasitoids in attempts at predicting population and damage levels (Koul and Cuperus, 2008).

The adoption of areawide mating disruption launched by The Agricultural Research Service, USA against the codling moth, a pest in apple and pear orchards, on 7700 acres (3100 ha) in the Pacific north-west reduced use of insecticide sprays targeting codling moth by 24 per cent (IPM Report, 2006). In 2002, ARS scientists in Stillwater, Oklahoma started an AWPM project on Russian wheat aphid and greenbug on wheat in the US Great Plains using customized cultural practices, pest-resistant cultivars, biological control agents and other biologically based pest control technologies. Also in 2002, an AWPM project began for tarnished plant bug on cotton in the delta of Mississippi and Louisiana using host destruction, host-plant resistance and remote-sensing technology.

Areawide management of oriental fruit fly and sweet potato weevil in Taiwan is well known. An areawide control programme for oriental fruit fly using methyl eugenol has been conducted for many years, but its population remains at a high level in the field. Culture and protein bait containing spinosad has also been used, while non-chemical efforts are concentrated on: (i) organizing farmers; (ii) establishing geological information; (iii) monitoring fruit fly density; (iv) providing fruit production information; (v) holding training programmes; and (vi) maintaining orchard sanitation (Koul and Cuperus, 2008).

AW-IPM requires multiyear planning, and an organization dedicated exclusively to its implementation, whereas conventional pest management involves minimal forward planning, tends to be reactive, and is implemented independently by individual producers, businesses, or households. AWIPM tends to utilize advanced technologies, whereas the conventional strategy tends to rely on traditional tactics and tools (Klassen, 2005).

15.4. CONCLUSIONS

The pest problems is agriculture are probably as old an agriculture itself. For a long time farmers could do nothing to contain these losses and slowly developed a number of cultural and mechanical control measures largely through experience. The introduction of synthetic chemical insecticides during the 1940s and 50s enabled us for the first time to gain an upper hand in the struggle against insects. However, the instensive and extensive use, misuse and abuse of insecticides during the ensuing decades caused widespread damage to the environment. In addition, the pest problems in some crops increased following intensive application of insecticides.

The concept of integrated pest management originated to minimize the use of chemical insecticides and to lay increased emphasis on biological control and other non-chemical tactics. But inspite of its inspirational value, most IPM programmes still include economic threshold level based application of chemical insecticides as a major input. There is now ample evidence of the adverse effects of these synthetic chemicals on the ecosystem, the environment and on human and animal health. Therefore, it is imperative to restrict and eventually eliminate the use of chemical pesticides in IPM programmes. In light of this, biointensive integrated pest management is considered the desirable path to sustainability in agriculture. As this chapter illustrates there is now a wide array of techniques available to replace the use of conventional chemical insecticides in IPM programmes. The challenge before applied entomologists is to develop, validate and disseminate the site-specific biointensive IPM technologies to the farmers. There is also a need to strengthen research in strategic areas like genetically improved bioagents/biopesticides and pest-resistant transgenics. Utilization of molecular approaches will go a long way in overcoming the major limitations of biological pest control. With these advancements, biointensive IPM appears poised to play a greater role as an environmentally benign alternative to chemical insecticide based IPM for sustainable insect pest management in future.

REFERENCES

Acharya VS and Bhargava MC (2007) Effect of chitin synthesis inhibitor, diflubenzuron against larva of Spodoptera litura Fab. Noctuidae: Lepidoptera. *Insect Environment* **13:** 90-91.

Afifi AI, El Arnaouty SA, Attia AR and Abd Alla Ael M (2010) Biological control of citrus mealybug, *Planococcus citri* (Risso) using coccinellid predator, *Cryptolaemus montrouzieri* Muls. *Pakistan Journal of Biological Sciences* **13:** 216-222.

Aggarwal N (2013) Botanical Pesticides in Insect Pest Management. In: Arora R, Singh B and Dhawan, AK (eds), *Theory and Practice of Integrated Pest Management*. Scientific Publishers, Jodhpur, pp. 120-134.

Ahmed I, Yokota A, Yamazoa A and Fujiwara T (2007) Proposal of *Lysinibacillus boronitolerans* gen. nov. sp. nov., and transfer of *Bacillus fusiformis* to *Lysimibacillus fusiformis* comb. nov and *Bacillus sphaericus* to *Lysinibacillus sphaericus* comb. nov. *International Journal of Systematic Evolutionary Microbiology* **57**: 1117-1125.

Aksoy S, Weiss B and Attardo G (2008) Paratransgenesis applied for control of tsetse transmitted sleeping sickness. In: Aksoy S (ed), *Transgenesis and the Management of Vector Borne Disease.* Landes Bioscience, Austin (TX), USA. pp 35-48.

Aller HE and Ramsay JR (1988) RH-5849- A novel insect growth regulator with a new mode of action. *Brighton Crop Protection Conference: Pests and Diseases* **2**: 511-518.

Alston DG, Lindstrom T and Murray MS (2008) Woolly and green apple aphid suppression on apple trees with spirotetramat 240 SC. In: *Proceedings of 82nd Annual Orchard Pest and Disease Management Conference*, January 9-11, 2007, Hilton, Portland, USA. p 36.

Ameta OP and Bunker GK (2007) Efficacy of NNI0001 (Flubendiamide) 480SC against diamond back moth, *Plutella xylostella* L. in cabbage and its effects on natural enemies under field condition. *Pestology* **31**: 21-24.

Anonymous (2013) *Package of Practices for Kharif Crops.* Punjab Agricultural University, Ludhiana.

Ant T, Koukidou M and Alphey L (2011) Genetic engineering of the olive fruit fly, *Bactrocera oleae*, for use in the sterile insect technique (SIT). *5th Meeting of the IOBC/WPRS WG Integrated Protection of Olive Crops Bulletin,* 15-20 May 2011, Jerusalem, Israel.

Arakawa T (2008) Synergistic effect of a fungicide containing polyoxin B with insect growth regulators (IGRs) in the killing of common cutworm, *Spodoptera litura* (Lepidoptera: Noctuidae). *Applied Entomology and Zoology* **43**: 167-171.

Arora R and Battu GS (2001) Biopesticides: An Indian perspective. *Souvenir-Conference on Biopesticides: Emerging Trends,* (BET, 2001). Society of Biopesticide Science. India, Jalandhar, pp. 17-24.

Arora R and Shera PS (2014) Genetic improvement of biocontrol agents for sustainable pest management. In: K Sahayaraj (ed), *Basic and Applied Aspects of Biopesticides.* Springer India, pp. 255-285. DoI 10.1007/978-81-322-1877-7-15

Arora R and Singh P (2011) Bioefficacy of indoxacarb against *Helicoverpa armigera* in Egyptian clover seed crop. *Journal of Research, Punjab Agricultural University* **48**: 109-111.

Arora R, Battu GS and Ramakrishnan N (2000) Microbial pesticides: Current status and future outlook. In: Dhaliwal GS and Singh B (eds), *Pesticides and Environment.* Commonwealth Publishers, New Delhi, pp. 344-395.

Arora R and Dhaliwal GS (1994) Botanical pesticides in insect management: Ecological perspectives. In: Dhaliwal GS and Kansal BD (eds), *Management of Agricultural Pollution in India*. Commonwealth Publishers. New Delhi, pp. 215-245.

Atkinson PW, Pinkerton AC and O'Brochta DA (2001) Genetic transformation systems in insects. *Annual Review of Entomology* **46**: 317-346.

Baer CF, Tripp DW, Bjorksten A and Antolin MF (2004) Phylogeography of a parasitoid wasp (*Diaeretiella rapae*): No evidence of host associated lineages. *Molecular Ecology* **13**: 1569-1589.

Balachowsky AS (1951) *La lute contre les insects*. Payot, Paris.

Balikai RA (2002) Bio-efficacy of buprofezin 25 SC against grape mealybug *Maconellicoccus hirsutus* (Green). *Pestology* **26**: 20-23.

Barclay HJ, Matlock R, Gilchrist S, Suckling DM, Reyes J, Enkerlinn WR and Vreysen MJ (2011) A conceptual model for assessing the minimum size area for an area-wide integrated pest management program. *International Journal of Agronomy* (http://dx.doi.org/10.1155/2011/409328).

Barrett AGM, Carr RA, Attwood SV, Finch MAW and Richardson G (1985) The application of novel carbanion chemistry in milbemycin-avermectin synthesis. In : James NF (ed.), *Recent Advances in the Chemistry of Insect Control*, Royal Society of Chemistry, London.

Battu GS and Arora R (1997) Insect pest management through microorganisms. In: Dadarwal KR (ed.), *Biotechnological Approaches in Soil Microorganisms for Sustainable Crop Production.* Scientific Publishers, Jodhpur, pp 223-246.

Battu GS, Arora R and Dhaliwal GS (2002) Prospects of baculoviruses in integrated pest management. In: Koul O and Dhaliwal GS (eds), *Microbial Biopesticides.* Taylor & Francis, London, pp. 215-238.

Baum JA (1998) Transgenic *Bacillus thuringiensis. Phytoprotection* **79**: 127-130.

Baum JA, Johnson TB and Carlton TB (1999) *Bacillus thuringiensis*: Natural and recombinant biopesticide products. *Methods in Biotechnology* **5**: 189-209.

Baumhover AH (2002) A personal account of programs to eradicate the screw worm, *Cochliomyia hominivorax*, in the United States and Mexico with special emphasis on the Florida program. *Florida Entomologist* **85**: 669–676.

Beard CB, Dotsona EM, Penningtona PM and Eichlera S, Rosalesa CC and Durvasula RV (2001) Bacterial symbiosis and paratransgenic control of vector-borne chagas disease. *International Journal of Parasitology* **31**: 621-627.

Beech CJ, Koukidou M, Morrison NI and Alphey L (2012) Genetically modified insects: science, use, status and regulation. *Collection of Biosafety Reviews* **6**: 66-124.

Benbrook CM (1996) *Pest Management at the Crossroads*. Consumers Union, Yonkers, NY. 272 pp.

Benedict M, Eckerstorfer M, Franz G, Gaugitsch H, Greiter A, Heissenberger A, Knols B, Kumschick S, Nentwig W and Rabitsch W (2010) Defining Environmental Risk Assessment Criteria for Genetically Modified Insects to be Placed on the EU Market. Report from EFSA-funded project CT/ EFSA/ GMO/2009/03. Available at http://www.efsa.europa.eu/en/supporting/doc/71e.pdf.

Benedict JH and Ring DR (2004) Transgenic crops expressing Bt proteins: current status, challenges and outlook. In: Koul O and Dhaliwal GS (eds), *Transgenic crop Protection: Concepts and Strategies,* Science publishers, Inc., Enfield (NH), US.

Benedict MQ and Robinson AS (2003) The first releases of transgenic mosquitoes: an argument for the sterile insect technique. *Trends in Parasitology* **19**: 349-35

Bentur JS (2004) Genetics of Insect resistance in Crop Plants. In: Dhaliwal GS and Singh R (eds), *Host Plant resistance to Insects: Concept and Approaches*. Panima Publishing Corporation, New Delhi, pp. 126-158.

Bhattacharjee C, Prasad DT, Manjunath NH, Sanya D, Zarga SM (2012) Exploring plant proteinase inhibitors. *Genomics and Applied Biology* **3**: 8-21.

Bhattacharjee C, Theertha Prasad D, Manjunath NH, Sanyal D and Zarga SM (2012) Exploring Plant Proteinase Inhibitors. *Genomics and Applied Biology* **3**: 8-21.

Biddinger David J, Weber Donald D, and Hull Larry A (2009) "Coccinellidae as Predators of Mites: Stethorini in Biological Control" Publications from USDA-ARS / UNL Faculty. Paper 387.

Biradar SK, Sundaram RM, Thirumurugan T, Bentur JS, Amudhan S, Shenoy VV, Mishra B, Bennet J and Sharma PN (2004) Identification of flanking SSR markers for a major rice gall midge resistance gene *Gm1* and their validation. *Theoretical and Applied Genetics* **109**: 1468–1473.

Bishnoi RC (1999) Bioefficacy of Rimon 10 EC against diamaondback moth *Plutella xylostella* affecting cabbage. *Pestology* **23**: 3-5.

Blackwell M (2011) The fungi: 1, 2, 3... 5.1 million species? *American Journal of Botany* **98**: 426-438.

Bloomquist JR (2001) GABA and glutamate receptors as biochemical sites for insecticide action. In: Ishaaya I (ed), *Biochemical Sites of Insecticide Action and Resistance,* Springer, Berlin, pp. 17-41.

Boets A, Arnaut G, Van Rie J and Damme N (2004) Toxins. *United States Patent* No. 6,706,860.

Bone LW and Singer S (1991) Control of parasitic nematode ova/larvae with a *Bacillus laterosporus*. *United States Patent* No. 5,045,314.

Bonning BC, Boughton AJ, Jin H and Harrison RL (2003) Genetic enhancement of baculovirus insecticides. In: Upadhyay RK (ed) Advances in microbial control of insect pests, Kluwer Academic/Plenum Publishers, New York.

Bonning BC, Possee RD and Hammock BD (1999) Insecticidal efficacy of a recombinant baculovirus expressing JHE-KK, a modified juvenile hormone esterase. J Invertebr Pathol 73 : 234-236.

Brar KS, Virk JS, Jalali SK, Sohi AS and Shenhmar M (2007) Evaluation of multiple insecticide tolerant strain of *Trichogramma chilonis* Ishii for the management of cotton bollworms. *Journal of Insect Science* **20**: 208-211.

Bretschneider T, Benet-Buchholz Jordi B, Reiner and Ralf N (2003) Spirodiclofen and spiromesifen—Novel acaricidal and insecticidal tetronic acid derivatives with a new mode of action. *Chimia* **57**:697–701.

Brookes G and Barfoot P (2013) *GM Crops: Global Socio-economic and Environmental Impacts 1996-2012.* PG Economics Ltd, Dorchester, UK. Available at: http://www.pgeconomics.co.uk/ pdf/ 2012globalimpactstudyfinal.pdf.

Bukovinszky T, Gols R, Posthumus MA, Vet LEM and Lenteren JC (2005) Variation in plant volatiles and attraction of the parasitoid *Diadegma semiclausum*

(Hellen). *Journal of Chemical Ecology* **31**: 461-80.

Butron A, Chen YC, Rottinghaus GE and McMullen MD (2010) Genetic variation at bx1 controls DIMBOA content in maize. *Theory and Applied Genetics* 120:721–734

Carbonell LF, Hodge MR, Tomalski MD and Miller LK (1988) Synthesis of a gene coding for an insect specific scorpion neurotoxin and attempts to express it using baculovirus vectors. Gene 73: 409-418.

Chazeau J 1985. Predaceous insects. In: Helle W and Sabelis MW (Eds), Spider Mites; Their Biology, Natural Enemies, and Control, vol. B. Elsevier, Amsterdam, pp. 211–246.

Chen M, Shelton A amd Ye G (2011) Insect resistant genetically modified rice in China. From research to commercialization. *Annual Review of Entomology* **56**: 81-101.

Choudhary B and Gaur K (2010) *Bt Cotton in India: A Country Profile*. ISAAA Series of Biotech Crop Profiles. ISAAA: Ithaca, NY.

Combs D and Reissig WH (2008) Ultor® 150 SC for the control of New York apple pests. In: *Proceedings of 82nd Annual Orchard Pest and Disease Management Conference,* January 9-11, 2007, Washington State University, Pullman, WA. 25 pp.

Cook SM, Khan ZR and Pickett JA (2007) The use of push-pull strategies in integrated pest management. *Annual Review of Entomology* **52**: 375–400.

Cork A, Alam SN, Rouf FMA and Talekar NS (2005) Development of mass trapping technique for control of brinjal shoot and fruit borer, *Leucinodes orbonalis* (Lepidoptera: Pyralidae). *Bulletin of Entomological Research* **95**: 589–596.

de Faria MR and Wraight SP (2007) Mycoinsecticides and mycoacaricides: a comprehensive list with worldwide coverage and international classification of formulation types. *Biological Control* **43**: 237-256.

Dent D (2000) *Insect Pest Management* 2nd Edition, CABI Publishing, UK. 399 pp.

Dhaliwal GS and Arora R (2001) Role of phytochemicals in integrated pest management. In: Koul O and Dhaliwal GS (eds), *Phytochemical Biopesticides.* Harwood Academic Publishera, Amsterdam, The Netherlands. pp. 97-117.

Dhaliwal GS and Arora R (2006) *Integrated Pest Management: Concepts and Approaches*. 2nd Revised edition, Kalyani Publishers, New Delhi, 369 pp.

Dhaliwal GS and Singh R (eds) (2004) *Host Plant Resistance to Insects*. Panima Publ. Corp., New Delhi, 578 pp.

Douglas AE (2007) Symbiotic microorganisms: untapped resources for insect pest control. *Trends in Biotechnology* **25**: 338-342.

Dufour R (2001) Biointensive Integrated Pest Management (IPM) Available at: *http://www.attra.ncat.org/attra-pub/PDF/ipm.pdf*

Dutky SR (1940) Two new spore forming bacteria causing milky diseases of Japanese beetle larvae. *Journal of Agricultural Research* **61**: 57-68.

Eilenberg J, Hajek AE and Lomer C (2001) Suggestions for unifying the terminology in biological control. *BioControl* **46**: 387-400.

Elliott NC, Onstad DW and Brewer MJ (2008) History and Ecological Basis for Areawide Pest Management. In: Koul O and Cuperus GW (eds), *Areawide Pest Management. Theory and Implementation*, CAB International, UK. pp 15-33.

El-Sayed AM, Byers JA, Manning M, Jürgens A, Mitchell VJ and Suckling M (2008) Floral scent of Canada thistle and its potential as a generic insect attractant. *Journal of Economic Entomology* **101**:720–727.

FAO (1990) *Food, Agriculture, Conservation and Trade Act of 1990*, Public Law 101-624, Title XVI, Subtitle A, Section 1603. Government Printing Office, Washington, DC.

Favert EM and Yousten AA (1985) Insecticidal activity of *Bacillus laterosporus. Journal of Invertebrate Pathology* **45**: 195-203.

Favia G, Ricci I, Marzorati M, Negri I, Alma A, Sacchi L, Bandi C and Daffonchio D (2008) Bacteria of the genus Asaia: A potential paratransgenic weapon against malaria. In: Aksoy S (ed), *Transgenesis and the Management of Vector Borne Disease.* Landes Bioscience, Austin (TX), USA. pp 35-48.

Felton GW, Danato K, Del VRJ and Duffet SS (1992) Impact of oxidized plant henolics on the nutritional quality of dietary protein to noctuid herbivore, *Spodoptera exigua. Journal of Insect Physiology* **38**: 277-285.

Ferror DN (2013) *Cultural Control.* In: Radcliffe EB, Hutchison WD and Canalado RE (eds), *Radliffe's IPM World Textbook.* http//ipmworld.umn. edu, University of Minnesota, St Paul, MN.

Ffrench–Constant RH, Dowling A and Waterfield NR (2007) Insecticidal toxins from *Photorhabdus bacteria* and their potential use in agriculture. *Toxicon* **49**: 436–451.

Fischer MH and H Mrozik (1989) Chemistry. In: Campbell WG (ed), *Ivermectin and abamectin,* Springer, Berlin.

Fischer V (2008) Control of pear psylla with Ultor®. In: *Proceedings of 82nd Annual Orchard Pest and Disease Management Conference,* January 9-11, 2007, Hilton, Portland, USA. 27 pp.

Flint HM and Doanne CC (2013) Understanding semiochemicals with emphasis on insect sex pheromones in integrated pest management programs. In: Radcliffe EB, Hutchison WD and Canalado RE (eds), *Radliffe's IPM World Textbook.* (http//ipmworld.umn.edu), University of Minnesota, St Paul, MN.

Foissac X, Loc NT, Christou P, Gatehouse AMR and Gatehouse JA (2000). Resistance to green leafhopper (*Nephotettix virescens*) and brown plant hopper (*Nilaparvata lugens*) in transgenic rice expressing snowdrop lectin (*Galanthus nivalis agglutinin*; GNA). *Journal of Insect Physiology* **46**: 573–583.

Fontes E.(2009) Risk assessment and risk management under the Cartagena Protocol on Biosafety. *Asian Pacific Journal of Molecular Biology and Biotechnology* **17**: 97-98.

Franz G and Robinson AS (2011) Molecular technologies to improve the effectiveness of the sterile insect technique. *Genetica* **139**: 1-5.

Fuog D, Fergusson SJ and Fluckiger C (1998) Pymetrozine: A novel insecticide affecting aphids and whiteflies. In: Ishaaya I and Degheele D (eds), *Insecticides with Novel Modes of Action: Mechanism and Application.* Springer, Berlin.

Gai C, Lacava P, Quecine M, Auriac M, Spotti LJ, Araújo W, Miller TA and Azevedo JL (2009) Transmission of *Methylobacterium mesophilicum* by *Bucephalogonia xanthophis* for paratransgenic control strategy of Citrus variegated chlorosis. *The Journal of Microbiology* **47**: 448-454.

Gant DB, Bloomquist JR, Ayad M and Chalmers AE (1990) A comparison of mammalian and insect GABA receptor chloride channels. *Pesticide Science* **30**: 355–356.

Gatehouse AMR and Boulter D (1983) Assessment of the antimetabolic effects of trypsin inhibitors from cowpea (*Vigna unguiculata*) and other legumes on development of the bruchid beetle *Callosobruchus maculatus. Journal of Science of Food and Agriculture* **34**: 345–350.

Gatehouse JA (2011) Prospects for using proteinase inhibitors to protect transgenic plants against attack by herbivorous insects. *Current protein Peptide Science* **12:** 409-416.

Gautam RD, Chi WD and Lessey M (1998). Preliminary studies on inoculative releases of Australian beetle *Cryptolaemus montrouzieri* Mulsant and another Indian ladybird, Scymnus coccivora Aiyar against pink mealybug, *Maconellicoccus hirsutus* Green at point Fortin. *Proceedings of the First Symposium on the Hibiscus Mealybug*, Centeno, Trinidad and Tobago, pp. 25–29.

Gershburg E, Stockholm D, Froy O, Rashi S, Gurevitz M and Chejanovsky N (1998) Baculovirus mediated expression of scorpion depressant toxin improves the insecticidal efficacy achieved with excitatory toxins. *FEBS letters* **422**: 132-136.

Glare T, Caradus J, Gelernter W, Jackson T and Keyhani N (2012) Have biopesticides come of age? *Trends in Biotechnology* **30**: 250-257.

Glare TR and O'Callaghan M (2000) *Bacillus thuringiensis: Biology, Ecology and Safety*. John Wiley & Sons, Chichester.

Gogi MD, Sarfraz RM, Dosdall LM, Arif MJ, Keddie AB and Ashfaq M (2006) Effectiveness of two insect growth regulators against *Bemisia tabaci* (Gennadius) (Homoptera: Aleyrodidae) and *Helicoverpa armigera* (Hübner) (Lepidoptera: Noctuidae) and their impact on population densities of arthropod predators in cotton in Pakistan. *Pest Management Science* **62**: 982-990.

Gonzalez J and Tronsco P (2007) The fruit fly exclusion programme in Chile In: Vreysen AS, Robinson AS and Hendrichs J (eds), *Area-Wide Control of Insect Pests*. Springer, Dodrecht, The Netherlands. pp 641-651.

Gordon RD (1985) The Coccinellidae (Coleoptera) of America North of Mexico. Journal *of the New York Entomological Society* **93**: 88–99.

Guiderdoni E, Cordero MJ, Vignols F, Garcia-Garrido JM, Lescot M, Tharreau D, Meynard D, Ferriere N, Notteghem JL and Delseny M (2002) Inducibility by pathogen attack and developmental regulation of the rice *Ltp1* gene. *Plant Molecular Biology* **49**: 679-695.

Gujar GT and Dhillon MK (2011) Status of Bt transgenic crop research in India. In: A.K. Dhawan AK, Singh B, Arora R and Bhullar MB (eds), *Recent Trends in Integrated Pest Management*. Invited Papers, 3rd Congress on Insect Science, Pest Management for Food Security and Environment Health. April 18-20, 2011, INSAIS, PAU, Ludhiana, India, pp. 1-11.

Hallman GJ (2001) Irradiation as a quarantine treatment. In: Molins R (ed), *Food Irradiation: Principles and Applications*. Wiley, New York, pp. 113-130.

Hammes GG, Sherrod D and Mitchell WH (1998) A five year summary of beneficial arthropod studies with "Steward" (DPX MP 062). In: Drugger CP and Richter DA (eds), *Proceedings of the Beltswide Cotton Conference,* National Cotton Council, Memphis, TN.

Hammock BD, Bonning BB, Possee RD, Hanzlik TN and Maeda S (1990) Expression and effects of juvenile hormone esterase in a baculovirus vector. *Nature* **344:** 458-461.

Hamon N, Shaw R and Yang H (1996) Worldwide development of fipronil insecticide. In: Drugger CP and Richter DA (eds.) *Proceedings of the Beltswide Cotton Conference,* National Cotton Council, Memphis, TN.

Harder HH, Riley SL, Mccann SF and Irving SN (1996) DPX-MPO62; a novel broad-spectrum, environmentally soft, insect control compound. In: *Proceedings of the 1996 Brighton Crop Protection Conference — Pests and Diseases*, Vol. 2, pp. 449–454.

Harrison RL and Bonning BC (2000) Use of scorpion neurotoxins to improve the insecticidal activity of *Rachiplusia ou* multicapsid nucleopolygedrovirus. *Biological Control* **17**: 191-201.

Harrison R and Hoover K (2012) Baculoviruses and other occluded insect viruses. In: Vega FE and Kaya HK (eds), *Insect Pathology*. Second Edition. Academic Press, London.

Hayakawa T, Shimojo EI, Mori M, Kaido M, Furusawa I, Miyata S, Sano Y, Matsumoto T, Hashimoto Y and Granados RR (2000) Enhancement of baculovirus infection in *Spodoptera exigua* (Lepidoptera : Noctuidae) larvae with *Autographa californica* nucleopolyhedrovirus or *Nicotiana tabacum* engineered with a granulovirus enhancing gene. *Applied Entomology and Zoology* **35**: 163-170.

Hendrichs J, Vreysen JB, Enkerlin WR and Cayol JP (2005) Strategic options in using sterile insects for area-wide integrated pest management. In: Dyck, VA, Hendrichs J. and Robinson, AS (eds), *Sterile Insect Technique: Principles and Practice in Area-Wide Integrated Pest Management*, Springer, Dordrecht, The Netherlands, pp. 563-600.

Hibbett DS, Binder M, Bischoff JF, Blackwell M and Cannon PF (2007) A higher-level phylogenetic classification of the fungi. *Mycological Research* **111**: 509-547.

Hilder VA, Gatehouse AMR and Boulter D (1992) Transgenic plants conferring insect tolerance: protease inhibitor approach. In: Kung S and Wu R (eds), *Transgenic Plants* Academic Press, New York, USA. pp 310-338.

Hilker M and McNeil J (2008) Chemical and behavioral ecology in insect parasitoids: how to behave optimally in complex odorous environment. In: Wajnberg, E; Bernstein C and Van Alphen, J (eds), *Behavioral Ecology of Insect Parasitoids: from Theoretical Approaches to Field Applications*. Blackwell Publishing Ltd. pp. 92-112.

Hoover K, Schultz CM, Lane SS, Bonning BC, Duffey SS, McCutchen BF and Hammock BD (1995) Reduction in damage of cotton plants by a recombinant baculovirus that knocks moribund larvae of *Heliothis virescens* off the plant. *Biological Control* **5**: 419-526.

Horowitz AR and Ishaaya I (2004) Biorational insecticides— mechanism, selectivity and importance in pest management. In: Horowitz AR and Ishaaya I (eds), *Insect Pest Management: Field and Protected Crops*. Springer, Berlin.

Hosokawa T, Kikuchi Y, Shimada M and Fukatsu T (2007) Obligant symbiont involved in the pest status of host insect. *Proceedings of Royal Society B: Biological Sciences* **284**: 1979-1984.

Hsu ACT (1991) 1,2-diacyl-1-alkyl hydrazins, a novel class of insect growth regulators. In: *Synthesis and Chemistry of Agrochemicals II*, ACS Symp. Ser Washington DC, *American Chemicals Society* **443**: 478-490.

Hu Z, Chen X and Sun X (2003) Molecular biology of insect viruses. In: Upadhyay R.K. (ed) *Advances in Microbial Control of Insect Pests*. Kluwer Academic/Plenum Publishers, New York.

Huber RT, Moore L and Hoffmann MP (1979) Feasibility study of area-wide pheromone trapping of male pink bollworm *Pectinophora gossypiella* moths in a cotton insect pest management program. *Journal of Economic Entomology* **72:** 222–227.

Hughes PR, Wood HA, Breen JP, Simpson SF, Duggan AJ and Dybas JA (1997) Enhanced bioactivity of recombinant baculoviruses expressing insect-specific spider toxins in lepidopteran crop pests. *Journal of Invertebrate Pathology* **69**: 112-118.

Hurwitz I, Hillesland H, Fieck A, Das P and Durvasula R (2011) The paratransgenic sand fly: A platform for control of Leishmania transmission. *Parasites and Vectors* **4**: 82

Hussain MA and Lal KB (1940) The bionomics of *Emproasca devastans* (Distant) on some varieties of cotton in the Punjab. *Indian Journal of Entomology* **2**: 123-136.

Husseneder C and Grace JK (2005) Genetically engineered termite gut bacteria deliver and transfer foreign genes in termite colonies. *Applied Microbiology and Biotechnology* **68**: 360-367.

Jackson TA (2007) A novel bacterium for control of grass grub. In: Vincent C, Goettel MS and Lazarovits G (eds), *Biological Control: A Global Perspective.* CABI, Wallingford, UK, pp. 160-168.

Jackson TA, Pearson JF, O'Callaghan M, Mahanty HK and Willcocks MJ (1992) Pathogen to product-development of *Serratia entomophila* (Enterobacteriaceae) as a commercial biological agent for the New Zealand grass grub (*Costelytra Zealandica*). In: Jackson TA and Glare TR (eds), *Use of Pathogens in Scarab Pest Management.* Intercept Press, Andover, pp. 191-198.

Jagginavar SB, Sunitha ND and Biradar AP (2009) Bioefficacy of flubendiamide 480 SC against brinjal fruit and shoot borer, *Leucinodes orbonalis* Guen. *Karnataka Journal of Agricultural Sciences* **22**: 712-713.

Jalali SK and Venkatesan T (2011) Development and use of multiple insecticide resistant strain of *Trichogramma chilonis* Ishii for the management of *Helicoverpa armigera* (Hubner) on tomato in India. *Global Conference on Entomology*, March 5-9, 2011, Chiang Mai, Thailand.

Jalali SK, Singh SP, Venkatesan T, Murthy KS and Lalitha Y (2006) Development of endosulfan tolerant strain of an egg parasitoid *Trichogramma chilonis* Ishii (Hymenoptera : Trichogrammaitidae). *Indian Journal of Experimental Biology* **44**: 584-590.

James C (2013) *Global Status of Commercialized Biotech/GM Crops*: 2013. ISAAA Brief No.46. ISAAA: Ithaca, NY.

Jansens S, Cornelissen M, Clercq R, Reynaerts A and Peferoen M (1995) *Phthorimaea opercullela* (Lepidoptera: Gelechidae) resistance in potato by expression of *Bacillus thuringiensis Cry1Ab* insecticidal crystal protein. *Journal of Economic Entomology* **88**:1469-1476

Jaworska K, Olszak RW, Labanowska BH and Korzeniowski M (2012) Efficacy of spirotetramat in the control of pear psylla (*Cacopsylla pyri* L.) on pear trees in Poland. *Journal of Fruit and Ornamental Plant Research* **20**: 91-106

Jayaraj S, Amanthakrishnan TN and Veeresh GK (1994) *Biological Pest Control in India, Progress and Perspe-ctives.* RGICS Project No. 2, Rajiv Gandhi Institute of Contemporary Studies, New Delhi.

Jena BC, Srihari B and Mahapatra R (2006) Pesticidal management practices to control brinjal shoot and fruit borer. *Indian Journal of Entomology* **68**:193-196.

Josan A and Singh G (2000) Sublethal effects of lufenuron on the diamondback moth, *Plutella xylostella* (Linnaeus). *Insect Science and its Application* **20**: 303-308.

Jurat-Fuentes J and Jackson TA (2012) Bacterial entomopathogens. In: Vega FE and Kaya HK (eds), *Insect Pathology*. Second Edition. Academic Press, London, pp. 265-350.

Kairo MTK, Pollard GV, Peterkin DD and Lopez VF (2000) Biological control of the hibiscus mealybug, *Maconellicoccus hirsutus* Green (Hemiptera: Pseudococcidae) in the Caribbean. *Integrated Pest Management Reviews* **5**: 241–254.

Karban R and Baldwin IT (1997) *Induced Responses to Herbivory*. University of Chicago Press, Chicago, IL.

Karihaloo JL and Kumar PA (2009) *Bt Cotton in India – A status Report.* Second Edition. Asia-Pacific Consortium on Agricultural Biotechnology (APCoAB), New Delhi, India.

Karimzadeh R, Hejazi MJ, Rahimzadeh Khoei F and Moghaddam M (2007) Laboratory evaluation of five chitin synthesis inhibitors against the Colorado potato beetle, *Leptinotarsa decemlineata*. *Journal of Insect Science* **7**: 50-55.

Karp DS, Mendenhall CD, Sandi RF, Chaumont M, Ehrlich PR, Hardly EA and Daily GC (2013) Forest bolsters bird abundance, pest control and coffee yield. *Ecol Letters (doi:10.1111/ele.12173)*

Kaur, H. and Virk, J.S. (2011) Feeding potential of *Cryptolaemus montrouzieri* against the mealybug *Phenacoccus solenopsis* *Phytoparasitica* DOI 10.1007/s12600-011-0211-3

Kaur S (2000) Molecular approaches towards development of novel *Bacillus thuringiensis* biopesticides. *World Journal of Microbial Biotechnology* **16**: 781-793.

Kaya HK, Aguilera MM, Alumai A, Chaoo HY and de la Torre M (2006) Status of entomopathogenic nematodes and their symbiotic bacteria from selected countries or regions of the world. *Biological Control* **38**: 134-155.

Kayzer H, Kaufmann L and Schurmann F (1994) Pymetrozine (CGA 215'944); a novel compound for aphid and whitefly control. An overview of its mode of action. In *Proceedings of the 1994 Brighton Crop Protection Conference - Pests and Diseases,* Vol.2, pp. 737–742.

Khan ZR, Ampong-Nyarko K, Chiliswa P, Hassanali A and Kimani S (1997) Intercropping increases parasitism of pests. *Nature* **388**: 631–632

Khan Z, Midega C, Pittchar J, Pickett J and Bruce T (2011) Push-Pull technology: a conservation agriculture approach for integrated management of insect pests, weeds and soil health in Africa. *International Journal of Agricultural Sustainability* **9**: 162-170.

Kirst HA, Michel KH, Mynderse JS, Chao EH, Yao RC, Nakatsukasa WM, Boeck LD, Occlowitz J, Paschel JW, Deeter JB and Thompson GD (1992) Discovery, isolation and structure elucidation of a family of structurally unique fermentation-derived tetracyclic macrolides. In: Baker DR, Fenyes JG and Steffens JJ (eds.), *Synthesis and Chemistry of Agrochemicals III*. Am. Chem. Soc., Washington, DC, pp. 214-225.

Klassen W (2000) Area-wide approaches to insect pest interventions: history and lessons. In : Tan KH (ed), *Proceedings of International Conference on Area-Wide Control of Insect Pests, and the 5th International Symposium on Fruit Flies of Economic Importance,* 28 May–5 June 1998, Penerbit Universiti Sains Malaysia, Pulau Pinang, Malaysia.

Klassen W (2005) Area-wide integrated pest management and the sterile insect technique. In: Hendrichs DJ and Robinson AS (eds.) *Sterile Insect Technique. Principles and Practice in Area-Wide Integrated Pest Management,* IAEA. Springer, The Netherlands.

Knapp SJ (1994) Mapping quantitative trait loci. In: Phillipes, R.L. and Vasil, I.K. (eds.) *DNA-based Markers in Plants.* Kluwer Academic Publ., Dordrecht. The Netherlands.

Knipling EF (1959) Sterile-male method of population control. *Science* **130**: 902–904.

Knipling EF (1979) *The Basic Principles of Insect Population Suppression and Management*. US Department of Agriculture, Agriculture Handbook no. 512, Washington, DC, 659 pp.

Koppenhofer AM and Kaya HK (2002) Entomopathogenic nematodes and insect pest management. In: Koul O and

Dhaliwal GS (eds), *Microbial Biopesticides*. Taylor & Francis, London, pp. 277-306.

Koul O and Cuperus GW (2008) Current approaches to areawide integrated pest management: obstacles and future directions. In: Koul O and Cuperus GW (eds), *Areawide Pest Management Theory and Implementation*, CAB International, UK. pp. 560-572.

Kramer KJ, Morgan TD, Throne JE, Dowell FE, Bailey M and Howard JA (2000) Transgenic avidin maize is resistant to storage insect pests. *Nature Biotechnology* **18**: 670–674.

Kramer MG 2004. Recent advances in transgenic arthropod technology. *Bulletin of Entomological Research* **94**: 95-110.

Kuhar TP, Walgenbach JF and Doughty HB (2010) Control of *Helicoverpa zea* in tomatoes with Chlorantraniliprole applied through drip chemigation. *Plant Health Progress doi:10.1094/PHP-2009-0407-01-RS.*

Kumar A, Jalali SK, Venkatesan T, Nagesh M and Lalitha Y (2008) Genetic improvement of egg parasitoid *Trichogramma chilonis* Ishii for combined tolerance to multiple insecticides and high temperature. *Journal of Biological Control* **22**: 347-356.

Kumar BV, Kuttalam S and Chandrasekaran S (2009) Efficacy of a new insecticide spirotetramat against cotton whitefly. *Pesticide Research Journal* **21:** 45-48.

Kumar H and Kumar V (2004) Tomato expressing Cry1A(b) protein from *Bacillus thuringiensis* protected against tomato fruit borer, *Helicoverpa armigera* (Hubner) (Lepidoptera: Noctuidae) damage in the laboratory, greenhouse and field. *Crop Protection* **25:** 119-124.

Kumar MG and Prakasam CB (1984) The recovery of *Cryptolaemus montrouzieri* Muls. on the coffee green scale, *Coccus viridis* (Green) on the Shevaroy Hills. *Journal of Coffee Research* **14**: 34–35.

Kumar M, Chimote V, Singh R, Mishra, GP, Naik, PS, Pandey, SK and Chakrabarti SK (2010) Development of Bt transgenic potatoes for effective control of potato tuber moth by using *cry1Ab* gene regulated by GBSS promoter. *Crop Protection* **29:** 121-127.

Lahm GP, Stevenson TM, Selby TP, Freudenberger JH, Cordova D, Flexner L, Bellin CA, Dubas CM, Smith BK, Hughes KA, Hollingshaus JG, Clark CE and Benner EA (2007) Rynaxypyr: a new insecticidal anthranilic diamide that acts as a potent and selective ryanodine receptor activator. *Bioorganic and Medicinal Chemistry Newsletters* **17**: 6274-6279.

Lapied B, Grolleau F and Sattelle DB (2001) Indoxacarb, an oxadiazine, blocks insect neuronal sodium channels. *British Journal of Pharmacology* **132**: 587–595.

Lasota JA and Dybas RA (1991) Avermectin, a novel class of compounds: implications for use in arthropod pest control. *Annual Review of Entomology* **36**: 96–117.

Latif M, Mahbubar Rahman M, Zinnatul Alam M and Muffazal Hussain M (2009) Evaluation of flubendiamide as an IPM component for the management of brinjal shoot and fruit borer, *Leucinodes orbonalis* Guenee. *Munis Entomology and Zoology* **4**: 257-267.

Laubach AC (1916) Studies on aerobic, sporebearing, non-pathogenic bacteria. Spore bearing organism in water. *Journal of Bacteriology* **1**: 505-512.

Lawrence PK and Koundal KR 2002, Plant protease inhibitors in control of phytophagous insects. *Electronic Journal of Biotechnology* **5:** 93-109

Leo F, De Volpicella M, Licciulli F, Liuni S, Gallerani R and Ceci R (2002). Plant- PIs: a database for plant proteinase inhibitors and their genes. *Nucleic Acid Research* **30**: 347-348.

Lewis EE and Clarke DJ (2012) Nematode parasites and entomopathogens. In: Vega FE and Kaya HK (eds.) *Insect Pathology*. Second Edition, Academic Press, London, pp. 395-424.

Liu D, Burton S, Glancy T, Li ZS, Hampton R, Meade T and Merlo DJ (2003) Insect resistance conferred by 283-kDa *Photorhabdus luminescens* protein TcdA in *Arabidopsis thaliana*. *Nature Biotechnology* **21**: 1222–1228.

Liu SS and Jiang LH (2003) Differential parasitism of *Plutella xylostella* larvae by the parasitoid *Cotesia plutellae* on two host plant species. *Bulletin of Entomological Research* **93**: 65-72.

Liu Su, Liang QM, Zhou WW, Jiang YD, Zhu QZ, Yu H, Zhang CX, Gurr GM and Zhu ZR (2014) RNA interference of NADPH-cytochrome P450 reductase of the rice brown planthopper, *Nilaparvata lugens*, increases susceptibility to insecticides. *Pest Management Science* (DOI: 10.1002/ps.3760)

Liu XM, Smith CM and Gill BS (2002) Mapping of microsatellite markers linked to the *Dn4* and *Dn6* genes expressing Russian wheat aphid resistance in wheat. *Theoretical and Applied Genetics* **104**: 1042–1048.

Lou YG, Ma B and Cheng JA (2005) Attraction of parasitoid *Anagrus nilaparvatae* to rice volatiles induced by rice brown plant hopper, *Nilaparvata lugens*. *Journal of Chemical Ecology* **31**: 2357-2372.

Maeda S (1989) Expression of foreign genes in insects using baculovirus vectors. *Annual Review of Entomology* **34**: 351-372.

Maeda S, Volrath SL, Hanzlik TN, Harper SA, Maddox DW, Hammock BD and Fowler E (1991) Insecticidal effects of an insect specific neurotoxin expressed by a recombinant baculovirus. *Virology* **184**: 777-780.

Maeyer L and Geerinck R (2009) The multiple target use of spirodiclofen (Envidor 240 SC) in IPM pomefruit in Belgium. *Communication in Agricultural and Applied Biological Sciences* **74:** 225-232.

Mallapur CP, Rabindra RJ, Tippannavar PS and Sharanabasappa Kambrekar DN (2008) Colonization and establishment of *Encarsia flavoscutellum* in Southern India. *Endure International Conference 2008: Diversifying crop protection*, 12-15 October 2008, France.

Mandaokar AD, Goyal RK, Shukla A, Bisaria S, Bhalla R, Reddy VS, Chaurasia A, Sharma RP, Altosaar I and Kumar PA (2000) Transgenic tomato plants resistant to fruit borer (*Helicoverpa armigera* Hubner). *Crop Protection* **19:** 307-312.

Mani M (1990) The grapevine mealybug. *Indian Horticulture* **35**: 28–29.

Mani M and Krishnamoorthy A (1990) Evaluation of the exotic predator, *Cryptolaemus montrouzieri* Muls (Coccinellidae, Coleoptera) in the suppression of green scale, *Chloropulvinaria psii* (Mask.) (Coccidae, Homoptera). *Entomon* **15**: 45–48.

Mani M and Krishnamoorthy A (2008) Biological suppression of the mealy bugs *Planococcus citri* (Risso), *Ferrisia virgata* (Cockerell) and *Nipaecoccus viridis* (Newstead) on pummelo with *Cryptolaemus montrouzieri* Mulsant in India. *Journal of Biological Control* **22**: 169–172.

Mani M and Thontadarya TS (1988) Field evaluation of *Cryptolaemus montrouzieri* Muls. in the suppression of grape mealybug, *Maconellicoccus hirsutus* (Green). *Journal of Biological Control* **2**: 14–16.

McCray AH (1917) Spore-forming bacteria in the apiary. *Journal of Agricultural Research* **8**: 399-420.

McCutchen BF, Choundary PV, Crenshaw R, Maddox D, Kamita SG, Palekar N, Volrath S, Flower E, Hammock BD and Maeda S (1991) Development of a recombinant baculovirus expressing an insect selective neurotoxin: Potential for pest control. *Biotechnology* **9**: 848-852.

McKenna C, Gaskin R, Horgan D, Dobson S and Jia Y (2013) Efficacy of a postharvest spirotetramat spray against armoured scale insects on kiwifruit vines. *New Zealand Journal of Crop and Horticultural Science* **41**: 105-116.

McMeniman M, Lane C, Cass R, Fong B, Sidhu M, Wang YF and O'Neill S (2009) Stable introduction of a life-shortening *Wolbachia* infection into the mosquito *Aedes aegypti. Science* **323**: 141-144.

Michaelides P and Irving SN (1998) Cotton insect pest control with indoxacarb: A novel insecticide. In: *Proceedings of the World Cotton Research Conference 2,* Athens, Greece, pp. 773–776.

Miller TA, Lauzon CR, Lampe D, Durvasula R and Matthews C (2006) Paratransgenesis applied to control insect-transmitter plant pathogens: The Pierce's disease case. In: Bourtzis K and Miller TA (eds.) *Insect Symbiosis.* Volume 2. Taylor and Frances, London, UK/CRC Press, Boca Raton (FL), USA.

Misra HP (2011) Bio-efficacy of chlorantraniliprole against shoot and fruit borer of brinjal, *Leucinodes orbonalis* Guenee. *Journal of Insect Science* **24**: 60-64.

Mols C and Visser M (2002) Great tits can reduce caterpillar damage in apple orchards. *Journal of Applied Ecology* **39**: 888–899

Mooney KA, Gruner DS, Barber NA, Van Bael SA, Philpott SM and Greenberg R (2010) Interactions among predators and the cascading effects of vertebrate insectivores on arthropod communities and plants. *Proceedings National Academy of Sciences,* USA **107**: 7335-7340.

Moraes MCB, Laurmann R, Sujii ER, Pires C and Borges M (2005) Induced volatiles in soybean and pigeon pea plants artificially infested with the neotropiocal brown stink bug, *Euschistus heros*, and their effect on egg parasitoid, *Telenomus podisi. Entomologia Exerimentalis et Applicata* **115**: 227-237.

Moran NA (2006) Symbiosis. *Current Biology* **16** : 866-871.

Morrison NI, Segura DF, Stainton KC, Fu G, Donnelly CA and Alphey LS (2009) Sexual competitiveness of a transgenic sexing strain of the Mediterranean fruit fly, *Ceratitis capitata. Entomologia Experimentalis et Applicata* **133**: 146-153.

Morton RL, Schroeder HE, Bateman KS, Chrispeels MJ, Armstrong E and Higgins TJV (2000) Bean alpha-amylase inhibitor 1 in transgenic peas (*Pisum sativum*) provides complete protection from pea weevil (*Bruchus pisorum*) under field conditions. *Proceedings National Academy of Sciences* USA **97**: 3820–3825.

Moscardi F (1999) Assessment of the applications of baculoviruses for control of Lepidoptera. *Annual Review of Entomology* **44**: 257-289.

Moses TK, Pollard GV, Peterkin DD and Lopez F (2002) Biological control of the hibiscus mealybug, *Maconellicoccus hirsutus* Green (Hemiptera: Pseudococcidae) in the Caribbean. *Integrated Pest Management Reviews* **5**: 241–254.

Muthukrishnan N, Manoharan T, Thirumalai Thevan PS and Anbu S (2005) Evaluation of buprofezin for the management of grape mealy bug *Maconellicoccus hirsutus* (Green), *Journal of Entomological Research* **29**: 339-344.

Nagadhara D, Ramesh S, Pasalu IC, Rao YK, Krishnaiah NV, Sarma NP, Bown DP, Gatehouse JA, Reddy VD and Rao KV (2003) Transgenic indica rice resistant to sap sucking insects. *Plant Biotechnology Journal* **1**: 231-240.

Naimov S, Dukiandjiev S and Maagd RS (2003) A hybrid *Bacillus thuringiensis* delta-endotoxin gives resistance against a coleopteran and a lepidopteran pest in transgenic potato. *Plant Biotechnology Journal* **1**: 51-57.

Nathan SS, Kalaivani K, Choi MY and Paik CH (2009) Effects of jasmonic acid-induced resistance in rice on the plant brownhopper, *Nilaparvata lugens* Stal (Homoptera: Delphacidae). *Pesticide Biochemistry and Physiology* **95**: 77-84.

Nauen R, Schnorbach HJ and Elbert A (2005) The biological profile of spiromesifen (Oberon) - A new tetronic acid insecticide/ acaricide. *Pflanzenschutz-Nachrichten Bayer* **58**: 417–440.

Nauen R and Konanz S (2005) Spiromesifen as a new chemical option for resistance management in whiteflies and spider mites. *Pflanzenschutz-Nachrichten Bayer* **58**: 485–502.

Nauen R, Bretschneider T, Elbert A, Fischer R and Tieman R (2003) Spirodiclofen and spiromesifen. *Pesticide Outlook* **14**: 243–246.

Nester EW, Thomashow LS, Metz M and Godon M (2002) 100 years of *Bacillus thuringiensis*: A Critical Scientific Assessment. *American Academy of Microbiology*, Washington, 22 pp.

O'Reilly DR and Miller LK (1991) Improvement of a baculovirus pesticide by deletion of *egt* gene. *Biotechnology* **9**: 1086-1089.

Oerke EC (2006) Crop losses to pests. *Journal of Agricultural Science* **144**: 31-43.

Oliveira EJ, Rabinovitch L, Monnerat RG, Passos LK and Zahner V (2004) Molecular characterization of *Brevibacillus laterosporus* and its potential use in biological control. *Applied Environmental Microbiology* **70**: 6657-6664.

Painter RH (1951) *Insect Resistance in Crop Plants*. University of Kansas Press, Lawrence, Kansas.

Palumbo JC (2007) Influence of spray timing on the efficacy of Movento (spirotetramat) against aphids in lettuce. In: *ESA Annual Meeting*, December 9-12, 2007, Department of Entomology, University of Arizona, AZ, USA, 1303 pp.

Perlak FJ, Stone TB, Muskopf YN, Peterson LJ, Parker GB, McPherson SA, Wyman J, Love S, Reed G, Biever D and Fischoff DA (1993) Genetically improved potatoes: protection from damage by colarado potato beetles. *Plant Molecular Biology* **22**: 313-321

Pittman G, Brumbley S, Allsopp P and O'Neill SL (2008) Assessment of gut bacteria for a paratransgenic approach to control *Dermolepida albohirtum* larvae. *Applied and Environmental Microbiology* **74**: 4036-4043.

Prikhodko GG, Robson M, Warmke JW, Cohen CJ, Smith MM, Wang PY, Warren V, Kaczorowski G, Vander Ploeg LHT and Miller LK (1996) Properties of three baculoviruses expressing genes that encode insect-selective toxins: Mu-Aga-IV, AsII, and SH1. *Biological Control* **7**: 236-244.

Purcell JP, Greenplate JT, Jennings MG, Ryers JS, Pershing JC, Sims SR, Prinsen MJ, Corbin DR, Tran M, Sammons RD and Stonard RJ (1993) Cholesterol oxidase: A potent insecticidal protein active against bollweevil larvae. *Biochemical and Biophysical Research Communication* **196**: 1406-1413

Purohit SS and Vyas SP (2004) *Medicinal plants cultivation, A Scientific Approach*. Agrobias, Jodhpur, India.

Rabindra RJ and Grzywacz DC (2010) Asia: India. In: Kabaluk JT, Svircev AM, Goettel MS and Woo SG (Eds), *The Use and Regulation of Microbial Pesticides in Representative Jurisdictions Worldwide*. IOBC Global, pp. 12-17. (Available online through www.IOBC-Global.org)

Rajavel DS, Mohamraj A and Bharathi K (2011) Efficacy of chlorantraniliprole (Coragen 20SC) against brinjal shoot and fruit borer, *Leucinodes orbonalis* (Guen.). *Pest Management in Horticultural Ecosystems* **17**: 28-31.

Rao KV, Rathore KS, Hodges TK, Fu X, Stoger E, Sudhakar D, Williams S, Christou P, Bharathi M and Bown DP (1998) Expression of snowdrop lectin (GNA) in transgenic rice plants confers resistance to rice brown planthopper. *Plant Journal* **15**: 469–477

Rashid MMU, Khattak MK, Abdullah K, Amir M, Tariq M and Nawaz S (2012) Feeding potential of *Chrysoperla Carnea* and *Cryptolaemus Montrouzieri* on Cotton Mealybug, *Phenacoccus solenopsis*. *The Journal of Animal and Plant Sciences* **22**: 639-643

Rawlings ND, Tolle DP and Barrett AJ (2004) MEROPS: the peptidase database, *Nucleic Acids Research* **32**: 160-164.

Reddy RA and Narayan LK (1986). Biology and control of the grape mealybug. *Indian Grape Journal* **2**: 30–39 .

Rigev A, Keller M, Strizhov M, Sneh B, Prudovsky E, Chet I, Ginzberg I, Koncz Z, Schell J and Zilberstein A (1996) Synergistic activity of a *Bacillus thuringinensis* d-endotoxin and bacterial endochitinase against *Spodoptera littoralis* larvae. *Applied Environmental Microbiology* **62**: 3581–3586.

Rivers L, Vann CN, Zimmack HL and Dean DH (1991) Mosquitocidal activity of *Bacillus laterosporus*. *Journal of Invertebrate Pathology* **58**: 444-447.

Roberts DW and Humber RA (1981) Entomogenous fungi. In: Cole GT and Kendrick B (eds), *Biology of Conidial Fungi*. Vol 2. Academic Press, New York, pp. 201-236.

Robinson AS (1998) Genetic Control of Insect Pests. In: Rechcigl JE and Rechcigl NA (eds), *Biological and Biotechnological Control of Insect Pests*, CRC Press, New York.

Rosell R, Quero C, Coll J and Guerrero A (2008) Biorational insecticides in pest management. *Journal of Pesticide Science* **33**: 103-121.

Ruiu L, Delrio G, Ellar DJ, Floris I, Paglietti B, Rubino S and Satta A (2006) Lethal and sub-lethal effects of *Brevibacillus laterosporus* on the housefly (*Musca domestica*). *Entomologia Experimentalis et Applicata* **118**: 137-144.

Ruiu L, Satta A and Floris I (2013) Emerging entomopathogenic bacteria for insect pest managemet. *Bulletin of Insectology* **66**: 181-186.

Saha P, Dasgupta I and Das S (2006b) A novel approach for developing resistance in rice against phloem limited viruses by antagonizing the phloem feeding hemipteran vectors. *Plant Molecular Biology* **62**: 735–752.

Saha P, Majumder P, Dutta I, Ray T, Roy SC and Das S (2006a) Transgenic rice expressing *Allium sativum* leaf lectin with enhanced resistance against sap sucking insect pests. *Planta* **223**: 1329–1343.

Saikia R, Gogoi DK, Mazumder S, Yadav A, Sarma RK, Bora TC and Gogoi BK (2011) *Brevibacillus laterosporus* strain BPM3, a potential biocontrol agent isolated from a natural hot water spring of Assam, India. *Microbiological Research* **166**: 216-225.

Saito T (2004) Insecticide susceptibility of the leafminer, *Chromatomyia horticola* (Goureau) (Diptera: Agromyzidae). *Applied Entomology and Zoology* **39**: 203–208.

Saker MM, Salama HS, Salama M, Banna AEl and Ghanv NMA (2011) Development of Bt transgenic potatoes for effective control of potato tuber moth by using *cry1Ab* gene regulated by GBSS promoter. *Journal of Genetic Engineering and Biotechnology* **9**: 149-155.

Sammour EA, Kandil MA and Abdel-Aziz NF (2008) The Reproductive Potential and Fate of Chlorfluazuron and Leufenuron

Against Cotton Leafworm *Spodoptera littoralis* (Boisd.). *American-Eurasian Journal of Agriculture and Environment Science* **4**: 62-67.

Sanahuja G, Banakar R, Twyman RM, Capell T and Christou P (2011) *Bacillus thuringiensis*: a centry of research, development and commercial applications. *Plant Biotechnology Journal* **9**: 283-300.

Sanchis V (2011) From microbial sprays to insect resistant transgenic plants: history of the biopesticides *Bacillus thuringiensis*: A Review. *Agronomy for Sustainable Development* **31**: 217-231.

Sandhu SK and Arora R (2013) Breeding for Insect Resistance in Crop Plants. In: Dhawan AK, Singh B, Bhullar M and Arora R (eds.) *Integrated Pest Management*, Scientific Publishers, Jodhpur, pp. 267-300.

Sankyo Co. (1997) *Technical Data Sheet on Milbemectin*. Sankyo Co, Tokyo.

Sawant SS, Patil RS and Mule RS (2004) Field efficacy of novaluron (Rimon 10 EC) against brinjal fruit and shoot borer, *Leucinodes orbonalis* Guenee. *Pestology* **28**:28-30.

Schetelig M, Caceres C, Zacharopolou A, Franz G and Wimmer E (2009) Conditional embryonic lethality to improve the sterile insect technique in *Ceratitis capitata* (Diptera: Tephritidae). *BMC Biology* **7**: 4.

Schneider JC (1989) Role of movement in evaluation of area-wide insect pest management tactics. *Environmental Entomology* **18**: 868–874.

Sekercioglu CH (2006) Increasing awareness of avian ecological function. *Trends in Ecology and Evolution* **21**: 469-471.

Shade RE, Schroeder HE, Pueyo JJ, Tabe LM, Murdock LL, Higgins TJV, Chrispeels MJ (1994) Transgenic pea seeds expressing α-amylase inhibitor of the common bean are resistant to bruchid beetles. *Biotechnology* **12**:793-796.

Sharma HC (2009) *Biotechnological Approaches for Pest Management and Ecological Sustainability*. CRC Press, Boca Raton, FL, USA. 548 pp.

Sharma HC (2012) Application of modern tools of biotechnology for pest management – prospects and limitations. In: Singh B and Arora R (eds), *Souvenir: National Seminar on Biotechnological Approaches in Pest Management*. May 4-5, 2012, Punjab Agricultural University, Ludhiana, pp. 1-13.

Sharma OP, Sehgal M, Jeswani MD, Trivedi TP, Garg DK, Srivastava CP, Benagi VI, Nigam GL and Lavekar RC (2004) Pigeonpea. In: Singh A, Sardana HR and Sabie N (eds), *Validated IPM Technologies for Selected Crops*, National Centre for Integrated Pest Management, New Delhi.

Shelton AM, Fuchs M and Shotkoshi F (2008) Transgenic vegetables and fruits for the control of insects and insect vectored pathogens. In: Romeis J, Shelton AM and Kennedy GG (eds.). *Integration of Insect Resistant Genetically Modified Crops within IPM Programs*. Springer, Berlin, Germany. pp. 249-272.

Shu Q, Ye G, Cui H, Cheng X, Xiang Y, Wu D, Gao M, Xia Y, Hu C, Sardana R and Altosaar I (2000) Transgenic rice with synthetic *cry1Ab* gene from bacillus thuringiensis is highly resistant to eight lepidopteran insects. *Molecular Breeding* **6**: 433-39.

Sidde Gowda DK, Suhas Y and Patil B (2004) Field efficacy of novaluron (Rimon 10 EC) against pigeonpea pod borer *Helicoverpa armigera* (Hubner). *Pestology* **28**: 16-18.

Simmons GS, McKemey AR, Morrison NI, O'Connell S, Tabashnik BE, Claus J, Fu G, Tang G, Sledge M, Walker AS, Phillips CE, Miller ED, Rose RI, Staten RT, Donnelly CA and Alphey L (2011) Field performance of a genetically engineered strain of pink bollworm. *PLoS ONE* **6**: doi:10.1371/journal.pone.0024110.

Singer S (1996) The utility of morphological group II *Bacillus. Advances in Applied Microbiology* **42**: 219-261.

Singh B (2012) Role of pesticides in management of crop pests. In: Arora R, Singh B and Dhawan AK (eds), *Theory and Practice of Integrated Pest Management.* Scientific Publishers (India), Jodhpur, pp. 76-83.

Singh G and Suri KS (2012) Potential of insect growth regulators in insect pest management. In: Arora R, Singh B and Dhawan AK (eds) *Theory and Practice of Integrated Pest Management.* Scientific Publishers, Jodhpur, pp. 110-119.

Singh SP (1978). Propagation of a coccinellid beetle for the biological control of citrus and coffee mealybugs. Scientific Conference CPA. Bangalore, India: Indian Institute of Horticultural Research

Singh S, Shenhmar M, Brar KS and Jalali SK (2007) Evaluation of genetically improved strain of *Trichogramma chilonis* for the management of sugarcane stalk borer (*Chilo auricilius*). *Indian Journal of Agricultural Sciences* **78**: 868-872.

Singh SP (2003) Role of predators and parasitoids in biological control of crop pests. In: Koul O, Dhaliwal GS, Marwaha SS and Arora JK (eds.). *Biopesticides and Pest Management.* Vol. 1, Campus Books International, New Delhi, pp.196-221.

Slesak E, Slesak M and Gabrys B (2001) Effect of methyl jasmonate on hydroxamic acid, protease activity, and bird cherry-oat aphid *Rhoplaosiphum padi* L. probing behavior. *Journal of Chemical Ecology* **12**: 2529–2543.

Smith CM (2005) Plant Resistance to Arthropods: Molecular and Conventional Approaches. Dordrecht, The Netherlands: Springer. 423 pp.

Smith CM (2004) Plant resistance against pests: issues and strategies. In: Koul O, Dhaliwal GS and Cuperus GW (eds). *Integrated Pest Management: Potential, Constraints and Challenges.* CABI, Wallingford, pp. 147-168.

Smith CM and Clement SL (2012). Molecular bases of plant resistance to arthropods. *Annual Review of Entomology* **57**: 309-328.

St. Leger RJ and Wang C (2009) Entomopathogenic fungi and the genomics era. In: Stock SP, Vandenberg J, Glazer I and Boemare N (eds) *Insect Pathogens: Molecular Approaches and Techniques.* CABI, Wallingford, pp. 365-400.

Stewart LMD, Hirst M, Ferber ML, Merryweather AT, Cayley PJ and Possee RD (1991) Construction of an improved baculovirus insecticide containing an insect specific toxin gene. *Nature* **352**: 85-88.

Sullivan MJ, Turnipseed SG and Robinson D (1999) Insecticidal efficacy against a complex of fall and beet armyworms and soybean looper in South Carolina Cotton. In: Dugger CP and Richte DA (eds.) *Proceedings of the 1999 Beltswide Cotton Production Conference,* National Council of America, Memphis, TN.

Sun LH, Wang CM, Su CC, Liu YQ, Zhai HQ and Wan JM (2006) Mapping and Marker-assisted selection of a brown plant hopper resistance gene bph2 in Rice (*Oryza sativa* L.). *Acta Genetica Sinica.* **33**: 717–723.

Suri KS (2013) Bioefficacy of some insecticides against rice stemborers and leaffolder infesting rice in Punjab. *Pesticide Research Journal* **23**: 190-193.

Thaler JS, Stout MJ, Karban R and Duffey SS (1996) Exogenous jasmonates simulate insect wounding in tomato plants (*Lycopersicon esculentum*) in the laboratory and field. *Journal of Chemical Ecology* **22**: 1767-1781.

Tomalski MD and Miller LK (1991) Insect paralysis by baculovirus mediated expression of a mite neurotoxin gene. *Nature* **352**: 82.

Treacy MF, Rensner PE and All JN (2000) Comparative insecticidal properties of two nucleopolyhydrovirus vectors encoding a similar toxin gene chimer. *Journal of Economic Entomology* **93**: 1096-1104.

Triplett S, Luck GW and Spooner P (2012) The importance of managing the costs and benefits of bird activity for agricultural sustainability. *International Journal of Agricultural Sustainability* **10**: 268-288.

Tu JM, Zhang GA, data K, Xu CG and He YQ (2000) Field performance of transgenic elite commercial hybrid rice expressing *Bacillus thuringiensis* delta-entotoxin. *Nature Biotechnology* **18**: 1101-1104.

Udikeri SS, Patil SB, Rachappa V, Khadi M and Panchabhavi KS (2004) Novaluron (Rimon 10 EC): an ideal IGR for management of cotton bollworms. *Pestology*. **28**: 28-31.

Vadodaria MP, Maisuria IM, Patel RB, Patel CJ and Patel UG (2000) Insect growth regulator (IGR) - a new tool in the management of *Helicoverpa* on cotton in Gujarat. *Pestology*. **24**:11-14.

Van Bael SA, Philpott SM, Greenberg R, Bichier P, Barber NA, Mooney KA and Gruner DS (2008) Birds as predators in tropical agroforestry systems. *Ecology* **89**: 928-934.

Van Driesche R, Hoddle M and Center T (2008) *Control of Pests and Weeds by Natural Enemies: An Introduction to Biological Control.* Blackwell Publishing, Malden, MA, USA.

Van Lenteran JC (2008) *Internet Book of Biological Control. 5th Edition*, Available at: www.IOBC-Global.org, Wageningen, The Netherlands.

Van Lenteran JC (2012) *Internet Book of Biological Control. 6th Edition*, Available at: www.IOBC-Global.org, Wageningen, The Netherlands.

Van Lenteren JC, Bale J, Bigler F, Hokkanen HMT and Loomans AJM (2006) Assessing risks of releasing exotic biological control agents of arthropod pests. *Annual Review of Entomology* **51**: 609-634.

Vargas RI, Ronald FL, Mau EB, Jang RM and Lyle W (2008) The Hawaii Fruit Fly Areawide Pest Management In: Koul O and Cuperus GW (eds.) *Areawide Pest Management. Theory and Implementation*, CAB International, UK. pp. 300-325.

Vargas RI, Jang EB and Klungness LM (2003) Areawide pest management of fruit flies in Hawaiian fruits and vegetables. In: Inamine K (ed.) *Recent Trends on Sterile Insect Technique and Areawide Integrated Pest Management.* Research Institute for Subtropics, Okinawa, Japan, pp. 37–46.

Vega FE and Kaya HK (eds.) (2012) *Insect Pathology.* Second Edition. Academic Press, London, 490 pp.

Venkatesan T, Jalali SK, Helen M, Lalitha Y and Kumar A (2011) Development of pesticide tolerant *Chrysoperla zastrowi arabica* Henry *et al.* and their efficacy against sucking pests of cotton. *Global Conference on Entomology*, March 5-9, 2011, Chiang Mai, Thailand.

Vincent C, Hallman G, Panneton B and Lessard FF (2003) Management of agricultural insects with physical control methods. *Annual Review of Entomology* **48**: 261-281.

Virk JS and Shera PS (2013) Behaviour modifying chemicals to enhance the effectiveness of natural enemies. In: Dhawan AK, Singh B, Bhullar M and Arora R (eds), *Integrated Pest Management*, Scientific Publishers, Jodhpur, pp. 325-350.

Wall C, Garthwaite DG, Blood Smyth JA and Sherwood A (1987) The efficacy of sex-attractant monitoring for the pea moth, *Cydia nigricana*, in England, 1980–1985. *Annals of Applied Biology* **110**: 223–229.

Wang Y, Gao C, Xu Z, Zhu YC, Zhang J, Li W, Dai D, Lin Y, Zhou W and Shen J (2008)

Buprofezin susceptibility survey, resistance selection and preliminary determination of resistance mechanism in *Nilaparvata lugens* (Homoptera: Delphacidae). *Pest Management Science* **64**: 1050-1056.

Wenny DG, Devault TL, Johnson MD, Kelly D, Sekercioglu CH, Tomback DF and Whelan CJ (2011) Perspectives in ornithology: The need to quantity ecosystem services provided by birds. *The Auk* **128**: 1-14.

Whelan CJ, Wenn DG and Marquis RJ (2008) Ecosystem services provided by birds. *Annals of the New York Academy of Sciences* **1134**: 25-60.

Whitman DW and Nordlund DA (1994) Plant chemicals and the location of herbivorous arthropods by their natural enemies. In: Ananthakrishnan TN (ed), *Functional Dynamics of Phytophagous Insects*. Oxford & IBH Publishing Co. Pvt. Ltd, New Delhi, pp. 133-159.

WHO (2009). *Progress and prospects for the use of genetically modified mosquitoes to inhibit disease transmission*. Report on planning meeting 1: Technical consultation on current status and planning for future development of genetically modified mosquitoes for malaria and dengue control. World Health Organization (WHO), Geneva, Switzerland. Available at http://whqlibdoc.who.int/publications/2010/9789241599238_eng.pdf.

Williams CM (1967) Third generation pesticides. *Scientific American* **217**: 13-15.

Wimme E (2003) Applications of insect transgenesis. *Nature Reviews Genetics* **4**: 225-232.

Xu X, Yan G, Kong R, Liu X and Hu Y (2000) Analysis of expression of the binary toxin genes from *Bacillus sphaericus* in *Anabaena* and potential in mosquito control. *Current Microbiology* **41**: 352 -356.

Yao J, Rotenberg D, Afsharifar A, Alviar KB and Whitefield AE (2013) Development of RNAi methods for *Peregrinus maidis*, the corn planthopper (*DOI: 10.1371/journal.pone.0070243*)

Yap HH (1990) Field trials of *Bacillus sphaericus* for mosquito control. In: de Barjac H and Sutherland DJ (eds), *Bacterial Control of Mosquitoes & Black Flies: Biochemistry, Genetics & Application of Bacillus thuringiensis & Bacillus sphaericus.* Rutzers University, Press, New Brunswick, pp. 307-320.

Yarom I, Blumberg D and Ishaaya I (1988) Effects of buprofezin on California red scale (Homoptera: Diaspididae) and Mediterranean black scale (Homoptera: Coccidae). *Journal of Economic Entomology* **81**: 1581-1585.

Ye XY, Ng TB, Tsang PW, and Wang J (2001) Isolation of a homodimeric lectin with antifungal and antiviral activities from red kidney bean (*Phaseolus vulgaris*) seeds, *Journal of Protein Chemistry* **20**: 367-375

Yencho GC, Cohen MB and Byrne PF (2000) Applications of tagging and mapping insect resistance loci in plants. *Annual Review of Entomology* **45**: 393–422.

Zhang P,Wang Y, Zhang J, Maddock S, Snook M and Peterson T (2003) Amaize QTL for silk maysin levels contains duplicated Myb-homologous genes which jointly regulate flavone biosynthesis. *Plant Molecular Biology* **52**:1–15

Zhang WL (2011) Pymetrozien has been a preferred product for controlling the rice planthoppers in China. *Pesticide Market News* **5:** 28–29.